Advances in Computer Communications and Networks
From Green, Mobile, Pervasive Networking to Big Data Computing

RIVER PUBLISHERS SERIES IN COMMUNICATIONS

Series Editors

ABBAS JAMALIPOUR
The University of Sydney
Australia

MARINA RUGGIERI
University of Rome Tor Vergata
Italy

HOMAYOUN NIKOOKAR
Delft University of Technology
The Netherlands

The "River Publishers Series in Communications" is a series of comprehensive academic and professional books which focus on communication and network systems. The series focuses on topics ranging from the theory and use of systems involving all terminals, computers, and information processors; wired and wireless networks; and network layouts, protocols, architectures, and implementations. Furthermore, developments toward new market demands in systems, products, and technologies such as personal communications services, multimedia systems, enterprise networks, and optical communications systems are also covered.

Books published in the series include research monographs, edited volumes, handbooks and textbooks. The books provide professionals, researchers, educators, and advanced students in the field with an invaluable insight into the latest research and developments.

Topics covered in the series include, but are by no means restricted to the following:

- Wireless Communications
- Networks
- Security
- Antennas & Propagation
- Microwaves
- Software Defined Radio

For a list of other books in this series, visit www.riverpublishers.com

Advances in Computer Communications and Networks
From Green, Mobile, Pervasive Networking to Big Data Computing

Editors

Kewei Sha

University of Houston
Clear Lake, USA

Aaron Striegel

University of Notre Dame, USA

Min Song

Michigan Tech, USA

River Publishers

Published, sold and distributed by:
River Publishers
Alsbjergvej 10
9260 Gistrup
Denmark

River Publishers
Lange Geer 44
2611 PW Delft
The Netherlands

Tel.: +45369953197
www.riverpublishers.com

ISBN: 978-87-93379-87-9 (Hardback)
    978-87-93379-88-6 (Ebook)

©2017 River Publishers

Contents

PART I: Data Center Computing

1 **Flyover: A Cost-Efficient and Scale-Out Data Center Network Architecture**   **3**
Sheng Xu, Binzhang Fu, Mingyu Chen
and Lixin Zhang

PART II: Mobile Computing

3 Mitigating Bufferbloat with Receiver-based TCP Flow Control Mechanism in Cellular Networks   **65**

Xiaolan Liu, Fengyuan Ren, Ran Shu, Tong Zhang and Tao Dai

7 Improving the Effectiveness of Data Transfers in Mobile Computing Using Lossless Compression Utilities   181

Armen Dzhagaryan, Aleksandar Milenković
and Martin Burtscher

PART III: Spectrum

8 Scheduling-Inspired Spectrum Assignment Algorithms for Mesh Elastic Optical Networks   225

Mahmoud Fayez, Iyad Katib, George N. Rouskas
and Hossam M. Faheem

Prosanta Paul, ChunSheng Xin, Min Song
and Yanxiao Zhao

PART V: Multimedia Networks

14 User Experience Awareness Network Optimization for Video Streaming Based Applications   395

Hengky Susanto, ByungGuk Kim and Benyuan Liu

15 METhoD: A Framework for the Emulation of a Delay-Tolerant Network Scenario for Media Content Distribution in Under-Served Regions   427

Adriano Galati, Sandra Siby, Theodoros Bourchas, Maria Olivares, Stefan Mangold and Thomas R. Gross

PART VI: Network Optimization

16 On the Routing of Kademlia-type Systems   461

Stefanie Roos, Hani Salah and Thorsten Strufe

Preface

Recent developments in computer communications and networks have enabled the deployment of exciting new areas such as Internet of Things and collaborative big data analysis. The design and implementation of energy efficient future generation communication and networking technologies also require the clever research and development of mobile, pervasive, and large-scale computing technologies. This book studies and presents recent advances in communication and networking technologies reflecting the state-of-the-art research achievements in novel communication technology and network optimization.

This book is organized into six parts. The first part includes two novel data center design techniques that aim to save cost and manage power usage. The second part of the book focuses on mobile network design and optimization. Topics cover flow control mechanism, network monitoring and topology generation, mobility management, and data management protocols. Spectrum is a very important topic in wireless communication and it is discussed in depth in the third part of the book. Internet of Things (IoT) are one of the most promising and emerging networked systems. In the fourth part of the book, solutions to important issues in IoT such as deployment, simulation, resource management, and applications are demonstrated. Part IV illustrates two important approaches to improve the performance of multimedia networks. Finally, five optimization mechanisms are proposed in the last part. It targets to optimize routing protocol, improve bloom filter design, maximize information, reduce noise, and balance the load.

Advances in Computer Communications and Networks is ideal for personnel in computer communication and networking industries as well as academic staff and collegial, master, Ph.D. students in computer science, computer engineering, electrical engineering and telecommunication systems.

Acknowledgments

First and foremost, we would like to express our warm appreciation to University of Houston – Clear Lake, University of Notre Dame, and Michigan Technological University. Special thanks go to our authors who contributed excellent book chapters. We would like to express our warm appreciation to the River Publisher staffs who allowed us to publish our work and gave their valuable time to review our book. We would also like to thank the reviewers who provided feedback and suggestions for our book. Finally, we want to thank our families who supported and encouraged us in spite of all the time it took us away from them. Last and not least, we beg forgiveness of all those whose names we have failed to mention. Any suggestions, comments, and feedback for further improvement of the text are welcome.

List of Contributors

A. Swami, *Army Research Laboratory, Adelphi, MD 20783, United States*

A. Wolf, *Department of Computing, Imperial College, London SW7 2AZ, United Kingdom*

Aamir Saeed, *Wireless Communication Networks, Department of Electronic Systems, Aalborg University, Denmark*

Abdollah Homaifar, *ECE Department, North Carolina A&T State University, Greensboro, NC 27411, USA*

Adriano Galati, *Disney Research, Zurich, Switzerland*

Akash Kapoor, *Department of Computer Science, University of Illinois at Urbana-Champaign, Champaign, 61801, Illinois, USA*

Alberto E. Schaeffer-Filho, *Institute of Informatics, Federal University of Rio Grande do Sul, BR*

Aleksandar Milenković, *Department of Electrical and Computer Engineering, The University of Alabama in Huntsville, Huntsville, AL 35899, USA*

Amir Rastegarnia, *Department of Electrical Engineering, Malayer University, Malayer 65719-95863, Iran*

Andrea Fumagalli, *Open Networking Advanced Research (OpNeAR) Lab, Department of Electrical Engineering, The University of Texas at Dallas, Richardson, TX, USA*

Andreas Mauthe, *School of Computing and Communications, Lancaster University, UK*

Armen Dzhagaryan, *Department of Electrical and Computer Engineering, The University of Alabama in Huntsville, Huntsville, AL 35899, USA*

Azam Khalili, *Department of Electrical Engineering, Malayer University, Malayer 65719-95863, Iran*

B. Holbert, *Department of Computer Science and Engineering, Pennsylvania State University, State College, PA 16801, United States*

Behcet Sarikaya, *Huawei Technologies Limited, Plano, TX, USA*

Benyuan Liu, *Computer Science Department, University of Massachusetts Lowell, Lowell, MA 01852, USA*

Biao Xu, *Institute of Computer Application Technology, Hangzhou Dianzi University, Hangzhou, China*

Binzhang Fu, *1. SKL Computer Architecture, ICT, CAS, Beijing, China 2. University of Chinese Academy of Sciences, Beijing, China*

Björn Richerzhagen, *Multimedia Communications Lab, Technische Universität Darmstadt, DE*

ByungGuk Kim, *Computer Science Department, University of Massachusetts Lowell, Lowell, MA 01852, USA*

Christopher Osiegbu, *ECE Department, North Carolina A&T State University, Greensboro, NC 27411, USA*

ChunSheng Xin, *ECE Department, Old Dominion University, 5115 Hampton Blvd, Norfolk, VA 23529, USA*

Daniel Limbrick, *ECE Department, North Carolina A&T State University, Greensboro, NC 27411, USA*

David Griffith, *The Communications Technology Laboratory (CTL), National Institute of Standards and Technology, Gaithersburg, Maryland 20899, USA*

Dominik Stingl, *Multimedia Communications Lab, Technische Universität Darmstadt, DE*

Fatemeh Afghah, *ECS Department, Northern Arizona University, Flagstaff, AZ 86011, USA*

Fengyuan Ren, *1. Tsinghua National Laboratory for Information Science and Technology, Beijing, China*
2. Department of Computer Science and Technology, Tsinghua University, Beijing, China

Franz Eichhorn, *LOVOO GmbH, 01069 Dresden, Germany*

George N. Rouskas, *1. King Abdulaziz University, Jeddah, Saudi Arabia*
2. North Carolina State University, Raleigh, NC 27695, United States

Gil Einziger, *Computer Science Department, Technion, Haifa 32000, Israel*

Gokarna Sharma, *Department of Computer Science, Kent State University, Kent, OH 44242, USA*

Hani Salah, *TU Darmstadt, Darmstadt, Germany*

Hari Krishnan, *School of Electrical Engineering and Computer Science, Louisiana State University, Baton Rouge, LA 70803, USA*

Hengky Susanto, *Computer Science Department, University of Massachusetts Lowell, Lowell, MA 01852, USA*

Hossam M. Faheem, *1. Fujitsu Technology Solution, Munich, Germany*
2. Ain Shams University, Khalifa El-Maamon St, Cairo, Egypt

Ilya Moiseenko, *Department of Computer Science, University of California at Los Angeles, Los Angeles, 90095, California, USA*

Iyad Katib, *King Abdulaziz University, Jeddah, Saudi Arabia*

Jens Myrup Pedersen, *Wireless Communication Networks, Department of Electronic Systems, Aalborg University, Denmark*

Jie Wu, *Department of Computer and Information Sciences, Temple University, Philadelphia, PA 19122, USA*

Jongdeog Lee, *Department of Computer Science, University of Illinois at Urbana-Champaign, Champaign, 61801, Illinois, USA*

Jyotirmoy Banik, *Open Networking Advanced Research (OpNeAR) Lab, Department of Electrical Engineering, The University of Texas at Dallas, Richardson, TX, USA*

Klara Nahrstedt, *Department of Computer Science, University of Illinois, US*

Li Xue, *Huawei Technologies Limited, Haidian, Beijing, China*

Lixin Zhang, *1. SKL Computer Architecture, ICT, CAS, Beijing, China
2. University of Chinese Academy of Sciences, Beijing, China*

Mahmoud Fayez, *Fujitsu Technology Solution, Munich, Germany*

Marco Tacca, *Open Networking Advanced Research (OpNeAR) Lab, Department of Electrical Engineering, The University of Texas at Dallas, Richardson, TX, USA*

Maria Olivares, *Disney Research, Zurich, Switzerland*

Martin Burtscher, *Department of Computer Science, Texas State University, San Marcos, TX 78666, USA*

Md Tanvir Al Amin, *Department of Computer Science, University of Illinois at Urbana-Champaign, Champaign, 61801, Illinois, USA*

Min Song, *CS Department, Michigan Technological University, 1400 Townsend Dr, Houghton, MI 49931, USA*

Mingyu Chen, *1. SKL Computer Architecture, ICT, CAS, Beijing, China
2. University of Chinese Academy of Sciences, Beijing, China*

Nada Golmie, *The Communications Technology Laboratory (CTL), National Institute of Standards and Technology, Gaithersburg, Maryland 20899, USA*

Nils Richerzhagen, *Multimedia Communications Lab, Technische Universität Darmstadt, DE*

P. Novotny, *Department of Computing, Imperial College, London SW7 2AZ, United Kingdom*

Paul Moulema, *Department of Computer and Information Science, Towson University, Towson, Maryland 21252, USA*

Peng Liu, *Institute of Computer Application Technology, Hangzhou Dianzi University, Hangzhou, China*

Prosanta Paul, *ECE Department, Old Dominion University, 5115 Hampton Blvd, Norfolk, VA 23529, USA*

Radhika Goyal, *Department of Computer Science, University of Illinois at Urbana-Champaign, Champaign, 61801, Illinois, USA*

Ralf Steinmetz, *Multimedia Communications Lab, Technische Universität Darmstadt, DE*

Ran Shu, *1. Tsinghua National Laboratory for Information Science and Technology, Beijing, China*
2. Department of Computer Science and Technology, Tsinghua University, Beijing, China

Rasmus Løvenstein Olsen, *Wireless Communication Networks, Department of Electronic Systems, Aalborg University, Denmark*

Rhaban Hark, *Multimedia Communications Lab, Technische Universität Darmstadt, DE*

Roy Friedman, *Computer Science Department, Technion, Haifa 32000, Israel*

S. Silvestri, *Department of Computer Science, Missouri University of Science and Technology, Rolla, MO 65409, United States*

Saeid Sanei, *Department of Computer Science, University of Surrey, Surrey GU2 7XH, UK*

Sandra Siby, *Department of Computer Science, ETH Zurich, Switzerland*

Seifemichael B. Amsalu, *ECE Department, North Carolina A&T State University, Greensboro, NC 27411, USA*

Sheng Xu, *1. SKL Computer Architecture, ICT, CAS, Beijing, China*
2. University of Chinese Academy of Sciences, Beijing, China

Sriharsha Mallapuram, *Department of Computer and Information Science, Towson University, Towson, Maryland 21252, USA*

Stefan Mangold, *LovefieldWireless, Liebefeld, Switzerland*

Stefanie Roos, *TU Dresden, Dresden, Germany*

T. La Porta, *Department of Computer Science and Engineering, Pennsylvania State University, State College, PA 16801, United States*

Tao Dai, *1. Tsinghua National Laboratory for Information Science and Technology, Beijing, China*
2. Department of Computer Science and Technology, Tsinghua University, Beijing, China

Tarek Abdelzaher, *Department of Computer Science, University of Illinois at Urbana-Champaign, Champaign, 61801, Illinois, USA*

Theodoros Bourchas, *Disney Research, Zurich, Switzerland*

Thomas R. Gross, *Department of Computer Science, ETH Zurich, Switzerland*

Thorsten Strufe, *TU Dresden, Dresden, Germany*

Tong Zhang, *1. Tsinghua National Laboratory for Information Science and Technology, Beijing, China*
2. Department of Computer Science and Technology, Tsinghua University, Beijing, China

Wael M. Bazzi, *Electrical Engineering Department, American University in Dubai, P. O. Box 28282, Dubai, UAE*

Waltenegus Dargie, *Chair of Computer Networks, Faculty of Computer Science, Technical University of Dresden, 01062 Dresden, Germany*

Wei Yu, *Department of Computer and Information Science, Towson University, Towson, Maryland 21252, USA*

Xiaolan Liu, *1. Tsinghua National Laboratory for Information Science and Technology, Beijing, China*
2. Department of Computer Science and Technology, Tsinghua University, Beijing, China

Yanxiao Zhao, *ECE Department, South Dakota School of Mines & Technology, 501 E St Joseph St, Rapid City, SD 57701, USA*

Zeyuan Zhang, *Department of Computer Science, University of Illinois at Urbana-Champaign, Champaign, 61801, Illinois, USA*

Zhehao Wang, *REMAP, University of California at Los Angeles, Los Angeles, 90095, California, USA*

Zhen Jiang, *Department of Computer Science, West Chester University, West Chester, PA 1938, USA*

List of Figures

List of Tables

List of Algorithms

List of Abbreviations

3G	Third Generation
3GPP	3rd Generation Partnership Project
ABRWDA	Available Bandwidth based Receiver Window Dynamic Adjustment
AC	Alternating Current
ACK	Acknowledgement
AFCT	Average Flow Completion Time
AL	Alarm
AMC	Adaptive Modulation and Coding
AMD	Advanced Micro Devices
AMI	Advanced Metering Infrastructure
ANN	Artificial Neural Networks
AR	Arrival
AOB	Always On – Balanced
AODV	Ad hoc On-Demand Distance Vector Routing Protocol for mobile ad hoc networks (MANETs)
AOI	Area of Interest
AP	Access Point
AQM	Active Queue Management
ARIMA	Autoregressive (AR) Moving Average (MA)
ARM	Advanced RISC Machine
ARPU	Average Revenue Per User
ARQ	Automatic Repeat reQuest
AS	Autonomous System
ASPL	Average Shortest Path Length
ATC	Adapt-then-Combine
AWGN	Additive white Gaussian noise
BalancedBF	Balanced Bloom filters
BDP	Bandwidth-Delay-Product
BGP	Border Gateway Protocol
BlockedBF	Blocked Bloom filters

BNG	Broadband Network Gateway
B-Router	Border Router
CAF	Cyclic autocorrelation function
CBF	Counting Bloom filters
CCDF	Complementary Cumulative Distribution Function
CDF	Cumulative Distribution Function
CDMA2000	Code Division Multiple Access 2000
CDN	Content Distribution Network
CF	Compressed file
CMOS	Complementary Metal-Oxide Semiconductor
CN	Core Network
CoDel	Controlled Delay
COSMO	Co-Simulation with Matlab and OMNET++
CPU	Central Processing Unit
CQI	Channel Quality Indication
CR	Compression ratio
CRB	Cramer-Rao bound
CRNs	Cognitive radio networks
CS	Compressed file size
CSD	Cyclic spectrum density
CTA	Combine-then-Adapt
CWDM	Coarse Wavelength Division Multiplexer
DAP	Data Aggregator Point
DBN	Dynamic Bayesian Network
DBN-R	Dynamic Bayesian Network Reinforcement learning
DC	Direct Current
DCN	Data Center Network
DBN	Dynamic Bayesian network
DBN-R	Dynamic Bayesian network reinforcement learning
DHT	Distributed Hash Table
DMM	Distributed Mobility Management
DP	Departure
DRWA	Dynamic Receiver Window Adjustment
DSSS	Direct-sequence spread spectrum
DTN	Delay Tolerant Networks
DTNRG	DTN Research Group
DTX	Discontinuous Transmission
DVFS	Dynamic Voltage and Frequency Scaling
DWPT	Discrete wavelet packet transform

DWT	Discrete wavelet transform
EC2	Elastic Computing
ECMP	Equal Cost Multi Path
EE.C	Compression energy efficiency
EE.C(0)	Compression overhead energy efficiency
EE.C.LL	Compression energy efficiency lower estimate
EE.C.UL	Compression energy efficiency upper estimate
EE.D	Decompression energy efficiency
EE.D(0)	Decompression overhead energy efficiency
EE.D.LL	Decompression energy efficiency lower estimate
EE.D.UL	Decompression energy efficiency upper estimate
EE.UDW	Uncompressed download energy efficiency
EE.UUP	Uncompressed upload energy efficiency
EMAMS	Exponentially moving average based multi-scale summation
EMR	Emergency
eNB	evolved Node B
EPC	Evolved Packet Core
EPOCHS	Electric Power and Communication Synchronizing Simulator
EPS	Electrical Packet Switching
ET.C	Total energy for compression
ET.C(0)	Overhead energy for decompression
ET.D	Total energy for decompression
ET.D(0)	Overhead energy for compression
ET.UDW	Total energy for download of the uncompressed file
ET.UUP	Total energy for upload of the uncompressed file
EWMA	Exponential Weighted Moving Average
FCFS	First-Come First-Serve
F-coverage	Focused Coverage
FFD	First Fit Decreasing
FFT	Fast Fourier transform
FIFO	First In First Out
FIR	Finite impulse response
FNCS	Framework for Network Co-Simulation
FSO	Free Space Optics
GECO	Global Event-driven Co-simulation framework
GPS	Global Positioning System
GridMat	Matlab Toolbox for co-simulation
GTP	GPRS Tunneling Protocol

HAEC	Highly adaptive and Energy Efficient Computing
HAEP	Hospital Assignment for Emergent Patients
HD	Hard Disk
HDMI	High-Definition Multimedia Interface
HPC	High Performance Computing
HT	Hough transformation
HVAC	Heating, Ventilation, and Air Conditioning
IA	Initial Attachment
IaaS	Infrastructure as a Service
ICN	Information-centric network
ICT	Information and Communication Technology
IDC	International Data Corporation
IDE	Integrated Development Environment
IDR	InfoMax data retrieval
IETF	Internet Engineering Task Force
I_{IDLE}	Idle current
IO	Input-Output
IoT	Internet-of-Things
IPNSIG	InterPlaNetary Internet Special Interest Group
ISP	Internet Service Provider
iSR	Iterative Service Redeployment
IT	Information Technology
JSON	JavaScript Object Notation
LCM	Look-Compute-Move
LEO	Low Earth Orbit
LLC	Last Level Cache
LMS	Least Mean Squares
LO	Load Only
LTE	Long Term Evolution
LUM	luminosity
MAC	Medium Access Control
MAC	Media Access Control
MANETs	Mobile Ad Hoc Networks
MB	Mother (Main) Board
MCC	Maximum Correntropy Criterion
MEE	Minimum Error Entropy
MEMS	Micro Electro Mechanical Systems
METhoD	Mobile Emulator Testbed for DTNs
MME	Mobility Management Entity

MMOG	Multiplayer Online Game Servers
MOSAIC 2B	Mobile empowerment for Socio-economic development in South Africa
MPTCP	Multi-Path TCP
MSE	Mean Square Error
MSNs	Mobile Sensor Networks
MVU	Minimum Variance Unbiased
NB	Node B
NDN	Named-data network
NFV	Network Functions Virtualization
NIC	Network Interface Card
NS-2	Network Simulator version 2
NS-3	Network Simulation version 3
NUM	Network Utility Maximization
OCS	Optical Circuit Switching
OMNET++	Objective Modular Testbed in C++, discrete event simulation framework
ON	Always ON
OpenDSS	Electric power Distribution System Simulator
OSA	Open System Authentication
OSM	Open Street Map
P2P	Peer-to-Peer
PaaS	Platform as a Service
PCO	Protocol Configuration Option
PDA	Personal Digital Assistant
PDN	Packet Data Network
P-GW	Packet Data Network Gateway
PiccSIM	Platform for Integrated Communications and Control design, Simulation, Implementation and Modeling
PIE	Proportional Integral controller Enhanced
PIR	Passive InfraRed sensor
PIRL	PIR Living room
PNNL	Pacific Northwest National Laboratory
PNSR	Peak Signal Noise Ratio
POI	Point of Interest
POP	Popularity
PPM	Prediction by Partial Matching
P-Router	Penultimate Router
PSD	Power spectral density

PSLF	Positive Load Flow Program, Power flow and dynamics simulation software
PSTN	Public Switch Telephone Network
PU	Primary user
PUTCO	Public Utility Transport Corporation
QoE	Quality of Experience
QoS	Quality of Service
RAM	Random Access Memory
RAN	Radio Access Network
RBF	Radial basis function
RCP	Reception Control Protocol
RDR	Reliable data retrieval
RF	Radio Frequency
RFC	Request for Comments
RFID	Radio-Frequency IDentification
RG	Residential Gateway
RI	Radio identification
RNC	Radio Network Control
ROI	Region of Interest
RSFC	Receiver-side Flow Control
RSSI	Received Signal Strength Indication
RTT	Round-Trip Time
SA	Safe Away
SDN	Software Defined Networking
SDR	Simple data retrieval
SF	Sampling frequency
S-GW	Serving Gateway
SH	Safe Home Activity
SINR	Signal to Interference plus Noise Ratio
SLA	Service-Level Agreement
SMS	Short Message Service
SNMP	Simple Network Management Protocol
SNR	Signal to noise ratio
SPEC	Standard Performance Evaluation Corporation
SPEED	Sequence Prediction via Enhanced Episode Discovery
SRM	Structural Risk Minimization
SS	Safe Sleeping
SSL	Semi-Supervised Learning
SU	Secondary users

S-VLAN	Service VLAN
SVM	Support Vector Machine
SWDC	Small-World Datacenter
T.C	Time to compress
T.D	Time to decompress
T.UDW	Time to download the uncompressed file
T.UUP	Time to upload the uncompressed file
TCP	Transmission Control Protocol
TCP-RRE	TCP Receiver-Rate Estimation
TEID	Tunnel Endpoint Identifier
Th.C	Compression throughput
Th.C.LL	Compression throughput lower estimate
Th.C.UL	Compression throughput upper estimate
Th.D	Decompression throughput
Th.D.LL	Decompression throughput lower estimate
Th.D.UL	Decompression throughput upper estimate
Th.UDW	Uncompressed download throughput
Th.UUP	Uncompressed upload throughput
ToR	Top-of-Rack
TRILL	Transparent Interconnection of Lots of Links
TT	Triangle Tessellation
TTL	Time-To-Live
UAV	Unmanned Aerial Vehicle
UDP	User Datagram Protocol
UDR	Unreliable data retrieval
UE	User Equipment
UF	Uncompressed file
UG-Router	Unified Gateway Router
UMass	University of Massachusetts
UMTS	Universal Mobile Telecommunications System
UPS	Uninterruptible Power Supply
US	Uncompressed file size
USB	Universal Serial Bus
VBS	Supply voltage
VGA	Video Graphics Array
VLAN	Virtual Local Area Network
VM	Virtual Machine
WiMAX	Worldwide Interoperability for Microwave Access
WLAN	Wireless Local Area Network

WSN	Wireless Sensor Networks
WT	Wavelet transform
WTMM	Wavelet transform modulus maxima
WTMP	Wavelet transform multiscale product
WTMS	Wavelet transform multi-scale summation
ZiSAS	ZigBee-based intelligent self-adjusting sensor

PART I

Data Center Computing

1

Flyover: A Cost-Efficient and Scale-Out Data Center Network Architecture

Sheng Xu[1,2], Binzhang Fu[1,2], Mingyu Chen[1,2] and Lixin Zhang[1,2]

[1]SKL Computer Architecture, ICT, CAS, Beijing, China
[2]University of Chinese Academy of Sciences, Beijing, China

Abstract

Torus, which is simple and incrementally expandable, is considered a good scale-out model for the large-scale computing systems, such as data centers. However, one downside of Torus is its long network diameter. A typical approach to address this problem is using random shortcuts. However, this approach does not consider the variety of data center traffic and can lead to severe non-uniform network performance. In this chapter, we propose *Flyover*, which exploits the flexibility of optical circuit switching to add on-demand shortcuts, as a cost-efficient and scale-out network architecture for DCNs. *Flyover* has three key features. First, it gives priority to the *serpent* flows, which have large size and long distance, over the elephant flows, which only have large size. It reduces the pressure on the electrical torus network drastically and thus improves the overall network performance. Second, *Flyover* generates region-to-region instead of the point-to-point shortcuts. It allows the valuable optical shortcuts to be efficiently utilized. Third, a semi-random heuristic algorithm is used to gain improved network performance with reduced computation time. In addition, a few extensions of *Flyover* are discussed and their evaluation is provided to demonstrate the scalability of *Flyover*. Finally, *Flyover* is extensively analyzed and compared with its counterparts using both simulators and prototypes. The experimental results show that Flyover could improve the network throughput by 135% and reduce the latency by 73%.

Keywords: Datacenter Networks, Scale-out, Hybrid Networks Architecture.

1.1 Introduction

The success of large-scale data center applications heavily depends on the underlying cluster system [1]. In general, incrementally expanding computing and storage components is not a big problem. However, the network part is a different story. For example, for the tree-like networks [2, 3], their size is bounded by the number of ports in each switch. Thus, the network cannot be freely expanded unless free ports are reserved. Random networks [4–6] are arbitrarily expandable, but they suffer from high cabling and routing complexity due to the fact that there is no regularity can be exploited. An *n*-D torus, where nodes are arranged in an *n* dimensional cube and form a ring in each dimension,[1] is simple and regular. And it is also arbitrarily expandable as nodes can be freely added in any one dimension. Due to these good features, *n*-D torus is widely adopted by supercomputers [7]. However, its main drawback is the long network diameter, which may lead to longer average shortest path length (ASPL) and increase the possibility of congestion.

To alleviate these problems, in [8] the authors proposed a way that adds static random shortcuts to the torus network architectures. These static random shortcuts follow a small-inspired [9] pattern.[2] A data center based on this random topology is called as Small-World Datacenter (SWDC). Generally, this approach is effective. As shown in Table 1.1, by adding two random shortcuts per node to a 32 × 32 2-D torus network, the diameter and ASPL are reduced by 4.6× and 3.6×, respectively. The reduced network diameter and ASPL indeed lead to improved network performance. As shown in Figure 1.1, the network throughput could be improved by 1.2× in average (the associated experimental setting is in Section 1.4). However, we could see from this figure that static random shortcuts generate severe non-uniform performance across those 11 traffic patterns. For example, under random_32 traffic pattern, the network throughput is as high as 0.69. However, under permutation traffic

Table 1.1 The reduction of network diameter and ASPL by adding random shortcuts

32 × 32 Torus	Diameter	ASPL
without shortcut	32	16.02
one shortcut per node	8	5.39
two shortcuts per node	7	4.46

---

[1]In this chapter, we use the terms "node" and "switch" interchangeably.

[2]Each node has multiple links to its nearby nodes and also has several links connecting to the distant nodes when needed.

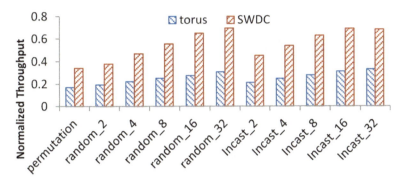

Figure 1.1   Performance comparison between 2-D torus and SWDC.

pattern, the network throughput is only 0.34. The main reason behind this phenomenon is that shortcuts are added randomly instead of on-demand.

The prerequisite for applying on-demand shortcuts is exactly knowing the traffic demand between any two nodes. Although there are SNMP and other popular tools, such as sFlow, could be utilized to monitor the network status, we believe that it makes sense, in certain circumstances, for high-priority applications to explicitly describe the bandwidth they need. The same philosophy was widely used by many solutions [10, 11] focused on improving QoS. In the following text of this chapter, we assume that the bandwidth demand between any two nodes is known in advance.

Applying the on-demand shortcuts means that the shortcuts should be added and removed dynamically. Obviously, this will lead to severe cabling problem. To address this problem, we exploit the optical circuit switching (OCS) to create and destroy the shortcuts dynamically. In other words, a shortcut could be made/removed by creating/destroying an optical circuit between the two nodes. Thus, we propose the *Flyover* network architecture. *Flyover* is a hybrid OCS/EPS (electrical packet switching) network architecture. The EPS layer is a 2-D torus network that connects all top-of-rack (ToR) switches and provides the basic connectivity. The OCS layer, on the other hand, provides the on-demand shortcuts for improved network performance. We should note that combining OCS network with EPS network is not a new idea. For example, the OCS network has been combined with the ToR switches in c-Through [12] to take over the job of transmitting elephant flows. Helios [13] did the similar work except that the OCS was introduced at the core level. However, the proposed *Flyover* is more than a straightforward use of the above idea on a 2-D torus EPS network. *Flyover* differs from them in the following three features.

First, *Flyover* defines a new type of flow, namely the *serpent* flow. Unlike the elephant flow, which only considers the size of flow, the *serpent* flow further takes the distance between source and destination into consideration. Existing solutions prioritize elephant flows; thus, those "not so fat but long distance" flows will be transported along the EPS network. On the other hand, as the distance increases, the possibility of congestion increases. Consequently, the network performance degrades. *Flyover*, on the other hand, utilizes the *serpent* flow to reflect the true requirement of flows on network resources. Therefore, the flows requiring more network resources are prioritized. In Section 1.4.2, we will show that the network throughput and latency could be improved by up to 13% and 22%, respectively, by taking the transportation distance into consideration.

Second, *Flyover* generates shortcuts in a region-to-region manner instead of the point-to-point manner. When generating the shortcuts, the traffic demand between two regions instead of two individual ToRs is considered. Once the shortcut is created, ToRs in the same region could share the shortcut. Furthermore, an individual ToR could belong to multiple regions. Thus, the possibility of using optical shortcuts is dramatically increased, and hence ASPL and the network performance could be significantly improved. As we will show in Section 1.4.3, by applying this technique, the network throughput and latency could be further improved by up to 35% and 47%, respectively.

Third, *Flyover* employs S-Random reconfiguration algorithm, which can reduce the computation time while improving the network performance. As we will show in Section 1.4.4, compared with Edmonds algorithm, S-Random algorithm could improve the network throughput and latency by up to 51% and 41%, and by 26% and 27% in average, respectively.

Finally, the proposed *Flyover* is highly scalable and cost efficient. As the network size increases, *Flyover* could always provide on-demand shortcuts by adding more OCS switches. To support this feature, three port partition algorithms are discussed and evaluated in this chapter. The results show that *Flyover* could always get satisfied performance. Furthermore, compared with the static shortcut solution, *Flyover* only adds small cost. For example, the overhead is smaller than 1% for a 1024 ToRs case.

We summarize our main contributions as follows.

1. A new hybrid OCS/EPS network architecture, *Flyover*, is proposed. *Flyover* adds on-demand shortcuts to 2-D torus and improves the network performance significantly.

2. *Serpent* flows, which takes both the flow size and transportation distance into consideration, are considered when utilizing the shortcuts for the flows that require more network resources.
3. *Flyover* utilizes the shortcuts in a region-to-region instead of the traditional point-to-point manner to improve the network performance.
4. *Flyover* utilizes semi-random heuristic algorithms that can reduce the computation time and improve the network performance simultaneously.

The rest of this chapter is organized as follows: Section 1.3 discusses the proposed *Flyover*. Section 1.2 is related works and motivation. Section 1.4 and Section 1.5 compare *Flyover* with its counterparts with results from both simulations and prototypes. Finally, Section 1.7 concludes this chapter.

1.2 Related Works and Motivation

1.2.1 Related Works

Recently, there are a body of works focusing on using novel technologies to design a flexible, cost-effective, and performance-efficient DCN. Such solutions include unstructured DCN topologies [4–6, 8, 14], wireless DCNs [15–19], and optical DCNs [12, 13, 20–24].

Unstructured DCN topologies: In previous works, the idea of unstructured topologies has been applied in both HPC systems [25–27] and data center systems [4–6, 8]. The traditional electrical switching-based DCNs, such as fat-tree [2] and VL2 [3], are limited to the regular structures, which are not easily and arbitrarily expandable. In order to resolve this problem, unstructured DCNs, such as Scafida [4], SWDC [8] and Jellyfish [5], are proposed to provide an incremental expansion capability. These unstructured network topologies are generated mainly by two methods: (1) by using random graphs [4, 5] and (2) by adding random shortcuts to regular topologies [8, 25, 27], such as rings and torus networks. These topologies achieve low diameter, low average shortest path length, and thus low end-to-end network latencies [26], so the performance of networks is greatly improved. However, all of them exploit only the static random shortcuts, which neglect the real demands of the applications' runtime traffics. Moreover, the complex wiring can also affect the realization of these topologies.

Wireless DCNs: Recently, some of the researchers are motivated by the advantages of wireless communication techniques in energy consumption, cabling complexity, and reconfigurable capacity. They proposed wireless DCNs as a flexible solution for high performance data centers. There are two wireless techniques adopted in previous works [15, 16–19]: radio-frequency (RF) and free-space-optics (FSO). In [15, 16], Kandual et al. proposed *flyways* which utilizes the 60 GHz wireless spectrum for providing on-demand links to meet the requirement of the hotspot communication. In [17], X. Zhou et al. put mirrors on the DC ceilings to reflect the incoming wireless signals (RF), and hence an indirect line-of-sight communication can be established between any two racks in a data center. Hamedazimi et al. proposed *firefly* [18, 19], a FSO-based wireless DCN, where FSO transceivers are placed on ranks and all inter-rank communications are performed using optical signals.

Optical DCNs: The optical DCNs can be divided into all-optical DCNs, such as OSA [21], WaveCube [22] and RODA [23], and hybrid electrical/ optical DCNs, such as c-Through [12], Helios [13], and Mordia [20]. OSA [21] shows the high flexibility as its topology architecture and link capacities can adapt to the changing traffic patterns, but its network scale is limited by the low port density of the MEMS (Micro-Electro-Mechanical Systems)-based switch. In order to overcome this problem, WaveCube and RODA, two MEMS-free optical DCN architectures, are suggested to achieve scalability, fault-tolerance, and high-performance simultaneously. The difference between them is that the RODA uses *tunable* transceivers which can switch in between different wavelengths, so RODA has more flexibility in using lightpaths. c-Through/Helios build single-hop lightpaths for the relatively steady and longer flows, whereas those delay-sensitive and bursty flows are transmitted through the electrical switches. Different from c-Through/Helios, Mordia could provide more fine-grained sharing of optical circuit by using the techniques of WDM and time slot by paying a higher implementation cost.

Flyover is closely related to two kinds of works, i.e., the random networks and hybrid optical/electrical networks. However, *Flyover* differs from them in three main features. First, *Flyover* adds shortcuts according to both the amount of data and distance of the transmissions. Second, *Flyover* allows ToRs to share shortcuts. Third, *Flyover* employs S-Random reconfiguration algorithm.

1.2.2 Motivation

As described above, SWDC can maintain the advantages of incremental expansion and high network performance (low diameter and ASPL). However, under different traffic patterns, SWDC generates severe non-uniform network performance, as shown in Figure 1.1. This drawback seriously affects the use of SWDC for the large scale of datacenter. Before addressing this problem, we first need to understand the main cause of this problem. Based on the analysis on the datacenter traffic patterns [15, 28, 29], only a few hotspot TORs in data centers communicate with a few other ToRs at the same time, where most elephant flows are generated. Reference [30] shows nearly 80% of elephant flows have a stable transmission time about 1.5–2.5 s. If elephant flows requiring high network bandwidth are transferred along the EPS network, they can easily lead to resources contention and hence increase congestion. In SWDC, the shortcuts are randomly generated and cannot satisfy the demand of elephant flows, particularly when these elephant flows are uneven distributed. To address this problem, one could utilize on-demand shortcuts to migrate parts of elephant flows to the exclusive high-speed network based on the use of MEMS-based optical circuit switching [12, 13]. This approach has two benefits: (1) increasing the network throughput and (2) reducing the EPS network congestion.

Although this idea of optimizing the elephant flows is straightforward, there are still several challenges that need to be addressed.

1. How to choose the appropriate elephant flows to be migrated from EPS network to OCS network.
2. How to use these shortcuts effectively.
3. How to build a large-scale data center by using optical switches with low port density.

All these challenges will be addressed in the following section.

1.3 The Flyover

1.3.1 Overview

As shown in Figure 1.2, *Flyover* is a two-level network. At the bottom, all servers are directly connected to ToR switches. Each ToR switch has four (groups) ports to connect its neighbors in four directions, i.e., east, west, south, and north. Particularly, the west neighbor of the ToR in the westmost column is the corresponding ToR in the eastmost column in the same raw. However, for simplicity, this kind of warp-around links are omitted in Figure 1.2. Obviously, the 2-D torus EPS network already provides the connectivity between any two ToRs. However, as we discussed in the above section, 2-D torus has a long network diameter.

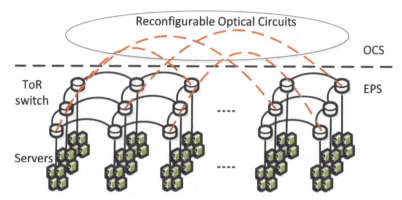

Figure 1.2   The overview of Flyover network architecture. Note that the wrap-around links of 2-D torus are omitted for simplicity.

Table 1.2   Table of notations

i, j	$\triangleq$	represent any two nodes.
N_i, S_i, E_i, W_i	$\triangleq$	represent the nodes neighboring with i in all four directions.
D	$\triangleq$	represents the bidirectional demand matrix.
MD	$\triangleq$	represents the Manhattan distance matrix.
D^S	$\triangleq$	represents the bidirectional demand matrix based on our defined *serpent* flow.
RD^S	$\triangleq$	represents the bidirectional regional demand matrix based on our defined *serpent* flow.
A_h	$\triangleq$	represents the amount of nodes with the distance h from each end of the shortcut.

To address this problem, *Flyover* exploits OCS switches (not shown in the figure) on top of ToRs to provide dynamically reconfigurable optical circuit. From the topology's point of view, these reconfigurable optical circuits are actually the shortcuts of the underlying 2-D torus EPS network. The bandwidth of optical shortcut could be different from the bandwidth of links connecting adjacent ToRs in the 2-D torus EPS network. However, in this work, we assume they are equal for simplicity. As for the routing algorithm, we treat the optical shortcuts as normal links. Thus, a shortest path first routing is utilized. Since on-demand shortcuts are created for flows with high demand of network resources, most traffic load is transported through shorter path and network performance is maximized. In the rest of this section, we will discuss the way to generate on-demand shortcuts and show how these shortcuts are utilized in a region-to-region manner.

1.3.2 Serpent Flow

Traditionally, shortcuts are generated according to the bidirectional demand matrix D. Particularly, each element of the matrix, such as $D_{i,j}$, reflects the amount of data that should be transmitted between the two nodes, say i and j, in both directions. Obviously, the bidirectional demand matrix D does not consider the factor of transportation distance of different flows. This principle works well in c-Through [12] and Helios [13] because they are based on the tree-like EPS networks. For example, for a multi-root tree network, the equal-cost multi-path routing (ECMP) is often utilized to balance the traffic. In this scenario, all flows will have a same transportation distance. Therefore, it is reasonable for them to ignore the factor of transportation distance. However, in 2-D torus network, this situation is quite different.

By taking the transportation distance into consideration, we define a new type of flow, namely the *serpent* flow. A *serpent* flow is big if it is "big" in both size and transportation distance. As shown in Equation (1.1), we define the *Serpent* of a flow as the product of the flow size ($Size\_f$) and Manhattan distance (MD_f) between the source and destination ToRs. Therefore, considering the transportation distance of flows, the bidirectional demand matrix D is updated to D^S that can be calculated as shown in Equation (1.2), where n is the number of flows, i and j are the ToRs in two ends. We do not consider the case that $i = j$, since we assume $D_{i,i} = 0$. We assume a minimal routing, such as the dimension-order routing, is utilized by the 2-D torus network. Therefore, the flows between i and j have the same transportation distance, i.e., $MD_{i,j}$.

$$Serpent\_f = Size\_f \times MD\_f \tag{1.1}$$

$$D^S_{i,j} = \sum_{f=1}^{n} Serpent\_f = D_{i,j} \times MD_{i,j} \tag{1.2}$$

1.3.3 Semi-Random Heuristic Algorithm

Based on the D^S, finding the optimal shortcuts can be translated into the classical maximum matching problem. Therefore, we could exploit the Edmonds algorithm [31] to generate the shortcuts. However, we find that there are two important problems in using Edmonds algorithm to generate the shortcuts. First, the complexity of Edmonds algorithm is O(N^3), where N is the number of ToRs. Therefore, for a large-scale network, the reconfiguration cost is high, especially for short-lived high-priority applications. Second, Edmonds algorithm calculates the optimal shortcuts according to serpent-flow traffic matrix, which easily results in the multiple shortcuts connecting two remote regions. As shown in Figure 1.3, the distance between i and j is 7 hops; according to the calculation method of *serpent* flow, the bandwidth demands between i and j is multiplied by 7. Similarly, the bandwidth demands between S_i and S_j, N_i and N_j, E_i and E_j, and W_i and W_j are all multiplied by 7. Since Edmonds algorithm only pursues the maximum point-to-point bandwidth demand and neglects the number of shortcuts between two regions, in the worst-case, shortcuts are established between i and j, S_i and S_j, N_i and N_j, E_i and E_j, and W_i and W_j. Thus, there are too many shortcut links being used for the connection between two regions; the benefits of using shortcuts are greatly reduced. The shortcuts used for connecting to other regions are reduced and, hence, potentially affects the performance negatively.

To resolve these two problems, we propose the use of semi-random heuristic algorithm (S-Random) to generate the shortcuts. This algorithm is easy to implement and effective. Generally, there are two basic steps to generate shortcuts for each ToR. As a first step, a ToR is chosen randomly, say ToR i. As a second step, for ToR i, a ToR j is selected, where ToR j fulfills the following two conditions. First, ToR j has free ports. Second, $D_{i,j}^S$ is not lower than $D_{i,k}^S$, where k is any ToR that has free ports. In the first step, by adding randomness to the choice of ToR, S-Random algorithm avoids the redundant shortcuts

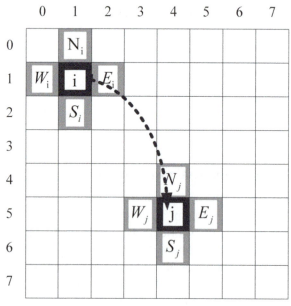

Figure 1.3   An illustrative example of using optical shortcuts in a region-to-region manner.

between two regions and allows better performance. Meanwhile, in the second step, we can make use of quick sort algorithm [32] to find the optimal ToR, and hence the complexity of the proposed algorithm can be reduced to $O(N \log N)$.

1.3.4 Region-to-Region On-Demand Shortcuts

To generate the region-to-region shortcuts, we first extend the bidirectional demand matrix D^S to be the bidirectional regional demand matrix RD^S as shown in Equation (1.3), where r indicates the maximal distance within which the ToRs could share the optical shortcut, A_h is the amount of ToRs with the distance h from each end of the shortcut, and i_a represents the a_{th} neighbor of i with distance h. In this way, the bidirectional demands between one end of the shortcut and the other end's neighbors are also considered. For example, if $r = 1$, only the direct neighbors are considered. As shown in Figure 1.3, the regional bidirectional traffic demand between i and j can be calculated as $RD^{i,j} = D^S_{i,j} + \frac{1}{4} \times \{D^S_{i,E_j} + D^S_{i,W_j} + D^S_{i,S_j} + D^S_{i,N_j} + D^S_{E_i,j} + D^S_{W_i,j} + D^S_{S_i,j} + D^S_{N_i,j}\}$. By replacing D^S with RD^S, the region-to-region shortcuts could be easily found using the semi-random algorithm discussed above.

$$RD^S_{i,j} = D^S_{i,j} + \sum_{h=1}^{r} \frac{\sum_{a=0}^{A_h} D^S_{i,j_a} + \sum_{a=0}^{A_h} D^S_{i_a,j}}{A_h} \qquad (1.3)$$

1.3.5 The Scalability of Flyover

Generally, one OCS switch could provide enough capacity to provide the on-demand shortcuts for a data center with a moderate size. For example, a 320-ports OCS switch could provide one optical shortcut for each of 320 ToR switches. Assuming each rack has 40 servers, the system could have 12800 servers. There are two situations where more OCS switches are needed. First, more than one shortcuts is required by each ToR switch. Second, the system is expanded with more servers. For any case, we could expand the *Flyover* by adding more OCS switches.

When there are multiple OCS switches, the ToRs should determine which switch they want to use to create shortcuts. We propose three algorithms. Before discussing the algorithms, we first index all ToRs with a two-element tuple in the traditional 2-D torus way. For example, the ToR i locating in column x and row y is indexed as (i_x, i_y). Furthermore, the number of OCS switches is defined as C. Therefore, the partition algorithms can be briefly described as follows.

Random: The OCS switch is randomly chosen by ToRs with equal possibility or different possibility according to the capacity of OCS switch.

Column-based: The OCS switch is selected based on the column the ToR belongs to. For example, we could assign ToR i to OCS $i_x \% C$.

XY-based: The OCS switch is selected based on the sum of coordinates of ToR in both row and column. For example, the OCS $(i_x + i_y) \% C$ will be used to select ToR i.

To compare the aforementioned partition algorithms, we use them to partition ToRs and compare the diameter and ASPL of the generated networks. As shown in Figure 1.4(a), where OCS_n_M indicates that n OCS switches and algorithm M are used, the column-based algorithm always generate higher network diameter than other two algorithms. This is expected since column distance increases as the number of OCS switches increases. As shown in Figure 1.4(b), the results of ASPL are consistent with the diameter results. Due to the good performance and simplicity, the *Random* algorithm is recommended and adopted by *Flyover*.

When multiple OCS switches are utilized, another problem is how to reconfigure them when the applications or applications' traffic changes. There are two options: i.e., *synchronous reconfiguration* and *pipelined (asynchronous) reconfiguration*. In the former, all optical switches are reconfigured in parallel. In other words, they have the same setup and stable periods. In setup periods, all optical switches are shut off and reconfigured simultaneously. In the latter, each optical switch has its own reconfiguration period. After one optical switch has finished its reconfiguration, another switch starts a new round of reconfiguration. Since the pipelined reconfiguration method can lead to a stable network performance during the period of reconfiguration (discussed in Section 1.4.6), it is adopted by the *Flyover*.

1.4 Simulations and Analyses

In this section, we first evaluate the impact of different parameters and configurations on the performance of *Flyover*. The parameters and configurations include the type of flow (*elephant flow* or *serpent flow*), the type of bandwidth demand (*point-to-point* or *region-to-region*), the type of reconfiguration strategy (*synchronous* or *asynchronous*), the type of reconfiguration algorithm (*Edmonds Algorithm* or *Semi-Random Heuristic Algorithm*), and the number of shortcuts (from 1 to 4). The purpose of these evaluations is to provide

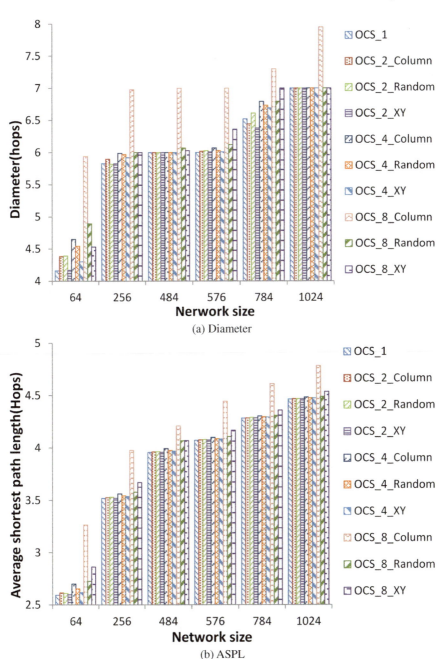

(a) Diameter

(b) ASPL

Figure 1.4   The performance comparison between different partition algorithms with different sizes.

a comparative view on the performance of *Flyover* with different parameters and configurations, and find the best configuration. We also present a comprehensive performance comparison between *Flyover* and *SWDC* to demonstrate that the on-demand shortcuts outperform the static random shortcuts. Furthermore, the evaluation of design space shows a good solution for the construction of large-scale data center network by using *Flyover* structure.

1.4.1 The Methodology

A cycle-accurate network simulator, BookSim2.0 [33], is used for simulations. We have extended the base simulator to support TCP flows and OCS switching techniques. Table 1.3 summarizes the base configurations we use for simulations. We simulate a data center network with 1024 ToR switches, and each ToR switch connects to 32 servers. As a result, the whole data center network contains 32768 servers. We use two different sizes of packet, 1 flit[3] and 20 flits, to represent the control packet (such as ACK) and data packet, respectively. The flow size is uniformly distributed between 500 and 10,000 packets. The ratio between the duration of a stable period and that of a setup period is 20:1. The same traffic patterns used by [34, 35] are adopted in the simulations, and the three traffic patterns are described as follows.

Permutation: Each ToR transfers data to one other ToR picked out randomly with the constraint that each ToR only receives one flow at any time. When a ToR finishes the data transfer, it re-selects a ToR with the same constraint.

Incast_N: Each ToR transfers data to N ToRs picked out randomly with the constraint that each ToR receives no more than $N + 1$ flows at any time. The N can be 2, 4, 8, 16, 32. When a ToR finishes one flow, it immediately chooses another ToR randomly to start a new flow with the same constraint.

Table 1.3   Simulation parameters

Parameter Description	Value
Servers Per ToR	32
Number of ToR	1024
Buffer Size	12.5 Data packets
Switch Delay (AS + CS)	2 cycles
Link Delay	1 cycles
Packet Size	1, 20

---

[3]The flit is the basic unit for flow control. It is widely used in the interconnection networks of supercomputers. We adopt this technique for its high performance.

Random_N: Each ToR transfers data to no more than N ToRs picked randomly with the constraint that each ToR receives no more than N flows at any time. The N can be 2, 4, 8, 16, and 32. When a ToR finishes one flow, it immediately chooses another ToR randomly to start a new flow with the same constraint.

1.4.2 Elephant Flow vs Serpent Flow

These experiments show the benefits of considering the distance of transmission. As shown in Figure 1.4(a), the "Serpent flow based" represents the *Flyover* network considering the distance of transmission. Compared with "Elephant flow based," which does not consider the distance, the network throughput is improved by up to 13% and 9.8% in average as shown in Figure 1.5(a). Correspondingly, the average packet latency is reduced by up to 22% and 17% in average. We should note that only one un-shared shortcut per ToR is used to highlight the effect of considering the distance factor. In the following experiments, we will find that the network performance could be further improved by sharing more shortcuts.

1.4.3 Point-to-Point vs Region-to-Region

This section discusses the effect of sharing shortcuts within regions. As shown in Figure 1.6, the "= 0" results represent the network that considers the factor of distance but does not allow ToRs to share shortcuts, and "$r = 1$" and "$r = 2$" results represent the networks that allows ToRs to share shortcuts within one and two hops respectively. Furthermore, only one shortcut per ToR is generated. Thus, the benefits are solely introduced by sharing shortcuts in regions.

As shown in Figure 1.6, by sharing the shortcuts within one hop, the network throughput is improved by 13% maximally and 8.9% in average, and the packet latency is improved by 22.2% maximally and 17.8% in average. By sharing the shortcuts within two hops, the network throughput is further improved by 18.8% maximally and 11.9% in average, and the packet latency is further improved by 22.2% maximally and 19% in average. As discussed before, the main benefit of sharing shortcuts is the increased utilization rate of valuable optical shortcuts. By sharing the shortcuts within one hop, the utilization of the shortcuts is improved by 26.8% maximally and 15% in average. By sharing the shortcuts with two hops, the utilization is further improved by 23.4% maximally and 15.4% in average. Considering the large overhead of further increasing the size of region, sharing the shortcuts within two hops is a good choice.

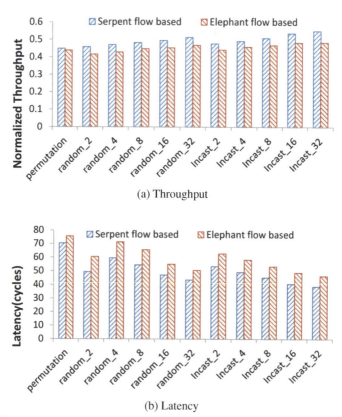

(a) Throughput

(b) Latency

Figure 1.5 The comparison between the networks generated with and without considering the transportation distance. Note that only one shortcut is generated for each ToR and the shortcuts are allowed to be shared to emphasize the effect of considering the transportation distance.

1.4.4 Edmonds Algorithm vs Semi-Random Heuristic Algorithm

This section discusses the network performance of *Flyover* with two different reconfiguration algorithms: Edmonds algorithm and semi-random heuristic algorithm. The results of simulation experiments are shown in Figure 1.7. Compared with Edmonds algorithm, S-Random algorithm could improve the network throughput and latency by 51% and 41% maximally, and by 26% and 27% in average, respectively. Moreover, the advantages of S-Random algorithm can also come from the small average diameter and smaller average shortest path length, and both of two metrics are strongly correlated with the performance of network. Compared with Edmonds algorithm, the average

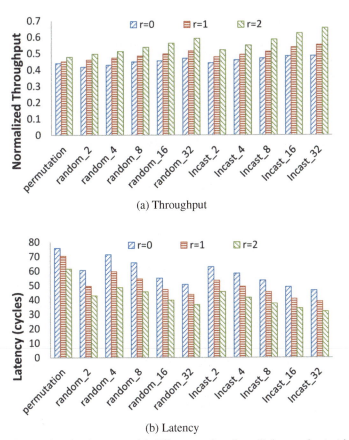

(a) Throughput

(b) Latency

Figure 1.6   Sharing the shortcuts with different region sizes. Only one shortcut is used.

network diameter is reduced by 13.5% maximally and 11.8% in average, and the average shortest path length is reduced by 8.2% maximally and 6.6% in average. The main reason for these results can be found in Section 1.3.3.

1.4.5 The Number of Shortcuts

Increasing the number of optical shortcuts not only increases the network performance, but also increases the system cost. As a trade-off between performance and cost, we utilize maximally 4 optical shortcuts for each ToR in our experiments. Furthermore, to understand the performance benefits further enabled by utilizing more shortcuts, we allow the ToRs to share the shortcuts within two hops according to the analysis above.

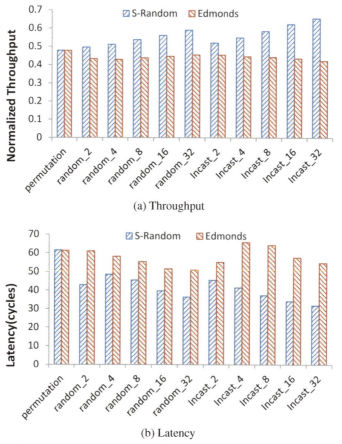

(a) Throughput

(b) Latency

Figure 1.7   Comparison between s-random algorithm and edmonds algorithm.

As shown in Figure 1.8, by using two shortcuts per ToR, the throughput is improved by 68% maximally and 30.3% in average, and the packet latency is improved by 145% maximally and 44.3% in average, compared with using only one shortcut. By adding two more shortcuts per ToR, the throughput is further improved by 17.3% maximally and 10% in average, and the packet latency is further reduced by 22% maximally and 15% in average. Considering the fact that doubling the number of shortcuts per ToR will significantly increase the cost, two shortcuts per ToR provide the better performance-cost. Another benefit of using more shortcuts per ToR is that more uniform network performance could be provided to different applications. By using two instead of one shortcut, the standard deviation of

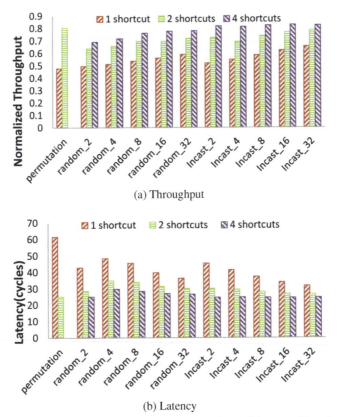

(a) Throughput

(b) Latency

Figure 1.8 The network performance with different numbers of shortcuts. Note that shortcuts within two hops could be shared in this experiment.

the throughput is reduced from 0.054 to 0.051, and the standard deviation of packet latency is reduced from 8.21 to 2.99. We can see that using more shortcuts leads to much more uniform latency performance across different traffic patterns. As expected, by further using more shortcuts, the standard deviation could be further reduced but the further improvement is no longer significant.

1.4.6 Reconfiguration Cost

This section discusses the methods of reconfiguring multiple OCS switches. Assume there are two OCS switches, namely OCS-1 and OCS-2. In the synchronous reconfiguration strategy, OCS-1 and OCS-2 are simultaneously reconfigured by every 8,000 cycles. On the other hand, in the pipelined reconfiguration strategy, OCS-1 and OCS-2 are iteratively reconfigured with an interval of 4,000 cycles. For example, OCS-1 is reconfigured at cycle-0, cycle-8,000, ..., and OCS-2 is reconfigured at cycle-4,000, cycle-12,000 , ...,. The reconfiguration cost of both switches are assumed as 400 cycles, and the simulations run 400,000 cycles.

As shown in Figure 1.9, compared with the synchronous manner, the pipelined reconfiguration strategy could improve the network throughput and latency by 7% and 7.8% maximally, and by 4.7% and 1.5% in average, respectively. We can see that the improvement is relatively modest. The reason is that the reconfiguration cost (400 cycles) is only 5% to the reconfiguration period (8000 cycles). However, if we look at the standard deviation of the packet latency, we can see that the pipelined strategy could get a lot more stable performance. With the synchronous reconfiguration strategy, the standard deviation of the packet latency under Random_8 traffic pattern is 35.1.

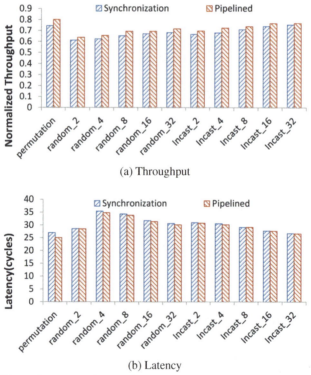

(a) Throughput

(b) Latency

Figure 1.9   The performance of networks with synchronous and asynchronous (Pipelined) reconfiguration algorithms.

However, with asynchronous reconfiguration strategy, the standard deviation is reduced to 31.1.

1.4.7 Flyover vs SWDC

In this section, we bring together all of *Flyover*'s unique features and compare *Flyover* with SWDC. Particularly, *Flyover* adopts the *serpent* flow-based solution to generate two shortcuts for each ToR, and the shortcuts are shared by ToRs within two hops. The SWDC, on the other hand, adopts the *elephant* flow-based solution to generate two shortcuts too, but shortcuts are not shared by ToRs. As shown in Figure 1.10, compared with SWDC, *Flyover* improves the network throughput and latency by up to 135% and 277%, and by 37.4% and 69% in average, respectively.

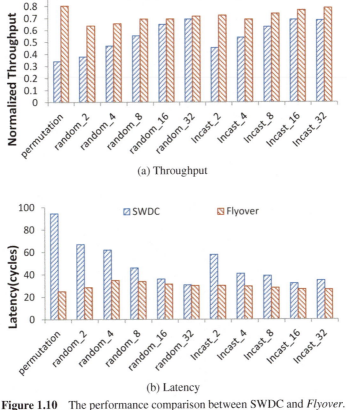

(a) Throughput

(b) Latency

Figure 1.10 The performance comparison between SWDC and *Flyover*.

In addition to improving network performance, *Flyover* can provide more uniform performance, which impacts not only the network performance under different traffic patterns, but also the performance of all nodes in the network. This benefit can be seen from Figure 1.11, which shows the cumulative

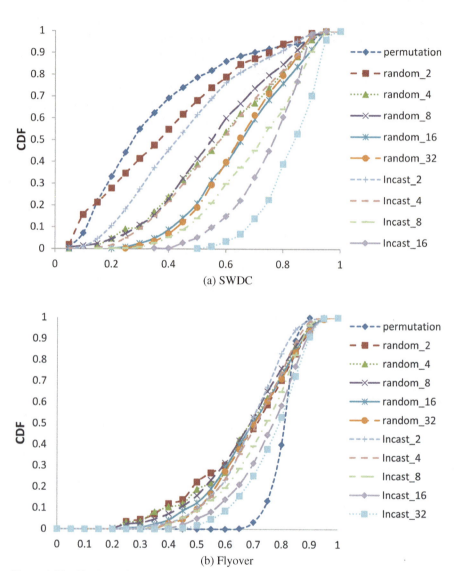

Figure 1.11   The CDF of the normalized throughput for both SWDC and Flyover. The network size is 1024, and each ToR has two shortcuts.

distribution function (CDF) of node's normalized throughput for both SWDC and *Flyover* under different traffic patterns. As shown in Figure 1.11(a), there are significant differences in the distribution of the normalized throughput under different traffic patterns. For example, under the permutation traffic pattern, about 80 percent of the ToRs' normalized throughput is from 0.1 to 0.85, while under incast_32 traffic pattern, over 80 percent of the nodes' normalized throughput is between 0.65 and 0.92. As shown in Figure 1.11(b), opposite to SWDC, the network throughput of *Flyover* is more concentrated. Under the permutation traffic pattern, each ToR switch has less number of aggregated traffics, so when these high-bandwidth flows are migrated from EPS network to OCS network, the network resources competition is greatly alleviated, resulting in more uniform network performance. For example, almost 90 percent of the nodes' normalized throughput is between 0.75 and 0.9. Generally speaking, for all traffic patterns we have evaluated, 80 percent of the nodes' normalized throughput is between 0.35 and 0.9 in *Flyover*, which is far better than SWDC.

The cost of *Flyover* over SWDC mainly depends on the cost of OCS switches. Generally, the cost of OCS switches can be normalized to the number of ports. In this chapter, we adopt the same assumption as [21] that the cost per OCS port is $500. Thus, to provide one shortcut per ToR for a network with 1024 ToRs, 1024 OCS ports are required. The added cost is about $512 k. Assuming the cost of a ToR switch with tens of 10 GE ports and a few 100 GE ports is $100 k, the cost overhead is smaller than 0.5%.

1.4.8 The Design Space

In this chapter, we have considered two methods to build a large-scale data center network by using *Flyover* structure. The first is to build a high-dimensional torus as the baseline network and add small amount of shortcuts on each ToR switch. The second method is to build a low-dimensional torus as the baseline network and add multiple shortcuts on each ToR switch. This section compares the performance and cost of two methodologies. In the following part of this section, we assume a datacenter network with 4096 ToRs, and each ToR connects with 32 servers. In Figure 1.12, nD indicates that n-D torus architecture is selected to build a data center network, and $nD + mshortcuts$ indicates that n-D torus is used as a baseline network architecture and each ToR switch has m ports connected to OCSs to create shortcuts.

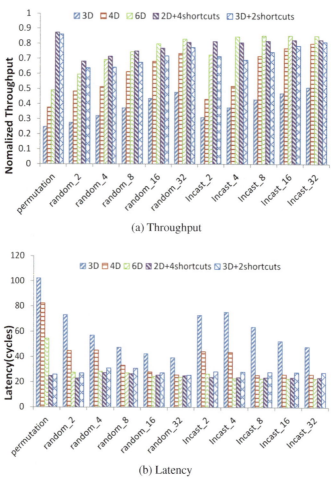

(a) Throughput

(b) Latency

Figure 1.12   Performance of building a large-scale data center network.

The experimental results show that the second method has the better performance in both network throughput and latency. As shown in Figure 1.12, compared with $3D + 2shortcuts$, $2D + 4shortcuts$ improves the network throughput and latency by up to 13.7% and 11%, and by 6.5% and 5.8% in average, respectively. Furthermore, we find that 3D + 2shortcuts and 2D + 4shortcuts cost the same number of ports as an 4D torus, but the performance of having shortcuts is much better. Similarly, $2D + 4shortcuts$ has the similar performance to $6D$ torus, while the number of ports required for connecting the other ToR switches is reduced by 40%. The cost of ToR switch is directly related to the number of the high-speed ports connected with other ToRs, so *Flyover* has higher performance/cost ratio and more stable performance than *nD* torus architecture.

1.4.9 The Cost Comparison

This section focuses on the cost comparison among traditional oversubscribed networks, *Flyover* and fat-tree. Table 1.4 shows the cost and power usage of different network elements, where some of the values are from OSA [21]. In *Flyover*, the connection between ToRs requires the high-speed port. We use 3 100 G-port [36] to replace a 320 G-port needed. We use CWDM-based transceiver ($w = 16$) which can provide high-bandwidth and multiple wavelength channels in a single piece of fiber [13].

To build a data center, which contains 1024 ToRs and each ToR connects with 32 servers, using three-tried 2:1:1 oversubscribed structure, we need 1024 48×10 G ToR switches, 512 64×10 G aggregation switches and 512 32×10 G core switches. To connect all switches, the number of fibers required is 32768. The bisection bandwidth that a 2:1:1 oversubscribed network can provide is 50% of the non-blocking network. The total cost of the structure is $21.3 million, and the power consumption is 295 KW. To construct a fat-tree data center network, we use 2048 64×10 G switches and 1024 32×10 G switches. The number of fibers required is 65536. The fat-tree structure costs is $36.1 million. The power consumption is 491.5 KW. In *Flyover*, each ToR switch has 32 10 G-port, 12 100 G-port, and 32 10 G optical transceivers. We use 1024 ToR switches and 2 1024-port optical circuit switches. The number of fibers required is 8192. The total cost is $35.5 million, with a power consumption of 170.9 KW. For all traffic patterns we have evaluated, *Flyover* achieved 63–81% of the non-blocking bandwidth. As shown in Table 1.5, *Flyover* consumes the lowest power and requires the fewest numbers of fibers. The use of the large number of expensive optical elements increases the cost

Table 1.4   Cost (USD) and Power (Watt) per port for different elements

Element	Cost ($)	Power (W)
MEMS[†]	0.5 K	0.24
ToR (100 G)	1.2 K	3.2
ToR (10 G)[†]	0.2 K	3
Transceiver (CWDM)[†]	0.4 K	1
Fiber[†]	0.05 K	0

[†]The values are from OSA [16].

Table 1.5   Cost, power, wiring and performance of different networks

Architecture	Cost ($)	Power (KW)	Wiring	% of Non-blocking
Traditional (2:1:1)	21.3 M	295	32768	50%
Fat-tree	36.1 M	491.5	65536	100%
Flyover	35.5 M	170.9	8192	63–81%

of *Flyover*, but it is expected that the cost of those optical elements will fall significantly in the future.

1.5 Prototype Evaluations

The testbed used in our experiments consists of 16 Tecal RH2285 servers with one 8-port 1 Gbps NICs and one S5710-52C-EI 48-port 1GigE Ethernet switch. Each of these 16 servers runs an Open vSwitch to emulate ToR switches. These Open vSwitches are directly connected to form a 2-D Torus network. We use the 2-D mesh network to simulate a subnetwork of a huge 2-D torus network. For example, these 16 switches could locate in the middle of a 32×32 2-D torus network. To ensure that the Open vSwitch does not become the bottleneck, each switch port is set to work at 100 Mbps. Within the ToR switch, one switch port is connected to the Ethernet switch, which emulates electrical or optical shortcut used for end-to-end data transfers. We configure routing tables for all ToR switches by using the OpenFlow controller.

In SWDC, the shortcuts connected between two ToRs are determined in advance and not changed. But in *Flyover*, all shortcuts are reconfigured periodically to meet the changing traffic demands. When reconfiguring shortcuts, the OpenFlow controller communicates with all ToR switches to update their routing table. The reconfiguring time is set to be 25 ms, and the stable time is set to be 1 s. Each server runs 2 VMs connected to an Open vSwitch. One VM is used for generating uniform background flow and the other one runs a MapReduce application, TritonSort [37]. As an ideal model, the full bisection bandwidth of the packet-switched network is implemented in the testbed by forwarding all traffic to the Ethernet switch. And background flows are removed to maximize the performance.

The TritonSort application uses 1 GB of random data as its input set, and the number of slavers is set to be 4. On the one hand, according to the distance between master and slavers, the distribution proportion of slavers can be approximately considered as 1:2:1 (close:middle:remote). On the other hand, the number of virtual machines is 32. Considering the fact that hotspot traffic in data center is no more than 10%, so that the 4 VMs used as slaver

Table 1.6 The average completion time

Network Architecture	SWDC	Flyover	Full Bisection Bandwidth
Completion Time (s)	239	151	132

is a rational choice for our protype. In both SWDC and *Flyover*, the slavers are randomly selected out of the 15 slavers. Table 1.6 shows the average completion time of the TritonSort task over 10 runs. The average completion time of *Flyover* is 36.7% shorter than that of SWDC, and is only 12.5% over the ideal network.

We also plot the TritonSort's execution timeline in Figure 1.13. The upper subfigure shows the CDF of the completion time of map tasks. We can see that there is a severe long tail problem in SWDC. The reason is that the background flows result in higher congestions in SWDC. With the help of variable shortcuts, *Flyover* eliminates the long tail and significantly reduces the completion time of the map phase. The lower subfigure shows the execution

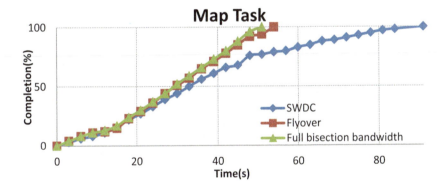

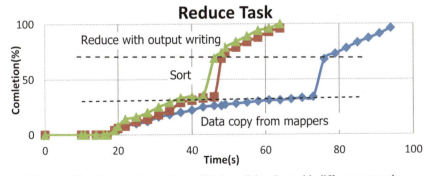

Figure 1.13 The completion time of Hadoop TritonSort with different networks.

timeline of the Reduce phase. It shows three steps: copying data from mappers, sorting, and the execution of the reduce function [12]. Both the first and third steps need high bandwidth to support the moving of a large amount of data. Flyover provides variable shortcuts for large flows and thus can accelerate these steps significantly.

1.6 Discussion

1.6.1 The Bandwidth of Shortcut

The shortcut's bandwidth is an important factor impacting the network performance of $Flyover$. If the shortcuts cannot provide sufficient bandwidth to meet the requirements of elephant flows, the advantages of using on-demand shortcuts to promote network performance will be greatly reduced. In order to avoid the shortcuts becoming the bottleneck of the network, one approach is to increase the shortcut's bandwidth to the peak bandwidth, which equals to five times the link's bandwidth of the EPS network (including one local injection port and four adjacent injection ports). However, in fact, the probability of requiring such high bandwidth is extremely low. In Section 1.4.3, we calculated the utilization of shortcuts for all traffic evaluated patterns. The average utilization of shortcuts fluctuates with the changing traffic patterns. On the one hand, under permutation traffic pattern, the average utilization of shortcuts is the highest but is still less than 67%[4]. Ever for single shortcut, the maximum utilization is not more than 75%. On the other hand, under all traffic patterns, the average utilizations are all greater than 50%. On average, the aggregated bandwidth requirements for the shortcuts are always higher than half of the link's bandwidth of the EPS network. We increase the shortcut bandwidth to two times the link's bandwidth of the EPS network and simulate the network performance. The results show that there is no noticeable performance improvement in most cases. As such, this chapter assumes that the shortcut's bandwidth equals to the link's bandwidth of the EPS network.

1.6.2 The Sharing Region of Shortcut

In this chapter, we increase the utilization rate of shortcuts by sharing shortcuts with adjacent nodes, which introduces an important question of the best way to set the size of sharing region. Although extending the sharing region of

---

[4]In this chapter, we assume that the bandwidth of optical shortcut equals to the bandwidth of links connecting adjacent nodes in the EPS network.

shortcuts can improve the utilization of shortcuts, it also causes new problems. When the bandwidth requirement of the aggregated traffics exceeds the shortcut's bandwidth, the shortcut will become the bottleneck. This problem can be addressed by increasing the shortcut's bandwidth. Another problem is that the large processing time required for the reconfiguration of shortcuts will severely affect the effectiveness of shortcuts. It is because that the large processing time can easily lead to outdated traffic demands and thus reduce the utilization rate of shortcuts. The processing operation consists of calculating traffic demand matrix, calculating configuration for shortcuts, calculating and updating routing table entries for all nodes. And these operations have to be executed serially and their cost is directly related to the size of sharing region. For example, when the size of sharing region increases from 2-hops to 3-hops, the number of routing table entries that the controller updates for each node increases from $13^2 \times 32 = 5408$ to $25^2 \times 32 = 20000$, which greatly increases the reconfiguration time of shortcuts. In order to avoid the large overhead caused by sharing region, we keep sharing shortcuts within two hops only.

In the future, the larger scale of *Flyover* is required to meet the burgeoning demand. To further optimize the network performance, we may need to increase the sharing region of shortcuts, and the shortcut bandwidth may also need to be increased accordingly. This will introduce some new challenges in the implement of *Flyover*, which is part of our future work.

1.7 Conclusion

In this chapter, we propose *Flyover*, a two-level hybrid OCS/EPS network architecture. *Flyover* exploits on-demand shortcuts to maximize the network performance. In *Flyover*, the shortcuts are added and removed based on the current traffic demands so that the high-bandwidth and long-distance flows are migrated from EPS network to OCS network when needed. The benefits of on-demand shortcuts are two folds. Firstly, by directly providing the high-speed optical circuits to the high-bandwidth flows, the network throughput can be improved substantially; secondly, by transferring flows with larger influences (i.e., flows with large amount of data and long forwarding distance) from EPS network to OCS network, the network resources competition and congestion can be greatly alleviated and better network performance is achieved.

Compared with other hybrid OCS/EPS architectures, *Flyover* has two distinct advantages. First, the *serpent* flow, which is big in the size of flow and the distance of data transmission, is prioritized. Therefore, the pressure on the underlying 2-D torus EPS network is significantly reduced. Second, the shortcuts are generated and used in a region-to-region manner, enabling high utilization of the valuable optical shortcuts. Furthermore, a number of expansions of *Flyover* are discussed and evaluated to show that *Flyover* is highly scalable. The extensive simulations and prototype experiments prove that the *Flyover* is quite suitable for current and future data centers.

Acknowledgment

This work was partially supported by the Strategic Priority Research Program of the Chinese Academy of Sciences under Grant No. XDA06010401, by National Natural Science Foundation of China (NSFC) under Grant No. 61202056, Grant No. 61331008, and Grant No. 61221062, and by Huawei Research Program YBCB2011030.

References

[1] Barroso, L. A., Dean, J., and Holzle, U. (2003). Web search for a planet: The Google cluster architecture. IEEE Micro.

[2] Al-Fares, M., Loukissas, A., and Vahdat, A. (2008). A scalable, commodity data center network architecture. In *ACM international conference on the applications, technologies, architectures, and protocols for computer communication*.

[3] Albert, G., James, H., Navendu, J., Srikanth, K., Changhoon, K., Lahiri, Parantap, David, M., Parveen, P., and Sengupta (2009). VL2: a scalable and flexible data center network. In *ACM International Conference on the applications, technologies, architectures, and protocols for computer communication*.

[4] Gyarmati, L., and Trinh T. A. (2010). Scafida: A scale-free network inspired data center architecture. In *ACM International Conference on the applications, technologies, architectures, and protocols for computer communication*.

[5] Singla, A., Hong, C., Popa, L., and Godfrey, P. (2012). Jellyfish: Networking data centers randomly. In *USENIX symposium on networked systems design and implementation*.

[6] Curtis, A., Carpenter, T., Elsheikh, M., Lpez-Ortiz, A., and Keshav, S. (2012). Rewire: An optimization-based framework for unstructured data

center network design. In *IEEE international conference on computer communications.*

[7] http://www.top500.org/

[8] Shin, J., Wong, B., and Sirer, E. (2011). Small-world datacenters. In *ACM symposium on cloud computing.*

[9] Watts, D. J., and Strogatz, S. H. (1998). Collective dynamics of 'small-world' networks. *Nature.*

[10] Guo, C., Lu, G., Wang, H. J., Helen, J., Yang, S., Chao, K., Sun, P., Wu, W., and Zhang, Y. (2010). SecondNet: a data center network virtualization architecture with bandwidth guarantees. In *ACM International Conference on emerging Networking EXperiments and Technologies.*

[11] Ballani, H., Costa, P., Karagiannis, T., and Rowstron, A. (2011). Towards predictable datacenter networks. In *ACM international conference on the applications, technologies, architectures, and protocols for computer communication.*

[12] Wang, G., Andersen, D., Kaminsky, M., Papagiannaki, K., TS, N., Kozuch, M., and Ryan, M. (2011). c-Through: part-time optics in data centers. In *ACM international conference on the applications, technologies, architectures, and protocols for computer communication.*

[13] Farrington, N., Porter, G., Radhakrishnan, S., Bazzaz, H., Subramanya, V., Fainman, Y., George, P., and Vahdat, A. (2011). Helios: a hybrid electrical/optical switch architecture for modular data centers. *ACM international conference on the applications, technologies, architectures, and protocols for computer communication.*

[14] Fu, B., Xu, S., Bao, W., Jiang, G., Chen, M., Zhang, L., Tao, Y., He, R., and Zhao, J. (2014). Dandelion: A locally-high-performance and globally-high-scalability hierarchical data center network. *IEEE Comput. Commun. Networks.*

[15] Kandula, S., Padhye, J., and Bahl, P. (2009). Flyways to de-congest data center networks. In *The Workshop on Hot Topics in Networks.*

[16] Halperin, D., Kandula, S., Padhye, J., Bahl, P., and Wetherall, D. (2011). Augmenting data center networks with multi-gigabit wireless links. In *International conference on the applications, technologies, architectures, and protocols for computer communication.*

[17] Zhou, X., Zhang, Z., Zhu, Y., Li, Y., Kumar, S., Vahdat, A., Zhao, B.Y., and Zheng, H. (2012). Mirror mirror on the ceiling: flexible wireless links for data centers. *International conference on the applications, technologies, architectures, and protocols for computer communication.*

[18] Hamedazimi, N., Qazi, Z., Gupta, H. et al. (2015). FireFly: a reconfigurable wireless data center fabric using free-space optics. In *International conference on the applications, technologies, architectures, and protocols for computer communication*.

[19] Hamedazimi, N., Qazi, Z., Gupta, H. et al. (2013). Patch panels in the sky: A case for free-space optics in data centers. In *The Workshop on Hot Topics in Networks*.

[20] George, P., Richard, S., Nathan, F., Alex, F., Pang, C., Tajana, R., Yeshaiahu, F., George, P., and Amin, V. (2013). Integrating microsecond circuit switching into the data center. In *ACM international conference on the applications, technologies, architectures, and protocols for computer communication*.

[21] Kai, C., Anubhav, S., Ashutosh, S., Kishore, R., Lei, X., Yueping, Z., Xitao, W., and Yan, C. (2014). OSA: an optical switching architecture for data center networks with unprecedented flexibility. In *Symposium on network system design and implementation*.

[22] Chen, K., Wen, X., Ma, X., Chen, Y., Xia, Y., Hu, C., and Dong, Q. (2015). WaveCube: A scalable, fault-tolerant, high-performance optical data center architecture. In *IEEE international conference on computer communications*.

[23] Pal, A., and Kant, K. (2015). RODA: A reconfigurable optical data center network architecture. *Local Computer Networks*.

[24] Ye, T., Lee, T., Ge, M., and Hu, W. (2016). Modular AWG-based interconnection for large-scale data center networks. In *IEEE transactions on cloud computing*.

[25] Yu, Y., and Qian, C. (2014). Space shuffle: A scalable, flexible, and high-bandwidth data center network. *IEEE Network Protocols (ICNP)*.

[26] Koibuchi, M., Fujiwara, I., Matsutani, H., and Casanova, H. (2013). Layout-conscious random topologies for hpc off-chip interconnects. *High Performance Computer Architecture*.

[27] Koibuchi, M., Matsutani, H., Amano, H., Hsu, D., and Casanova, H. (2012). A case for random shortcut topologies for HPC interconnects. *International symposium on computer architecture*.

[28] Benson, T., Akella, A., and Maltz, D. A. (2010). Network traffic characteristics of data centers in the wild. In *Internet Measurement Conference*.

[29] Kandula, S., Sengupta, S., Greenberg, A., Pateland P., and Chaiken, R. (2009). The nature of data center traffic: measurements & analysis. In *Internet Measurement Conference*.

[30] Benson, T., Anand, A., Akella, A. and Zhang, M. (2011). MicroTE: fine grained traffic engineering for data centers. In *Emerging Networking Experiments and Technologies*.

[31] Paths, J. E. (1965). Trees and flowers. *Canadian J. Mathemat.*

[32] Yueying, P., Shicai, L., and Miao, L. (2007). Quick sorting algorithm of matrix. In *International Conference on Electronic Measurement and Instruments*.

[33] Jiang, N., Becker, D. U., Michelogiannakis, G., Balfour, J., Towles, B., Shaw, D. E., Kim, J. and Dally, W. J. (2013). A detailed and flexible cycle-accurate network-on-chip simulator. In *Performance Analysis of Systems and Software*.

[34] Cao, Y., Xu, M., Fu, X., and Dong, E. (2013). Explicit multipath congestion control for data center networks. In *Emerging Networking Experiments and Technologies*.

[35] Benson, T., Akella, A., and Maltz, D. A. (2010). "Network traffic characteristics of data centers in the wild. In *Proceedings of the 10th ACM SIGCOMM conference on internet measurement. ACM*, pp. 267–280.

[36] http://gb.gigalight.com.cn/products_list/pmcId=171.html.

[37] Rasmussen, A., Porter, G., Conley, M., Harsha, V., Mysore, R. N., Pucher, A., and Vahdat, A. (2011). TritonSort: A balanced large-scale sorting system. In *Symposium on network system design and implementation*.

2

Dynamic Power Management in Data Centres

Waltenegus Dargie[1] and Franz Eichhorn[2]

[1]Chair of Computer Networks, Faculty of Computer Science, Technical University of Dresden, 01062 Dresden, Germany
[2]LOVOO GmbH, 01069 Dresden, Germany

Abstract

The amount of data that are computed, stored, and communicated by data centres has been increasing for years and will keep on increasing. One of the consequences of this phenomenon is the deployment of a considerable number of high-capacity servers each year. In contrast, several independent studies reveal that resources (including energy) are not optimally utilised in most existing data centres. The introduction of server virtualisation and cloud computing promises an efficient resource utilisation because they enable the dynamic consolidation of workloads. This chapter discusses and experimentally demonstrates the scope of different dynamic power management strategies (dynamic voltage and frequency scaling, load balancing, and workload consolidation) in data centres. It also introduces the HAECubie demonstrator we developed to quantitatively analyse the relationship between energy/power consumption and the utility (performance) that can be achieved through workload consolidation.

Keywords: Adaptation, demonstrator, dynamic power management, energy-efficient computing, energy-efficient data centres, workload consolidation, video hosting platform, video streaming.

2.1 Introduction

A data centre is a collection of physical or virtual servers which are interconnected with each other with high-speed links to compute, store, and communicate a large amount of data. The efficient set-up and management of data centres are critical aspects for achieving high performance, cost-effectiveness, resilience, scalability, and adaptiveness (most importantly, in terms of supporting new applications). Some of the key features pertaining to data centre set-up are port density, access layer uplink bandwidth, true server capacity, and the capability to accommodate and deal with over-subscription [1].

Different data centres have different size but the average size of typical data centres has been steadily increasing over the years. The estimated worldwide server deployment in 2010 has been 40 million units [2], but additional servers have been steadily deployed since then. According to the International Data Corporation (IDC),[1] the total cloud IT infrastructure spending (server, storage, and Ethernet switch) forecast for 2015 was a growth of 26.4% when compared to the spending of 2014, 'accounting for a third of all IT infrastructure spending, up from 28.1% in 2014'. As far as capacity per individual servers is concerned, systems with operation capacity in the range of exaflops and an optical link number of around 10^8 are expected by the year 2020 [3].

Likewise, the amount of data hosted and processed by data centres has been increasing over the years. The trend will remain the same in future mainly for two reasons. Firstly, the improved capacity of smart phones, digital cameras, and tablet computers has increased the ease with which multimedia data can be generated and shared on the spur of the moment. Secondly, the introduction of private and public cloud computing environments is accelerating the transfer and storage of a large amount of data from terminal devices to back-end servers. According to the recent Cisco Global Cloud Index, by 2019, global data centre traffic will grow nearly by 3-fold (compared to the annual traffic in October 2014), reaching at 10.4 zettabytes per year. Moreover,

- 83% of all data centre traffic will come from the cloud, and
- 4 out of 5 data centre workloads will be processed in the cloud [4].

As the number, functions, and complexity of Internet-based data centres grow, their power consumption grows with comparable magnitude. Unfortunately, independent studies reveal that presently computing resources are not efficiently utilised in many existing data centres and server clusters. For example,

---

[1]http://www.idc.com/getdoc.jsp?containerId=prUS24476413 (last visited May 4, 2016).

in a typical Twitter server, the CPU utilisation is reported to be less than 20% and the RAM utilisation is between 40 and 50% [5]. Likewise, in a typical Google server, the CPU utilisation is reported to be between 25 and 35% whereas the RAM utilisation is approximately 40% [6]. In Amazon's EC2 cloud environment, the CPU utilisation per server is reported to be between 3 and 17% [7]. At the same time, the idle power consumption of existing servers is between 50 and 60% of their peak power consumption [8–10].

Embedding dynamic power management strategies at various abstraction layers in the computational hierarchy of large-scale data centres is crucial to attain efficient and sustainable computing. Depending on the relative 'nearness' of dynamic power management strategies to actual hardware subsystems, they can have from short-term to long-term scope. Those strategies directly dealing with hardware subsystems usually have short-term scope (typically, in millisecond range). As the abstraction layer in which a dynamic power management strategy resides moves away from the hardware layer, its data gathering and computational complexity increase, which means the time it requires to reach at a decision increases, and therefore, its scope becomes from mid-term (in the range of a few hundred seconds) to long term (in the range of hours).

2.2 Analysis of DC Power

In order to identify the suitable strategies for managing the power consumption of the different hardware subsystems of a physical server, it is useful to first examine (1) the approximate amount of power each subsystem consumes and (2) the dynamic range of the power consumption. As an example, consider Figures 2.1 and 2.2 where we plot pie charts to account for the portion of power consumed by the different subsystems of a small-scale server (built on a D2461 Siemens-Fujitsu motherboard, integrating a 2 GHz AMD Athlon 64 dual core processor) running two different applications (workloads). One of the applications was a CPU-bound video transcoder and the other an IO-bound video streaming application. Depending on the load on the power supply unit, between 25 and 35% of the AC power is dissipated in the form of heat during the AC to DC conversion. We believe there is at least a 10% additional power loss on the DC power due to a further DC to DC conversion at the various voltage regulators. We were not able to measure this loss, however, as this meant essentially modifying the motherboard structure. Therefore, our measurement was limited to the circuits between the output of the power supply unit and the voltage regulators.

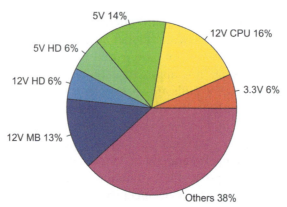

Figure 2.1   The portion of the DC power consumed by the various subsystems in the AMD server through different power lines when an IO-intensive application (Apache video streaming server) was running. MB: Motherboard. HD: Disk drive. The 5 V and 3.3 V power lines are used by various subsystems including by the memory bank [9].

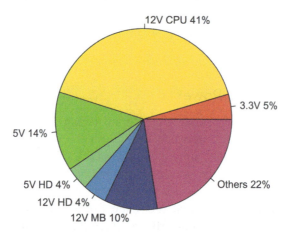

Figure 2.2   The portion of the DC power consumed by the various subsystems in the AMD server through different power lines when a CPU-intensive application (transcoder) was running. MB: Motherboard. HD: Disk drive. The 5 V and 3.3 V power lines are used by various subsystems including by the memory bank [9].

As can be seen, in both charts, a considerable amount of the AC power is lost in the form of heat dissipation (labelled as others in the figure). The average AC power consumption of the server when the IO-bound application was running was 53 W and the combined average DC power consumption of the different subsystems was 33 W. Hence, 20 W (about 38%) was lost in the form of heat dissipation due to the inefficiency of the power supply unit. The power efficiency improved to 78% with the CPU-bound application.

2.3 Dynamic Voltage and Frequency Scaling

One of the widely used dynamic power management strategies is dynamic voltage and frequency scaling (DVFS). The motivation for employing this strategy is that when compared to the power consumption of all the other active subsystems (RAM, NIC, disk drive, etc.), the power consumption of the processor is the greatest; and by adapting the supply voltage and clock frequency of the processor, it is possible to adapt its power consumption according to the required operation speed (task deadline) and the expected idle time of the processor. For a given operation frequency (f) and a supply voltage (V), the power consumption of the processor can be approximated by:

$$P \approx g\left(f, V^2, C\right) \tag{2.1}$$

where C is a capacitance associated with the CMOS technology. Hence, increasing (or decreasing) the switching frequency of the processor by one step results in a proportional increment (or reduction) in its power consumption, but increasing (or decreasing) the core voltage by one step results in an exponential increment (or decrement) of its power consumption. This does not mean, however, that the operation frequency and voltage of processors can be varied arbitrarily. Most existing processors have limited operation frequencies and there is a strong dependency between these frequencies and the permissible processor voltages (however, the possible number of discrete operation frequencies has been steadily increasing with each processor generation; moreover, modern multicore processors provide separate phase locked loops to individual cores, so that each core can be scaled independently [11–14]).

The appropriate speed of a processor depends on two important factors, namely (1) the deadline set by a scheduler for the completion of a task and (2) the expected time interval between the arrival of two consecutive tasks. If a processor completes processing a task before the deadline or if the inter task interval is considerably long, then the processor becomes idle even though it is technically active and consumes power. In order to avoid idle power consumption, the processor's operation frequency/voltage can be reduced, so that it can process tasks slowly, thereby reducing the inter task interval as well as its power consumption. Figure 2.3 illustrates the main idea of DVFS.

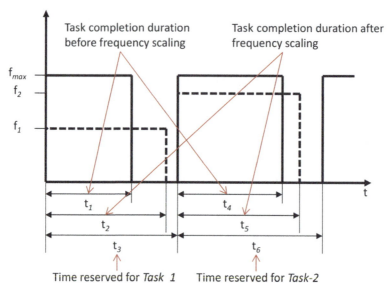

Figure 2.3   The essential aspect of dynamic voltage and frequency scaling.

The solid lines illustrate the task execution frequency before dynamic frequency scaling and the corresponding task completion durations whereas the dashed lines illustrate the extended execution durations when dynamic frequency scaling is applied. Ideally, these extensions are still within the bounds of the allocated durations, but in reality it is difficult to reduce the power consumption of the processor and at the same time avoid the violation of task completion deadlines, in which case, dynamic voltage and frequency scaling entail performance degradation.

The amount of power that can be saved by DVFS depends mainly on four factors: (1) The number of discrete voltages and frequencies that are permissible to scale (the resolution of scaling), (2) the estimation accuracy of the task arrival rate (or the inter task interval) which in turn depends on the statistics gathered to estimate inter task interval, (3) the nature of the CPU load, and (4) the scaling policy or algorithm.

2.4 Workload Consolidation

The introduction of server virtualisation has enabled a more efficient resource utilisation (and, as a result, a more efficient energy consumption) because multiple virtual machines belonging to different owners can share or run in a single physical server without interfering with each other's privacy (though it must be remarked that the overhead of virtualisation must be compensated by the number of active virtual machines a physical server hosts). Virtualisation enables also the existence of virtual data centres and separates the provision of physical infrastructures (infrastructure as a service, IaaS) from the provision of service computing platforms (platform as a service, PaaS) [15–17]. Its additional advantages include the following:

1. Consolidating virtual machines (workload) at runtime to switch off idle and underutilised servers; this strategy avoids the 33 to 45% power loss due to heat dissipation and the additional cost arising from the power consumption of idle subsystems (refer to Figures 2.1 and 2.2);
2. Distributing virtual machines to balance the workload of the data centre (for example, to relieve overloaded servers or to deal with an unforeseen surge in workload) and to dynamically accommodate new customers.

As far as dynamic power management is concerned, the online decisions are the most relevant ones.

A virtual machine typically encapsulates an operating system and one or more applications. When an application executes, a multi-dimensional communication takes place between the processor and the memory unit; the memory unit and the network interface card; and the memory unit and the storage system. As a result, the memory content of the application is steadily modified. During the live migration of a virtual machine, all these activities must be coordinated. Considering (i) the time it takes to migrate virtual machines, (ii) the complexity of the coordination required to ensure that the performance of application execution does not significantly deteriorate during migration, and (iii) the resources required for the coordination process, workload consolidation should have a long-term power management scope. Some of the important steps in workload consolidation are the following:

- The detection of overloaded and underutilised situations;
- the selection of candidate virtual machines for migration; and,
- the identification of target servers to which virtual machines can be migrated.

Beloglazov et al. [18] express the consolidation of virtual machine as a bin packing problem, where the virtual machines represent the items to be allocated, the bin sizes are the servers' CPU capacities, and the price is the power consumption of the servers. It is solved by first sorting virtual machines in decreasing order of their CPU utilisations and then by allocating to a destination server virtual machines which cost the lowest price (the lowest increment of power consumption due to the VM allocation). Once overloaded servers are dealt with, the strategy addresses underutilised servers. Virtual machines from these servers are migrated to other servers as long as this can be done without overloading the latter. The performances of the proposed heuristics were evaluated using the CloudSim simulation framework.[2] The simulated IaaS cloud environment consisted of 800 heterogeneous servers, of which half of them were HP ProLiant ML110 G4 servers with dual core Intel Xeon processors 3040 (1860 MHz) and the other half were HP ProLiant ML110 G5 servers with dual core Intel Xeon processors 3075 (2660 MHz). Both types of servers were assigned 4 GB RAM and 1 GBps network bandwidth. They consider more than 1000 virtual machines which resembled Amazon's EC2 virtual machine instance types[3] with the only difference that each virtual machine was assigned a single core and an appropriately scaled amount of RAM. The authors reported that the combination of the regression and the minimum migration time approaches produced the best results when both energy saving and minimisation of SLA violation were considered at the same time. The combination of the fixed threshold overloading heuristic with the minimum migration time produced the highest energy saving but at the price of a higher SLA violations.

Verma et al. [19] employ bin-packing to consolidate virtual machines by assuming that (1) a mechanism exists for predicting the optimal minimum size of a VM to meet its SLA goals, and (2) servers can be monotonously ordered according to their energy efficiency, meaning, if some server s_i is more energy-efficient for some application a_k than another server s_j, then s_i is more energy-efficient than s_j for *any* other application. To place and re-locate virtual machines in a cluster, they first sort all servers in decreasing order of their energy efficiency and all virtual machines in decreasing order of the optimium size (i.e. necessary allocation of resources). Secondly, they compute a mapping of virtual machines in such a way that starting with a theoretical utilisation of 0 for all servers, they allocate as many virtual machines from the sorted list to the most-energy efficient server until its capacity is filled. The process is continued with the second most energy-efficient server and the remaining virtual machines, until all virtual machines are mapped to some server. Further iterations are made to efficiently utilise the resources.

---

[2]https://code.google.com/p/cloudsim/
[3]http://aws.amazon.com/de/ec2/instance-types/

Xu et al. [20] propose a virtual machine consolidation strategy which aims to avoid SLA violation while minimising energy consumption. This is done by minimising the interferences during migration and co-location of virtual machines. The authors employ a multi-dimensional supply–demand model to compute the interference of a migrated virtual machine on CPU, network I/O, CPU cash, and memory bandwidth. Based on this quantity, candidate virtual machines are identified. The authors tested their approach using different workloads (including, SPEC CPU2006,[4] netperf,[5] Hadoop,[6] and SPECweb2005[7]). The proposed consolidation strategy was tested on 10 homogeneous physical servers each having two quad-core Intel Xeon E5620 2.40 GHz processors, 12 MB shared LLC, 24 GB memory, and 800 GB NFS storage. The physical servers were connected via 1 Gbps Ethernet switches. The servers as well as 50 virtual machines were running CentOS5.5 with Linux 2.6.18.8 kernel. Small VM instances were allocated with 1 virtual CPU core and 1.7 GB RAM, while large VM instances were allocated 4 virtual CPU cores and 7.5 GB RAM. Experimental results show that the proposed strategy can improve the performance of CPU and memory intensive workloads by 16%–28% and network-intensive workloads by 45–65% in comparison with First Fit Decreasing (FFD) [19] and Sandpiper [21] consolidation strategies. Even though the proposed approach is self-sufficient, it can be integrated into existing load balancing (Sandpiper [21]) or power-aware (FFD [19]) VM consolidation strategies.

Zhu et al. [22] proposed a consolidation framework to improve the energy efficiency by placing together virtual machines which have complementary resource consumption characteristics. They define a distance metric that quantifies the peak resource demand of applications for all resources (CPU, memory, storage, network) together with an affinity score that favours applications with a direct communication link between them. Virtual machines are placed together if they are 'near' to each other according to the distance metric. The authors evaluated their approach with two clusters each comprising of 64 servers. In one of the clusters, the servers integrate AMD Opteron 250 processors and in the other Intel Xeon E5345 processors. The authors claim that their approach was able to reduce the overall energy consumption by up to 55% compared to when the virtual machines were executed separately.

---

[4] https://www.spec.org/cpu2006/

[5] http://www.netperf.org/netperf/

[6] https://hadoop.apache.org/

[7] https://www.spec.org/web2005/

The associated cost is increased execution time. When virtual machines with dissimilar resource consumption characteristics were co-located, execution times increased by almost 10%. The effect was pronounced when virtual machines with similar characteristics were executed. For example, for two CPU-intensive applications, execution times doubled, and for two memory-intensive applications, execution times increased by 41%.

One of the challenges in studying the consequences of dynamic workload or virtual machine consolidation is the difficulty of testing algorithms with real-world platforms and workloads. Most of the proposed approaches were tested with simulated platforms, even though in some cases traces of real-world workloads have been obtained. As an example, Beloglazov et al. employed real workload traces obtained from PlanetLab [23] to carry out statistical analysis on the CPU utilisation of the workload. To complement ongoing efforts, we developed a demonstrator which emulates an Internet video hosting platform. In the next sections, we shall introduce this demonstrator and, using the demonstrator, quantitatively compare the performance of different dynamic workload consolidation strategies.

2.5 The HAECubie Demonstrator

The HAECubie[8] demonstrator (see Figure 2.4) [24] consists of 30 Cubieboard2[9] nodes, each node representing a single server in a data centre. The purpose of the data centre is to host video streaming (similar to the servers introduced in Section 2.3). Each Cubieboard2 has 1 GB memory, dual-core ARM coretex-A7 processor, 100 Mbps Ethernet, and 32 GB storage on SD Card. The nodes are organised into six clusters and each cluster is connected to a 1 Gbps switch, which is in turn connected to a master switch with the same capacity. The master switch is connected to a router so that video streaming requests can be received via the Internet. The organisation of nodes into clusters enables dynamic power management at core, node, cluster, and data centre levels. One extra node, the *master node*, is directly connected to the master switch and is in charge of managing the system as a whole. All incoming requests are accepted by the master node which then forwards the requests to one of the nodes. Individual nodes are directly responsible for

---

[8]The acronym HAEC stands for Highly Adaptive Energy Efficient Computing; it is the title of our project which is funded by the German Research Foundation. See the ACKNOWLEDGEMENT section for detail.

[9]http://cubieboard.org/

Figure 2.4 The HAECubie demonstrator for hosting video streaming.

streaming videos to users. Figure 2.6 displays a partial view of the hardware architecture of the video hosting platform.

We deployed 1800 videos on the HEACubie which have different length and duration, the shortest video being 14 KB and the longest 50 MB (Figure 2.5 shows the probability density function of the size of the videos). In the beginning, the videos are randomly placed on all nodes. They gain popularity credits every time they are requested and this way accumulate popularity over time. In order to accommodate sudden traffic surges, each node can utilise only up to 40% of its network bandwidth.[10] The cluster to which a node belongs depends on the popularity of the videos it hosts. Nodes hosting less popular videos spend much of their time sleeping to save energy.

At the software level, the video hosting platform consists of four services running on each node and five services on the master node. The services running on each node are a video management service, a monitoring service, an

---

[10]The amount of bandwidth a data streaming server should support depends, among others, on the expected round trip time of a flow passing through the link and the available buffer size (in case of congestion). Generally, the buffer size is directly proportional to the flow rate [26, 27]. For an exponential workload (video size) with an average streaming duration per request of 1 s and 0.4 request per second per board (assuming that the board experiences idle state, which is the basis for consolidation), the bandwidth required to stream the video without any congestion would be 40% of the 1 Gbps data rate.

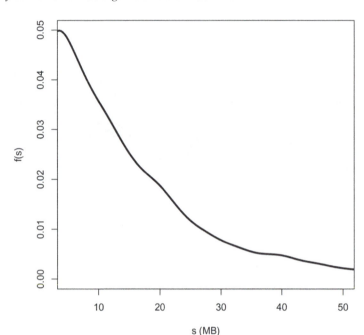

Figure 2.5 The 1800 videos (in MB) deployed on the demonstrator have an exponential probability density function. A review of literature on video size distribution in Internet video hosting platforms suggests that the videos hosted have a long-tailed power distribution which can be approximated by exponential density function [25].

activity service, and a migration service. Each of these services have instances running on the master node as well. The master node in addition runs a power management service. Figure 2.7 displays the software architecture of the video hosting platform.

At the node level, the video management service is responsible for streaming videos to users and for tracking and reporting the status of a video to the video management service running on the master node. The video management service at the master node is responsible for receiving and forwarding user requests and for calculating and updating the popularity of a video. It is also responsible for predicting the popularity of a video, which is useful for determining to which cluster a video should belong (to be discussed in the next subsection). The monitoring service gathers statistics pertaining to the resource and power consumption of a node and reports them to the master. The monitoring service at the master collects and aggregates these statistics and stores them into the database. All tasks pertaining to adaptation,

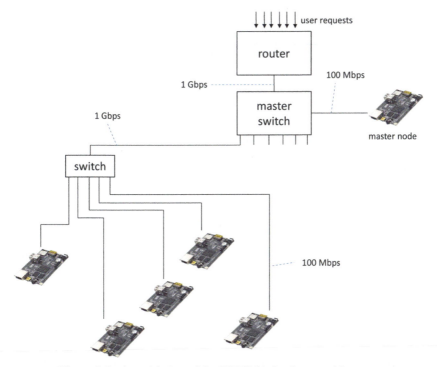

Figure 2.6   A partial view of the HAECubie hardware architecture.

migration, and power management inside the demonstrator are asynchronous tasks; therefore, a mechanism is required to ensure that these tasks are executed to their completion and all deadlines are respected. The activity service will be notified when a task is issued and completed, so that it ensures these tasks are executed as intended. The activity service at the master overlooks the activities of individual nodes.

The migration service at the node level is responsible for migrating videos between nodes, in a distributed manner, but the migration strategy comes from the migration service residing on the master node, which notifies nodes of these strategies. Finally, the power management service is responsible for turning on and off nodes and switches. The services interact with each other and with the database to exchange the relevant information. In addition, the software platform provides a web-based administration support, so that the status of the entire data centre can be viewed at runtime and parameters which are relevant for adaptation (for example, popularity threshold) can be modified and adaptation policies can be manually selected.

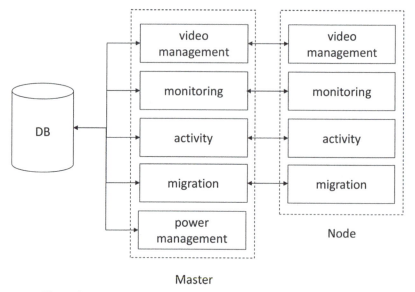

Figure 2.7   The software architecture of the HAECubie demonstrator.

2.5.1 Consolidation

The purpose of workload consolidation is to dynamically balance the demand for and the supply of resources inside HAECubie, so that underutilised nodes can be switched off and the workload of overloaded nodes can be reduced to avoid a resource bottleneck. Because of the limited computing and communication resources in our demonstrator, we migrate data (videos) at runtime instead of virtual machines. Apart from the abstraction of the entities being migrated (video versus virtual machines), the migration process in both cases is similar, since the process in both cases is predominantly IO-bound. The only differences between the two is that (a) compared to the videos we migrate, virtual machines are typically larger in size and require a longer migration time, and (b) since virtual machines are migrated while the services they abstract still execute, the memory content of the source machine is continuously updated during migration. As a result, multiple iterations and a 'stop condition' are required during virtual machine migration [28, 29] while this is not the case for video migration. In this chapter, our focus is on the comparison between different consolidation strategies with the emphasis being on the balance between the supply of and the demand for resources. Consequently, the impact of migrating video data as opposed to virtual machines on the efficiency of the consolidation strategies should be minor.

Our demonstrator supports four types of consolidation strategies:

- All On (ON). This strategy turns on all nodes without performing any migration or load-balancing. It serves as a reference to all the other strategies.
- All On – Balanced (AOB). Uses all nodes, but tries to improve throughput and latency by distributing popular videos (by migrating them at runtime) on all nodes, so that bandwidth utilisation in the cluster is optimised.
- Load Only (LO). Activates the number of nodes that are necessary to meet the demands of the short term load of the cluster (which is estimated based on the statistics of recently accessed videos). It uses load-balancing to distribute popular videos among active nodes. The basic assumption behind this strategy is that recently streamed videos are more likely to be requested.
- Popularity (POP). This strategy classifies videos as popular and unpopular and nodes as class *A* and *B*. Class *B* nodes host predominantly unpopular videos and they are candidates to be switched off. They will be turned on *on demand*, when unpopular videos they are hosting are requested. In addition, the strategy balances the load of all active nodes.

To quantify the energy-utility trade-off arising from dynamic workload consolidation, we defined streaming latency and throughput as utility and evaluated the relationship between these two utilities and the power and energy consumption of the entire cluster.

2.5.2 Workload Prediction

The video management service at the master node uses a mean square estimation filter to predict the popularity of each video (the number of views) for the next time slot based on which it determines the resource demand of the HAECubie. Hence, at time $t - 1$ the video management service estimates the popularity of video x for the time slot $[t - 1, t]$ based on the statistics it has up to that point in time. This quantity is denoted by $p_e(t)$. Meanwhile, between $t - 1$ and t, it counts (measures) the number of requests for video x and this quantity is denoted by $p_m(t)$. Then, the estimated popularity of video x for the time slot $[t, t + 1]$ is given as:

$$p_e(t + 1) = p_e(t) + \alpha_x[p_m(t) - p_e(t)], \ 0 < \alpha_x < 1 \qquad (2.2)$$

The expected error between the predicted and the actually observed popularity (the term inside the square brackets in Equation (2.2)) determines the value of α_x. If this error is small (i.e. if the measurement is correlated with the estimation), then α_x will be small, otherwise, α_x will be large. Note that Equation (2.2) can be rewritten as:

$$p_e(t + 1) = \alpha_x p_m(t) + (1 - \alpha_x)p_e(t) \tag{2.3}$$

$$
\begin{aligned}
p_e(t + 1) &= \alpha_x p_m(t) + (1 - \alpha_x)\left[\alpha_x p_m(t - 1) + (1 - \alpha_x)p_e(t - 1)\right] \\
&= \alpha_x p_m(t) + \alpha_x(1 - \alpha_x)p_m(t - 1) + (1 - \alpha_x)^2 p_e(t - 1) \quad (2.4) \\
&= \alpha_x p_m(t) + \ldots + \alpha_x(1 - \alpha_x)^{t-1}p_m(1)
\end{aligned}
$$

For most practical cases, the autocorrelation of p falls to zero after the k-th sample. Hence, only the recent k samples are considered for estimation:

$$p_e(t + 1) = \alpha_x p_m(t) + \ldots + \alpha_x(1 - \alpha_x)^{t-k}p_m(k) \tag{2.5}$$

Once the master's video management service estimates the popularity of each video for the next time slot, it computes the overall expected workload by aggregating the video popularities. This is useful to estimate the number of nodes required to handle the workload.

For the POP strategy, the master node defines a popularity threshold based on a specified energy-utility function. Those videos which have popularities above this threshold belong to Class A nodes, which should be always on. Those videos which have popularities below this threshold belong to Class B nodes, which should be switched off but can be switched on *on demand*, when unpopular videos are requested. Furthermore, the master notifies each node to which Class it belongs along with the popularity of the videos the node hosts. Based on this information each node migrates those videos it hosts which do not belong to its Class. Migration takes place on peer-to-peer basis, without the participation of the master node. Once Class B nodes are free of Class A videos, then the power management service puts them all into a sleep state, i.e. they will be switched off. Consequently, the access time for unpopular videos is relatively high, because of the booting time required to make a Class B node active.

2.5.3 Workload Generation

Video streaming requests have two statistically independent components, namely the request arrival rate (**r**) and the streaming duration (or video size) of each requested video (**s**). For most practical scenarios, both components are random variables. Thus, a workload **w** can be expressed as **w** = **rs**. We expressed the workload of our demonstrator by the joint probability density function $f(\mathbf{w}) = f(\mathbf{r}, \mathbf{s})$, which can be expressed as the convolution of the probability density functions of **r** and **s**:

$$f(w) = \int_0^{w/s} f(w/s) f(s) ds \qquad (2.6)$$

The most plausible way of generating a realistic workload for our demonstrator is by reusing traces and log files from actual video hosting platforms. Unfortunately, existing platforms do not make traces and log files available for the public nor do they permit web crawler due to non-disclosure agreements and privacy concerns. Nevertheless, in the literature there exist several stochastic models that can estimate particular aspects of real-world workloads based on already available traces and log files. For example, some of these models estimate the distribution of file size and video popularity growth in YouTube, Youku, Dailymotion, and Metacafe [30–33]. Hence, we convolved existing models, to generate the workload of our demonstrator. For the detailed analysis of our workload generation strategy, we refer the reader to [25].

2.6 HAECubie Evaluation

To evaluate the energy-utility trade-off that can be achieved by the different consolidation strategies, we generated a stochastic workload and supplied it to the HAECubie demonstrator. But the statistical parameters (the density functions of the workload size and the workload arrival rate) are the same for all the strategies. For each experiment, the demonstrator streamed videos as per user requests for one hour. We used an extra 32-core server (with Intel Xeon E5-4603 processors and 16 GB RAM) to generate and feed the workloads to our HAECubie. In our evaluation, we considered the overall energy consumption and the average power consumption of the system (the HEACubie only), the average throughput in terms of the number of bits per second the demonstrator was able to stream, and the video streaming latency (the time between the reception of a request and the beginning of streaming).

2.6.1 Dealing with Underutilisation and Overloading Conditions

Except for the always on strategy, all the other strategies dealt with overloading conditions by enabling dynamic load-balancing through the migration of videos at runtime. The only server which does not support load-balancing was the master node. For our scenario, however, it was rarely overloaded, since its task was limited to the forwarding of incoming requests and the computation of relevant statistics which are used for dynamic workload consolidation. Hence, its average CPU utilisation was below 12%. Similarly, except the always on strategy, all the others dealt with underutilisation condition by trying to strike a balance between the demand for computational resources and the supply of resources and by switching off excess resources. Dealing with both conditions introduces an adaptation cost in terms of additional energy consumption, reduction of throughput, and increased latency. This shall be discussed in the subsequent subsections.

2.6.2 Energy and Power Consumption

Whereas the energy consumption is a measure of the cost of energy incurred to run the system, the power consumption is a measure of how much electrical load the system introduces on the power supply line. The amount of load the system should produce is specified by the contract between the client and the power supplier because violating this agreement may make the power supply line unstable and disturb the normal operation of other systems which share the same line. Since for our case each experiment was conducted for one hour, it is possible to directly infer the energy consumption of the system from its average power consumption.

Figure 2.8 displays the density function of the power consumption of the HAECubie for the different strategies. The always on strategy (ON) consumed the largest amount of power as well as energy. In the absence of load balancing, some nodes received more streaming requests than others as a result of which these nodes often experience congestion. The always on with load-balancing strategy (AOB) has consumed the next largest power (energy). Perhaps this is because the gain in power consumption as a result of load-balancing has been counterbalanced by the adaptation cost that was needed to migrate the videos between nodes.

For the POP strategy, we considered two popularity thresholds: 0.25 (POP) and 0.125 (POP2). In the first, videos, the popularity of which is below the 0.25 quantile in the popularity distribution belong to Class B nodes which are normally switched off. In the second, the quantile threshold is 0.125. In other words, more popular videos are stored in Class B servers in the POP strategy than in the POP2 strategy. Consequently, the POP strategy resulted in a larger amount of power and energy consumptions compared to the POP2 strategy,

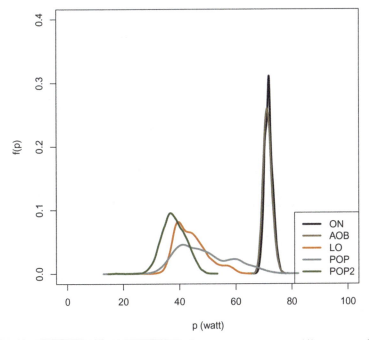

Figure 2.8 The density function of the power consumption of the different consolidation strategies.

since in the POP strategy, the Class B servers had to be frequently turned off and on to stream the popular videos they host. This indicates that switching off nodes in which a large number of popular videos are stored does not save power significantly in the long term. The LO strategy, which keeps all nodes active on which recently accessed videos reside and turn off all other nodes except when they are demanded, performed better than the POP strategy in terms of power and energy consumption.

2.6.3 Throughput

Throughput is a measure of the rate at which the HAECubie streams videos to the users. Interestingly, the ON strategy is the one which performed the poorest (as can be seen in Figure 2.9). One of the plausible reasons for this is that since the strategy does not load balance, some of the boards hosting popular videos were congested as a result of which overall streaming rate was slow. This confirms to the significance of load balancing in server clusters and data centres. Nevertheless, load-balancing alone does not produce the highest throughput

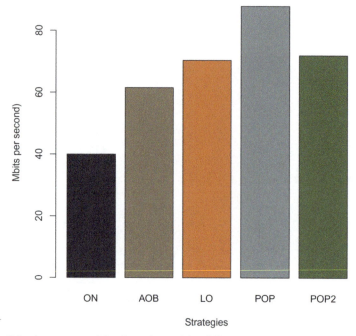

Figure 2.9   A summary of the throughput of the different strategies in Mbits per second.

even if there is a surplus of computing and communication resources. In fact, all the consolidation strategies yield relatively higher throughputs compared with the ON and AOB strategies. The highest throughput was achieved when the POP strategy was used followed by the POP2 strategy, confirming to the fact that accurate estimation of the videos popularity enables to accurately balance the demand for and the supply of resources in video hosting platforms.

2.6.4 Latency

The challenge with dynamic workload consolidation is dynamic resource-pool sizing, particularly when more resources are demanded. For our HAECubies, sometimes up to several tens of seconds are required to make videos available from nodes which are completely switched off. The time required to turn on a sleeping node depends on the workload of the system and how busy all the management services are. As a result it is a random variable.

As can be seen in Figure 2.10, the ON strategy (top) has the smallest latency whereas all the rest resulted in some amount of latency. However, the average latency for each strategy is below 3 s. The POP strategy resulted in the highest average latency compared to the LO and POP2 strategies. One of

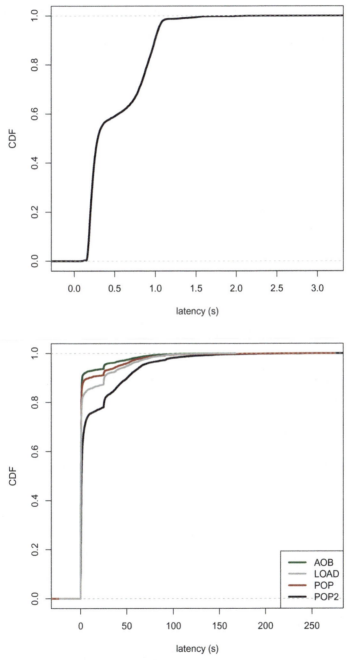

Figure 2.10 The CDF of the latency of the ON strategy (top) and the AOB, LO, POP, and POP2 strategies (bottom).

Table 2.1   Summary of energy-utility trade-off for different consolidation strategies

Strategy	ON	AOB	LOAD	POP	POP2
Energy consumption (Watts-hour)	71.626	71.3	48.3	56.1	45
Mean power consumption (Watt)	71.8	71.4	44.0	48.5	38.0
Throughput (Mbps)	40.0	61.6	70.4	88.0	72.0
Gain in % (Throughput)	0	54	76	120	80
Gain in watt (Avg. Power)	0.0	56	38.7	32.5	47.1

the reasons for the significant latency in the POP strategy is that more popular videos (for example, compared to POP2) are likely to be assigned to cluster B boards which should be started *on demand*. The AOB is the worst strategy in terms of latency.

2.6.5 Summary

Our experiments show that there is no single strategy that can outperform all the others by all accounts. Table 2.1 summarises the energy-utility trade-off that can be achieved by the different strategies. Whereas the POP strategy produces the largest throughput for a small amount of average power consumption, it is also the one which produces the largest delay. On the other hand, the POP2 strategy has the smallest amount of average power consumption and the third best in average latency. It has also the second best in throughput. From this it is possible to conclude that the threshold of the popularity strategy can be adjusted to achieve the optimal energy-utility trade-off for a named utility. The optimal threshold, however, should take the hardware and software architectures as well as the workload of the system into consideration.

In terms of gain (by taking the outcomes of the ON strategy as a baseline), the POP2 improved throughput by 120% whereas it reduces the power consumption by 32.5% only. The highest gain in terms of saving power is achieved by the AOB strategy, but when the corresponding throughput gain is considered, the POP2 has the highest gain.

2.7 Conclusion

In this chapter, we experimentally investigated the scope and usefulness of different dynamic power management strategies (DVFS, load balancing, and dynamic workload consolidation) to achieve high-performance and energy-efficient computing in data centres. Our experiment with dynamic DVFS consisted of different servers and workload types. Similarly, our experiment with load-balancing and dynamic workload consolidation was carried out on a custom-made demonstrator consisting of 30 Cubieboards which are organised into six clusters. For all our experiments, we used video streaming and video transcoding as the preferred applications.

We demonstrated that the scope of a dynamic power management strategy depends on its relative 'nearness' to actual hardware components or sub-systems. DVFS typically has a scope of milliseconds (short term) whereas dynamic workload consolidation has a scope of several hundred seconds to hours (long term). The usefulness of DVFS depends on (1) the number of permissible operation frequencies and core voltages, (2) the rate at which the CPU utilisation is sampled to estimate inter task intervals, (3) the workload type, and (4) the algorithm of scaling.

We implement three types of workload consolidation strategies. In the AOB strategy, all nodes are active but load-balancing is used to avoid resource bottleneck. In the LOAD strategy, all nodes from which videos are requested in the recent past remain active. The POP strategy uses the popularity distribution of all nodes to determine the number of nodes that should be active to handle the incoming workload. Our experiment results confirm that all strategies performed well when compared with the baseline in which all nodes were active and no load-balancing was employed. However, they also show that there was no single strategy that can achieve high energy-efficiency, low streaming latency, and high throughput at the same time. The POP strategy, with the selection of the appropriate threshold, can achieve most of these goals. Because we used relatively simple boards to construct our demonstrator, these boards experienced failure when they were frequently switched on and off (this was due to an unfortunate bug we discovered in the Linux operating system for the specified boards). Consequently, some of our measurements did not make sense to us during statistical analysis. We rejected them as outliers. We are currently building the second generation demonstrator using more powerful and more stable ODROID boards (ODROID-XU4, each board consisting of octa core CPU).

Acknowledgement

This work has been partially funded by the German Research Foundation (DFG) under the project agreement SFB 912/2 2015.

References

[1] Cisco Inc. (2013) Cisco virtualized multi-tenant data center, version 2.2.
[2] Dreslinski, R. G., Wieckowski, M., Blaauw, D., Sylvester, D., and Mudge, T. N. (2010). Near-threshold computing: Reclaiming moore's law through energy efficient integrated circuits. *Proc. IEEE*, 98 (2), 253–266.
[3] Benner, A. (2012). Optical interconnect opportunities in supercomputers and high end computing. In *Optical Fiber Communication Conference and Exposition and the National Fiber Optic Engineers Conference (OFC/NFOEC)*, pp. 1–60.
[4] Cisco Inc. (2015). Cisco global cloud index: Forecast and methodology. 2015–2019.
[5] Delimitrou, C., and Kozyrakis, C. (2014). Quasar: Resource-efficient and qos-aware cluster management. In *ASPLOS '14*, pp. 127–144, ACM.
[6] Reiss, C., Tumanov, A., Ganger, G. R., Katz, R. H., and Kozuch, M. A. (2012). Heterogeneity and dynamicity of clouds at scale: Google trace analysis. In *SoCC '12*, pp. 7:1–7:13, ACM.
[7] Liu, H. (2012). Host server cpu utilization in amazon ec2 cloud.
[8] Dargie, W. (2015). A stochastic model for estimating the power consumption of a processor. *IEEE Trans. Comput.* 64, 1311–1322, May.
[9] Brihi, A., and Dargie, W. (2013). Dynamic voltage and frequency scaling in multimedia servers. *2013 IEEE 27th international conference on advanced information networking and applications (AINA)*, vol. 0, pp. 374–380.
[10] Mobius, C., Dargie, W., and Schill, A. (2014). Power consumption estimation models for processors, virtual machines, and servers. *IEEE Trans. Parallel Distr. Syst.* 25, 1600–1614, June.
[11] Kalla, R., Sinharoy, B., Starke, W. J., and Floyd, M. (2010). Power 7: IBM's next-generation server processor. *IEEE Micro.* 30 (2), 7–15.
[12] Sinharoy, B., Kalla, R., Starke, W. J., Le, H. Q., Cargnoni, R., Van Norstrand, J. A., Ronchetti, B. J., Stuecheli, J., Leenstra, J., Guthrie, G. L., Nguyen, D. Q., Blaner, B., Marino, C. F., Retter, E., and Williams, P. (2011). IBM POWER 7 multicore server processor. *IBM J. Res. Dev.* 55 (3), 1:1–1:29.
[13] Gerards, M. E. T., Hurink, J. L., and Kuper, J. (2015). On the interplay between global DVFS and scheduling tasks with precedence constraints. *IEEE Trans. Comput.* 64, pp. 1742–1754.

[14] Maiti, S., Kapadia, N., and Pasricha, S. (2015). Process variation aware dynamic power management in multicore systems with extended range voltage/frequency scaling. In *IEEE 58th international midwest symposium on circuits and systems (MWSCAS)*, pp. 1–4.

[15] Jadeja, Y., and Modi, K. (2012). Cloud computing-concepts, architecture and challenges. In *2012 International Conference on Computing, Electronics and Electrical Technologies (ICCEET)*, pp. 877–880, IEEE.

[16] Seinturier, L., Merle, P., Rouvoy, R., Romero, D., Schiavoni, V., and Stefani, J.-B. (2012). A component-based middleware platform for reconfigurable service-oriented architectures. *Software: Practice Exp.* 42 (5), 559–583.

[17] Abrishami, S., Naghibzadeh, M., and Epema, D. H. (2013). Deadline-constrained workflow scheduling algorithms for infrastructure as a service clouds. *Future Gen. Comput. Syst.* 29 (1), 158–169.

[18] Beloglazov, A., and Buyya, R. (2012). Optimal online deterministic algorithms and adaptive heuristics for energy and performance efficient dynamic consolidation of virtual machines in cloud data centers. *Concurr. Comput. Practice Exp.* 24 (13), 1397–1420.

[19] Verma, A., Ahuja, P., and Neogi, A. (2008). pmapper: Power and migration cost aware application placement in virtualized systems. In *Proceedings of the 9th ACM/IFIP/USENIX International Conference on Middleware*, Middleware '08, New York, NY, USA, pp. 243–264, Springer-Verlag New York, Inc.

[20] Xu, F., Liu, F., Liu, L., Jin, H., Li, B., and Li, B. (2014). iaware: Making live migration of virtual machines interference-aware in the cloud. *IEEE Trans. Comput.* 63, 3012–3025, Dec.

[21] Wood, T., Shenoy, P., Venkataramani, A., and Yousif, M. (2007). Black-box and gray-box strategies for virtual machine migration. In *Proceedings of the 4th USENIX conference on Networked systems design and implementation*, NSDI'07, Berkeley, CA, USA, p. 17, USENIX Association.

[22] Zhu, Q., Zhu, J., and Agrawal, G. (2010). Power-aware consolidation of scientific workflows in virtualized environments. Technical report, Ohio State University.

[23] Park, K., and Pai, V. S. (2006). Comon: A mostly-scalable monitoring system for planetlab. *SIGOPS Oper. Syst. Rev.* 40, 65–74, Jan.

[24] Eichhorn, F., Dargie, W., Mobius, C., and Rybina, K. (2015). Haecubie: A highly adaptive and energy-efficient computing demonstrator.

In *2015 24th International Conference on Computer Communication and Networks (ICCCN)*, pp. 1–8, Aug.

[25] Moebius, C., and Dargie, W. (2014). Statistical analysis of the workload of a video hosting server. In *Analytical and Stochastic Modeling Techniques and Applications* (B. Sericola, M. Telek, and G. HorvÃąth, eds.), vol. 8499 of *Lecture Notes in Computer Science*, pp. 223–237, Springer International Publishing, Switzerland.

[26] Wu, H., Feng, Z., C. Guo, and Y. Zhang (2013). Ictcp: Incast congestion control for tcp in data-center networks. *IEEE/ACM Trans. Netw.* 21, 345–358, Apr.

[27] Enachescu, M., Ganjali, Y., Goel, A., McKeown, N., and Roughgarden, T. (2005). Part iii: Routers with very small buffers. *SIGCOMM Comput. Commun. Rev.* 35, 83–90, July.

[28] Rybina, K., Dargie, W., Strunk, A., and Schill, A. (2013). Investigation into the energy cost of live migration of virtual machines. In *Sustainable Internet and ICT for Sustainability (SustainIT), 2013*, pp. 1–8, Oct.

[29] Strunk, A., and Dargie, W. (2013). Does live migration of virtual machines cost energy? In *2013 IEEE 27th International Conference on Advanced Information Networking and Applications (AINA)*, pp. 514–521, IEEE.

[30] Cheng, X., Dale, C., and Liu, J. (2008). Statistics and Social Network of YouTube Videos. *2008 16th Interntional Workshop on Quality of Service*, pp. 229–238, June.

[31] Cha, M., Kwak, H., Rodriguez, P., Ahn, Y.-Y., and Moon, S. (2009). Analyzing the video popularity characteristics of large-scale user generated content systems. *IEEE/ACM Trans. Network.* 17 (5), 1357–1370.

[32] Zink, M., Suh, K., Gu, Y., and Kurose, J. (2009). Characteristics of YouTube network traffic at a campus network – Measurements, models, and implications. *Computer Networks*, 53 (4), 501–514.

[33] Mitra, S., Agrawal, M., Yadav, A., Carlsson, N., Eager, D., and Mahanti, A. (2011). Characterizing web-based video sharing workloads. *ACM Transactions on the Web*, 5 (2), pp. 8:1–8:27.

PART II

Mobile Computing

3

Mitigating Bufferbloat with Receiver-based TCP Flow Control Mechanism in Cellular Networks

Xiaolan Liu[1,2], Fengyuan Ren[1,2], Ran Shu[1,2], Tong Zhang[1,2] and Tao Dai[1,2]

[1]Tsinghua National Laboratory for Information Science and Technology, Beijing, China
[2]Department of Computer Science and Technology, Tsinghua University, Beijing, China

Abstract

The symbol of network congestion is the packet dropping event detected by TCP congestion control mechanism. But in cellular networks, the packet dropping event is always concealed by a large buffer deployed necessarily to absorb the traffic burst and to achieve the retransmission mechanism of link layer for every user in the base station. Excessive packets beyond link capacity are blocked in the large buffer, so sender cannot detect the network congestion and will increase its sending window consistently. And this may trigger excessively long delays and network jitter without contribution to the improvement of throughput at the same time. This intractable phenomenon is named bufferbloat. More seriously, the dynamic variation of wireless link bandwidth can deteriorate such circumstance and severely degrade the network performance and user's QoE. Wireless link is always the bottleneck link of a TCP connection in cellular networks, and its available bandwidth is that of the connection. Wireless link is also the last hop of the connection, and it connects the mobile terminal directly. Thus, the mobile terminal can get all of the channel state information such as signal strength. So it is more feasible to resolve the bufferbloat at receiver side with the help of channel state information; nevertheless, which is ignored in previous studies. So in this chapter, we unveil the root cause of bufferbloat is the in-adaptation between the adjustment of sending windows and the dynamic variation of wireless link bandwidth at the existence of large buffer. And we present a receiver-based flow control strategy named ABRWDA that retrieves the available bandwidth of a wireless link at receiver side directly and uses it to dynamically calculate the receiving window (*rwnd*) and then adjusts sending windows to mitigate the bufferbloat in cellular networks. Because of the dynamics of wireless link bandwidth, we introduce kalman filter algorithm to maintain the system stability. We test our approach with the NS-2 simulation, and the results indicate that ABRWDA achieves 0.1 times and 0.4 times shorter queues, 0.5 times and 0.8 times lower latency, while still maintaining the same high throughput as that of Newreno and a previous solution DRWA, respectively.

Keywords: Bufferbloat, TCP, Flow control, Cellular networks.

3.1 Introduction

With the development of mobile communication technologies, the cellular network is becoming a main media for us to access the Internet. The data traffics account for an increasing proportion annually in mobile traffic. In the first half year of 2014, for example, the mean monthly usage of mobile access has made a 16.6% jump than that of last year, while it is only 13.4% for fix access [1, 2]. So it is a major mission of advancing QoS to make sure the data transmission quality in cellular networks for network operators.

In the Internet, the network layer only supports the best effort service. The end-to-end reliability of data transmission is ensured by TCP [3]. ACK packet acknowledges correct data receiving. Network congestion is detected by packet dropping event indicated by three duplicated ACK or sending timeout, and mitigated by sending window adjustment controlled by congestion control mechanism and flow control mechanism. But in the cellular networks, wireless link is the unique way to access the cellular network for mobile user. Because of the dynamic variation of wireless signal quality, the bandwidth of wireless link varies dynamically, which causes the transmission errors and decline of link utilization. So a large buffer is deployed in base station for each user to absorb burst traffic coming from wired network and to achieve retransmission mechanism of link layer, however, which may cause TCP performance degradation, because the maximum accepted by TCP sending window is the link capacity, also named as bandwidth-delay-product (BDP).

When the size of sending window is beyond the link capacity, network congestion may cause packet dropping, and sender can adjust the size of sending window accordingly. The packets exceeding link capacity are blocked in buffer when there is a large buffer and there will be no packet dropping. So the sender cannot detect the network congestion and increase the size of sending window persistently. With the increasing queue length of blocked packets in buffer, the latency is increased dramatically, which deteriorates the network reliability and performance severely, and which is the so-called bufferbloat [4]. So bufferbloat is happened when input rate is bigger than output rate of packet in large buffer. The input transmission rate is decided by the size of sending window, while the output transmission rate is determined by the available bandwidth. Wireless link is always the bottleneck link of TCP connection compared to the wired link [5], so the available bandwidth of wireless link is that of TCP connection. Thus, the key to solve bufferbloat in the cellular networks is to make the size of sending window adapt to the available bandwidth of wireless link.

The research of bufferbloat initially begins from Internet. This term is proposed by J. Gettys [5], and corresponding research approaches such as AQM, CoDel and PIE [6–8] are presented, and followed with researches in cellular networks [5, 9, 10], accordingly. But approaches proposed in these researches need either to alter the function of sender or to adjust the deployment of the medium nodes along the network link. These approaches are not deployed widely because of the expensive price of adjustment. The approach based on receiver, however, is a light selection. Being a receiver, the mobile terminal is changeable to the user. DRWA [5], for example, is a receiver-based window adjustment mechanism. But the performance of DRWA degrades greatly when propagation latency increases.

In this chapter, we present an approach named as Available Bandwidth-based Receiver Window Dynamic Adjustment (ABRWDA) to dynamically mitigate bufferbloat in the cellular networks. The cellular networks have special characteristics that the wireless link is always the last hop of downlink and bottleneck link of the TCP connection, and the available bandwidth of wireless link is related to the channel state closely, and UE can get all these channel state information because of the direct connection with it. ABRWDA retrieves the bandwidth at receiver side directly by tracking the channel state information. Meanwhile, RTT is also estimated at the receiver-side. By multiplying the bandwidth and RTT, the receiving window (*rwnd*) is obtained. To control sending window at receiver-side, ABRWDA modifies the TCP flow control mechanism with dynamically adjusted *rwnd* calculated by previous method. We insist that our solution is easy to deploy because it can be functioned as either a kernel module or a net-filter, while requires no modifications to the network or application software, and no support from ISPs.

To evaluate the performances of ABRWDA, several simulations on NS2 platform are conducted. The experimental results prove that ABRWDA can reduce the latency while maintaining the appropriate throughput of TCP. Further experimental results show that the average flow completion time (AFCT) is reduced by 25–78% compared to TCP NewReno and almost 50% improvements compared to the previous solutions. Results also reveal better network environment adaptability of ABRWDA.

The rest of the chapter is organized as follows. We present the background, related work and motivation in Section 3.2. Detailed algorithm design and analysis of ABRWDA are presented in Section 3.3. Section 3.4 outlines the experimental configurations thoroughly, followed by the experimental results and performance comparison with that of other approaches in Section 3.5. We conclude our chapter and discuss our future work in Section 3.6.

3.2 Background, Related Work and Motivation

In this section, we describe the background, related work and motivation of this chapter.

3.2.1 Background

3.2.1.1 Cellular networks

The cellular network is a heterogeneous network. It possesses all characteristics of high rate and fix bandwidth of wired network and varied bandwidth and high error rate of wireless network. The architecture of cellular networks is illustrated in Figure 3.1, mainly consisted by core network (CN), radio access network (RAN) and user equipment (UE).

CN is the central part of cellular networks. It is in charge of switching of voice service to the public switch telephone network (PSTN), 3G or 4G system and routing of data service, as well as managing authority, charging, QoS, security, etc. RAN connects UE and CN, and schedules radio resources. The structure of RAN tends to simplification with the development of wireless communication technologies. For example, the RAN in 3G is simplified as Node B (NB) and radio network control (RNC) model, while all functions of RAN are focused on evolved Node B (eNB). The simplification decreases the

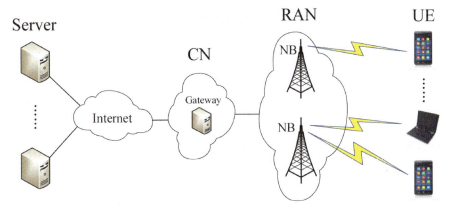

Figure 3.1   The architecture of cellular networks.

response time of RAN and enhances the network function accordingly. UE is the user platform to access the cellular networks. Mobility management, connectivity management and radio resource control is controlled by UE and CN. The usage of UE is spurred by the update of hardware and mobile applications.

The radio channel is the only link between UE and NB. Due to the dynamic variation of radio signal, there is a high error rate within the data transmission in wireless link. The link layer retransmission mechanism, such as Automatic Repeat reQuest (ARQ) and error correction technology, is introduced in the cellular networks. To achieve this mechanism, a large buffer must be deployed in NB for each user. And large buffer also can contribute to absorb the burst traffic from wired network.

The bandwidth of wireless link is varied dynamically because of the dynamic characteristics of radio channel state. Wireless link is always the bottleneck link of a TCP connection in cellular networks [5]. The reliability of data transmission is ensured by TCP in cellular networks, which is similar to the wired networks. But there will be a conflict between the congestion control mechanism of TCP and dynamic characteristic of wireless link, which will affect the performance of TCP partially.

3.2.1.2 Bufferbloat

Bufferbloat [4, 5, 7, 9] is an intractable phenomenon in both the Internet and cellular networks, which may cause excessively long delays without contribution to the improvement of throughput at the same time. The problem matters more for the latter, since large buffer is the inevitable choice due to the demand of link layer retransmission mechanism as well as the need of mitigating traffic burst.

In a bufferbloated circumstance we consider in this chapter, the source increases its sending window persistently because the packets dropping is concealed by the large buffer. So the RTT latency increases sharply when packets are blocked in large buffer.

3.2.1.3 TCP flow control mechanism

The TCP receive window is originally designed to prevent a fast sender from overwhelming a slow receiver with limited buffer space, used in TCP flow control mechanism [3]. This mechanism governs the size of a sending window together with the congestion control mechanism. It reflects the available buffer size on the receiver side so that the sender will not send more packets than the receiver can accommodate.

3.2.1.4 Available bandwidth

The available bandwidth of a connection is the difference of the link capacity and the data rate of cross traffic. End-to-end available bandwidth estimation is important for a range of applications such as network management, flow control and rate adaptation in real-time multi-media streaming [11]. It depends on that of the bottleneck link. In a practical cellular network communication, the wireless link forms the last hop from the base station to the mobile clients, which is usually the bottleneck of the routes in terms of congestion [5]. Therefore, the available bandwidth of the wireless link is usually that of the whole connection.

3.2.2 Related Works

3.2.2.1 Bufferbloat

The issue about bufferbloat was first exposed by Dave P. Reed in [12]. He found large RTTs along the routes but without packet loss. Kathleen Nichols and Van Jacobson proposed a modern AQM, CoDel [7], to solve the bufferbloat by gauging the packet-sojourn time through the queue with the help of the timestamp. Rong Pan et al. issued a latency-based design to control bufferbloat in the Internet in [8], which could effectively cap the average queueing latency with a reference value.

All these works are about tackling bufferbloat in the Internet, and bufferbloat also exists in the cellular networks. DRWA, a dynamic receive window adjustment approach, was proposed in [5] to tackle the bufferbloat by estimating the *cwnd* with the received data at receiver side. In order to mitigate the ACK delays by eliminating TCP ACK clocking, a new TCP variant TCP-RRE was presented in [13]. It used TCP timestamp to estimate the receive rate at receiver side and determines the sending rate accordingly, which can keep the occupancy of the downlink buffer low. In [10] Yung-Chih, Chen et al. researched the bufferbloat's effect on multi-path TCP (MPTCP) within WiFi and cellular networks. They showed that MPTCP might suffer from bufferbloat when there is another long-lived WiFi flow, and the severer the bufferbloat is, the more harm will be done on the performance of MPTCP.

3.2.2.2 Receiver-side flow control in cellular networks

TCP exploits the sliding window mechanism to adjust the sending rate at the sender side. The sending window size is the minimum of the congestion window *cwnd* and the advertised receiver window *rwnd*. Wireless link is always the bottleneck link of the TCP connection in cellular networks. It is vital to abstain the state information of wireless link for analyzing the TCP performance. Being a receiver, UE can conveniently retrieve the state information of wireless link because of the direct connection with the wireless link. So many researches about TCP performance are based on TCP receiver. RCP [14] is a TCP clone that exchanges the functions of sender and receiver thoroughly. The receiver in RCP controls all the key functionalities of protocol such as congestion control, flow control and reliability. CLAMP [6] establishes queue length model in buffer along the bottleneck link with the rate control of link layer. And it allocates fairly the radio resource and maintains the link utility through the flow control of receiver side. RSFC [15] is receiver-side flow control scheme used to monitor the available upload capacity and adjusts the TCP *rwnd* dynamically in a two-way TCP circumstance. Ahmed Mansy et al. showed that a single video stream can cause significant delays to other ongoing application in a bufferbloat circumstance. They propose a technique [16] that enables a video client to smoothly download video segments without causing significant delays.

3.2.3 Motivation

From the above subsection, we know many works have been performed in cellular networks to tackle bufferbloat. Seminal as they are, all these approaches do not uncover the root cause of bufferbloat, which is the mismatching between the adjustment of sending windows and the dynamic variation of available bandwidth at the existence of large buffer. In this case, the key to solving bufferbloat is to make the sending rate accommodate the available bandwidth and its fluctuations.

In the cellular networks, the wireless link is always the bottleneck link of the network, so its available bandwidth is usually that of the connection. The UE acts as receiver for data sent from the server in the wired data network, which consist of the downlink of the connection. Being adjacent to the wireless link, which is the last-hop of the connection, UE obviously can obtain first-hand information of the wireless link [14] to determine the rate at which packets should be sent by the server [17]. This information (such as CQI or SINR) have congruent relationship for certain network type, as stated in [18, 19]. Accordingly, UE can get the variable bandwidth value with the variation of wireless channel states.

In a word, we can retrieve the wireless link bandwidth at UE directly with the help of the channel information, estimate RTT with the same method in [5], calculate *rwnd* with the former two and adjust the sending rate with calculated *rwnd* finally under TCP flow control mechanism.

3.3 Algorithm Analysis and Design

As previously mentioned, seminal as these researches about bufferbloat in cellular networks are, they either need adjustment of deployment of medium nodes, or need to change the configuration of server or base station. The cost is huge. And they do not uncover the root cause of bufferbloat, which is the mismatching between the adjustment of sending window and the variation of available bandwidth. So the key to solve the bufferbloat is the adaption of the send rate to available bandwidth.

In the cellular networks, more than 30% bytes flows are served by well-known CDN servers. And these servers are located on the network nodes of operators to reduce latency and improve user's QoS. Compared with wired link with fix bandwidth, the wireless link becomes the bottleneck link in the cellular networks, and its available bandwidth is that of a TCP connection. UE can obtain the state information of wireless link via direct connection [17, 20]. So UE can retrieve available bandwidth directly [18, 21]. In this chapter, we propose a new congestion control scheme named ABRWDA to mitigate bufferbloat. ABRWDA retrieves the available bandwidth at UE directly, calculates the *rwnd* after the RTT estimation and adjusts the size of sending window together with TCP congestion control mechanism. The framework of ABRWDA is illustrated in Figure 3.2.

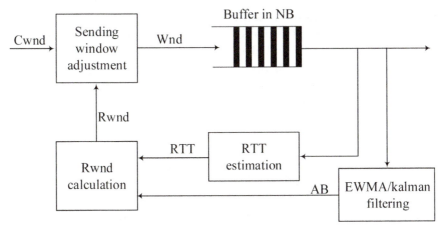

Figure 3.2   The framework of ABRWDA, comprising three components, available bandwidth disposing, RTT estimating and *rwnd* calculating.

3.3.1 Retrieving the Available Bandwidth

The UE can obtain the overall information about the physical channels, and the information can be used to calculate the available bandwidth [18, 19, 21] despite different state parameters owned by different type of networks. In work [18], for example, UE selects a node with the biggest SNR from two relays and NB as its data transmission node. That node transmits data with suitable data rate under certain adaptive modulation and coding (AMC) selected with SNR afterwards. Similarly, the available bandwidth is calculated after CQI and DTX are measured directly by UE in a new flow control scheme in [21]. Thus, our approach is based on the assumption that UE can directly retrieve the available bandwidth of wireless link.

In the Internet, there may be several flows at the same time. So the bandwidth is shared by these flows. But the research result in [22] shows that in LTE there is only one TCP flow actively downloading data in 72.1% of the time, and this percentage might be even larger for smart phone users because a fraction of users that using LTE data cards on their laptops may be included in their data set, which may have high TCP flow concurrency. That is to say, in most cases there is only one TCP flow in a short time scale in the cellular networks. And according to the findings in [23], a single large buffer is deployed for each user in cellular networks, so there will be no interplay between data flows of different users. Hence, we insist on that the link bandwidth is used by one flow in the cellular networks in one time, and the available bandwidth of a connection is exactly that of the wireless link.

The bandwidth of wireless link is dynamically varied in cellular networks. The network fluctuation is harmful to TCP performance. So we introduce exponential weighted moving average (EWMA) algorithm in ABRWDA to maintain the system stability, illustrated in Equation (3.1)

$$Dbw_i = \alpha \cdot Dbw_i + (1 - \alpha) \cdot bandwidth \qquad (3.1)$$

where $0 < \alpha < 1$, $bandwidth$ is the available bandwidth of wireless link, Dbw is smoothed bandwidth. But the performance of EWMA is related to the selection of parameter closely. The value of parameter is different for different contexts, which is inconvenient in the application. Therefore, Kalman filter algorithm is introduced [24, 25] in our chapter. The Kalman filter is a recursive approach to estimate the state of a process, in a way that minimizes the mean of the squared error. It can be used either for stable process or for unstable process. It retrieves the optimal value of system state with the priori state estimate at a previous time step and the current measurement when there is a minimum for mean square error.

The variation of bandwidth is a time-varying linear process in the cellular networks. Being a measured variable, the time update equation and measurement update equation can be expressed with Equations (3.2) and (3.3) without thinking about the control input [24, 25]:

$$B_k = AB_{k-1} + w_{k-1} \qquad (3.2)$$

$$z_k = HB_k + v_k \qquad (3.3)$$

where A is the state transition matrix of state equation and H is the measurement matrix of measurement equation. w is the process noise, and v is the and measurement noise. They are assumed to be white and independent noise with normal probability distributions:

$$p(w) \sim N(0, Q) \qquad (3.4)$$

$$p(v) \sim N(0, R) \qquad (3.5)$$

where the Q and R represent the process and measurement noise covariance matrices. They are constant here. The so-called optimal estimation of Kalman filter is that in the sense of the minimum of mean square error (MSE) for bandwidth estimation, represented in Equation (3.6)

$$J(\widehat{B}) = min(E[(\widehat{B}_k - \widehat{B}_k^-)(\widehat{B}_k - \widehat{B}_k^-)^T]) \tag{3.6}$$

The time update process of the Kalman filter is to predict the bandwidth $\widehat{B}_k^-$ and MSE P_k^- at time step k with the estimation of bandwidth $\widehat{B}_{k-1}$ and MSE P_{k-1} at time step $k-1$ respectively:

$$\widehat{B}_k^- = A\widehat{B}_{k-1} \tag{3.7}$$

$$P_k^- = AP_{k-1}A^T + Q \tag{3.8}$$

The estimation of bandwidth $\widehat{B}_k^-$ with the $\widehat{B}_k^-$ and measurement residual $z_k - H_k B_k^-$ and MSE P_k at time step k is obtained in measurement update process:

$$K_k = P_k^- H_k^T (H_k P_k^- H_k^T + R)^{-1} \tag{3.9}$$

$$\widehat{B}_k = \widehat{B}_k^- + K_k(z_k - H_k\widehat{B}_k^-) \tag{3.10}$$

$$P_k = (I - K_k H_k)P_k^- \tag{3.11}$$

where K_k is the Kalman gain, which is weighting matric for measurement residual $z_k - H_k\widehat{B}_k^-$ at time step k.

3.3.2 RTT Estimation

We get RTT estimation by averaging the RTT samples obtained from the timestamp within the last RTT when timestamp option is on. Otherwise, we estimate the RTT value with the time interval between a first acknowledged byte and a byte receipted whose sequence number is at least one window size forward, the same method in [26].

3.3.3 Rwnd Calculation

We calculate *rwnd* with the retrieved available bandwidth and estimated RTT and adjust the sending window with TCP flow control mechanism. The detailed realization of window acknowledgement mechanism needs a RTT period. The adjustment effect to sending window of *rwnd* will be seen at least after a RTT. But in a RTT, the bandwidth may be varied from a small value to a big one, which can causes the tendency to zero buffer and throughput decreasing. Thereafter, we set a scaling factor λ to handle this case.

Algorithm 3.1   ABRWDA

Initialization:

1:   $wind\_ \Leftarrow CWND\_INIT$;
2:   $last\_wind\_ \Leftarrow CWND\_INIT$;
3:   $data\_cumd\_ \Leftarrow 0.0$;
4:   $rtt\_min\_ \Leftarrow a\ large\ enough\ value$;
5:   **if** $data\_cumd\_ == 0.0$ **then**
6:     $start\_time\_ \leftarrow now$;
7:   **end if**
8:   $data\_cumd\_+ = PacketSize$;
9:   **if** $data\_cumd\_ > last\_wind\_$ **then**
10:    $rtt\_est\_ \leftarrow now - start\_time\_$;
11:    **if** $rtt\_min\_ > rtt\_est\_$ **then**
12:      $rtt\_min\_ \leftarrow rtt\_est$;
13:    **end if**
14:    $last\_wind\_ \leftarrow wind\_$;
15:    $Dbw\_ \leftarrow (1 - \alpha) * Dbw\_ + \alpha * bandwidth\_$;
16:    $wind\_ \leftarrow \lambda * Dbw\_ * rtt\_min\_$;
17:    $rwnd\_ \leftarrow wind\_ > last\_wind\_?wind\_ : last\_wind\_$;
18:    $data\_cumd\_ = 0.0$;
19:   **else**
20:    $rwnd\_ \leftarrow wind\_$;
21:   **end if**

Because of the dynamic variation of wireless link bandwidth, the size of current *rwnd* window may be smaller than that of the previous one when there is a sudden increase in the wireless link bandwidth, which could cause the abnormal phenomenon that the estimated RTT value is smaller than the theoretical minimum (propagation delay only). So in our design, we select the bigger one between the current window and the previous one as new *rwnd*.

The pseudocode of ABRWDA is presented in Algorithm 3.1.

3.4 Simulation Configurations

In light of the limitations stated in aforementioned literatures, we propose a method named ABRWDA to mitigate the bufferbloat in cellular networks. We perform several simulation experiments to test the effectiveness of our approach. The simulations are performed using NS-2 version 2.30, which is running on the Ubuntu 10.04 LTS with 2.6.32-21-generic Linux kernel. And the simulation topology is shown in Figure 3.3.

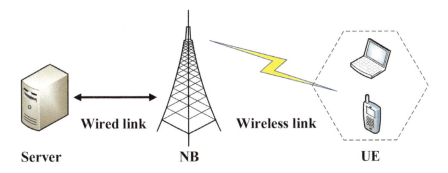

Figure 3.3 Network topology in simulation.

There are three kinds of nodes in the simulations, denoting the remote server, the base-station (NB) and the UE, respectively. The remote server lies in the wired network. The link bandwidth between the server and NB is 1 Gbps, and the propagation delay is 50 ms. In a real experiment in [5], they observed more than 800 KB (almost 560 packets with 1500 B packet size) packets in flight for a certain carrier. The minimum BDP is 58 KB for this network state. So in our simulation we set the buffer size of NB to be 500 packets, far greater than the minimum BDP. The queue management mechanism is drop-tail. The UE can be a mobile phone or a laptop.

In NS-2, any two nodes are connected by an *OTcl* object [27]. The link bandwidth and propagation delay are stored in two member variables of *OTcl* class *DelayLink,* respectively. In our simulation, we obtain the available bandwidth in file *tcp_sink.cc* using a customized static member variable. It inherits the member function *bandwidth* of its parent class, which returns the link bandwidth value bound with it given in *Tcl* file. We determine the configurations of the wireless link according to the UMTS specifications, and the model presented in [28] is also referred to. The bandwidth is set to be the theoretical value of the CDMA2000. All the parameters are listed in Table 3.1.

In real network circumstances, wireless physical link rate is dynamically variable due to the variation of signal strength and other radio environment [29]. In our simulation, the raw TCP version is NewReno. We simulate the dynamism of the wireless link bandwidth with random variables in *Tcl* file. We use two random variables to reappear the dynamical variation of the wireless link bandwidth. The first random variable simulates a different

Table 3.1   Simulation parameters

Parameter	Value
Packet size	1500 B
Queue management	Drop Tail
Buffer size in NB	500 packets
Bandwidth between sender and NB	1000 Mbps
Bandwidth between NB and UE: downlink	(3.1~9.3) Mbps
Bandwidth between NB and UE: uplink	(1.8~5.4) Mbps
Link delay between server and NB	50 ms
Link delay between NB and UE	25 ms
CWND_INIT	10 packets

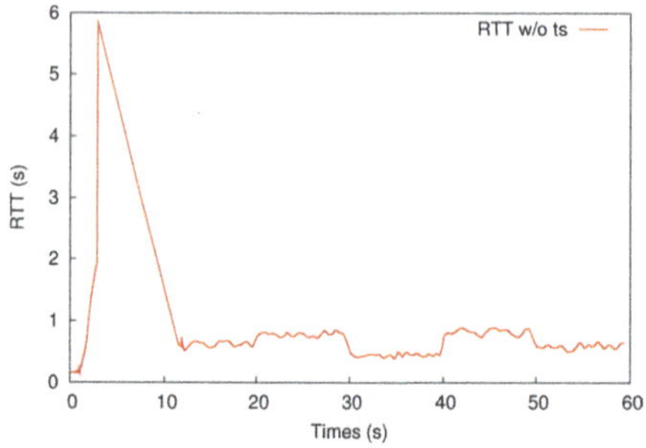

(a) RTT

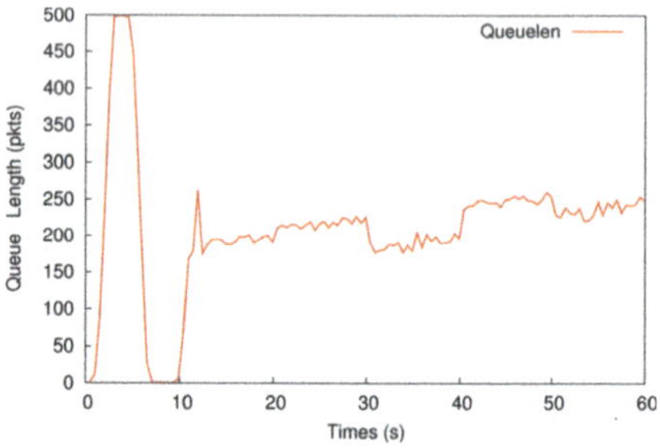

(b) Queue Length

Figure 3.4   The RTT and queue length of NewReno in a bufferbloated circumstance.

experimental circumstance with good or weak signal condition. And the second one simulates the dynamic variation of wireless link bandwidth. Our simulation results in Figure 3.4 agree with that in [5], which are measured in real networks. So configurations in our simulation can depict the practical networks conditions perfectly.

3.5 Experimental Results and Performance Evaluation

In this section, we analyze the performance of ABRWDA and make a contrast with that of DRWA and some other TCP versions. There have been many works about TCP performance in wired networks, but what it is like in bufferbloated circumstance is still undiscovered. In this section, we present the performance evaluation of ABRWDA, DRWA and Vegas in a bufferbloated cellular network.

3.5.1 TCP Performance Analysis Under Bufferbloated Circumstance

3.5.1.1 TCP performance of ABRWDA in a bufferbloated circumstance

The buffer size is excessively bigger than BDP in a bufferbloated network circumstance. The sender will persistently send packets when there is no packet dropping. So the queue length in buffer will grow up quickly, as shown with red line in right picture of Figure 3.4. The queue size reaches to the maximum at 3.51 s, and it begins to decrease after a stable state because of the packet dropping. It decreases to the zero at 8.92 s. Though after a new window evolvement, buffer size still keeps in a great value in the simulation.

The big buffer size can cause extra-long delay. The red line in left picture of Figure 3.4 stands for the RTT in a bufferbloated circumstance, estimated with the approach in the Algorithm 3.1. We can see that RTT in real networks with NewReno is greater than that with ABRWDA, represented with green line in right picture of Figure 3.5. The great spike of RTT is caused by the absence of packets which should be received normally in standard TCP. The zero buffer size harms the throughput seriously just as the full buffer does to RTT. The throughput collapses steeply when buffer size decreases to zero. We do not present throughput for NewReno due to the space limitation.

The essence of ABRWDA is the adaption of the sending rate to the bandwidth variation of wireless links at the existence of large buffer. In Figure 3.5, the green line stands for the performance of ABRWDA. The pink line in the left picture represents the variation of bandwidth. We can see that ABRWDA can adapt well to the variation of bandwidth dynamically.

The appropriate queue length ensures the perfect throughput as well as low delays. The green line in the middle picture presents the measured queue length of ABRWDA in a 60 s simulation. We can see that the maximum queue length does not exceed 63 packets, being bigger than BDP calculated with minimum bottleneck bandwidth (about 40 packets size). The RTT measurements are represented in the right picture of Figure 3.5. We can find that its variation is in a opposite tendency to that of bandwidth, and the RTT values are steadily varied in the range of 0.154 (the delay without the queuing delay) to 0.33 s, which is a great improvement contrasted to the standard TCP without ABRWDA, shown in the left picture of Figure 3.4.

From Figure 3.5, we can see that ABRWDA keeps an appropriate queue length that can maintain the high throughput and low delay at the same time.

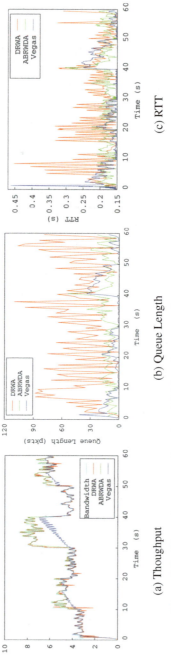

(a) Thoughput

(b) Queue Length

(c) RTT

Figure 3.5   Throughput, queue length and RTT performance of ABRWDA, DRWA and Vegas in a 60 s simulation.

3.5.1.2 TCP performance of DRWA in a bufferbloated circumstance

The performances of DRWA are presented with red line in Figure 3.5. DRWA estimates the *cwnd* with received data at receiver side and restrains the sending rate with *rwnd* calculated with the estimated *cwnd* and RTT.

Since the estimation of *rwnd* is only based on the transport layer information, DRWA has no knowledge of the real link available bandwidth. So the sending rate cannot adapt to the variation of link bandwidth, which can explain the several collapses and the dramatic fluctuation in queue length, as shown in middle picture. And RTT plotted in right picture fluctuates accordingly.

From Figure 3.5 we can find that ABRWDA's adaptability to the variation of network states is superior to that of DRWA. ABRWDA can perfectly adapt to the variation of network states, keep queue length within an appropriate range and reduce the RTT naturally.

3.5.1.3 TCP performance of vegas in a bufferbloated circumstance

Vegas is a TCP variant that confines all the changes at the sending side [30]. It adjusts the *cwnd* with the difference of expected sending rate and actual sending one. When there is congestion in network, the difference will increase; then, Vegas decrease the sending rate to avoid congestion. Thus, the congestion avoiding is achieved in return for the decrease in throughput. We can see in the left picture of Figure 3.5 that Vegas's adaptability to the variation of bandwidth is severely poor. The throughput collapses severely in several periods of times and does not vary with the variation of bandwidth, which can be seen at 10.001, 18.109, and 40.001 s, respectively. The average queue length is only 6 packets, with the maximum of 84 packets size at the beginning, as shown in the middle picture of Figure 3.5. In 40.76% of simulation time, the queue length is zero, under which the RTT is relatively small. The average value is 0.166 second, except for a 0.449 second of maximum at the very beginning.

3.5.2 The Effect on TCP Performance Caused by Parameters' Selection

The essence of ABRWDA is the adaption of sending rate to the variation of wireless link state at the existence of large buffer. In the left picture of Figure 3.5, the pink line indicates the variation of throughput well keeps up with the variation of bandwidth. Several little jitters triggered by bandwidth variation can be found, but they dissipate very soon.

There are two parameters to be configured in Algorithm 3.1: one is α, which is the coefficient of the moving average filter, and the other is λ, the scaling factor to adapt to the bandwidth variation. We seek for their optimal values in 2 steps: first, we set λ as 1, and test optimal α with different values; second, we test the optimal λ value with the optimal α value found in the former step.

In Algorithm 3.1, we dispose the variation of bandwidth with a smooth average filter. We test the appropriate α in experiments and calculate both the CDF of link utilization and RTT under different α values, as shown in Figure 3.6. We can see that compared with other α values, the link utilization is significantly lower when α is 1. All the α values less than 1 produce similar utilizations, and the value of 1/4 produces the best utilization.

In Figure 3.6, though we can see that there is little difference in RTT CDF in the cases of different α values, we can still find that the smaller α is, the less RTT varies. Combining the CDF of link utilization with that of RTT, we choose the 1/4 as the optimal value of α.

We test different values of λ with the optimal α value selected in former experiment; the results are shown in Figure 3.7. In the left picture, we can see that the link utilizations are all close to 1 with different λ values, except for the case when λ is 1.0. Furthermore, as is shown in the right picture of Figure 3.7, the bigger λ is, the greater RTT varies. Taking into account both the link utilization and the RTT variation, we set the optimal value of λ as 1.2.

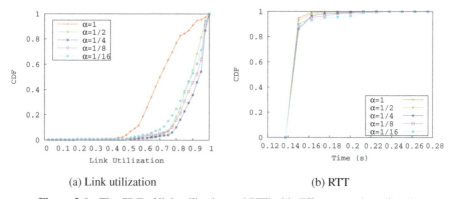

(a) Link utilization   (b) RTT

Figure 3.6   The CDF of link utilization and RTT with different α values ($\lambda = 1$).

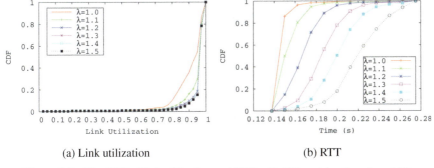

(a) Link utilization   (b) RTT

Figure 3.7   The CDF of link utilization and RTT with different λ values ($\alpha = 1/4$).

3.5.3 The Improvement in User Experiences

Nowadays, mobile applications (such as Web browsing, file downloading and online game playing, etc.) have flooded into people's lives. When it comes to network service, different applications vary greatly in data volumes. A typical mobile scenario is simultaneous large file downloading and Web browsing. In this case, short flows and long flows will coexist on the same network path. In a bufferbloated circumstance, the packets of the long flow will occupy most of the buffer space, causing great queueing delay to short ones. The consequence of the bufferbloat problem could be the slow response to a delay-sensitive application such as simple Web page request, which seriously impacts the user experience. So the completion time of short flows is a vital merit to evaluate the user experience in this situation. In this experiment, we compare the averaged short flow completion times (AFCT) among ABRWDA, DRWA, NewReno and Vegas. We start 100 flows in each test—one is long and the others are short. The long flow starts at the beginning of the test, and the short ones start at different time in turn. We emphasize that none of the short flows overlap with each other on the time dimension to avoid the inter-influence among different short flows. The outcome is shown in Figures 3.8 and 3.9.

In this experiment, we set three kinds of flow size (1, 30 and 200 packets) to represent clicking the mouse, small Web pages and large Web components separately. For each flow size, we change the distance between the UE and the server by setting different propagation delays, which are selected according to the settings in [5]. The values in Figure 3.8 are the AFCTs of short flows in each scenario. Comparing the three subgraphs, we can see that as a delay-based protocol acting on the source, Vegas achieves the shortest AFCT when

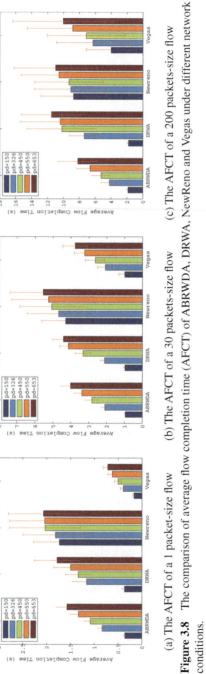

(a) The AFCT of a 1 packet-size flow

(b) The AFCT of a 30 packets-size flow

(c) The AFCT of a 200 packets-size flow

Figure 3.8   The comparison of average flow completion time (AFCT) of ABRWDA, DRWA, NewReno and Vegas under different network conditions.

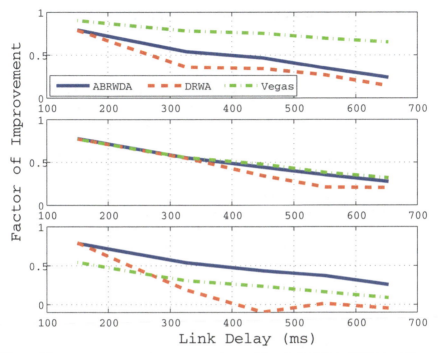

Figure 3.9 The performance improvement about AFCT of ABRWDA, DRWA and Vegas with different flow size and network circumstance (Top: 1 packet-size; Middle: 30 packets-size; Bottom: 200 packets-size).

the flow size is 1 packet. However, when as the flow size becomes larger, the advantage of Vegas becomes less obvious, which can be seen from the second and third subgraph of Figure 3.8. In addition, the low delay of Vegas always comes with the cost of throughput loss. DRWA and ABRWDA are both receiver-based algorithms, and they both perform well with the flow size being 1 or 30. Nevertheless, when the flow size is 200 packets, DRWA loses effect, and in some cases its performance is even poorer than NewReno. In the four algorithms, only ABRWDA keeps good performance with three flow sizes. What's more, AFCT's growth of ABRWDA is very steady in each network circumstance. Compared with NewReno, The performance improvement factor of AFCT for ABRWDA, DRWA and Vegas are shown in Figure 3.9. We can find that the performance of ABRWDA is very steady; however, neither of Vegas and DRWA can maintain the performance as the flow size increases.

3.5.4 The Improvement of System Performance from Kalman Filter

To test the improvement of system performance from Kalman filter, we use the real trace file of bandwidth measured in [31] in this experiment. That is to say that the value of z_k is the sum of the bandwidth in trace file and Gaussian white noise. The bandwidth of current time is the correction of precious time, and the measured value is that in trace file. So the values of coefficient matrix A and H are one.

The trace file of bandwidth of [31] is the data sets of the HTTP-based mobile video stream received by receiver (video client) in 3G. The sender and receiver are located in the same place, which can reduce the effect of wired link to bandwidth measurement. We test all the trace files and only illustrated with trace-7 because of the limited space. The bandwidth fluctuation is large, ranging from 1 $kbps$ to 4.96 $Mbps$, and with a median of 1.88 $Mbps$.

The performance comparison between Kalman filter and ABRWDA is illustrated in Figure 3.10. From subgraph (a), the CDF of RTT, we can see that the Kalman filter is stabler than ABRWDA in real network environment. The jitter of RTT of ABRWDA is great, while the RTT jitter of the Kalman filter is small and centralized. The 95th percentile of RTT of Kalman filter reduces 19% compared to ABRWDA. We also can see that there is a heavy tail of RTT for them all. The reason for tail increase is the fluctuation of bandwidth.

Both the estimation of previous time and the measurement of current time are used in bandwidth estimation in the Kalman filter. But they are not averaged by a constant parameter. The Kalman gain is dynamically adjusted with

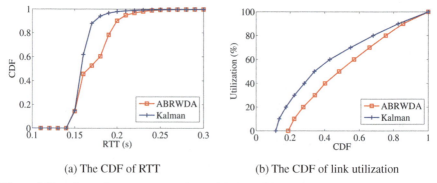

(a) The CDF of RTT   (b) The CDF of link utilization

Figure 3.10   The performance comparison of Kalman filter and ABRWDA with real bandwidth trace file. The subgraph (a) represents the cdf of RTT and (b) represents the cdf of link utilization.

different measurement errors. So the estimated bandwidth of the Kalman filter is more approximate to real value. From the subgraph (b) we can clearly see that the link utilization of the Kalman filter is greater than that of ABRWDA.

From above experiment, we can find the fluctuation of bandwidth heavily affects TCP performance. RTT increases dramatically and the throughput degrades greatly when there is a bandwidth burst, which deteriorates the network performance severely.

3.6 Conclusion

Cellular networks deploy large buffers to absorb traffic bursts and to achieve the link layer retransmission mechanism for better performance. As a result of the existence of link layer reliable retransmissions and large buffers, the packet dropping, which is referred to as the signal of the congestion by TCP, is totally concealed. Without the packet-loss-driven decrease in the sending window, excessively long delays are caused and then deteriorate the TCP performance. This phenomenon is named as bufferbloat. In this chapter, we propose ABRWDA, a receiver-based solution to mitigate bufferbloat in the cellular networks. ABRWDA obtains the wireless link bandwidth by means of the relationship between bandwidth and SINR, and then uses it and the estimated RTT to calculate the receiving window which will be used to adjust the sending window. We implement ABRWDA with NS-2 and compare its performance with the current receiver-based solution DRWA and the delay-based TCP version Vegas. The experimental results manifest that ABRWDA can achieve considerable performance benefits compared with other approaches.

Acknowledgment

The authors gratefully acknowledge the anonymous reviewers for their constructive comments. This work is supported in part by National Basic Research Program of China (973 Program) under Grant No. 2012CB315803 and National Natural Science Foundation of China (NSFC) under Grant No. 61225011.

References

[1] "GLOBAL INTERNET PHENOMENA REPORT 1h 2014." https://www.sandvine.com/downloads/general/global-internet-phenomena/2014/1h-2014-global-internet-phenomena-report.pdf, 2014. [Online; accessed 19-July-2014].

[2] "GLOBAL INTERNET PHENOMENA REPORT 1h 2013." https://www.sandvine.com/downloads/general/global-internet-phenomena/2013/sandvine-global-internet-phenomena-report-1h-2013.pdf, 2014. [Online; accessed 19-July-2014].

[3] Paxson, V., Allman, M., and Stevens, W. (1999). TCP Congestion Control. *RFC2581, April.*

[4] Gettys, J., and Nichols, K. (2011). Bufferbloat: Dark buffers in the internet. *Queue*, 9(11), 40.

[5] Jiang, H., Wang, Y., Lee, K., and Rhee, I. (2012). Tackling bufferbloat in 3G/4G networks, in *Proceedings of the 2012 ACM Conference on Internet Measurement Conference*, pp. 329–342, ACM.

[6] Andrew, L. L., Hanly, S. V., and Mukhtar, R. G. (2008). Active queue management for fair resource allocation in wireless networks. *IEEE Trans. Mobile Comput.* 7(2), 231–246.

[7] Nichols, K., and Jacobson, V. (2012) Controlling queue delay. *Communications of the ACM*, 55(7), 42–50.

[8] Pan, R., Natarajan, P., Piglione, C., Prabhu, M. S., Subramanian, V., Baker, F., and VerSteeg, B. (2013). PIE: a lightweight control scheme to address the bufferbloat problem, in *2013 IEEE 14th International Conference on High Performance Switching and Routing (HPSR),* pp. 148–155, IEEE.

[9] Jiang, H., Liu, Z., Wang, Y., Lee, K., and Rhee, I. (2012). Understanding Bufferbloat in Cellular Networks. In *Proceedings of the 2012 ACM SIGCOMM workshop on cellular networks: operations, challenges, and future design*, pp. 1–6, ACM.

[10] Chen, Y.-C., and Towsley, D. (2014). On bufferbloat and delay analysis of multipath TCP in wireless networks, in *Networking Conference, 2014 IFIP*, pp. 1–9, IEEE.

[11] Li, M., Wu, Y.-L., and Chang, C.-R. (2013). Available bandwidth estimation for the network paths with multiple tight links and bursty traffic. *J. Network Comput. Appl.* 36(1), 353–367.

[12] Reed, D. (2014). What's wrong with this picture. http://mailman.postel.org/pipermail/end2end-interest/2009-September/007742.html, 2014. [Online; accessed 09-July-2014].

[13] Leong, W. K., Xu, Y., Leong, B., and Wang, Z. (2013). Mitigating egregious ACK delays in cellular data networks by eliminating TCP ACK clocking, in *ICNP*, pp. 1–10.

[14] Hsieh, H.-Y., Kim, K.-H., Zhu, Y., and Sivakumar, R. (2003). A receiver-centric transport protocol for mobile hosts with heterogeneous wireless

interfaces, in *Proceedings of the 9th annual international conference on Mobile computing and networking*, pp. 1–15, ACM.

[15] Xu, Y., Leong, W. K., Leong, B., and Razeen, A. (2012). Dynamic regulation of mobile 3G/HSPA uplink buffer with receiver-side flow control. In *2012 20th IEEE International Conference on Network Protocols (ICNP)*. pp. 1–10, IEEE.

[16] Mansy, A., Ver Steeg, B., and Ammar, M. (2013). A client based technique for mitigating the buffer bloat effect of adaptive video flows, in *Proceedings of the 4th ACM Multimedia Systems Conference*, pp. 214–225, ACM.

[17] Spring, N. T., Chesire, M., Berryman, M., Sahasranaman, V., Anderson, T. and Bershad, B. (2000). Receiver based management of low bandwidth access links. In *INFOCOM 2000. Nineteenth Annual Joint Conference of the IEEE Computer and Communications Societies. Proceedings. IEEE*, 1, pp. 245–254.

[18] Hu, H., Yanikomeroglu, H., Falconer, D. D., and Periyalwar, S. (2004). Range extension without capacity penalty in cellular networks with digital fixed relays. http://www.sce.carleton.ca/faculty/yanikome roglu/Pub/gc04_hh.pdf, 2004. [Online; accessed 15-July-2014].

[19] Average Cell Throughput Calculations for LTE. http://www.ray maps.com/index.php/average-cell-throughput-calculations-for-lte/, 2011. [Online; accessed 15-July-2014].

[20] Kim, K.-H., Zhu, Y., Sivakumar, R., and Hsieh, H.-Y. (2005). A receiver-centric transport protocol for mobile hosts with heterogeneous wireless interfaces. *Wireless Networks* 11(4), 363–382.

[21] Lu, F., Du, H., Jain, A., Voelker, G. M., Snoeren, A. C., and Terzis, A. (2015). Cqic: Revisiting cross-layer congestion control for cellular networks. In *Proceedings of the 16th International Workshop on Mobile Computing Systems and Applications*, pp. 45–50, ACM.

[22] Huang, J., Qian, F., Guo, Y., Zhou, Y., Xu, Q., Mao, Z. M., Sen, S., and Spatscheck, O. (2013). An in-depth study of lte: effect of network protocol and application behavior on performance, in *ACM SIGCOMM Computer Communication Review*, Vol. 43, pp. 363–374, ACM.

[23] Winstein, K., Sivaraman, A., Balakrishnan, H., *et al.* (2013). Stochastic forecasts achieve high throughput and low delay over cellular networks. In *NSDI*, pp. 459–471.

[24] Welch, G., and Bishop, G. (2001). An introduction to the Kalman filter. *Proceedings of the Siggraph Course, Los Angeles*.

[25] An Introduction to the Kalman Filter. https://www.cs.unc.edu/welch/media/pdf/kalman_intro.pdf, 2014. [Online; accessed 20-Oct-2014].

[26] Feng, W.-C., Fisk, M., Gardner, M. and E. Weigle (2002). Dynamic right-sizing: an automated, lightweight, and scalable technique for enhancing grid performance, in *Protocols for High Speed Networks*, pp. 69–83, Springer.

[27] Issariyakul, T., and Hossain, E. (2011). *Introduction to network simulator NS2*. Springer, Berlin.

[28] Gurtov, A., and Floyd, S. (2004). Modeling wireless links for transport protocols. *ACM SIGCOMM Comput. Commun. Rev.* 34(2), 85–96.

[29] Ren, F., and Lin, C. (2011). Modeling and improving TCP performance over cellular link with variable bandwidth. *IEEE Trans. Mobile Comput.* 10(8), 1057–1070.

[30] Brakmo, L. S., and Peterson, L. L. (1995). TCP Vegas: End to end congestion avoidance on a global internet. *IEEE J. Selected Areas Commun.* 13(8), 1465–1480.

[31] Riiser, H., Vigmostad, P., Griwodz, C. and Halvorsen, P. (2013). Commute path bandwidth traces from 3G networks: analysis and applications. In *Proceedings of the 4th ACM Multimedia Systems Conference*, pp. 114–118, ACM.

4

Adaptive Monitoring for Mobile Networks in Challenging Environments

Nils Richerzhagen[1], Björn Richerzhagen[1], Rhaban Hark[1], Dominik Stingl[1], Andreas Mauthe[2], Alberto E. Schaeffer-Filho[3], Klara Nahrstedt[4] and Ralf Steinmetz[1]

[1]Multimedia Communications Lab, Technische Universität Darmstadt, DE
[2]School of Computing and Communications, Lancaster University, UK
[3]Institute of Informatics, Federal University of Rio Grande do Sul, BR
[4]Department of Computer Science, University of Illinois, US

Abstract

The increasing capabilities of mobile communication devices are changing the way people interconnect today. Similar trends in the communication technology domain are leading to the expectation that data and media are available everywhere and anytime. The consequence is an increasing load on communication networks. Such data requirements paired with environmental conditions such as mobility or node density increase the risk of network failure, especially in dynamic mobile networks that particularly rely on wireless communication. Consequently, monitoring is crucial in mobile networks to ensure reliable and efficient operations. Current monitoring mechanisms mostly rely on a static architecture and unveil problems to handle the changes of mobile networks, applications, and environmental conditions over time. In this chapter, an adaptive monitoring mechanism is presented to overcome these limitations. The mechanism exploits the connectivity and resource characteristics of mobile communication devices to (i) reconfigure the monitoring topology and (ii) adapt to changes of mobile networks and environmental conditions. Through evaluations we show that our proposed solution reduces the achieved relative monitoring error by a factor of six. It also represents a highly robust and reliable monitoring mechanism for these challenging environments.

Keywords: Adaptive monitoring, Mobile networks, Dynamic environment, Data collection.

4.1 Introduction

Well-equipped handheld communication devices, such as smartphones or tablet PCs, paired with a growing availability of wireless broadband access over cellular networks [1] enable the use of applications and services anytime and anywhere. Due to this, data-intensive traffic is massively increasing, which has to be primarily handled by cellular networks. Especially in crowded areas (e.g., tourist attractions or transportation and communication hubs), or during popular events, the resulting traffic can exceed the capacity of cellular networks. This results in a degraded network performance noticed by the users and negatively affecting their experience [2]. To overcome these problems several solutions have been proposed (i) to offload the resulting traffic of data-intensive applications over Wi-Fi ad hoc [3] or (ii) to exploit the locality of interaction as, for instance, in location-based services [4]. One of them is the system TopT [5]. TopT describes a video streaming mechanism that is able to adapt its streaming scheme used for delivery of the streams. The adaptation depends on the number of nodes joining or leaving the network. In the extremes, so-called flash crowds, the mechanism shows the benefit of adapting the used streaming scheme. In the context of locality for user interaction, Bypass [6] represents an example for an adaptive system, which benefits from offloading by exploiting the locality of content and interactions. Depending on the current state of the network and the environment, Bypass adapts between global and local communication strategies to react on these fluctuating conditions. For instance, the direct exchange of information between the devices may be preferred if connections over a cellular network exhibit degrading performance. Consequently, Bypass executes a transition from a centralized information exchange between the devices and a central coordinating service/content provider to a decentralized exchange between the devices themselves in addition to the exchange with the coordinating service/content provider.

To react on fluctuating conditions it is crucial for adaptive systems, such as Bypass, that the current state of the network, the applications, and the environment is known. This state is retrieved by monitoring. Based on the obtained monitoring information, adaptive systems are able to decide on the appropriate strategies for adaptation. As a result, the necessity to collect the monitored data from the devices at the central coordinating service/content provider becomes obvious. However, harnessing the density of mobile devices for the direct data exchange between the devices leads to a decentralization, where not every device may be connected to the central coordinating service/content provider. Whereas this decentralization is beneficial for offloading cellular networks and exploiting the locality of content and interactions, it strongly influences the collection of monitored data. As a coordinating service/content provider may only be connected to a small number of devices, the collection of monitored data from these devices leads to an incomplete and consequently imprecise view on the current state of the network and the environment.

To address this problem and to accurately monitor the whole network with its participating devices, we introduce CRATER. The presented approach constitutes an adaptive monitoring solution for dynamic, i.e., challenging environments. It particularly targets crowded places where the number of devices and the resulting traffic exceed the maximum capacity of cellular networks. Depending on the current network state, CRATER facilitates *centralized* as well as *decentralized monitoring* to gather relevant data from all participating devices. It relies on centralized monitoring if all devices obtain a direct connection to a service/content provider over the cellular network. Decentralized monitoring is deployed if direct communication between the mobile devices is used. In the latter case CRATER autonomously identifies devices that have (i) a cellular network connection and (ii) sufficient energy resources. These devices serve to collect monitoring data from other devices without an active cellular connection. To this end, ad hoc connections are established among nearby devices to forward the collected data to devices with a cellular network connection. Subsequently, the collected information is transmitted over the cellular network to a service/content provider to enable an accurate and complete view on all participating devices in the considered environment.

Apart from its adaptive design to facilitate centralized as well as decentralized monitoring, CRATER additionally tackles the challenges that arise from decentralized monitoring in mobile ad hoc networks (MANETs) [7, 8]. More precisely, it (i) operates on top of mobile devices (which leads to a constantly changing communication topology with intermittent connection), (ii) handles the error-prone and wireless communication medium with a limited communication range, and (iii) considers the limited resources of the mobile devices (e.g., energy) as well as the limited capacity of the shared communication medium. Using a model of a train station that represents an example for the envisaged challenging environments, the evaluation illustrates the advantages of CRATER's adaptive and flexible design. Furthermore, the evaluation outlines that CRATER represents a robust solution for decentralized monitoring tackling the challenges, which arise from its deployment in MANETs.

This chapter is based on the initial publication of CRATER [9] and provides more detailed insights into the system and monitoring of dynamic mobile networks. The remainder of the chapter is structured as follows. The background on monitoring in mobile networks is given in Section 4.2. Afterwards, Section 4.3 deals with the related work on decentralized monitoring and data collection in mobile networks. Section 4.4 describes the targeted scenario with the corresponding assumptions. The system design of CRATER is introduced in Section 4.5. Section 4.6 shows the simulation-based evaluation of the system and Section 4.7 concludes the chapter.

4.2 Background on Monitoring in Mobile Networks

This section deals with the details about monitoring in mobile networks. Initially, the arising challenges of mobile networks are described, which have to be considered and tackled by a monitoring mechanism that is deployed on top of these networks. Subsequently, the common functional and non-functional requirements of a monitoring mechanism for mobile networks are presented. Given the arising challenges as well as relevant requirements, the process of monitoring in mobile networks is discussed, detailing the relevant phases of a monitoring mechanism for mobile networks.

Previous works [10–12] have already identified some of the most prevalent challenges a monitoring mechanism in heterogeneous mobile networks has to handle. They include the following:

- **Limited resources**: Nodes[1] in mobile networks most often have limited resources such as energy. These limitations may be observed constantly or are transient if, e.g., a user enters an area with poor network coverage, limiting the available bandwidth significantly [10].
- **Highly dynamic networks**: Mobile networks are often characterized as highly dynamic, i.e., a highly dynamic network topology is observable. Due to their mobility nodes might frequently enter or leave the network, thereby changing its topology [11]. Thus, monitoring mechanism must cope with situations, where monitoring tasks fail due to node movement or complete disappearance of nodes. Monitoring mechanisms for mobile networks must handle such dynamics also including the rapidly

---

[1]Throughout this chapter the terms *node* and *device* will be used interchangeably for a mobile handheld communication device in a mobile network.

increasing or decreasing numbers of nodes in certain sections of the network.

- **Heterogeneous structure**: Mobile networks may consist of a great variety of devices. Those devices not only dispose of limited energy, bandwidth, and computational power, but might also vary greatly in their respective contribution. The potential usage of different network protocols adds even more complexity to the monitoring process [12].
- **Wireless communication**: Communication in mobile networks is realized using the wireless medium. That involves that (monitoring) mechanisms have to deal with short-lived, bi-directional, and error-prone connections [13].

Apart from the aforementioned challenges the monitoring mechanisms must satisfy a number of *functional* and *non-functional* requirements that have been identified in past research [14, 15]. One of the most important functional requirements is that the monitoring mechanism has to be able to accurately reproduce network, application, and environmental state, which is represented by a set of distinct attributes [14]. As highlighted in [14], "*the collected data may not tolerate errors and the success of the monitoring application depends on its accuracy.*" Further, such a mechanism also needs to adhere to a set of non-functional requirements that further specify its behavior as identified in [15]. Consequently, the delivered monitoring information shall be as *accurate* as possible, which implies that the reported state must not deviate from the actual state. A monitoring mechanism has to measure, process, and provide information ideally immediately—the *timeliness* is important. The monitoring mechanism has to be able to cope with a very large number of network participants; thus, it has to provide *scalability* in multiple dimensions. Those dimensions include not only the vast amount of network participants but also the various environmental and topological changes. Monitoring must also guarantee for scalability concerning the stakeholder of the monitoring mechanism; thus, it must be generic to provide monitoring for a broad range of possible use cases. Especially in mobile networks, a monitoring mechanism needs to be able to handle the arrival and departure of network participants. Furthermore, as described above, a monitoring mechanism has to exhibit a certain degree of *robustness* to handle node movement as well as the failure of individual nodes.

To overcome the above challenges and meet the functional as well as non-functional requirements, four different phases have been identified that a monitoring mechanism in (mobile) networks has to implement for its successful mode of operation. These four phases comprise (i) the *measurement*, (ii) the *data collection*, (iii) the *data analysis*, and finally (iv) the *information distribution* phase, which are detailed in the following.

4.2.1 Measurement

The data measurement phase is usually overlooked in the monitoring process as most mechanisms assume data to be available on the network nodes. With monitoring ultimately aiming at improving user experience, measurements should be focusing on a minimal set of attributes determining the state of the network, the application, and/or the environment. This minimal set of attributes must be chosen sensible as with more information that is measured, the produced overhead of the monitoring is also increasing. However, mostly individual attributes are not sufficient to retrieve the needed information value [16]. Today's networks consist of multiple layers. In each of those layers, specific protocols exist that fulfill certain functions. Monitoring must thus be able to measure the relevant attributes across multiple layers from a large amount of protocols. There exist multiple classes of measurements on mobile nodes: (i) sensor measurements (e.g., video, audio, accelerometer, gyrometer, GPS, WiFi), (ii) network performance measurements (e.g., bandwidth, delay), (iii) protocol metadata measurements (e.g., TCP/IP source, destination, checksums), and (iv) application-specific measurements (e.g., stalling rate/time, subscriptions per second). These parameters can, if not available locally, be measured via *active* or *passive* measurement approaches.

Active measurement approaches gather metrics in an active requesting fashion. It often implies additional generated application or network traffic. Due to the active procedure the information value is often significantly higher compared to passive measurement approaches. In contrast, passive approaches are based on the surveillance of standard user and application traffic and behavior—either on the node itself or on the network. A passive approach might even deploy network nodes exclusively dedicated to measurement, as proposed in [17] for MANETs. The advantage of dedicated monitoring nodes is to preserve potentially scarce resources of nodes for application use. Passive measurement solutions, however (e.g., [17, 18]), are not commonly used as the information value is crucial for the performance of a monitoring mechanism [14]. In [19], the active measurement framework Proton is proposed. The framework is able to measure the requested data locally in a cross-layered fashion. Using so-called monitoring access points, Proton is able to access different protocols from different layers.

4.2.2 Data Collection

Subsequent to the local measurement of monitoring data on the nodes, the data must be collected at distinct points in the network, where it may be analyzed or where it has been requested. Due to the dynamic nature of mobile networks, potential bandwidth constraints and the fact that monitoring traffic is often considered as overhead, this represents no trivial task. In data collection approaches the selection of the right set of sensors is a very important and not trivial [20]. The data collection mechanisms are often characterized by three distinct attributes [15, 21]: (i) the *monitoring topology* used to transmit the measured data, (ii) the data *exchange strategy* determining how the data is transmitted over the topology, and (iii) the *exchange trigger* mechanism specifying when the data exchange is initiated. On a coarse level, the exchange strategy can be divided into push- and pull-based approaches. In a pull-based approach a node actively requests and receives monitored data from a set of other nodes. Contrary to that, in a push-based approach, a single node sends its measured data to a set of other nodes. The exchange trigger can either be event-driven, e.g., an attribute exceeds a certain threshold or periodic if the data collection is always triggered after a fixed time interval. More detailed information about state-of-the-art collection approaches is given in Section 4.3.

4.2.3 Data Analysis

On a higher level, monitoring may be seen as an important aspect to manage and adapt networks, providing the required information and data basis to make decisions. The measurement and collection of raw data is only a first step towards this goal. A higher-level representation of network, application, and environment state (derived from the collected data through the process of data analysis) is missing. This process is done by developing analysis functions to infer the network state from the collected data. Due to the fact that even small networks generate large amounts of data, analyzing the measured data is a complex task [22]. Furthermore, due to the very high traffic overhead real-time analysis is rendered nearly impossible [23]. Lee et al. [24] name the following six common analysis functions in their recent survey. (i) *General-purpose traffic analysis* is mostly summarizing bare measurement data to provide useful summary information. Methods such as aggregation, filtering, or clustering are used to gain higher level information. (ii) *Estimation of traffic demands* is provided by additional predictive analysis of traffic demand [24]. This view can help, e.g., in traffic engineering and anomaly detection.

Anomalies will be detected by deviations of actual traffic from estimated demand. (iii) *Traffic classification per application* is an analysis function closely linked to applications. However, once approaches take payload signatures into account [25] the achieved result is highly dependent on the used encryption technique. (iv) *Mining of communication patterns* describes the process when operators apply classical data mining techniques on the raw measurement data to identify the traffic patterns and the applications they are linked to [26]. (v) *Fault management* is a traditional application of monitoring and data analysis. It can help in both fault identification [27] and fault localization [24]. (vi) *Automatic updating of network documentation* is already a common practice in large networks via a so-called discovery process. This applies to topology, routing policies and versions of software components. For mobile networks it needs to be extended to the spatial distribution of endpoints across the network.

4.2.4 Information Distribution

Once information is obtained from the analysis, the distribution of that information back to interested nodes is important in distributed systems. The nodes in a distributed system and in a mobile network must be aware of the network, application, and environmental state. As the location of the measured data, the collection and the analysis of the data do not necessarily match with the locations where the gained information is needed, the differentiation of collection and distribution of monitoring data are important. Furthermore, the distribution of information may be to all network nodes or only to a few. Those challenges must be respected and fulfilled by information distribution mechanisms.

4.3 Related Work: Data Collection in Mobile Networks

In the area of network monitoring the general approaches are either centralized or decentralized. Since centralized approaches (e.g., [28, 29]) are deployed in wired networks their suitability as comparison of benchmark for CRATER (addressing wireless networks such as MANETs) is limited. The dynamic conditions in MANETs (such as node density and movement speed) have a significant impact on the performance of the monitoring mechanism. Accordingly, the remainder of the related work focuses on (i) decentralized monitoring mechanisms and (ii) bio-inspired routing approaches. These routing approaches are examined since they constitute an essential ingredient for data collection in the hybrid topology of CRATER.

An approach to reduce the overhead of flooding is presented by DAMON [13]. DAMON relies on a distributed monitoring architecture for multi-hop mobile networks that uses an agent-sink topology for data collection. The agents control the flooding of the network, which leads to less overhead and reduces the possibility of collisions. Though, sinks are static and predefined by the network operator. This is infeasible for heterogeneous dynamic networks. Considering a detailed local and a sparse global network view as presented by Nanda and Kotz [30] is helpful to gain a more precise monitoring result. Nonetheless, the used hierarchy consisting of static mesh nodes and mobile nodes cannot be maintained in a dynamic MANET environment. Beside that Mesh-Mon is evaluated in a small-scale scenario with less than 25 nodes, which is not appropriate for heterogeneous network scenarios as presented in this work. Load balancing is an important factor in resource-constrained environments such as the envisaged scenario. HMAN by Battat and Kheddouci [31] establishes a three-tiered topology based on weights of nodes. Those weights incorporate factors, such as energy consummation, the distance to other nodes, and the storage capacity left. However, HMAN is dependent on the used routing protocol to manage and maintain the topology. Not separating the data and management communication may reduce the overhead but can render monitoring (a core network service) useless for example in overloaded situations. Tree-based topologies organize nodes into a hierarchical tree structure where data collection is done in a bottom-up fashion. BlockTree [15] describes a fully decentralized monitoring approach for MANETs, which establishes a hierarchy build by location-aware nodes. BlockTree is capable to provide location-aware monitoring information. Flat approaches, such as Mobi-G [21], show improved performance in sparsely populated areas, as a topology has not to be established and maintained. However, both approaches require detailed information about the nodes' location, which are provided by additional services (e.g., GPS) rendering such approaches useless in indoor scenarios or when the localization is not as accurate as needed. CrowdWatch [32] describes a scalable distributed energy-efficient crowd-sourcing framework. By introducing a three-tiered hierarchical structure and by not incorporating all nodes in the data collection approach, using grouping CrowdWatch is able to work energy efficiently. However, with less nodes included into the data collection process, the precision of the collected information is reduced. Barbera et al. [33] propose a grouping approach that considers social network properties. In their work so-called VIPs are elected in the network, which represent nodes with many social connections. The main challenge in this approach is given with its strength, as social connections of users are needed directly on the nodes to determine their importance in the social graph. Obtaining such social characteristics online may stress the resources of the nodes too much leading to higher energy consumption.

Dealing with bio-inspired routing approaches the issue of single-point of failure is discussed by Kiri et al. [34]. In such multi-cluster topologies, identification and separation of the individual clusters are essential. While a sink failure is handled in the approach, only a predefined set of sinks is available, which gives the approach a maximum lifetime. The benefit of using bio-inspired pheromone values for the routing process is demonstrated by Zhu et al. in [35]. Using pheromone values allows an adaption on resources and environmental changes is performed. But, using a single sink configuration can render the approach useless especially in networks with resource-constrained devices.

4.4 Scenario

The scenario used in this work models a populated place in an urban area that is subject to high dynamics, which we consider being a challenging environment. In our scenario, the dynamics become particularly apparent by the arrival and departure rate of users leading to crowds that alter the current user density, ranging from a sparsely to a densely populated place. Dependent on the user density in the considered area we assume that the present cellular network is able to handle the resulting traffic to a certain degree but operates unreliably, once a threshold is exceeded [3]. The varying user density accompanied by the problems of the communication network requires transitions from one operating mode to another. In the following we present the details of the scenario and outline our assumptions regarding the location, the users, and the utilized communication devices.

The considered place in our scenario is represented by an urban railway station. The railway station is an abstracted model of Waterloo Station, London, where people arrive and leave by train or foot. Figure 4.1(a) shows both a sketch of the levels of Waterloo Station in Figure 4.1(a) and the possible, directed flow of people in and around the station in Figure 4.1(b). Waterloo Station is the busiest railway station in Europe and is prone to fluctuations. In 2011/2012 more than 94,045,510 passengers passed through the station, resulting in an average of 257,659 passengers a day. Considering the opening times from 4:30 am to 01:05 am this yields to 11,984 passengers in an hour. During rush hour in the morning (highest peak between 8:00 and 8:59 am) 103,000 passengers are reported. The afternoon rush hour is

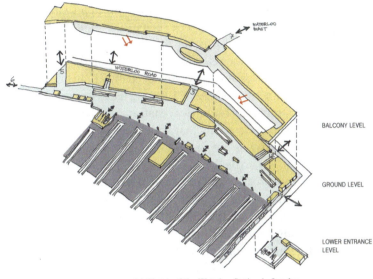

(a) Sketch of the Waterloo Station in London

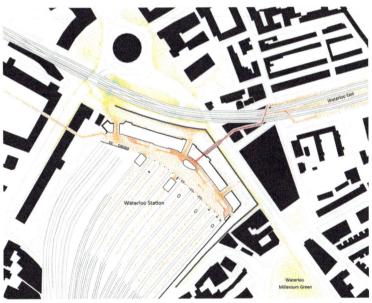

(b) Exemplary people flow in Waterloo Station

Figure 4.1 Sketch of Waterloo Station showing the different station levels in (a). An exemplary people flow is shown in (b) with different highlighted colors for the different levels in the station. For the scenario a sub-part of the station is modeled with focus on the ground floor including the platforms. The images were provided by Dr.-Ing. M.Sc. Nebojsa Camprag and his Ph.D. student Yang working at TU Darmstadt.

more spread out between 16:00 h and 18:59 h with approximately 87,000 passengers leaving via Waterloo [36]. The scenario used in this work models a sub-part of the station with focus on the ground floor of Figure 4.1(a) including the platforms. With the selection of a railway station for our scenario we follow the suggestions from Badonnel et al. [37]. In their work they investigate the establishment of a MANET among users to extend the existing communication infrastructure of a railway station. Furthermore, the railway station is populated with pedestrians that walk around with crowds forming during busy times. The crowds are formed through passengers that (i) arrive by foot to catch a train, (ii) arrive by train to leave the railway station, or (iii) arrive by train and try to catch another train. The peaks created by people periodical arriving and leaving are used to model the varying user density with the resulting overload of the cellular communication network. With respect to the communication infrastructure it is assumed that the railway station is covered with a cellular network that is accessible from anywhere within the station and can serve a maximum number of users, which we will elaborate later in this work. Furthermore, the station might be covered with Wi-Fi access points. However, they may not be accessible for everybody but only grant access for a limited number of users. Consequently, the presented version of CRATER does not consider Wi-Fi access points as additional means to communicate.

The considered users in the scenario have handheld communication devices, such as smartphones or tablet PCs, and move through the railway station. With their communication devices, users may consume different types of applications and services. In addition to the applications and services a monitoring mechanism is deployed that monitors different attributes to characterize the state of the communication network and devices. The corresponding monitoring entities are assumed to be located on the users' devices to locally monitor relevant attributes. The measured monitoring data is subsequently collected by a distributed infrastructure to a central entity (e.g., a cloud) which provides the required information to applications and services for adaptation. The cloud entity is assumed to have full scalability possibilities. The mechanism assumes that provider equipment, e.g., cell towers, cannot be utilized for deployment of the monitoring solution.

4.5 CRATER: Design of an Adaptive Monitoring Solution

CRATER is an adaptive monitoring mechanism targeted at mobile networks with handheld communication devices that have different communication interfaces, i.e., for cellular and wireless networks. It is designed as a self-contained service [13], since it does not rely on any other services or applications but operates independently to provide the required monitoring information even if other services or application fail. Consequently, we differentiate between a *data-* and *management-plane*, where the monitoring data is separated from the overall data and transmitted over the management-plane. As a result, the monitoring mechanism still uses the same communication interfaces of a device similar to applications or services. However, it operates on top of its own flexible topology and uses its own tailored routing mechanisms to transmit the monitoring data. CRATER consists of two different types of devices, comprising (i) the handheld devices of end-users, referred to as *nodes* and (ii) a server realized in cloud infrastructure, the *cloud component*. The mobile nodes periodically measure a set of predefined monitoring attributes, serving as basis to determine the current state of the network and nodes. The cloud component establishes and maintains a connection to every node to collect the locally measured monitoring data. Based on this data it determines the current state of the network and nodes, which serves other applications and services as knowledge base.

As illustrated by our scenario, the server may just maintain a direct connection to a fraction of nodes, because the cellular network is overloaded and does not grant access for every node in the network. Consequently, the server is not able to collect the required monitoring data from all nodes, which leads to an incomplete view on the current state of the network and nodes. To enable continuous and complete monitoring in spite of the arising problems in challenging environments, CRATER is designed to monitor the network even if only a fraction of nodes is connected to the server. Thus, the monitoring mechanism operates on top of a flexible topology, which is adapted depending on the environmental conditions. As depicted in Figure 4.2, it has two different topology structures, comprising a *uniform* and a *hybrid* topology structure. In the uniform topology structure all nodes maintain a direct link over the cellular network to the cloud component, which is able to collect the locally measured monitoring attributes from the nodes. In the hybrid topology structure, as depicted on the right-hand side in Figure 4.2, only a set of nodes has these direct links. We refer to these nodes as *sinks*. The remaining nodes, which are not able to connect to the server due to the overloaded cellular network, are denoted as *leaves*. As leaves do not have the possibility for a direct upload of their collected monitoring data, they must identify and affiliate to nearby sinks. Consequently, leaves and sinks directly exchange information with each other without the need for a prevailing communication infrastructure, using for instance Wi-Fi ad hoc or Bluetooth. As a result, sinks serve two purposes:

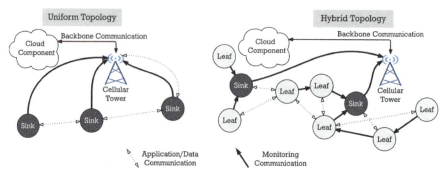

Figure 4.2   The uniform and hybrid topology structures of CRATER.

(i) they collect monitoring data from leaves and (ii) upload the collected as well as own monitoring data over the cellular network to the cloud component. To obtain collected monitoring data from leaves, which are currently not located in the direct communication range of a sink, CRATER applies concepts from mobile ad hoc networks (MANETs). The leaves simultaneously act as sources and forwarding nodes of a message so that nearby as well as distant leaves are able to transmit their collected monitoring data to sinks. As shown in Figure 4.2, CRATER facilitates that in both network states, uniform and hybrid, monitoring data can be gathered from all nodes, despite being a sink or a leaf.

Based on the presented design two main challenges arise that stem from the (i) adaptation between two different topology states and (ii) application of concepts from MANETs. With respect to the first challenge CRATER must detect when the cellular network is overloaded to execute a transition from a uniform to a hybrid topology structure. On the contrary, it must detect if resources are released to execute a transition from the hybrid to the uniform network structure. The second challenge arises from the direct information exchange between nodes to collect monitoring data from nodes, which are not directly connected to the cloud component. Consequently, CRATER must deal with the peculiarities related to MANETs, covering (i) node mobility, (ii) the limited communication range, (iii) the resulting short-lived connections between nodes, and (iv) the resource-constrained devices.

To implement the presented concepts of CRATER as well as to tackle the resulting two major challenges, CRATER consists of three basic components that are deployed on the mobile nodes (cf. Figure 4.3). These components

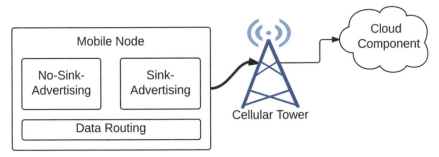

Figure 4.3   Components of the mobile CRATER nodes and the cloud component.

serve to (i) detect the changes in the environment, (ii) react on these changes by switching to the appropriate topology, and (iii) continuously monitor the current network state. The *No-Sink Advertising* component is introduced in Section 4.5.1 and used to react on missing connectivity and to identify the potential sinks for the collection of monitoring data. For the affiliation of leaves to sinks, CRATER introduces the *Sink Advertising* component, as detailed in Section 4.5.2. Finally, the *Data Routing* component is explained in Section 4.5.3 and serves for the collection of locally measured monitoring data from the leaves at sinks.

4.5.1 No-Sink Advertising

The corresponding procedure from the No-Sink Advertising component is executed by leaves, which lost their connection or could not yet connect to the cloud component over the cellular network. The procedure is performed to trigger a transition of CRATER from the uniform to the hybrid topology structure, if new nodes try to join the network but are not able to as the infrastructure entities may be overloaded. This transition from a uniform to a hybrid structure has to be executed to enable the affiliation of leaves to the sinks and to incorporate and collect their monitoring data. Furthermore, the No-Sink Advertising procedure is performed by leaves to keep CRATER in the hybrid topology structure due to the continuous overload of the cellular network. Consequently, the procedure is responsible to ensure that leaves advertise themselves if they have no direct upload to the server and are not yet connected to at least one sink. During the remainder of this section we detail the corresponding state chart with the related actions of the No-Sink Advertising procedure.

As depicted on the left-hand side in Figure 4.4, a leaf starts in the idle state. If it is not aware of any sink, either in its communication range or several hops away, the leaf executes transition $a1$, enters the *active* state, and broadcasts a *No-Sink-Messages*. The transmission of No-Sink-Messages for a prospective affiliation with a sink is always executed by leaves to avoid a proactive advertisement by a sink even if it is not required. Leaves may stay in the active state and periodically broadcast No-Sink-Messages as shown by the state transition a 2. CRATER uses a contention-based sending scheme [38] to

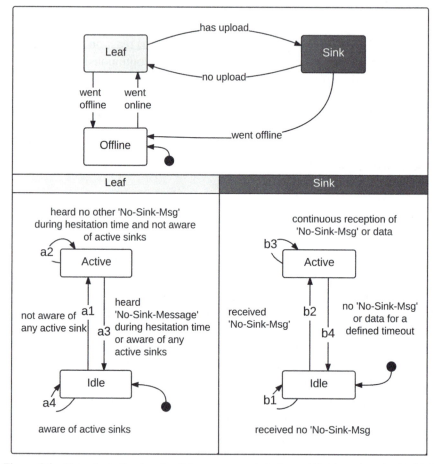

Figure 4.4 State chart of main advertising components in the mobile CRATER node: No-Sink Advertising and Sink Advertising combined with the role according to the upload management performed by the Sink Advertising.

prevent the whole network from unnecessary broadcasts, if multiple leaves are in an active state. Contention-based sending or forwarding describes a robust communication scheme where multiple nodes prepare the transmission of the same message, while only a subset of nodes executes the transmission. For the selection of these nodes a *hesitation time* T_h is introduced, which consists of a fixed maximum time T_{h_max} and a hesitation factor between 0 and 1 to delay the transmission of that message. The factor is called *hesitation factor* h_f and may depend on multiple weighted attributes or on a single attribute. Equation (4.1) shows the basic formula for the hesitation time. On reception of a message, nodes calculate T_h and delay the forwarding of that message correspondingly. If nodes overhear the forwarded message from another node (due to its smaller T_h), they do not forward but discard the message. Consequently, the resulting traffic is reduced, because only a few nodes forward messages in the network.

$$T_h = (1 - h_f) \cdot T_{h\_max} \quad \text{with} \quad 0 \leq h_f \leq 1 \qquad (4.1)$$

Thus, leaves only broadcast a No-Sink-Message when they did not overhear any other No-Sink-Message during their hesitation time and are still not aware of any sink. The hesitation factor h_f for a No-Sink-Message is computed by the normalized battery capacity as shown in (4.2). Consequently, leaves with higher energy resources will broadcast earlier. In cast that a leaf (i) overhears a No-Sink-Message or (ii) becomes aware of an active sink, transition $a3$ is executed and the node enters the idle state again. Otherwise, it remains in the active state and keeps broadcasting No-Sink-Messages. Assuming the leaf became aware of a sink or still overhears No-Sink-Messages, it stays in idle state, which is shown by transition $a4$. Otherwise, it executes transition $a1$ and enters the active state.

$$h_f = \frac{bat_{cur\_cap}}{bat_{max\_cap}} \qquad (4.2)$$

To prevent nodes from remaining leaves and staying either in the idle or active state, nodes periodically check if they may become sinks. This check is performed by the *Sink Advertising* mechanism, as described in the following section.

4.5.2 Sink Advertising

The *Sink Advertising* component provides two important procedures for the successful deployment of CRATER. The first procedure is responsible to determine if a node is either leaf or sink. This determination is periodically checked by trying to connect to the cloud component over the cellular network in case that no connection could be established so far. In Figure 4.4, the corresponding state chart is depicted on top of the figure. According to the resulting state, nodes become either leaves or sinks. Based on this procedure, CRATER ensures that a transition from the hybrid to the uniform topology structure is executed if the cellular network has sufficient capacity to serve all nodes. The second procedure of the Sink Advertising component targets the advertising of sinks so that leaves can affiliate to an advertised sink. As indicated before, sinks are important nodes, because they collect the monitoring data from affiliated leaves and have a direct connection to the cloud component to upload the collected data. Consequently, the Sink Advertising component must ensure that leaves (i) are aware of sinks, (ii) affiliate to one sink, and (iii) successfully transmit the collected data. In general, CRATER uses a multiple-sinks approach, because single-sink mechanisms show a degrading performance with an increased risk of failure. To distinguish between multiple sinks, every sink obtains a unique *sinkID*. The ID is used (i) to determine the affiliation from a leaf to a sink and (ii) to forward the monitoring data to the correct sink. CRATER adapts bio-inspired routing schemes from wireless sensor networks to establish and maintain the paths between leaves and sinks. To reduce the negative influence of node mobility on stored paths, it relies on loose paths where leaves decide themselves if they are suitable to perform forwarding. As a result, a gradient value, similar to pheromones or steepness indicators [34, 35], is introduced that decreases with an increasing distance to a sink. The collected monitoring data travels along an increasing gradient to reach the respective sink. In the following, we present the state chart for the advertisement of sinks, which is depicted on the right-hand side of Figure 4.4. Afterwards, we describe the structure of a *sink table* that is used to manage the information about sinks and detail the corresponding *Sink-Advertising-Messages*.

After becoming a sink, the node starts in the *idle* state of the corresponding state chart. As long as a sink does not receive a No-Sink-Message, it remains in the idle state, as depicted by transition $b1$. The reception of a No-Sink-Message indicates that at least one leaf is trying to affiliate to a sink. As indicated by transition $b2$ the sink enters the *active* state, where it periodically advertises itself by broadcasting *Sink-Advertising-Messages*. A sink remains active as long as it receives either successive No-Sink-Messages or data from leaves, which prove that its active presence is still required (state transition $b3$). Otherwise, transition $b4$ is executed and the sinks become idle, stopping the proactive Sink Advertising. To reduce the overhead and the probability of collisions, a contention-based sending scheme similar to the one for the No-Sink Advertising is used. Advertising messages are only broadcasted when no other nearby sink advertises itself during the delayed transmission of the message. The hesitation factor is computed on the weighted battery status similar to (4.2). On reception of a Sink-Advertising-Message during the delayed transmission of the own Sink-Advertising-Message sinks execute transition $b4$ and go back to the idle state. Received Sink-Advertising-Messages are stored in the local sink table of sinks and leaves. Sink tables are maintained by both sinks and leaves to manage information about advertised sinks. The sink table consists of (i) the sinkID, (ii) the received gradient, (iii) the number of updates received by that sink, and (iv) a timestamp for the last update. The timestamp is used to modify the current value of the gradient for the corresponding sink. With an older timestamp, the value of the gradient is reduced to account for the mobility of nodes. Older entries for a sink are potentially less reliable due to the constantly changing topology as well as the arrival of new sinks and the disappearance of old ones. Furthermore, the number of updates from a sink is an indicator for stable sink-leaf connections. This column of the sink table represents the main criterion for the selection of the own sink. In CRATER leaves select stable sinks, where multiple consecutive Sink-Advertising-Messages have been received in the past. Based on this decision CRATER prevents leaves from selecting sinks that just pass by leading to a frequent selection of new sinks instead of a few but constant sinks. If multiple sinks exhibit the same value for the number of updates, sinks with a higher gradient are preferred.

A Sink-Advertising-Message, which is initiated by an active sink, contains (i) a *messageID* for the unique identification, (ii) the sinkID to identify the corresponding sink, (iii) the hop count that represents the aforementioned gradient and is decreased with every hop to build a relative topology around sinks, (iv) the time-to-live (TTL) to adjust the spatial size of the region the sink may be responsible for, and (v) the sink quality, which corresponds to the battery status and is calculated as in (4.2). On reception of a Sink-Advertising-Message, leaves just forward the message if different conditions are met. A received message is only considered to be forwarded if (i) the TTL permits a further hop and (ii) the message with the related messageID has not been processed before. After both criteria have been met the message is forwarded if either (i) the sink table is empty, (ii) the received sinkID corresponds to the currently best sinkID according to the sink table, or (iii) the received gradient is greater or equal to the own best sink in the sink table. If at least one of these criteria is fulfilled, the node forwards the message using the same contention-based forwarding scheme as the sink.

4.5.3 Data Routing

To conclude the section about the system design, we describe the methods to route the collected monitoring data from leaves to sinks. These methods are used if CRATER is in the hybrid topology, where leaves are aware of at least one sink. The forwarding of data from leaves to sinks is mainly based on the concept that only local information is used to determine the path towards a sink. Consequently, the next hop is determined without a route discovery mechanism. Instead, as introduced in the beginning of this section, CRATER adapts bio-inspired routing schemes. Nodes decide based on their current value of the gradient if messages should be forwarded.

Whenever a leaf starts to send monitoring data to its respective best sink, it broadcasts a *Data-Message* containing the data as payload. If another Data-Message is received during a given time window, which must be forwarded, the leaf uses this message to piggyback its data. A Data-Message includes field for the unique sinkID and the gradient. The sinkID is used for the identification of the correct sink. If a leaf receives a Data-Message with a sinkID that differs from the ID of the affiliated sink, the message is dropped. The gradient is used to head the message into the right direction, since messages are routed towards the highest gradient value. If the gradient of the message is equal or smaller than the stored gradient in the sink table, the message is dropped as well. Both sinkID and gradient fields limit the number of potential forwarders. To forward the messages after both criteria have been met, the contention-based forwarding scheme from the Sink Advertising mechanism is used. The hesitation factor is based on the current gradient and the battery as shown in Equation (4.3). This computation favors leaves that have the highest gradient and a high battery capacity left.

$$h_f = w_{bat} \cdot \frac{bat_{cur\_cap}}{bat_{max\_cap}} + w_{grad} \cdot \frac{grad_{own\_grad}}{grad_{max\_grad}} \qquad (4.3)$$

On reception of Data-Messages, sinks and leaves buffer and process the collected monitoring data according to the type of the data, as discussed below. A leaf subsequently forwards the data towards its sink, whereas a sink uploads the monitoring data to the cloud component. CRATER distinguishes three different types of data: *normal uncompressed data, duplicate sensitive aggregates,* and *duplicate insensitive aggregates*. Especially for uncompressed data and duplicate sensitive aggregates, the rate of received and undetected duplicates at intermediate leaves as well as sinks should be small. The normal uncompressed data describes simple raw data points. The aggregated data in CRATER is according to its sensitivity to duplicates, following the classification of Madden et al. [39]. Duplicate insensitive aggregates may comprise minimum or maximum, whereas duplicate sensitive aggregates cover sums, averages, or counts, which are affected by duplicate processing at nodes.

4.5.4 CRATER Cloud Component

The cloud component in the CRATER design is fairly simple. It is used as central gathering point of all data, which is uploaded from the sinks. With the received information from the sinks in the network the cloud component computes the global view on the network and nodes. In the current design of CRATER the cloud component is not responsible for any other task except acting as the central data sink. It does not maintain any topology structures in the network or affects the selection of sinks.

4.6 Evaluation

The simulation-based evaluation of the proposed adaptive monitoring mechanism CRATER is split into three parts. Currently no prototypical deployment of the system exists. First, a system evaluation is conducted to examine how the different changeable parameters influence the performance of the monitoring mechanism. Second, the robustness of the system is evaluated under different environmental conditions. Third, a comparison of the proposed adaptive monitoring mechanism with a static centralized solution was made. The second evaluation, examining the robustness of CRATER under different environmental settings, targets four main scenario configurations: (i) changing the maximum number of base station connections $|S|_{\mathrm{max}}$, to prove the ability of CRATER to provide monitoring from best to worst connected environments, (ii) changing the ratio of background to main traffic nodes $R_{\mathrm{N_b}/\mathrm{N_m}}$, thus the impact of the node churn, (iii) changing the movement speed of the nodes, causing for example shorter inter-connection times, (iv) changing the density of nodes in the network from very sparse populations to an high populated area. The scenarios have been chosen, as they represent typical challenges from MANETs that must be tackled. In terms of the static centralized solution, used in part three of the evaluation, mobile devices can only connect to the cloud component over the cellular network. In the following, we detail (i) the modeling of the scenario and the used evaluation parameters, (ii) the evaluation of the robustness of CRATER, and the (iii) comparison between CRATER and the static centralized monitoring mechanism.

4.6.1 Modeling of the Scenario and Evaluation Setup

The modeled environment in the scenario corresponds to a railway station, as proposed by [37]. As depicted in Figure 4.5, we model a part from a railway station that comprises tracks and shopping facilities and is populated with user that move according to different movement models. The different movement models are used to model (i) continuously present users in the railway station and (ii) the arrival or departure of users by incoming or outgoing trains. In Figure 4.5, the continuously present users are represented by the black dots. They move through the station according to the Steady-state Random Waypoint Mobility model [40]. These users are seen as *noise* in the environment and may represent people waiting or going through the station

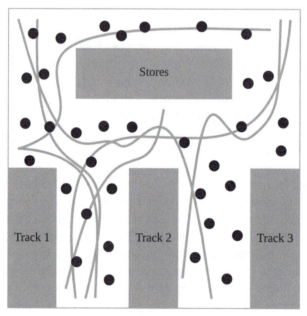

Figure 4.5   Modeling of the railway station with obstacles.

without a direct aim. The second movement model depicts the arrival and departure of groups of users. It is indicated by the solid lines, which sketch the potential paths through the station. Along these paths groups of nodes arrive, move along the paths, and subsequently leave the railway station either by train or by one of the exits. The groups of nodes form around so-called *attraction points*. The attraction points follow multiple paths along the map leading to the groups of nodes moving along the map accordingly. For the simulation-based evaluation comprising the model of the environment and the user mobility, we rely on the Simonstrator Platform [41] that comprises a model from ns-3 [42] of the IEEE 802.11g standard to simulate the Wi-Fi ad hoc communication between the devices. Three hours of operation are simulated where the first hour is used to reach a steady state of the simulated scenario, while measurements are taken during the remaining two hours. Five different seeds are used for repeating the experiments. Bar charts show the average with the 95% confidence interval. For a better understanding of the distribution of results box plots are utilized. Boxes represent the lower and upper quartile and the median is depicted by the solid line inside the box. Whiskers show the upper (lower) data point within 1.5 of the interquartile range. Outliers are represented by crosses.

Table 4.1 summarizes the simulation setup. Furthermore, it outlines the different settings of the movement speed as well as node densities. The underlined configurations represent the default configurations. The movement speed of nodes is uniformly distributed in the given intervals. The density in the network is varied by changing the number of nodes while keeping the simulated area constant.

Table 4.1 Scenario and simulation setup. Default values are underlined

Simulated Area	$2000\,m \times 2000\,m$
Max. Wi-Fi Comm. Range	$129\,m$
Max. Base Station Connections	585
$|S|_{max}$	315 (always hybrid 70%), 450 (100%), 495 (110%), <u>585 (130%)</u>, 750 (always uniform)
R_{N_b/N_m}	10%/90%, 20%/80%, 40%/60%, 50%/50%, <u>60%/40%</u>, 70%/30%
Network Density $\left[\dfrac{nodes}{km^2}\right]$	$36.3, 92.2, 181.5, \underline{273.7}, 363, 544.5$
Movement Speed $\left[\dfrac{m}{s}\right]$	<u>1–2</u>, 2–4, 4–8

4.6.2 System Parameter Configurations

To understand how the designed system works and how different changeable parameters influence, the system a parameter evaluation is performed. For that purpose the changeable system parameters and the respective variations are summarized in Table 4.2. The default configurations are underlined in each column. The system parameter evaluation is conducted by changing one parameter at a time while all others are fixed with the respective default values. For the system parameter evaluations the default scenario configuration, as presented in Section 4.1, is used. In this scenario CRATER does not face the extremes of the scenario as these are considered later in this section.

A variation of the Sink Advertising interval I_{sa} between 15 s and 60 s is assumed to have an impact on multiple metrics. The completeness shows a strong dependency on I_{sa}. In Figure 4.6(a) the box plot of the completeness for the three variations is visible. First of all, it is clearly visible that the median completeness, indicated by the red middle stroke in the box plots, remains above 98% even for the worst case. However, with larger I_{sa} the distribution of the completeness broadens, but even with I_{sa} equal to 60 s 99.3% of the completeness values range between 92.5% and 100%. The CDFs of the relative error, as depicted in Figure 4.6(b), show that the error is larger for higher I_{sa}. Furthermore, it is visible that CRATER is likely to underestimate the current system state as over 90% of the error values are less or equal to zero. The prominent sharp bend of the CDFs going positive is based on the fact that the movement of the main traffic nodes is designed in the way that when they are leaving, they are leaving nearly at the same point in time. The other case, when they go online is designed in a smoother way resulting in a more rounded graph. From the graph it becomes apparent that the relative errors range mostly between −10% and +5%. For the No-Sink Advertising interval no significant changes were observed.

When examining the data upload interval of sinks, $I_{d\uparrow\uparrow}$, the main thing to remember is that the data sampling interval on the cloud component is

Table 4.2 Overview of the changeable system parameters used for the evaluation of CRATER

Parameter	Symbol	Variations
Sink Advertising Int.	I_{sa}	<u>15s</u>, 30s, 60s
No Sink Advertising Int.	I_{nsa}	<u>10s</u>, 20s, 40s, 80s
Data Upload Int.	$I_{d\uparrow\uparrow}$	30s, <u>60s</u>, 120s
Data Sending Int.	$I_{d\rightleftarrows}$	15s, <u>30s</u>, 60s
Cloud Entity Sampling Int.	$\triangle$	30s, <u>1m</u>, 2m, 3m
Initial TTL of Sink-Adv-Mesgs	$TTL_{initial}$	1, <u>3</u>, 6

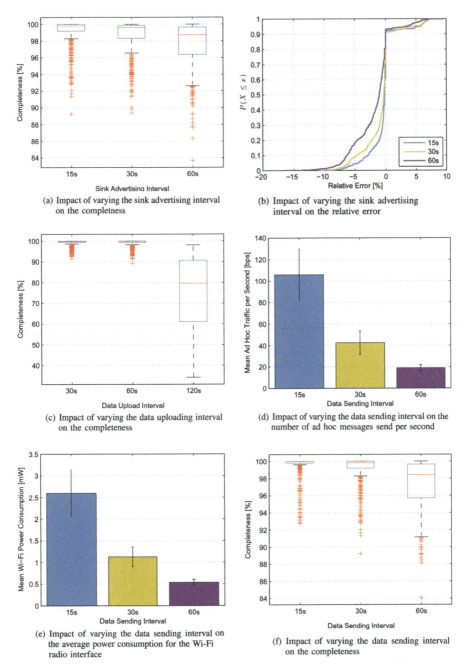

(a) Impact of varying the sink advertising interval on the completness

(b) Impact of varying the sink advertising interval on the relative error

(c) Impact of varying the data uploading interval on the completness

(d) Impact of varying the data sending interval on the number of ad hoc messages send per second

(e) Impact of varying the data sending interval on the average power consumption for the Wi-Fi radio interface

(f) Impact of varying the data sending interval on the completness

Figure 4.6 System parameter evaluation of CRATER showing the impact of (i) different sink advertising, (ii) data uploading, and (iii) data sending interval configurations.

set to 60 s. Figure 4.6(c) shows the completeness of the different uploading intervals. While there is no significant difference between 30 and 60 s, a large drop for 120 s is observable. In fact that drop is induced in combination with the data sampling interval Δ. As the data is sampled twice as much as it is uploaded the completeness drops, because some data is missing at the cloud component. As a result the data upload interval should always be chosen equal or lower to the sampling interval at the cloud component.

The data sending interval $I_{d\rightleftarrows}$ showed some previously examined effects. For example the delivery ratios at the cloud component only get better because less in-network traffic is present with higher $I_{d\rightleftarrows}$. With less in-network traffic the data added and sent by sinks gains on influence and changes the delivery ratios. The duplicate ratios at the sinks shows that similar change as less often messages are send in the network which reduces the loss which directly affects the contention-based forwarding scheme used by CRATER. In general the ad hoc as well as cellular traffic rises with lower $I_{d\rightleftarrows}$. Obviously, with more data sent, up to four times in the minimal $I_{d\rightleftarrows}$ case compared to the maximum, the traffic rises accordingly. Figure 4.6(d) shows that rising trend towards lower $I_{d\rightleftarrows}$. It is visible that the $I_{d\rightleftarrows}$ and the send ad hoc traffic behave proportional for sending intervals of 20 and 40 s, but with a smaller $I_{d\rightleftarrows}$ of 15 s the send ad hoc traffic is by a factor of 2.5 times larger compared to the 30 s upload interval measurement. This effect is due to leaves overhearing more messages as a result of congestion and increased loss in the network. Accordingly, the contention-based routing scheme suffers from this network utilization. A major impact of varying $I_{d\rightleftarrows}$ becomes apparent by observing the energy consumption. Figure 4.6(e) shows the Wi-Fi ad hoc energy consumption/cost relative to the data sending interval. Therefore, regarding the small difference between $I_{d\rightleftarrows}$ set to 15 or 30 s with respect to the completeness (cf. Figure 4.6(f)) but the significant impact on the occurring traffic and energy consumption of the nodes, $I_{d\rightleftarrows}$ should only be set to very low values when it is somehow needed, for example, in a system where very frequent updates are desired. In other cases where data freshness is not the main objective, $I_{d\rightleftarrows}$ can be chosen higher but should not be set to close to the sampling interval Δ of the cloud component.

Varying the sampling interval Δ of the cloud component while keeping $I_{d\rightleftarrows}$ and especially $I_{d\uparrow\uparrow}$ fixed shows the expected impact on the completeness and error estimates. With $\Delta < I_{d\uparrow\uparrow}$ the completeness drops significantly and errors become very large. The completeness becomes better with larger Δ, which is obvious as leaves and sinks have even more possibilities to success-fully send data to the cloud component with longer Δ. While the completeness only improves with larger sampling intervals the errors suffer from a too large Δ and become bigger again for sampling intervals of two and three minutes.

Changing the maximal number of hops Sink-Advertising-Messages are able to take is done with TTL_{initial}. With TTL_{initial} set to 1 there are no duplicates occurring in the network as each leaf that is aware of a sink uses the largest gradient leaves can have. Thus, other leaves hearing that messages do, by design of CRATER, not accept the messages as they do not have a larger gradient. With this configuration there is no chance that duplicates can occur; however, the spanned craters are very small in size leading to a lower coverage of CRATER. Between TTL_{initial} set to three and six, there is no significant difference visible in none of the analyzed metrics. From this observation it can be noted that it is very likely that the sink coverage in the used scenario is good enough that nodes do not forward the Sink-Advertising-Messages any further than three hops as leaves that far away are already aware of other more adjacent sinks.

Summarizing the main findings from the system parameter evaluation, CRATER tends to underestimate the network state. This was especially observable with variation of the Sink Advertising interval I_{sa} and the data sending interval $I_{d \rightleftarrows}$. Furthermore, CRATER suffers from outdated data when metrics are computed based on this data. The $I_{d \rightleftarrows}$ showed that impact very well. The freshness of the data received by the cloud component is mainly affected by the upload interval of the sinks $I_{d \uparrow\uparrow}$. As a result the following condition should be met by CRATER $I_{d \uparrow\uparrow} \leq \Delta$. To get (i) the best compromise of energy usage, (ii) the low power requirement, (iii) and the best coverage requirement the upload interval $I_{d \uparrow\uparrow}$ should be equal to the sampling interval of the server Δ. Whenever $\Delta < I_{d \uparrow\uparrow}$ the achieved coverage drops significantly and errors become very large as not all nodes are considered in the collected attributes.

4.6.3 Robustness

A cellular coverage with consistent connectivity characteristics is still not achieved in current networks. Signal strength, latency, bandwidth and the available cellular communication technologies, i.e., EDGE, UMTS, HSDPA, 4G, undergo fluctuations. Thus, evaluating the system performance of CRATER under different connectivity features is essential to obtain the robustness of the system. Therefore, the impact of changing the maximal possible number of connections from nodes to base stations $|S|_{\max}$ is evaluated. Variations of $|S|_{\max}$ range from scenarios in which the system stays in hybrid topology mode during the whole simulation time to scenarios in which the system stays in the uniform topology. Furthermore there are three cases in which the adaptivity of the system is evaluated as the scenario forces CRATER to adapt.

The achieved completeness with varying $|S|_{\max}$ improves with the number of available sinks in the network. This is a logical consequence from the wider coverage gained with more potentially available sinks in the network. However, it must be stated that CRATER is able to capture more than 96% of all nodes even in scenarios with poor connectivity characteristics. This effect is visible in Figure 4.7(a), which shows the box plots for the variations of $|S|_{\max}$.

The active ratio of nodes and sinks is also affected by variations of $|S|_{\max}$. Figure 4.7(b) shows a relatively constant leaf active ratio up to the point where enough sinks are present in the network that the few leaves left do not have to wait long to get connected. But with leaves being on average 9% of their online time active, thus requiring a connection to sinks, is a quite good evidence that leaves in the network searching for connectivity do not burden the wireless medium for too long and regain network connectivity fast. The sink active ratio R_{sa} shows significant changes nearly halving from the always hybrid to the transitional topology. With an improving coverage of sinks the percentage of sinks that need to get active is reduced. Finally, nearly zero percent for the uniform case in which joining nodes did request for connectivity to sinks before they were able to establish a connection to the base station.

Networks are very dynamic. In the considered scenario the dynamic environment is enabled by entities joining and leaving the area throughout the time with different arrival and joining rates. Those dynamics are modeled in the evaluation by varying the ratio from background to main traffic nodes $R_{\mathrm{N_b}/\mathrm{N_m}}$ which drastically changes the impact of incoming and outgoing trains in the scenario. Furthermore, by keeping the number of base station connections fixed at 110% of the present background nodes less sinks are available when $R_{\mathrm{N_b}/\mathrm{N_m}}$ is changed so that less background nodes are available. This shows to what extent of sink-to-node ratio CRATER is able to deliver the monitoring effort. The variations of $R_{\mathrm{N_b}/\mathrm{N_m}}$ show the expected impact on the achieved completeness (cf. Figure 4.7(c)), which increases with more background nodes and accordingly more present sinks in the network. The relative error, which is shown in Figure 4.7(d), shows similar results, and it increases with more background nodes and narrows in its range. A thing to highlight here is that CRATER underestimates the system state. In mobile scenarios the movement speed is an important factor as it may reduce the system performance significantly. Reasons for that are short-lived links, unstable topology structures, and fluctuating link qualities. Hence, the robustness of

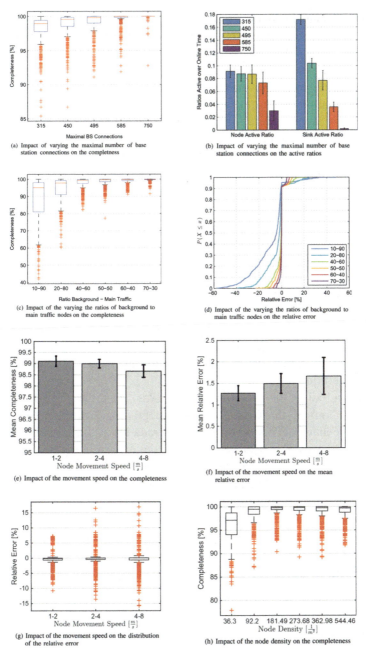

Figure 4.7 Evaluation of the robustness of CRATER for a varying (i) maximal number of base station connections, (ii) ratio of background to main traffic nodes, (iii) varying movement speed, and (iv) node density.

CRATER is evaluated under three different movement speed intervals 1–2, 2–4, and 4–8 *m/s*. Those intervals have been chosen as comfortable walking speed averages at around 1.4 *m/s* for people above their thirties, whereas 2.5 *m/s* is possible for younger people [43, 44]. However, as running people must also be considered, a movement speed up to to 8 *m/s* is also considered. For the evaluation of CRATER's performance the relative error as well as the completeness of the monitoring data are examined. For the relative error, the duplicate sensitive monitoring attribute *node count* has been chosen.

On average CRATER is able to deliver the monitoring data from at least 98.5% of the nodes irrespective of the movement speed interval. Changing the movement speed does not affect the performance of CRATER significantly as visible in Figure 4.7(e). The narrow confidence intervals underpin the performance and the robustness CRATER achieves. This is achieved while keeping the relative error very low at the same time. On average the relative error is lower than 2.5% and is not significantly affected by variations of the movement speed as shown in Figure 4.7(f). Taking a look at the distribution of the relative error, it becomes apparent that CRATER basically underestimates the system state (cf. Figure 4.7(g)), because the boxes stay below zero. An overestimation due to duplicate processing of data applies only for a fraction of the results. The quartile and whisker range remain lower than two percent, thus underpinning the very accurate monitoring results even under highly dynamic conditions.

Taking a look at the performance of CRATER for different node densities, Figure 4.7(h) unveils that CRATER exhibits a degrading performance in sparsely populated scenarios. This performance degradation is attributable to natural limitations, because the distance between nodes and the communication range impede the establishment of the hybrid topology. However, the probability that networks are overloaded in sparsely populated areas is less, reducing the need to operate on the hybrid topology. As depicted in Figure 4.7(h) CRATER only suffers from the intermittent connection between nodes in very sparsely populated scenarios, however, still exhibiting a completeness of over 85%.

4.6.4 Static Monitoring vs. CRATER

To estimate the benefit of using adaptive monitoring mechanisms such as CRATER, a comparison against a static approach is conducted. The static monitoring approach is only able to serve mobile nodes with direct cellular upload. Figure 4.8(a) shows the comparison of both mechanisms during one run to highlight the differences between both approaches over time. The plot shows the current number of present nodes and the maximum amount of connections that can be established over the cellular network. The results outline that both topology states of CRATER, uniform and hybrid, are used over time. The static approach reaches 100% of completeness whenever the number of nodes is lower than the maximal number of cellular connections. Once the number of nodes exceeds that threshold, the static monitoring approach cannot deliver accurate results any more. In contrast, CRATER is able to achieve a constantly high completeness, meaning more than 95% of the nodes are included in the monitoring results despite of the environmental conditions. Furthermore, in normal situations, where no overload of the base stations is present, the performance of CRATER is as good as the of the static approach. As shown in Figure 4.8(b) CRATER achieves significantly more accurate results with a mean relative error below 1%.

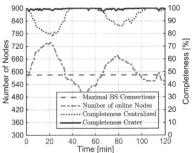 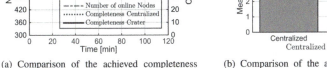 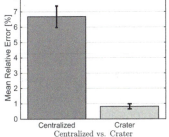

(a) Comparison of the achieved completeness over time for one run

(b) Comparison of the achieved mean relative error

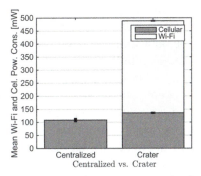

(c) Comparison of the power consumption for the monitoring data exchange

Figure 4.8 Comparison between CRATER and the static centralized monitoring mechanism.

However, the higher performance and the more accurate results are not for free. The additional cost added by CRATER become apparent in Figure 4.8(c), which represents the mean power consumption, using a component-based energy consumption model for smartphones [45]. Since the static approach does not rely on Wi-Fi, only cellular communication burdens the battery, leading to a mean power consumption of approximately 108 mW. In comparison, CRATER needs approximately 485 mW on average, due to (i) an increased cellular traffic and (ii) the Wi-Fi ad hoc communication to incorporate all nodes in the network. While the additional cellular traffic accounts for less than 30 mW of the mean power consumption compared to the static approach, the main impact arises from the utilization of Wi-Fi ad hoc. The reason for the high mean power consumption does not arise from an extensive data exchange but from the idle state of the ad hoc mode, which accounts for 353 mW. The additional burden, added by the Wi-Fi ad hoc communication, accounts for 0.35 mW on average.

4.7 Conclusion

In this chapter, an adaptive monitoring mechanism called CRATER especially designed for challenging environments is introduced. CRATER exploits the connectivity and resource characteristics of mobile handheld devices to facilitate continuous monitoring of the network and nodes. CRATER is part of a much larger research initiative exploring the potential of transitions between communication mechanisms to improve networking and applications in the future Internet, called MAKI [46]. By reconfiguring the monitoring topology structure, the system is able to adapt to a wide range of environmental and network-specific changes, providing a significantly increased performance despite such changes. At the same time, the system is robust against MANET-specific challenges such as (i) cellular connectivity changes, (ii) different churn rates, (iii) varying node movement speeds, and (iv) different node densities. Comparing the system with a non-adaptive monitoring mechanism unveils the potential of CRATER since it provides complete and accurate results. Nevertheless, the increased performance comes at increased cost. The power consumption of the nodes is significantly higher in contrast to the static approach, because more data is transmitted in the network and two communication interfaces of the device are used.

The presented system is currently concentrated on gathering data from mobile nodes in the network to gain a global network state. Analyzing subsets of the monitored data on, e.g., sinks to gain a local and regional knowledge may facilitate local or regional adaptations to improve the performance or reduce the cost. For instance, the periodic transmission of CRATER's different message types may be adapted according to the number of affiliated leaves to a sink. Furthermore, using No-Sink-Messages and Sink-Advertising-Messages deliver monitoring information during the topology structuring phase may improve the response time of the monitoring mechanism and provide a more recent view on the network state. Incorporating infrastructure devices like Wi-Fi access points can improve the systems performance as such devices constitute fixed sinks with less energy constraints compared to mobile user devices and have a direct server connection. Examining the use of different communication technologies for the direct communication, e.g., Wi-Fi ad hoc and Bluetooth, is part of our future work, because it may offer further possibilities to reduce the cost or increase the performance.

Acknowledgment

This work has been funded by the German Research Foundation (DFG) as part of projects B1, C2, and C3 within the Collaborative Research Centre (CRC) 1053–MAKI.

References

[1] Ericsson. (2013). Ericsson Mobility Report—On the Pulse of the Networked Society. pp. 1–32, 2013. [Online]. Available: http://www.ericsson.com/res/docs/2013/ericsson-mobility-report-november-2013.pdf

[2] Shafiq, M. Z., Ji, L., Liu, A. X., Pang, J., Venkataraman, S., and Wang, J. (2013). A first look at cellular network performance during crowded events, in *ACM SIGMETRICS*.

[3] Bao, X., Lin, Y., Lee, U., Rimac, I., and Choudhury, R. R. (2013). DataSpotting: Exploiting Naturally Clustered Mobile Devices to Offload Cellular Traffic, in *IEEE INFOCOM*.

[4] Ott, J., Hyytiä, E., Lassila, P., Vaegs, T., and Kangasharju, J. (2011). Floating Content: Information Sharing in Urban Areas, in *IEEE PerCom*.

[5] Rückert, J., Richerzhagen, B., Lidanski, E., Steinmetz, R., and Hausheer, D. (2015). TOPT: supporting flash crowd events in hybrid overlay-based live streaming, in *IFIP Networking*, pp. 1–9.

[6] Richerzhagen, B., Stingl, D., Hans, R., Gross, C., and Steinmetz, R. (2014). Bypassing the cloud: Peer-assisted event dissemination for augmented reality games, in *IEEE P2P*.

[7] Chlamtac, I., Conti, M., and Liu, J. J.-N. (2003). Mobile Ad Hoc Networking: Imperatives and Challenges. *Ad Hoc Networks* 1(1), 13–64.

[8] Sengul, C., Viana, A. C., and Ziviani, A. (2012). A survey of adaptive services to cope with dynamics in wireless self-organizing networks, *ACM Comput. Surveys* 44(4), 1–35.

[9] Akkaya, K., and Younis, M. (2005). A survey on routing protocols for wireless sensor networks. *Ad Hoc Networks* 3(3), 325–349.

[10] Kush, S. T., and Ashwani (2010). A survey of routing protocols in mobile ad hoc networks. *Int. J. Innov. Manage. Technol.* 1(3), 279–285.

[11] Aggarwal, V., Halepovic, E., Pang, J., Venkataraman, S., and Yan, H. (2014). Prometheus: toward quality-of-experience estimation for mobile apps from passive network measurements. *ACM HotMobile.*

[12] Ramachandran, K. N., Belding-Royer, E. M., and Almeroth, K. C. (2004). DAMON: A distributed architecture for monitoring multi-hop mobile networks, in *IEEE SECON.*

[13] Battat, N., Seba, H., and Kheddouci, H. (2014). Monitoring in mobile Ad Hoc networks: A survey. *Comput. Networks* 69, 82–100.

[14] Stingl, D., Gross, C., Nobach, L., Steinmetz, R., and Hausheer, D. (2013). BlockTree: Location-aware Decentralized Monitoring in Mobile Ad Hoc Networks, *IEEE LCN.*

[15] Sonntag, S., Manner, J., and Schulte, L. (2013) Netradar—measuring the wireless world, in *IEEE WiOpt.*

[16] Kazemi, H., Hadjichristofi, G., and DaSilva, L. A. (2008). MMAN— A monitor for mobile ad hoc networks: design, implementation, and experimental evaluation, *ACM WiNTECH.*

[17] Michelinakis, F., Bui, N., Fioravantti, G., Widmer, J., Kaup, F., and Hausheer, D. (2015). Lightweight mobile bandwidth availability measurement, in *IFIP Networking.*

[18] Richerzhagen, N., Li, T., Stingl, D., Richerzhagen, B., Steinmetz, R. and Santini, S. (2015). A step towards a protocol-independent measurement framework for dynamic networks, in *IEEE LCN.*

[19] Nahrstedt, K., Li, H., Ngyuen, P., Chang, S., and Vu, L. (2016). Internet of mobile things: mobility-driven challenges, designs and implementations, in *IEEE IoDT.*

[20] Stingl, D., Retz, R., Richerzhagen, B., Gross, C., and Steinmetz, R. (2014). Mobi-G: Gossip-based Monitoring in MANETs, in *IEEE/IFIP NOMS.*

[21] Keys, K., Moore, D., and Estan, C. (2005). A robust system for accurate real-time summaries of internet traffic, in *ACM SIGMETRICS Performance Evaluat. Rev.* 33(1), 85–96.

[22] Wang, X., Abraham, A., and Smith, K. A. (2005). Intelligent web traffic mining and analysis. *J. Network Comput. Appl.* 28(2), 147–165.

[23] Lee, S., Levanti, K., and Kim, H. S. (2014). Network monitoring: present and future. *Computer Networ.* 65, 84–98.

[24] Moore, A. W., and Papagiannaki, K. (2005). Toward the accurate identification of network applications, in *Springer PAM*.

[25] Nguyen, T. T., and Armitage, G. (2008). A survey of techniques for internet traffic classification using machine learning, *IEEE Commun. Surveys Tutor.* 10(4), 56–76.

[26] Katzela, I., and Schwartz, M. (1995). Schemes for fault identification in communication networks. *IEEE/ACM Trans. Network.* 3(6), 753–764.

[27] Boulon, J., Konwinski, A., Qi, R., Rabkin, A., Yang, E., and Yang, M. (2008). Chukwa: A Large-scale Monitoring System, in *CCA*, 8.

[28] Park, K., and Pai, V. S. (2006). CoMon: A mostly-scalable monitoring system for planetlab. *SIGOPS Operat. Syst. Rev.* 40(1), 65–74.

[29] Soumendra, N., and Kotz, D. (2008). Mesh-mon: A multi-radio mesh monitoring and management system. *Computer Commun.* 31(8), 1588–1601.

[30] Battat, N., and Kheddouci, H. (2011). HMAN: hierarchical monitoring for ad hoc network. In *IFIP EUC*.

[31] Kravets, R., Alkaff, H., Campbell, A., Karahalios, K., and Nahrstedt, K. (2013). CrowdWatch: Enabling In-Network Crowd-sourcing, *ACM SIGCOMM MCC*, 57–62.

[32] Barbera, M. V., Viana, A. C., De Amorim, M. D., and Stefa, J. (2014). Data offloading in social mobile networks through VIP delegation. *Ad Hoc Networks*, 19, pp. 92–110.

[33] Kiri, Y., Sugano, M., and Murata, M. (2007) Self-organized data-gathering scheme for multi-sink sensor networks inspired by swarm intelligence, in *IEEE SASO*, 2007.

[34] Zhu, X. (2007). Pheromone based energy aware directed diffusion algorithm for wireless sensor network. In *Advanced Intelligent Computing Theories and Applications. With Aspects of Theoretical and Methodological Issues*, D.-S. Huang, L. Heutte, and M. Loog, Eds., pp. 283–291.

[35] D. for Transport, "Rail Passenger Numbers and Crowding on Weekdays in Major Cities in England and Wales: 2012," 2013. [Online]. Available:

https://www.gov.uk/government/uploads/system/uploads/attachment_
data/file/252516/rail-passengers-crowding-2012-revised.pdf

[36] Badonnel, R., State, R., and Festor, O. (2005). Self-Organized Monitoring in Ad-Hoc Networks. *Telecommun. Syst.* 30, 143–160.

[37] Zorzi, M., and Rao, R. R. (2003). Geographic Random Forwarding (GeRaF) for Ad Hoc and Sensor Networks: Multihop Performance. *IEEE Trans. Mobile Comput.* 2(4), 337–348.

[38] Madden, S., Franklin, M. J., Hellerstein, J. M., and Hong, W. (2002). TAG: A tiny aggregation service for ad-hoc sensor networks. *SIGOPS Operating Syst. Rev.* 36, 131–146.

[39] Navidi, W., and Camp, T. (2004). Stationary distributions for the random waypoint mobility model. *IEEE Trans. Mobile Comput.* 3(1), 99–108.'

[40] Richerzhagen, B., Stingl, D., Rückert, J., and Steinmetz, R. (2015). Simonstrator: simulation and prototyping platform for distributed mobile applications, in *SIMUTOOLS*.

[41] Henderson, T. R., Roy, S., Floyd, S., and Riley, G. F. (2006). ns-3 Project Goals, in *ACM WNS2*.

[42] Bohannon, R. W. (1997). Comfortable and maximum walking speed of adults aged 20–79 years: reference values and determinants. *Age Ageing*, 26(1), 15–19.

[43] Himann, J. E., Cunningham, D. A., Rechnitzer, P. A., and Paterson, D. H. (1988). Age-related changes in speed of walking. *Medicine Sci. Sports Exercise*, 20(2), 161–166.

[44] Gross, C., Kaup, F., Stingl, D., Richerzhagen, B., Hausheer, D., and Steinmetz, R. (2013). EnerSim: an energy consumption model for large-scale overlay simulators, in *IEEE LCN*.

[45] "MAKI – Multi-Mechanisms Adaptation for the Future Internet," 2015. [Online]. Available: http://www.maki.tu-darmstadt.de/sfb_maki/ueber_
maki/index.en.jsp

[46] Richerzhagen, N., Stingl, D., Richerzhagen, B., Mauthe, A., and Steinmetz, R., "Adaptive Monitoring for Mobile Networks in Challenging Environments," in *IEEE ICCCN*, 2015.

5

Inferring Network Topologies in MANETs: Application to Service Redeployment

S. Silvestri[1], B. Holbert[2], P. Novotny[3], T. La Porta[2], A. Wolf[3] and A. Swami[4]

[1]Department of Computer Science, Missouri University of Science and Technology, Rolla, MO 65409, United States
[2]Department of Computer Science and Engineering, Pennsylvania State University, State College, PA 16801, United States
[3]Department of Computing, Imperial College, London SW7 2AZ, United Kingdom
[4]Army Research Laboratory, Adelphi, MD 20783, United States

Acknowledgement

This work was supported by the Defense Threat Reduction Agency under the grant HDTRA1-10-1-0085 State Estimation and Optimal Recovery in Networks with Massive Cascading Failures, and by the U.S. Army Research Laboratory and the U.K. Ministry of Defense agreement number W911NF-06-3-0001.

Abstract

The heterogeneous and dynamic nature of tactical coalition networks poses several challenges to common network management tasks, due to the lack of complete and accurate network information. We consider the problem of redeploying services in mobile tactical networks. We propose M-iTop, an algorithm for inferring the network topology when only partial information is available. M-iTop initially constructs a virtual topology that overestimates the number of network components and then repeatedly merges links in this topology

127

to resolve it towards the structure of the true network. We also propose an iterative service redeployment (iSR) algorithm for service redeployment in tactical networks. Extensive simulations show that M-iTop and iSR allow an efficient redeployment of services over the network despite the limitation of partial information.

Keywords: Topology inference, Partial information, Service redeployment.

5.1 Introduction

Tactical coalition networks are typically Mobile Ad hoc Networks (MANETs) composed of devices carried by soldiers or on vehicles from different nations and coalitions [1]. Some of these devices provide services, such as positioning and map services, to other devices often over multiple hops [2, 3]. As a consequence of device mobility, a static service deployment may result in poor performance and even complete lack of service. For this reason, services need to be redeployed as the topology of the network changes over time.

Previous works on service redeployment generally assume full knowledge of the service interconnections [4] and of the network topology [5–12]. This knowledge, however, may not be available in tactical coalition networks for several reasons. First, some coalitions may not share information such as the structure of their services or the connectivity of the devices. Second, wireless communications can easily experience transmission errors, especially in critical battlefield scenarios, preventing information collection. Finally, nodes are mobile and the collected information rapidly becomes stale. As a result, only partial and possibly inaccurate information on the network is available, which makes the service redeployment task challenging.

In order to address the lack of complete topology information, we propose a topology inference algorithm called M-iTop. M-iTop infers the network topology and its deployment in the physical space given the partial collected information. To the best of our knowledge, this is the first time that topology inference techniques are applied to service redeployment in MANETs.

M-iTop periodically probes the network to determine its topology. Due to the different coalition partners in the network and possible transmission errors, some nodes may not participate in the probing or may fail to provide their information. As a result, the learned topology shows missing nodes and links. Based on the collected information, M-iTop first generates a *virtual topology* by augmenting the learned topology with virtual links and nodes to restore known connectivity. It then repeatedly merges links in this topology according to consistency rules which are derived from an analysis of the collected information. The merging process resolves the virtual topology towards the structure of the true network. The inferred topology is then finally mapped in the physical space by inferring the geographical positions of the nodes.

The ability of M-iTop to accurately infer the network topology is limited by two main factors: the observability of the network (not all paths are exercised, and all links may not be covered by the probes), and the dynamics in the network while the algorithm is running. Our results show that for the typical size of a tactical network, M-iTop runs fast enough to accurately infer the current network topology even though nodes may have moved since the inference process started.

We also consider the problem of redeploying multiple service replicas in tactical networks. We formalize the problem as a multi-objective mixed integer linear programming problem, formulated using scalarization techniques. The objective is to maximize the number of nodes receiving service as well as the overall Quality of Service (QoS). In order to efficiently solve this problem, we propose the Iterative Service Redeployment (iSR) algorithm. Our results show that the short execution time of M-iTop and the accuracy of the inferred topology enable efficient service redeployment by iSR over the network, despite the limitations of partial information.

In summary, our main contributions are the following:

- We propose M-iTop to infer the network topology of mobile coalition networks.
- We formulate a mixed integer linear programming problem for multiple service redeployment.
- We propose the iSR to efficiently solve the redeployment problem.
- We show that M-iTop has a short execution time for typical coalition networks.
- We show that M-iTop and iSR provide an accurate estimate of the actual topology which enables efficient service redeployment.

5.2 Related Work

The problem of service redeployment has been recently considered in both wired networks with static topology [5–7, 13] and wireless networks with dynamic topology [8–11, 14, 15]. These methods assume full knowledge of the network topology and would not be able to operate in a tactical coalition network characterized by partial topology information.

The problem of inferring network topology in the presence of non-cooperating nodes which leads to partial information has been considered in several recent works [16–19]. The problem was first introduced in [16] and extended in [17] to scenarios with less information. Reference [19] identifies patterns in the traces that indicate certain structures in real network. While the approach in [19] has lower runtime complexity than [17], it is not as accurate. The approach proposed in [18] supplements trace information with additional data from the *record route*. This results in a more accurate inferred topology, but requires information which may not be available. Finally, in [20] we propose an algorithm called iTop to infer static network topologies. This algorithm provides higher inference accuracy than previous approaches.

The above-cited works only focus on static Internet-like topologies. As a consequence, they are not directly applicable to the context of MANETs. In this chapter, for the first time, we apply topology inference techniques to service redeployment in MANETs. We introduce the algorithm M-iTop, which extends our previous algorithm iTop [20] by specifically taking into account features of dynamic networks such as MANETs (hence the name Mobile-iTop). In particular, M-iTop addresses the node mobility and infers the network deployment in the physical space.

5.3 Network Model

We consider a network of nodes which form a MANET. Nodes are mobile and able to communicate with each other. We consider a binary disk communication model, i.e., two nodes are able to communicate if their physical distance is less than the fixed *transmission range* R_{tx}. Our approach can be easily extended to anisotropic and heterogeneous communication environments. We assume the MANET is composed of two coalition partners. Hence, nodes belong to two teams, to which we refer as *owned network* C_{own} and *allied network* C_{ally}. We assume the availability of an underlying routing algorithm in which all the nodes participate for communications[1]. We further assume that pairs of nodes in the owned network have a mechanism to measure the hop distance of the path between them, for example by analyzing the routing tables.

---

[1]In the following we assume that the routing algorithm is based on shortest path. Our approach can be easily extended to different routing strategies.

We refer to the real topology as the *Ground Truth* topology. We represent the network at time t by an undirected graph $G_{GT}^t = (V_{GT}, E_{GT}^t)$. V_{GT} is the set of nodes. There is an edge in E_{GT}^t between two nodes if they can directly communicate at time t, i.e., they are within physical distance R_{tx}.

Periodically, nodes in the owned network probe the network in order to acquire information about the current topology. In particular, each owned node sends a probe to every other node in C_{own}. We refer to the topology generated by the union of the probed paths as the *observable topology* $G_{obs}^t = (V_{obs}, E_{obs}^t)$. This topology is a subgraph of the G_{GT}^t topology as, in general, routing paths between owned nodes may only partially cover the network. Owned nodes reply to the probing with their information such as ID and geographic position. On the contrary, allied nodes do not provide any information even though they forward probes. We refer to the partial (or total) path information acquired by an owned node as a *trace*.

Traces are centrally collected and combined at the Network Operating Center (NOC). Since allied nodes do not provide their information when probed, the acquired information only partially represents the observable topology G_{obs}^t. The NOC executes our topology inference algorithm in order to estimate the observable topology. A service redeployment algorithm is then executed on the inferred topology.

5.4 M-iTop Approach

In this section we describe the operation of M-iTop, our topology inference approach with partial information. In the following we consider the operations performed by M-iTop at a given time t, but for ease of presentation the superscript t is omitted.

M-iTop operates in four phases. In the first phase it analyzes the traces and constructs the *virtual topology* $G_{VT} = (V_{VT}, E_{VT})$, which is a vastly overestimated topology compared to the observable topology. During the construction of the virtual topology, M-iTop classifies nodes in the virtual topology on the basis of their observed behavior with respect to the probes. In the second phase, M-iTop determines the *merge options* for each link in the virtual topology. These options indicate pairs of links in the virtual topology that can be merged while preserving consistency with respect to the observed characteristics of the ground truth topology and the node classifications assigned in the previous phase. In the third phase, M-iTop infers the merged topology $G_{MT} = (V_{MT}, E_{MT})$ by iteratively merging pairs of links based on their merge options and removing any merge options that are made invalid as a result.

The first three phases of M-iTop estimate the topological structure, i.e., the network graph, of the observable topology. As a result, these phases make use of the *hop distance* between nodes. The fourth phase infers the geographical locations of nodes of the allied network by exploiting the known locations of owned nodes and their connectivity in G_{MT}. For this reason, this phase makes use the *physical distances* between nodes.

The goal of the algorithm is to infer a G_{MT} that is as close as possible to G_{obs} given the partial information collected by the owned nodes.

5.4.1 Virtual Topology Construction

The NOC collects the information gathered by all nodes in the owned network and constructs the virtual topology as follows. Consider two owned nodes m_1 and m_2 connected by the path $m_1, v_1, v_2, \ldots, v_{n-1}, m_2$. Without loss of generality, assume that node m_1 initiates a probe to node m_2, to estimate their mutual hop distance $d(m_1, m_2)$ and collect the path information. The gathered information and hop distances are sent to the NOC by m_1, which analyzes them to infer the virtual topology and partitions the nodes into two classes. These classes are introduced in order to guide the merging process of M-iTop and reflect the node behavior as observed in the traces.

We define two classes of nodes. R is the class of *responding* nodes and A is the class of *anonymous* nodes. Responding nodes belong to the owned network while anonymous nodes belong to the allied network. When processing the collected traces, the NOC marks each node as belonging to one of these classes as follows.

Consider the case in which a node m_1 probes the path to m_2 and successfully receives a response from m_2. Since a reply was received, the NOC can conclude that these two nodes are connected. The path information is of the form $(m_1, x_1, \ldots, x_{n-1}, m_2)$ where $d(m_1, m_2) = n$ and each x_i either identifies a node v_i that is responding, or is a $*$ to denote no response. All nodes corresponding to a $*$ are anonymous. The NOC adds a node in the virtual topology for each responding node observed in the trace and connects them accordingly. It marks these nodes as responding and combines multiple instances of the same responding nodes reported by other owned nodes to avoid duplication of observed components. The anonymous nodes and the links connecting them are also added to the virtual topology. A virtual node is added for each ∗ in the trace and it is marked as anonymous.

The constructed G_{VT} topology overestimates the network because it may contain multiple anonymous nodes which are the same node in the G_{GT} topology. Since allied nodes do not provide their information, there is no simple way to determine which ones are the same. Therefore they are assumed to be distinct nodes until merged in the third phase of M-iTop.

Figure 5.1 shows an example scenario. The G_{GT} topology is shown in Figure 5.1(a). Dark nodes are the owned nodes, which take part and respond to the probing, while white nodes belong to the allied network and do not take part in the probing nor provide their information. In this example we assume shortest path routing for probing. As a result, nodes F, G, and H do

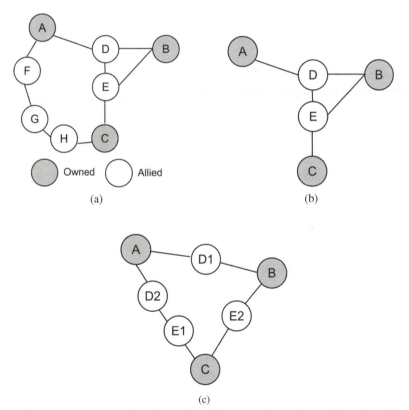

Figure 5.1   Example: Ground truth topology G_{GT} (a), observable topology G_{obs} (b) and virtual topology G_{VT} (c).

not appear in the observable topology G_{obs} as they are not covered by any shortest path between owned nodes as shown in Figure 5.1(b). Figure 5.1(c) shows the corresponding virtual topology. Multiple occurrences of the same anonymous node in the traces are represented as different nodes in the virtual topology, resulting in an overestimation of the observable topology G_{obs}. These occurrences are then merged in the subsequent phases of M-iTop.

5.4.2 Merge Options

In order to infer G_{MT} from G_{VT}, M-iTop identifies the valid merge options for each link e_i in E_{MT}, i.e., the set of links with which e_i can be merged. Initially, $E_{MT} = E_{VT}$. We introduce three conditions which have to be satisfied for a merge option to be valid. These conditions check the consistency of a merge option with the information gathered from the traces and with the node classification provided in the previous phase.

The set M_i denotes the set of links which are valid merge options for link e_i. A link e_i has a valid merge option with e_j if all the following conditions are verified. The sets of merge options are then used during the merging phase to determine which merges occur and in what order.

Trace Preservation: Since paths do not contain loops, a link will never appear twice in the same path. A merge option between two links satisfies trace preservation if these links do not appear together in any path.

Hop Distance Preservation: The hop distance between any two owned nodes in G_{VT} is consistent with the G_{obs} topology. A merge option is valid if the hop distance between owned nodes in G_{MT} is the same as that in G_{VT}. For example, in Figure 5.1(c), A-D1 can be merged with A-D2, C-E1 with C-E2, but not B-D1 with B-E2.

Link Endpoint Compatibility: The node classification gives us additional information on valid merging options. In particular, a node can be either responding or anonymous. A merge option between two links is valid if there is a way to combine their endpoints without violating their classification. In particular, a merge option between two links e_i and e_j is valid in two cases: (1) the endpoints of e_i and e_j are all classified as anonymous, or (2) for both e_i and e_j one endpoint is classified as anonymous and the other as responding. This second option is valid only if the responding node is the same for both links.

5.4.3 Merging Links

The next phase merges the links in the virtual topology to derive the M-iTop topology G_{MT}. Initially, $G_{MT} = G_{VT}$. The merging phase reduces G_{MT} by iteratively merging pairs of links based upon the existing merge options. Each merge combines two links in E_{MT}, combining their endpoints and reducing the number of components in the network accordingly. When no merge options remain, the merging phase is complete and the M-iTop topology is finalized.

Algorithm 5.1 shows the pseudo-code of the merging phase. In each step, M-iTop chooses two links and attempts to merge them. Several alternatives are possible to determine the order in which links are merged, which influences the resulting final topology. Since links with few merging options have fewer merging possibilities, they are more likely to be the same link in the ground truth topology. On the basis of this observation, we first select the link e_i with the fewest merging options and then link e_j which has the fewest merging options out of the links with which e_i can be merged (Algorithm 5.1, lines 2–3). We experimented with several alternative heuristics and the one described above provides the best results.

Link merging

Two links e_i and e_j are merged by the function $Merge(e_i, e_j)$ (Algorithm 5.1, line 4). For ease of exposition we describe the merging of e_j into e_i, as the same result would be obtained by the opposite merging. All paths containing e_j are modified to contain e_i in its place and the set M_i is changed so that $M_i = M_i \cap M_j$. Any links which could have been merged with both e_i and e_j retain their merge option with e_i and the merge option for e_j is removed. Any links which had a merge option with either e_i or e_j but not both have that option removed. This ensures that e_i can only be further

---

Algorithm 5.1   M-iTop merging phase

Input: Initial Virtual Topology $G_{VT} = (V_{VT}, E_{VT})$, Merge Options M
Output: Merged M-iTop Topology $G_{MT} = (V_{MT}, E_{MT})$
1 $G_{MT} = G_{VT}$;
2 **while** $\exists e_i \in E_{MT} \wedge M_i \neq \emptyset$ **do**
3 $\quad e_i = argmin_{e_i \in E_{MT}} |M_i|$;
4 $\quad e_j = argmin_{e_j \in M_i} |M_j|$;
$\quad$ // Link merging
5 $\quad$ Merge(e_i, e_j);
6 **return** $G_{MT} = (V_{MT}, E_{MT})$

---

merged with links that before were valid merges for both e_i and e_j. The link e_j is then removed from E_{MT}.

5.4.4 Inference of Nodes Physical Locations

The previous phases of M-iTop enable us to infer the structure of the network topology. In this phase we infer the geographical locations of nodes in the inferred topology. Owned nodes provide their location information when the network is probed. On the contrary, allied nodes do not, and thus their positions need to be inferred. In the following we describe the procedure M-iTop adopts for this purpose.

Let s be a node of the allied network and let $N_{loc}(s)$ be the set of its one-hop neighbors whose positions are known; this set may be empty. For each allied node s, we calculate the area $\mathcal{R}(s)$ defined as follows:

$$\mathcal{R}(s) = \left(\bigcap_{p \in N_{loc}(s)} R_{tx}(p) \right) \setminus \left(\bigcup_{p \in (C_{own} \setminus N_{loc}(s))} R_{tx}(p) \right) \qquad (5.1)$$

where $R_{tx}(p)$ is the transmission area of node p, i.e., the circle centered at p and with radius R_{tx}. $\mathcal{R}(s)$ defines the area in which node s can lie given the physical distance constraints imposed by the inferred connectivity. In order to determine the location of allied nodes, we rely on graph drawing algorithms modified to our needs. In particular, we fix the position of owned nodes and we constrain the position of each allied node s to $\mathcal{R}(s)$. We further modify the drawing algorithm by imposing the physical distance between two neighboring nodes in G_{MT} to be at most R_{tx}.

Given the location constraints, we execute a graph drawing algorithm until it converges to a stable configuration. We tried several graph drawing algorithms and adopted Force Atlas [21] as it provides good performance in the considered scenarios.

Note that, if owned nodes can estimate the physical distance to neighboring nodes, for example using the received signal strength [22], the above process can be further improved by including triangulation.

5.5 Iterative Service Redeployment (iSP) Algorithm

In this section we present the formalization of the multiple service replicas redeployment problem as well as of the Iterative Service Redeployment (iSR) algorithm to efficiently solve such problem.

5.5.1 Formalization of the Multiple Service Replicas Deployment Problem

As a consequence of node movements, the network is partitioned into m connected components $G_1, \ldots, G_m$. We represent each component G_i as a graph, i.e., $G_i = (V_i, E_i)$. The primary goal of the problem is to maximize the number of users that can access a service. To this purpose, up to R service replicas can be deployed on the nodes of the owned network. The secondary goal is to optimize the overall QoS provided by such services. We assume that, for each node v in the owned network, we know a QoS metric (e.g., delay, packet loss) $q(v, u)$ for any other node u in the entire network. Note that the QoS metric may be only partially available when service redeployment is applied to an inferred topology. Nevertheless, it can be estimated using the nodes' geographical locations inferred in the fourth phase of M-iTop [23].

We formulate the problem as a multi-objective optimization problem, since we aim to optimize both the number of served users and the overall QoS. The problem is expressed as a mixed integer linear programming problem using scalarization [24]. Let $V_i^{own} = V_i \cap C_{own}$ be the set of owned nodes in the G_i component. We formally define the problem as follows.

$$\max_{x_u} \sum_{i=1}^{m} z_i |G_i| + \beta \sum_{i=1}^{m} \frac{1}{|V_i|} \sum_{u \in V_i^{own}} \sum_{v \in V_i} y_{u,v} q(u, v) \qquad (5.2)$$

$$\text{subject to } z_i \leq \sum_{u \in V_i^{own}} x_u \quad i = 1, \ldots, m \qquad (5.3)$$

$$\sum_{u \in V_i^{own}} y_{u,v} \leq 1 \; \forall v \in V_i \qquad (5.4)$$

$$y_{u,v} \leq x_u \; \forall u, v \qquad (5.5)$$

$$\sum_{i=1}^{m} \sum_{u \in V_i^{own}} x_u \leq R \qquad (5.6)$$

$$x_u, z_i, y_{u,v} \in \{0, 1\} \qquad (5.7)$$

where x_u is equal to 1 if a service is deployed on node u, and to 0 otherwise; $y_{u,v}$ is equal to 1 if node v receives service from node u, to 0 otherwise; finally, z_i is equal to 1 if the connected component G_i contains at least one service, and to 0 otherwise. Table 5.1 summarizes the meaning of variables and constants used in the optimization problem.

Table 5.1   Summary of variables and constants used in the optimization problem (5.2)

Symbol	Meaning
$G_i = (V_i, E_i)$	i-th connected component
x_u	Equals 1 if a service is deployed on node u, 0 otherwise
$y_{u,v}$	Equals 1 if node v receives service from node u, 0 otherwise
z_i	Equals 1 if the component G_i contains at least one service, 0 otherwise
R	Number of services replicas to be deployed
$q(u, v)$	QoS metric between node u and node v

The goal is to jointly maximize the number of nodes served by at least one service as well as the average QoS in each component. We introduce the parameter $\beta \in \mathbb{R}^+$ to prioritize these objectives. In the remainder of the chapter, we assume that β is sufficiently small, such that having more nodes receiving service is always prioritized with respect to improving the QoS.

The constraint in Equation (5.3) ensures that a component is counted in the objective function only if at least one service is deployed in that component. Equation (5.4) constrains a node to receive service from at most one node. Equation (5.5) enables a node v to receive service from u, only if a service is deployed on u. Finally, Equation (5.6) ensures that at most R replicas are deployed.

The optimization problem may be computationally too intense to be solved exactly in a timely manner. For this reason, we propose the algorithm iSR that efficiently finds an approximate solution.

5.5.2 The iSR Algorithm

The Iterative Service Redeployment (iSR) algorithm is executed by the NOC and it is designed to efficiently find an approximate solution to the problem described in the previous section. The algorithm outputs, for each component G_i, the set of nodes $S_i \subseteq V_i^{own}$ on which service replicas are deployed. The pseudo code of the algorithm is shown in the algorithm iSR. In the pseudo-code, we explicitly omit the components with no owned nodes.

iSR iteratively deploys one service at a time. Initially, it considers the largest component that does not contain a service, and selects the node that optimizes the average QoS to all other nodes in that component (lines 4–7). Subsequently, when all components have at least one service (line 8), it identifies the node that would provide the best improvement in each component. To this purpose it calculates for each component G_i, the node u_i^* that provides the best average QoS improvement Δ_i^* in G_i (lines 9–11). It then selects the node with the highest Δ_i^* (lines 12–13).

Algorithm 5.2 Iterative service redeployment algorithm (iSR)

Input: Topology $G = (V, E)$, distance metric function
$q \colon V \times V \to \mathbb{R}^+$, number of services to be deployed R
Output: Set of selected nodes $S_1, \ldots, S_m$ in each component
for service redeployment

1 $r = 0$;
2 $S_i = \emptyset \ \forall i = 1, \ldots, m$;
3 **while** $r \leq R$ **do**
4   **if** $\exists \ G_i \ s.t. \ S_i = \emptyset$ **then**
      // Select largest connected component
5        $G^* = (V^*, E^*) = \mathrm{argmax}_{G_i \ s.t. \ S_i = \emptyset} |G_i|$;
      // Select largest best node
6        $u^* = \mathrm{argmax}_{u \in V^*} \frac{1}{|V^*|} \sum_{v \in V^*} q(u, v)$;
7        $S_i = S_i \cup \{u^*\}$;
8   **else**
9   **for** $i = 1$ *to* m **do**
10       $u_i^* = \mathrm{argmax}_{u \in V_i \setminus S_i} \frac{1}{|V_i|} \sum_{v \in V_i} \max_{\hat{u} \in S_i \cup \{u\}} q(\hat{u}, v)$;
11       $\Delta_i = \frac{1}{|V_i|} \Big(\sum_{v \in V_i} \max_{\hat{u} \in S_i \cup \{u\}} q(\hat{u}, v) - \sum_{v \in V_i} \max_{\hat{u} \in S_i} q(\hat{u}, v) \Big)$;
12   Let G^* be the component with maximum Δ_i and u^*
    the selected node;
13     $S^* = S^* \cup \{u^*\}$;
14   r++;
15 **return** $S_1, \ldots, S_m$;

---

iSR has complexity $O(m \log m + R \times m \times \max_i |V_i|^2)$. $m \log m$ is necessary to sort the m components by size, and for each of the R service replicas deployed, it is necessary to find the node with best improvement, which is proportional to the square of the number of nodes in the component.

5.6 Results

In this section we evaluate the performance of iSR when applied to the topology inferred by M-iTop. To this aim we developed a simulator based on the Wireless module of the Opnet environment [25].

When iSR is applied to an inferred topology, several sources of errors may occur in selecting nodes for service deployment. First, the topology inference algorithm may have a non-negligible execution time. As a result, nodes in the current topology might have moved, possibly making the inferred topology inaccurate by the time it is produced. Second, the nodes selected in the inferred topology may not be the actual best nodes in the G_{GT} topology, due to inaccuracies of the considered QoS metric when calculated on the inferred topology. Finally, the actual connected component sizes in the G_{GT} topology may not correspond to the sizes in the inferred topology. This is due to the inherent partial observability of the G_{GT} topology and to the possible inaccuracies in the inferencing process.

We use the following simulation setting. We randomly deploy nodes on a bounded square area of size 500 m × 500 m. Nodes move at a maximum speed of 0.5 m/s according to a local random way point model. In particular, each node selects a random position in a circle of radius 100 m around the initial location in which it was deployed. Nodes have a waiting time between consecutive movements chosen randomly in the interval $[0, 300]$ s. This movement strategy reflects mission-oriented local movements typical of military MANETs. We also assume shortest path routing and hop distance as the QoS metric for the service redeployment algorithm. Communication delays are integrated using the realistic wireless communications module of the Opnet simulator [25]. We use different numbers of nodes, communication ranges and percentage of nodes in the owned coalitions for different scenarios.

We first study the execution time of M-iTop by increasing the network size. For this experiments we use a transmission range of 200 m, which results in a highly connected network, and we assumed that 50% of the nodes are in the owned coalition. We averaged the results over 100 initial random deployments. Results are shown in Figure 5.2. Although the execution time of M-iTop rapidly increases with the network size, it is reasonably low for networks of 60 nodes, and in particular on the order of few seconds for networks with 40 nodes or less. The short execution time of M-iTop for small networks makes it suitable for inferring the topologies of tactical networks, whose size is generally on the order of a few tens of nodes.

In order to study the performance of iSR on the topology inferred by M-iTop, we performed four sets of experiments. In the first set we focus on the effects of the execution times of M-iTop, and of the inaccuracies of the inferred topologies, on the performance of iSR. In the second set, we study how the performance is affected by the percentage of nodes in the owned coalition. In the third set, we perform a sensitivity analysis to the setting of the transmission range and the nodes' speed. In the first three sets we consider the deployment of a single service replica, i.e., $R = 1$, and in the fourth set we consider the deployment of multiple replicas.

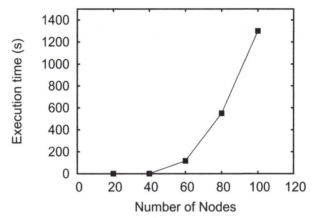

Figure 5.2 Execution time of M-iTop with increasing network size.

Before showing the results, we show an example of iSR applied to the G_{GT} topology, the observable topology G_{obs} and the topology inferred by M-iTop, to deploy one service replica. We consider a network with 40 nodes and transmission range 100 m. The G_{GT} topology is shown in Figure 5.3(a). Dark nodes belong to the owned network while white nodes to the allied network. The observable topology G_{obs} obtained by probing the network between owned nodes is shown in Figure 5.3(b). As expected this topology is a subset of the G_{GT} topology. Figure 5.3(c) shows the topology inferred by M-iTop. It closely matches G_{obs}, showing the effectiveness of the M-iTop algorithm.

In Figure 5.3(a–c), the nodes highlighted with a larger size are selected by the service redeployment algorithm. The node that belongs to the largest

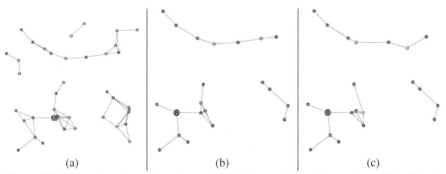

| (a) | (b) | (c) |

Figure 5.3 Example of ground truth topology G_{GT} (a), observable topology G_{obs} (b) and topology inferred by M-iTop G_{MT} (c).

component and minimizes the average hop distance to the other nodes is 18, as selected in the G_{GT} topology. In the G_{obs} topology, the largest component can still be correctly identified, but the partial information given by probing results in the selection of node 8. The topology inferred by M-iTop closely matches G_{obs}, which enables it to correctly determine the largest component and to select node 8. In the following experiments we show that the nodes selected from the inferred topologies provide good service when compared to the best choice that would be obtained with complete information.

5.6.1 First Set of Experiments

In these experiments we assume an equal number of allied and owned nodes. We periodically execute iSR every 500 s. This period provides good performance considering the mobility model adopted in this chapter. We apply iSR to deploy a single service replica, i.e., $R = 1$, to G_{GT}, G_{obs} and to the topology inferred by M-iTop G_{MT}. The case of multiple service replicas is studies in Section 5.6.4. For G_{GT} and G_{obs} the service redeployment takes effect instantly. On the contrary, for M-iTop the service redeployment is delayed by the time required for probing the network, collecting the information at the NOC and executing M-iTop. In particular, as soon as a new slot for service redeployment starts, the network is probed and M-iTop is executed. The node selected in the previous slot still provides service until M-iTop terminates and a new node can be selected. For comparison, we also consider the ideal unrealistic case in which the service is placed on the optimal node at each instant in time. This case is referred to OPT in the figures.

In the experiments, given a node u selected by iSR, we consider as a performance metric the actual average hop distance of u to the nodes belonging to its connected component in the G_{GT} topology.

5.6.1.1 Single connected component

We initially use a large transmission range of $200\ m$, which always results in a single connected component. This experiment enables us to focus on the effects of the execution times of M-iTop and of the inaccuracies of the inferred topology.

Figure 5.4(a) shows the average hop distance over time of the node selected by iSR, for a network of 40 nodes. Vertical lines indicate the times at which iSR is executed. In the case of M-iTop, these are the moments at which the network is probed and inferred. Once the inference is completed the service redeployment algorithm is then executed.

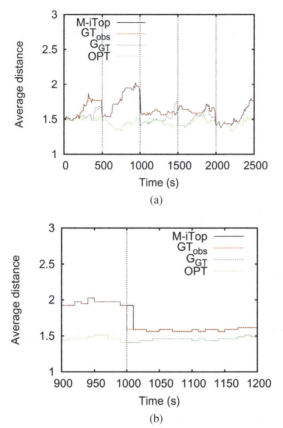

Figure 5.4 Network with 40 nodes, single connected component. Average hop distance over time (a), zoom on the service redeployment at 1000 s (b).

The distances under G_{GT}, G_{obs} and M-iTop generally get worse with respect to OPT as the amount of time since the last execution of the service redeployment algorithm increases. This is due to node mobility, which changes the structure of the network and causes the service redeployment decision to become stale. The vertical performance gap of M-iTop with respect to G_{GT} in Figure 5.4(a) shows the performance worsening due to the partial information available and to the possible inaccuracies of the inference process. The horizontal gap shows instead the delay which is incurred by the execution time of M-iTop. The results highlight that M-iTop has an average distance close and often equal to that of G_{GT} and G_{obs}. The inferred topology enables proper selection of a node for service redeployment even in the presence of partial information.

In order to highlight the effects of the execution times of M-iTop, we zoom on the results of the previous experiment around 1000 s in Figure 5.4(b). As the service redeployment algorithm is executed at time $t = 1000$, G_{GT} and G_{obs} instantly switch to the new node since no execution delay is considered for these approaches. M-iTop terminates in less than 10 s for a network of 40 nodes. As a result, after a short time a new node is selected on the basis of the inferred topology. This is the same node that is selected by G_{obs}, highlighting that the inferred topology closely matches the observable topology. The G_{GT} topology enables the selection of a slightly better node, but requires full knowledge of the network topology that is not available in a coalition network.

Figures 5.5(a) and (b) show the average hop distance over time for a 60-node network. The execution time of M-iTop is around 120 s for sixty nodes. Also in this case, M-iTop achieves the performance close to and often coinciding with G_{GT} and G_{obs}, with a slightly longer delay due to the algorithm execution.

5.6.1.2 Multiple connected components

We now consider a reduced setting of the transmission range to 100 m, which generates a network partitioned into several connected components over time. These experiments enable us to also show the possible inaccuracies incurred due to an incorrect selection of the largest connected component. We consider a network of 40 nodes.

Figures 5.6(a) and (b) show the average hop distance between the selected node and the number of nodes receiving service. In this case, M-iTop is able to provide the performance close to that of G_{GT} and G_{obs} most of the time. In the time interval between $t = 1000$ s and $t = 1500$ s, M-iTop has a lower average distance than do G_{GT} and G_{obs}. In this case, iSR applied to G_{MT} fails to select the actual largest connected component in G_{GT}. As a result, the average distance is lower but at the expense of fewer nodes receiving service. It should be noted that the network partition allows only a subset of the nodes to be connected to the service node. Figure 5.6(b) highlights that under M-iTop the selected node belongs to the largest component in most cases.

Overall, the topologies inferred by M-iTop enable the service redeployment algorithm to select a node which provides performance close to that in the case of complete knowledge, even in the presence of partial information, network partitions and node mobility.

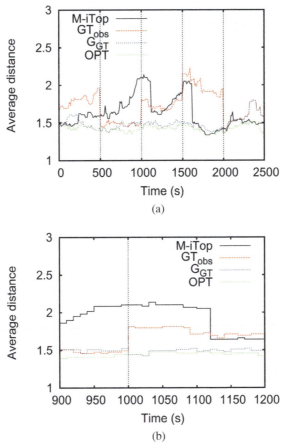

Figure 5.5 Network with 60 nodes, single connected component. Average hop distance over time (a), zoom on the service redeployment at 1000 s (b).

5.6.2 Second Set of Experiments

In this set of experiments we vary the percentage of owned nodes and study the performance of iSR applied to the inferred topologies. Also in this case, iSR deploys a single service replica ($R = 1$). Intuitively, a smaller number of owned nodes reduces the extent of the observable topology G_{obs}, and thus it also provides less information for the topology inference algorithm. On the contrary, when a large fraction of nodes belongs to the owned network, more information is available and thus the inferred topology more accurately represents the ground truth topology. In these experiments we set $R_{tx} = 200\ m$ which ensures the presence of a single connected component.

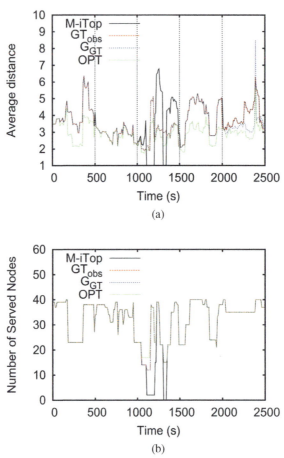

Figure 5.6   Network with 40 nodes, multiple connected components. Average hop distance over time (a), number of served nodes over time (b).

We study the average hop distance over time of the node selected by the service redeployment algorithm by increasing the percentage of nodes belonging to the owned network C_{own}. The distance is averaged over a period of 3000 s. and iSR is executed every 500 s. We compared the performance of iSR when applied on the full topology G_{GT}, on the observable topology G_{obs} and on the topology inferred by M-iTop. Also in these experiments we consider the ideal unrealistic case (OPT) in which the service is placed on the optimal node at each instant in time.

Figure 5.7(a) shows the results for a network of 40 nodes. When the full knowledge of G_{GT} is available, the service deployment algorithm achieves the best performance. The gap with respect to OPT is due to the time gap between subsequent executions of the deployment algorithm, during which nodes move and thus may result in suboptimal performance. When the percentage of owned nodes is small, G_{obs} only partially represents G_{GT}. As a result, the performance of the service deployment algorithm is inevitably poor. On the contrary, as the percentage of owned node increases, the performance improves and eventually converges to the case in which full knowledge is available. For a network of 40 nodes, M-iTop is able to accurately and quickly infer a

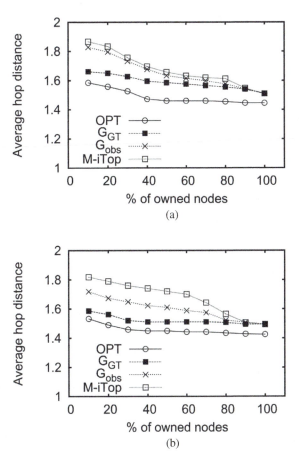

Figure 5.7 Average hop distance vs. the percentage of owned nodes. Network with 40 nodes (a), and network with 60 nodes (b).

topology close to the observable topology. As a result, the service deployment algorithm performs similar to the case in which the observable topology is perfectly known.

Figure 5.7(b) shows the results for a network of 60 nodes. The performance of G_{GT} and G_{obs} are similar to that in the case of 40 nodes. M-iTop has slightly worse performance than in the case of a 40-node network, due to the longer execution times (approximation 120 s. for a 60-node network). Such execution times delay the service redeployment, thus increasing the average distance (see also Figure 5.5). Nevertheless, M-iTop is always within 10% from the case in which the observable topology is perfectly known.

5.6.3 Third Set of Experiments

In this set of experiments we perform a sensitivity analysis of the performance of M-iTop to the setting of the communication radius R_{tx}. Intuitively, a smaller radius results in a higher number of small components, while a longer radius reduces the number of components and increases their size. In the experiments we consider a network of 60 nodes and we apply iSR every 500 s to deploy a single service replica. We increase the transmission radius from 25 to 250 m.

Figure 5.8(a) shows the average hop distance from the selected service nodes, while Figure 5.8(b) shows the number of nodes that receive service. When the transmission range is small (≤ 50 m) there are several small components, and additionally such components are constantly changing over time due to node mobility. OPT continuously redeploys the service, thus being able to always correctly select the largest component. Conversely, with G_{GT}, G_{obs} and M-iTop, iSR is executed every 500 s, penalizing the accuracy in selecting the largest component. As a result, when the radius is small, G_{GT}, G_{obs} and M-iTop show a lower average distance than OPT, which is the result of the fewer number of nodes receiving service.

As Figure 5.8(b) shows, there is a *percolation transition* as we increase the transmission range [26]. In particular, when R_{tx} increase from 50 to 100 m there is a sudden creation of a single *giant component*. Therefore, when $R_{tx} \geq 100$ m, all nodes in the network receive service under all approaches. This explains the peak around $R_{tx} = 100$ m in Figure 5.8(a). This value is close to the *percolation threshold;* therefore, there is a single, sparse, giant component. The sparsity of the component results in a higher hop distance to reach the service. As we increase the transmission radius, the density of

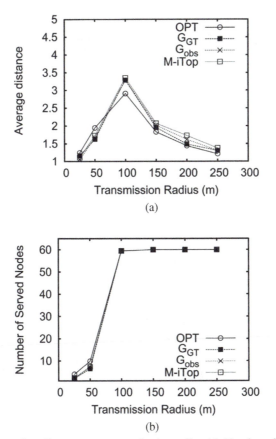

Figure 5.8 Average hop distance vs. communication radius (a). Number of served nodes vs. communication radius (b).

the component increases, and consequently the average distance decreases. Overall, M-iTop performs closely to G_{GT}, which however assumes perfect knowledge of the topology.

We further study the performance sensitivity to the nodes' speed. We set the transmission radius to 200 m, which always results in a single connected component, and we vary the speed from 0.1 to 3 m/s. Results are shown in Figure 5.9. When the speed is very low, the graph is more static; therefore, all approaches allow iSR to correctly estimate the largest component. M-iTop achieves a slightly higher average distance due to the non-negligible execution time. All approaches, including OPT, show an increase in the average hop

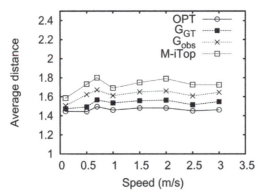

Figure 5.9   Average hop distance vs. nodes' movement speed.

distance as we increase the speed (0.5–0.75 m/s), after which they stabilize around that value, showing very little sensitivity to the speed parameter. This is due to the fact that, when the speed is sufficiently high, the nodes moving farther from the service are balanced by the nodes getting closer to it. Also in this case, M-iTop performs always within 10% from the case in which the observable topology is perfectly known.

5.6.4 Fourth Set of Experiments

The last set of experiments studies the performance of iSR when multiple services are deployed. We set the transmission radius to 200 m to have a single connected component in the network and a speed of 0.5 m/s. We assume that, when multiple services are present, a node receives service from the closest service replica. We compare the average hop distance achieved by iSR when applied to the topologies G_{GT}, G_{obs} and M-iTop. iSR is executed every 500 s.

Figures 5.10(a) and (b) show the average hop distance in a network of 40 and 60 nodes, respectively. As expected, in all approaches, the average distance decreases as we increase the number of replicas deployed. When the network is composed of 40 nodes, M-iTop can be executed in a short time and it is able to accurately infer the observable topology. Therefore, it achieves results similar to those in the case in which the observable topology is perfectly known. When 60 nodes are present, M-iTop is slightly penalized by the running time. Nevertheless, even in this case it performs within 10% of the case in which G_{obs} is known.

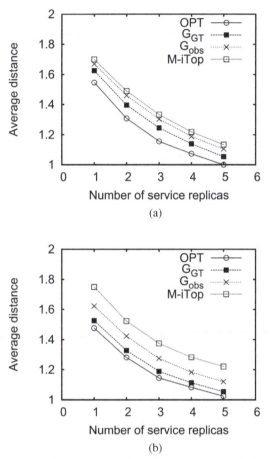

Figure 5.10 Average hop distance vs. number of service replicas deployed. Network with 40 nodes (a), and network with 60 nodes (b).

5.7 Discussion and Future Research Directions

The M-iTop algorithm provides a good approximation of the observable topology, which enables iSR to select nodes that provide good performance. We next discuss additional aspects which may lead to further performance improvements and will be investigated in our future works.

The M-iTop algorithm mainly focuses on network connectivity and hop distance metrics to perform the merging. The probing phase may provide additional information such as quality of service information on nodes and links. This information can be included in the merging process to perform more accurate merging. Furthermore, M-iTop can be extended in order to return quality of service information on the unobservable part of the network which can be used to improve the service redeployment algorithm.

The complexity of M-iTop can be reduced by considering constraints, such as the nodes' geographical position, that can significantly reduce the size of the merging options sets. The resulting shorter execution time will enable the applicability of the algorithm to large-scale networks.

Finally, movement prediction schemes can be included in the service redeployment algorithm to base the selection on the future expected position of nodes, improving the performance over time.

5.8 Conclusions

In this chapter, for the first time, we apply topology inference techniques to the service redeployment problem in tactical coalition networks. Service redeployment requires accurate topology information which is challenging in coalition networks, where the presence of allied nodes prevents the collection of full topology information. We present M-iTop, an algorithm for topology inference in the presence of partial information, and iSR, an iterative algorithm for multiple service redeployment. We show that M-iTop has a short execution time when applied to networks with the typical size of coalition networks. We perform extensive simulations showing that M-iTop, coupled with iSR, achieves the performance close to that of the optimal redeployment strategy based on complete information.

References

[1] Cirincione, G., and Gowens, J. (2007). The international technology alliance in network and information science. *IEEE Commun. Mag.* 6 (3), 14–18.

[2] Lund, K., Eggen, A., Hadzic, D., Hafsoe, T., and Johnsen, F. T. (2007). Using web services to realize service oriented architecture in military communication networks. *IEEE Commun. Mag.* 45 (10), 47–53.

[3] Lund, K., Skjervold, E., Johnsen, F. T., Hafsøe, T., and Eggen, A. (2010). Robust web services in heterogeneous military networks. *IEEE Commun. Mag.* 48 (10), 78–83.

[4] Novotny, P., Wolf, A. L., and Ko, B. J. (2013). Discovering service dependencies in mobile ad hoc networks. In *IFIP/IEEE IM*, pp. 527–533.

[5] Heydarnoori, A., and Binder, W. (2011). A graph-based approach for deploying component-based applications into channel-based distributed environments. *J. Softw.* 6, (8).

[6] Kalyvianaki, E., Wiesemann, W., Vu, Q. H., Kuhn, D., and Pietzuch, P. (2011). Sqpr: Stream query planning with reuse. In *IEEE ICDE*.

[7] Ooi, B.-Y., Chan, H.-Y., and Cheah, Y.-N. (2012). Dynamic service placement and replication framework to enhance service availability using team formation algorithm. *J. Syst. Softw.* 85 (9), 2048–2062.

[8] Ali, S., Mitschele-Thiel, A., Diab, A., and Rasheed, A. (2010). A survey of services placement mechanisms for future mobile communication networks. In *ACM FIT*.

[9] Herrmann, K. (2010). Self-organized service placement in ambient intelligence environments. *ACM Trans. Autonomous Adapt. Syst.* 5 (2), 6:1–6:39.

[10] Maia, M. E. F., Fonteles, A., Neto, B., Gadelha, R., Viana, W., and Andrade, R. M. C. (2013). Loccam – loosely coupled context acquisition middleware. In *ACM SAC*.

[11] Vega, D., Medina, E., Messeguer, R., Royo, D., and Freitag, F. (2011). A node placement heuristic to encourage resource sharing in mobile computing. In *Springer-Verlag ICCSA*.

[12] Tati, S., Silvestri, S., He, T., and Porta, T. L. (2014). Robust network tomography in the presence of failures. *IEEE ICDCS*.

[13] Famaey, J., Wauters, T., Turck, F. D., Dhoedt, B., and Demeester, P. (2011). Network-aware service placement and selection algorithms on large-scale overlay networks. *Elsevier Comput. Commun.* 34 (15), 1777–1787.

[14] Chadha, R., Poylisher, A., Deb, B., Littman, M., and Sabata, B. (2005). Adaptive dynamic server placement in manets. *IEEE MILCOM*.

[15] Graupner, S., Andrzejak, A., Kotov, V., and Trinks, H. (2005). Adaptive service placement algorithms for autonomous service networks. *Springer Eng. Self-Org. Syst.* 3464, 280–297.

[16] Yao, B., Viswanathan, R., Chang, F., and Waddington, D. (2003). Topology inference in the presence of anonymous routers. *IEEE INFOCOM*.

[17] Jin, X., Yiu, W.-P., Chan, S.-H., and Wang, Y. (2006). Network topology inference based on end-to-end measurements. *IEEE J. Selected Areas in Commun.* 24 (12), 2182–2195.

[18] Sherwood, R., Bender, A., and Spring, N. (2008). Discarte: A disjunctive internet cartographer. *ACM SIGCOMM*.

[19] Gunes, M. H., and Sarac, K. (2008). Resolving anonymous routers in internet topology measurement studies. In *IEEE INFOCOM*.

[20] Holbert, B., Tati, S., Silvestri, S., La Porta, T., and Swami, A. (2015). Network topology inference with partial path information. *IEEE ICNC*.

[21] Kobourov, S. G. (2012). Spring embedders and force directed graph drawing algorithms. *arXiv preprint arXiv:1201.3011*.

[22] Li, X. (2007). Collaborative localization with received-signal strength in wireless sensor networks. *IEEE Trans. Vehic. Technol.* 56 (6), 3807–3817.

[23] Belghith, A., and Belhassen, M. (2011). Routing enhancements in dynamic manets with obstacles. *IFIP WMNC*, 1–5, Oct.

[24] Jahn, J. (1985). *Scalarization in multi objective optimization*. Springer.

[25] Riverbed Technologies. http://www.riverbed.com/.

[26] Newman, M. (2010). *Networks: An introduction*. Oxford University Press.

6

Towards Unified Wireless Network: A Software Defined Architecture based on Network Virtualization and Distributed Mobility Management

Jyotirmoy Banik[1], Marco Tacca[1], Andrea Fumagalli[1], Behcet Sarikaya[2] and Li Xue[3]

[1]Open Networking Advanced Research (OpNeAR) Lab, The University of Texas at Dallas, Richardson, TX, USA
[2]Huawei Technologies Limited, Plano, TX, USA
[3]Huawei Technologies Limited, Haidian, Beijing, China

Abstract

In this chapter, we propose a software-defined networking (SDN)-based unified wireless network architecture, which enables distributed mobility management. The proposed approach envisions a virtualized control plane, placed in the cloud. The user plane is free of typical tunneling mechanisms and utilizes layer 2 and layer 3 traffic forwarding. In the first part, our work describes the network architecture for cellular users. We discuss about traffic flow in both upstream and downstream directions, different mobility scenarios, and an efficient address assignment scheme. We then use an analytical model to estimate the signaling overhead in evolved packet core (EPC), and an OpenFlow-based test bed to demonstrate the handover performance. The architecture is further extended to support WLAN users in fixed networks, with a description of the signaling schemes required in a number of mobility events.

Keywords: Software Defined Networking, Network Functions Virtualization, Unified Wireless Network, Fixed Mobile Convergence, Distributed Mobility Management, 5G Mobile Network, Evolved Packet Core.

6.1 Introduction

Cellular data traffic is growing exponentially with time. The increasing popularity of smartphones, tablets, and wearable gadgets (machine-to-machine communication) is primarily dictating this growth. A recent study predicts a tenfold increase in mobile data traffic, between 2014 and 2019 [1]. Long-term evolution (LTE)—an all IP wireless networking technology—is playing a key role in keeping pace with this increasing demand. However, expanding the network capacity proportionately with the anticipated demand growth is becoming a challenge for the network operators, especially considering projected average revenue per user (ARPU). The concept of separating the control plane and the user plane is gaining friction as resources can be provisioned efficiently and independently in each plane.

Another looming challenge for the control plane is the signaling overhead in the mobile core network (also known as evolved packet core or EPC). IP does not provide any native support for mobility. To address this issue, several solutions have been proposed by different standard bodies such as, IETF and 3GPP. For cellular networks, the most popular solution is the one provided by the 3GPP named as GPRS tunneling protocol (GTP). By using tunnel endpoint identifier (TEID), GTP tunnels enable a cellular user to move seamlessly between different eNodeBs, while allowing the user equipment (UE) to retain the same IP address. To keep track of the mobile user, GTP TEIDs are updated after every handover. With the rapid increase in cellular traffic demand, small cells are gaining momentum as they offer increased network capacity through space diversity. A side effect of this approach is that the frequency of handovers between cells tends to increase, and so does the number of GTP updates. This procedure imposes a significant signaling overhead on EPC. An additional drawback of using GTP at the UDP/IP layer is the header overhead of 36 bytes for IPv4 packets and 56 bytes for IPv6 packets. This overhead can be as high as 14% and 22%, for IPv4 and IPv6, respectively [2].

To overcome these challenges, we propose an alternative architecture in this chapter. The primary feature of our proposed architecture is to deploy distributed mobility management (DMM), instead of tracking mobility in a centralized way. This idea can be augmented by eradicating the notion of tunnels in the user plane and ensuring minimum participation of EPC in mobility management. Software-defined networking (SDN) plays a crucial role in this case by realizing the DMM scheme as well as separating the control plane from the user plane.

Network function virtualization (NFV) adds significant flexibilities in this architecture. Some of the important advantages are as follows:

- *Multitenancy*: It allows multiple operators to share the same physical infrastructure.
- *Easier Network Operation and Management*: Control functions are implemented in the cloud instead of using vendor-specific proprietary hardware. As a result, addition or deletion of features can be less complicated. Similarly, adding or removing resources can also be done on demand.
- *Cost Optimization*: Network virtualization allows to implement network functions in the cloud, which makes use of standard switches, routers, and servers instead of vendor-specific hardware. Operators have the opportunity to grow their network infrastructure as they need, instead of having all their resources provisioned in a predetermined fashion.

However, NFV imposes some challenges as well [3], especially in terms of security, computational performance, assuring carrier grade service, portability etc., which needs to be considered during any practical deployment.

Replacing vendor-specific gateways, switches, and routers by SDN-capable switches, from both core and backhaul networks, and placing the control plane functions of core network in the cloud transform the traditional network into a stateless and flat network. This essentially allows the architecture to accommodate a broader range of services, especially WLAN users from fixed network. This is particularly achievable as the control functions specific to the fixed broadband network can be placed in the cloud, as a separate module, and packet forwarding in the user plane is agnostic of the control states.

Notice that the proposed approach can coexist with GTP control (GTP-C) tunnels, thus allowing all the GTP-C functionalities to remain available and compliant with the 3GPP standard. Another desirable goal is to contain the number of changes required by this approach in evolved packet system to a minimum, thus simplifying its practical deployment.

Some works have already addressed these issues individually and come up with various solutions. References [2] proposes an architecture which leverages an enhanced Ethernet technology, referred to as transparent interconnection of lots of links (TRILL), and distributed hash table (DHT). However, in the TRILL-based approach, some Ethernet broadcasting functionalities are still required and can lead to undesirable signaling overhead. Another approach is presented in [4], where SDN/OpenFlow is applied to achieve a semidistributed mobility management scheme. The proposed method assigns additional responsibilities to the SDN controller, e.g., the handling of 3GPP control packets, differentiating between them, and taking actions accordingly. However, this solution affects both user (GTP-U) and control (GTP-C) planes. In this case, the mobility management entity (MME) must be aware of service VLAN (S-VLAN) assignment. As a result, the solution in [4] requires some changes in the 3GPP standard procedures as well. The solution in [5] makes use of a flat architecture and a virtualized core network. Traffic in the user plane is routed through border gateway protocol (BGP). However, this method does not address some scalability issues. The authors of this chapter have also done some work, which introduces DMM in mobile backhaul network, using SDN [6]. The method proposed in [6] eradicates tunnels from the user plane of backhaul network. Another work [7] from the authors of this chapter extends the DMM solution from the backhaul to the EPC.

The objective of this chapter is to describe an end-to-end DMM scheme. We start by describing a generic network architecture, which is followed by a description of the packet flow in both directions, i.e., uplink and downlink. We elaborately explain the mobility events and how the proposed scheme can be scaled up to support a very large network. This chapter also contains the validation of the claims made by the authors. An analytical model is presented to demonstrate the improvement in the signaling load on the EPC. Furthermore, experimental results are shown to prove the viability of the proposed architecture in terms of the handover performance. Finally, this chapter discusses how the architecture can be extended to support fixed wireless local area network (WLAN) users.

6.2 Software-Defined Distributed Mobility Management

This section describes the proposed SDN- and NFV-based distributed mobility management scheme. At first, this section describes how the proposed scheme is deployed in the backhaul network, which is followed by a description of how DMM is realized in the EPC.

6.2.1 Distributed Mobility Management in Mobile Backhaul Network

The mobile backhaul network is based on carrier Ethernet technology. The design eliminates the GTP tunnels from the user plane, and in order to get the location information, it uses the eNodeB medium access control (MAC) address. In conjunction with UE IP address, the eNodeB MAC address gives complete information about the user. This information is disseminated to other network entities by OpenFlow. The IEEE 802.1ad standard of virtual local area network (VLAN), commonly known as Q-in-Q, is used to differentiate between different eNodeB groups and differentiate between different user traffic types. IEEE 802.1ad provides an outer VLAN tag, referred to as S-VLAN, and an inner VLAN tag, referred to as C-VLAN.

The reference architecture is based on a layer 2 backhaul network. As Figure 6.1 shows, a group of eNodeBs is connected to an Ethernet switch, which is referred to as egress switch. An S-VLAN ID is assigned to the group of eNodeBs, which belong to the same egress switch. At the other end of the backhaul network, the switches that are adjacent to serving gateway(s) (S-GW) are referred to as ingress switches. Ingress and egress switches are connected via a network of switches to form a connected graph. Switches are controlled by one or more OpenFlow controllers. One S-VLAN tag is assigned to create one or more VLAN paths to offer layer 2 transport capabilities between every pair of egress switch and ingress switch. The route for each S-VLAN path can be computed accounting for predefined cost functions, e.g., shortest path (latency, hop count), load balancing, and link disjoint. The actual computation of such routes is performed by the SDN controller, taking into account the aforementioned parameters. Interested readers can find further details of the packet flow in backhaul network in [6].

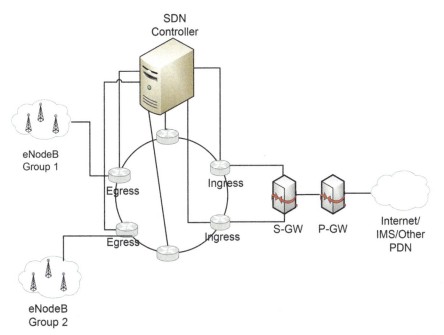

Figure 6.1 Network architecture with distributed mobility management in backhaul.

6.2.2 Virtualized Core Network Architecture

The solution described in Section 6.2.1 is further extended to develop an end-to-end system. In the data plane of the proposed architecture, conventional EPC gateways are replaced by the SDN capable switches. The control entities are moved to the northbound interface of the SDN controller; i.e., control functions of the nodes such as S-GW, packet data network gateway (P-GW), and MME are implemented as virtualized applications, which communicate with the SDN controller using the northbound API. The controller communicates with the switches using the southbound API, e.g., OpenFlow. In this chapter, we assume that the control plane remains intact, i.e., GTP-C tunnels are established between eNodeBs and S-GW, and between S-GW and P-GW. All the signaling procedures are identical to the 3GPP-defined standard. The end-to-end reference architecture is depicted in Figure 6.2.

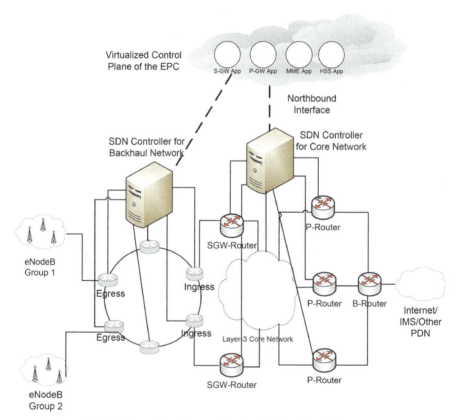

Figure 6.2   End-to-end architecture for the distributed mobility management.

6.2.3 Path Establishment and Packet Flow in the Core Network

We assume a layer-3 network in the core network. Recall that the routing in traditional IP network is a distributed process, where participating nodes exchange control messages among themselves to build routing tables. Since we are using controller-based approach, it is evident that the controller possesses the global knowledge of the entire network. As a result, the route computation is more efficient, and it can accommodate more sophistication.

In the data plane of the core network, S-GWs are implemented using OpenFlow-enabled switches. We assume that for a reasonably large network, there can be two or more S-GW routers geographically dispersed. A border router (B-Router) acts as point of access for the entire network, possibly being seen as an autonomous system (AS). Any data traffic leaving or entering the AS must go through the B-Router. A group of routers in the AS that neighbor the B-Router are referred to as penultimate routers (P-Router). A number of routers constitute the core network interconnecting S-GWs and P-Routers.

For ease of description, we assume that a distinct controller is deployed to control the core network entities, rather than using the same controller already defined for the backhaul network. These two controllers (the one for the core and the one for the backhaul) communicate with one another directly via customized interfaces [8].

The events concerning an LTE session are described in the following section, where the signaling and the data packet flow in both upstream and downstream directions are discussed.

6.2.3.1 Initial attachment and session establishment

When a UE joins the network, it exchanges several signaling messages with the mobile network as defined in the 3GPP standard. Initial attachment procedures are handled by the virtualized instances of the control plane in the cloud. Ultimately, the UE receives the IP address as a *protocol configuration option* (PCO).

6.2.3.2 Uplink (UE → PDN) packet flow

As the design of our previous work [6] dictates, a user packet from the UE, destined toward a packet data network (PDN), reaches the ingress switch according to any operator-defined criteria, e.g., the shortest path between the ingress and egress switches, and congestion along the paths. The ingress switch forwards the packet to its designated S-GW.

We divide the entire UE IP range to form a number of blocks that matches the number of P-Routers. Each UE IP block is exclusively assigned to one P-Router, i.e., each P-Router only handles the traffic from its designated UE IP block. As the IP forwarding follows the longest prefix mechanism, we can utilize the longer prefixes to handle the node failure issue. However, by looking into the source (UE) IP address field of the packet destined to a PDN, S-GW router forwards the packet to the designated P-Router. Before forwarding it to the P-Router, the controller notes down the S-GW router interface IP address in a table. The controller also notes down the in-port ID of the S-GW router (or the MAC address), through which it (S-GW router) receives the packet from the ingress switch. These two pieces of information will be used to route packets in the reverse direction. A default route to the B-Router is configured in the P-Router(s), and the B-Router eventually forwards the packet beyond the operator domain. Packet forwarding in the uplink direction is summarized in Table 6.1.

6.2.3.3 Downlink (PDN → UE) packet flow

A packet destined for the UE reaches the B-Router. The B-Router looks into the destination (UE) IP address and forwards it to the designated P-Router. This is relatively straightforward since each P-Router is statically configured to handle a certain block of UE IP addresses. The P-Router, upon consulting the SDN controller, learns about the S-GW IP address, to which it should route the packet. As the S-GW router receives the packet, it forwards the packet to the ingress switch interface, through which it receives the packet during the uplink transmission. Packet forwarding in downlink direction is summarized in Table 6.2.

6.2.3.4 Some key points of design

There are some key points to mention in this architecture.

- The B-Router does not require to be controlled by the SDN controller.
- The intermediate routers between the P-Routers and the S-GW routers do not need to be controlled by the SDN controller.

Table 6.1   Uplink packet forwarding at different segments of the network

Backhaul Network (Egress → Ingress)	Ingress → SGW-Router	SGW-Router → P-Router	P-Router → B-Router
VLAN based forwarding	Directly connected (default path)	Standard IP routing, considering the UE IP address	Default Route

Table 6.2 Downlink packet forwarding at different segments of the network

B-Router → P-Router	P-Router → SGW-Router	SGW-Router → Ingress	Backhaul Network (Ingress → Egress)
Standard IP routing, considering the UE IP address	Standard IP routing	Based on the stored information at the mobility table in the controller for the core network	VLAN-based forwarding

- B-Router contains only N number of specific routes, where N is the number of the P-Routers. However, the actual number of routes will vary, since we have to configure several other routes with longer prefixes, to handle the node failure scenarios.
- The theoretical number of states in any S-GW router is the number of subscribers (UE) connected through that particular S-GW router, added with the number of P-Routers, i.e., the number of routes to reach the P-Routers.
- For simplicity, we assume a single controller-based network. However, multiple controllers can be deployed to make this solution scalable. Reference [9] discusses about such a distributed controller scheme.

6.2.3.5 Mobility events

After a successful handover, when the UE moves to a different eNodeB, the SDN controller for the backhaul networks informs the SDN controller for the EPC, using intercontroller communication messages [8]. The controller for the core network holds a customized table, for keeping track of UE mobility. The table fields are presented in Table 6.3. Mobility scenarios when DMM in backhaul has been implemented, with possible transient events, have been discussed in [6]. A similar signaling sequence for mobility events is presented in Figure 6.3. It shows transient events that can possibly occur during a handover.

6.2.3.6 Multiple border router scenario

So far, we have considered only the scenario with one B-Router. However, in a countrywide large network, it is possible to have multiple B-Routers. We propose to deploy IP anycast to address such scenario.

Table 6.3 Mobility table stored in the SDN controller for the core network

UE IP address	S-GW IP address	S-GW in-port

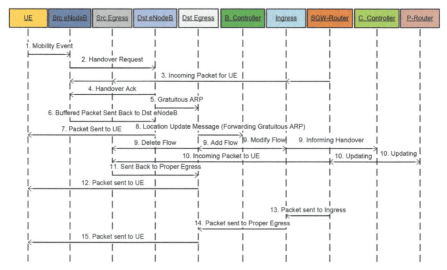

Figure 6.3   Mobility diagram for downlink direction.

As described in the previous section, each of the P-Routers is designated for a particular block of UE IP address range. There are several sets of P-Routers, each for a particular B-Router. One P-Router from each of these sets, handling the same UE IP blocks, forms an anycast group, which essentially belongs to the same anycast IP address. Whenever a user moves to such a location, which is closer to a new B-Router, rather than the existing one, it starts to forward traffic toward that B-Router and P-Routers. In such a way, a seamless mobility can be ensured within the operator domain.

The idea of mobility between multiple border routers is explained with an example. The scenario is depicted in Figure 6.4. For simplicity, we assume IPv4 addresses. This example holds true for IPv6 as well. In Figure 6.4, we can see two B-Routers. Each of the B-Routers has several designated P-Routers. In this simple example, we assume that the B-Router1 announces the IP prefix 192.168.1.0/24; on the other hand, the B-Router2 announces the prefix 192.168.2.0/24. A UE initially connects to the network, from a location closer to the B-Router1, and hence gets an IP address, 192.168.1.10; as we described earlier, P-Router1 and P-Router1' both hold the same anycast address, and both are designated to handle the 192.168.1.0/24 block. Since P-Router1 is geographically closer, it receives the traffic and forwards it to the B-Router1. As the B-Router1 is announcing the subnet prefix of this UE, the downlink traffic from the Internet will follow the same path in the reverse direction.

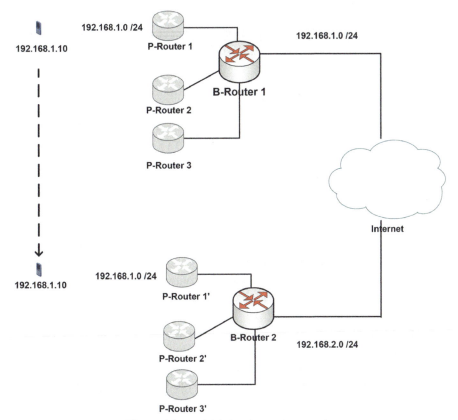

Figure 6.4 Multiple border router scenario.

Let us assume that the user moves to such a location, where P-Router1′ is closer to the UE, rather than P-Router1. This eventually means that B-Router2 will receive the traffic in uplink and forward it to the PDN. The B-Routers can be configured in such a way that it can forward the traffic from any IP block. However, the scenario is little tricky for downlink. All B-Routers cannot be configured to provide the best path for all the subnets. Hence, the downlink traffic would still arrive at B-Router1, and B-Router1 forwards that traffic to P-Router1. P-Router1 will consult the SDN controller and forward the traffic to the proper router. This suboptimal routing can add some additional latency to the user packets. However, this is not a usual scenario, and this asymmetric path does not impose any challenge for BGP.

6.2.3.7 IP address assignment considerations

There is no strict requirement of having any type of customization in IP address assignment. However, few points are suggested, which should enhance the performance.

- A spatial or zonal distribution of IP addresses is important. In the case of multiple B-Routers, it is evident that any one of the B-Routers cannot offer the best path for all the subnets. In other words, each of the B-Routers provides the best path to a particular number of IP subnets. Hence, a user location agnostic IP assignment can result in suboptimal routing of the packet. We can avoid this by adding some sort of location information in the IP address assignment application. This can be achieved in several ways. If we assume a geographically distributed SDN controller system, each controller can be assigned a particular set of subnets or IP address blocks. The B-Router corresponding to a particular controller should provide the best route for those subnets. However, if a user mobility results in a change in the assigned B-Router, the user traffic is routed through a suboptimal path. Such a scenario has been described in Section 2.3.6.
- The method we propose for the load balancing, by introducing the P-Routers, assumes a static/fixed configuration. It is also assumed that IP addresses will be equally divided between the P-Routers to distribute the load. To yield a better performance, a more dynamic approach can be adopted. Consider a hypothetical scenario, where the address assignment is agnostic of the network configuration. Since the user behavior and mobility pattern are largely dynamic, it is possible to face a situation where any particular P-Router is handling significantly higher traffic than the other P-Routers and eventually becomes overloaded. If the address assignment application has the intelligence to assign addresses considering the load-balancing problem, this particular scenario can be avoided. Every time a user requests for an IP address, the application should try to assign an address in round robin fashion, which balances the load among different P-Routers. While it may sound unrealistic in existing network infrastructure, thanks to the SDN paradigm, this kind of programmability is practically possible to attain.

6.3 Analytical Modeling of Signaling Load on EPC

This section provides a comparison between existing architecture and our proposed architecture, in terms of control message load on EPC. Signaling overheads can be very expensive for EPC gateways. A mobility management system that imposes less signaling traffic is always desired. Hence, it is imperative to compare the signaling load of the proposed system with the typically deployed system. A simple analytical model to compute signaling load on EPC is described in [4]. The model is modified to fit the proposed architecture, and the performances of different schemes are compared.

This model assumes a circular shaped cell and approximates the user mobility using Gauss–Markov mobility model [10]. It also assumes that the time is slotted. Average load on LTE system, L_I (packets/second), is defined as the addition of two components: packets sent to the EPC due to initial attachment procedure, L_{IA}, and packets sent to the EPC due to handover, L_{HO},

$$L_I = L_{IA} + L_{HO} \tag{6.1}$$

Let N be the number of the UEs, and the average time slots a UE stays in a cell before handover are denoted by k. The time interval between initial attachment (IA) and the network detachment (and vice versa) is modeled as an exponential random variable, with parameter λ. As described in [4],

$$L_I = L_{IA} + L_{HO} = \left(\lambda + \frac{1}{2k}\right) N \tag{6.2}$$

k can be determined from the Gauss–Markov model [11]. According to this model, speed and direction of the UE at n-th time slot is modeled respectively as follows:

$$S_n = \alpha S_{n-1} + (1 - \alpha) \mu + \sqrt{1 - \alpha^2} S_{x_{n-1}} \tag{6.3}$$

$$\theta_n = \alpha \theta_{n-1} + (1 - \alpha) \theta + \sqrt{1 - \alpha^2} \theta_{x_{n-1}} \tag{6.4}$$

α indicates the memory ($0 \leq \alpha \leq 1$), S_n and θ_n are speed and direction of the UE at n-th slot, and μ and θ are mean speed and direction of the UE. $S_{x_{n-1}}$ and $\theta_{x_{n-1}}$ are two independent and identically distributed (i.i.d) Gaussian random variables with means, μ and θ, respectively, and standard deviations, σ_μ and σ_θ, respectively. The speed can be expressed as follows [10]:

$$S_n = \alpha^n S_0 + (1 - \alpha^n) \mu + \sqrt{1 - \alpha^2} \sum_{i=0}^{n-1} S_{x_{n-1}} \alpha^{(n-i-1)} \tag{6.5}$$

where S_0 is the initial speed distribution. An approximation for speed and direction at n-th instant is given by:

$$S_n \sim N \left[\mu + \sqrt{\frac{1 - \alpha}{1 + \alpha}} (1 - \alpha^n) \mu, \frac{\alpha^{2n} \alpha_{\mu 0}^2}{3} + (1 - \alpha^{2n}) \sigma_\mu^2 \right] \tag{6.6}$$

$$\theta_n \sim N\left[\theta + \sqrt{\frac{1-\alpha}{1+\alpha}}\left(1-\alpha^n\right)\theta, \frac{\alpha^{2n}\alpha_{\theta_0}^2}{3} + \left(1-\alpha^{2n}\right)\sigma_\theta^2\right] \qquad (6.7)$$

If the UE is inside a given cell at the time slot, n, then p indicates the probability that the UE will remain in the same cell at the next time slot, $n+1$. As a result, the number of time slots an UE stays in the same cell can be modeled as a geometric random variable, with the parameter $(1-p)$. However, p can be defined as follows [4]:

$$p = \frac{4}{\pi r^2} \int_0^r \int_0^{\sqrt{r^2-x^2}} \int_{-\pi}^{\pi} P(S_n \le d) f_{\theta_n}(\phi)\, d\phi\, dy\, dx \qquad (6.8)$$

where r denotes the radius of the cell and d is the distance from any point in the circle to a point on the circle, which makes an angle ϕ with the x-axis. If we consider that *(x, y)* is the point in the circle, then d is defined as follows:

$$d = x\cos\phi + y\sin\phi + \sqrt{r^2 - (x\sin\phi - y\cos\phi)^2} \qquad (6.9)$$

Since the number of time slots the UE remains in the same cell before handover is distributed as a geometric random variable with parameter $(1-p)$, the average duration of staying in the same cell, k, is $1/(1-p)$.

In [6], where SDN-based distributed mobility management is limited in mobile backhaul network, it is required to inform the core network about a handover, only when the handover involves S-GW relocation. By modifying Equation (6.2), we can yield the signaling load on the EPC, L_{II}, as per our proposed scheme.

$$L_{II} = L_{IA} + L_{HO} = \left(\lambda + \frac{P_{SGW-Reloc}}{2k}\right)N \qquad (6.10)$$

$P_{SGW-Reloc}$ denotes the probability of S-GW relocation, given that a handover has taken place. However, for the work presented in this chapter, mobility management is entirely distributed in backhaul and EPC. As a result, Equation (6.10) is modified as follows:

$$L_{III} = L_{IA} + L_{HO} = \lambda N \qquad (6.11)$$

L_{III} denotes the signaling load on EPC for our proposed scheme. In Figure 6.5, comparison between different mobility management schemes have been presented. This analytical model is evaluated with the cell radius of 10 *km* and the user velocity of 4 *m/s*. It clearly shows that the proposed method outperforms both typical LTE system and our previous work where DMM is realized only in the backhaul.

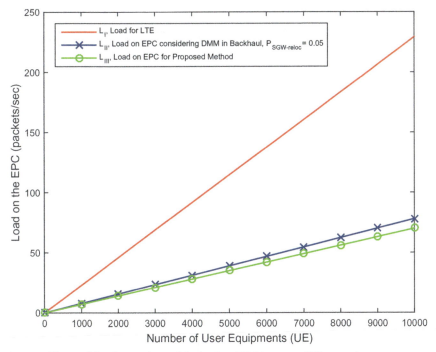

Figure 6.5 Comparison of the load on EPC between different schemes.

6.4 Experiments and Results

6.4.1 Test Bed Description

This section describes the details of the experimental setup that is used to collect the presented results. This experiment aims to measure the handover performance with distributed mobility management deployed in the backhaul network [6].

The test bed developed for this research consists of 1) four physical hosts running Open vSwitch, 2) a PC running the floodlight SDN controller, and 3) two hosts, one acting as UE and another as destination node. These results are also presented in [6].

Figure 6.6 shows the test bed layout. The test bed consists of two isolated networks: a data network and a management network. Data network links are solid lines in the figure, while management network links are dotted lines in the figure. The management network is used to connect (SSH) to client H1 and client H2, as well as OpenFlow traffic between the switches and the controller.

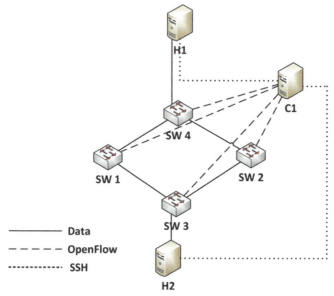

Figure 6.6   Test bed layout.

The test bed is used to emulate the behavior of the real network. Switch SW1 represents eNodeB group 1, while switch SW2 represents eNodeB group2. Switch SW3 represents the S-GW. Client H1 acts like a mobile user (UE). Mobility is emulated by having switch SW4 to send/receive traffic to either SW1 or SW2. Client H2 emulates a destination server in the core network or beyond. C1 is the SDN controller. The link between SW1 and SW2 is used to emulate the X2 interface tunnel.

Readers should keep in mind that this test bed represents a simplified LTE network. However, in an operational network, several additional nodes are present, and the obtained numbers may differ with the one yielded through the experiment.

6.4.2 Experiment Setup

The experiment is intended to present handover performance. Different transient situations may arise due to user mobility. As a result, it is important to ensure the handover performance of a mobility management system. There are several indexes to measure the handover performance. In this work, packet loss (for UDP traffic) and total transmission time (for TCP traffic) are selected. Similar experiments for WiFi and Wi-MAX networks are presented in [12].

Conventional hard handover is emulated in this experiment for two cases, namely 1) without an X2 interface tunnel and 2) with an X2 interface tunnel.

The first case is one where there is no X2 interface tunnel. Figure 6.7 illustrates the sequence of network states and directions of each flow in the system. The starting state (left side) is referred to as VLAN X state. As the UE moves, uplink traffic uses SW2, while downlink traffic is lost (center). After a delay D_1, the network updates the UE position (right side), and the state is referred to as VLAN Y state.

The second case, where there is an X2 interface tunnel, uses a similar approach. Figure 6.8 illustrates the sequence of network states. The only

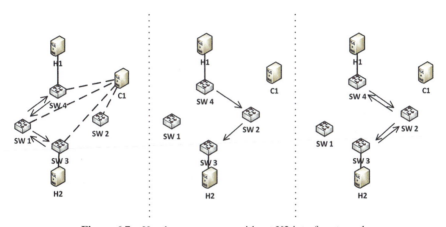

Figure 6.7 Handover sequence without X2 interface tunnel.

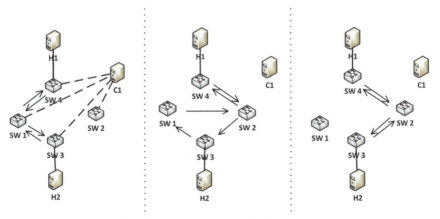

Figure 6.8 Handover sequence with X2 interface tunnel.

difference is—during handover when the UE moves to SW2—the downlink traffic is redirected from SW2 to SW1, using the X2 interface tunnel.

This tunnel is not available immediately. Instead, it becomes available for downlink traffic after a delay D_2. After a delay $D_1 > D_2$, the network updates the UE position (right side). The experiment consists of a sequence of cycles: Each cycle has a period of 10 seconds, followed by a transition from VLAN X state to VLAN Y state (or vice versa).

6.4.3 Results

This section reports the experimental results obtained with a number of scenarios for both the cases with and without the X2 interface tunnels. Results are obtained using two different types of traffic: UDP traffic and TCP traffic. In order to obtain network performance indicators under a variety of conditions, the values for D_1 and D_2 are arbitrarily changed in the experiments.

6.4.3.1 UDP traffic

UDP is a unidirectional protocol. As a result, the uplink and downlink traffic are tested independently. UDP traffic is generated using NetPerf [13].

Figure 6.9(a) and (b) reports the obtained results for uplink and downlink UDP traffic, respectively, without the X2 interface tunnel. Figure 6.9(a) (uplink traffic) shows the baseline loss of the system, while Figure 6.9(b) (downlink traffic) shows the packet loss that grows proportionally to the delay time D_1. This is as expected, as the UE can always connect to the server, no matter which eNodeB it is on. Figure 6.9(c) shows the results obtained with downlink UDP traffic using the X2 interface tunnel and $D_2 = 0$ s. As expected, the packet loss stays relatively constant, averaging close to 1%.

6.4.3.2 TCP traffic

Unlike UDP, TCP is a bidirectional protocol, which offers better reliability with the cost of additional transmission time used to retransmit the lost segments. Thus, the total transmission time is used as a metric to evaluate the performance of the proposed solution. The D-ITG (distributed internet traffic generator) [14] tool is used to generate TCP traffic. Tests are run to compare the trend of the transmission duration in the system with and without X2 interface tunnels. Two sets of data were collected for each experiment. The value for delay D_1 is varied in the interval $[0 \ s, 8 \ s]$ in $1 \ s$ steps. Three values for delay D_2 are considered: $D_2 = 0 \ s$, $D_2 = 0.25 \ s$, and $D_2 = 0.5 \ s$.

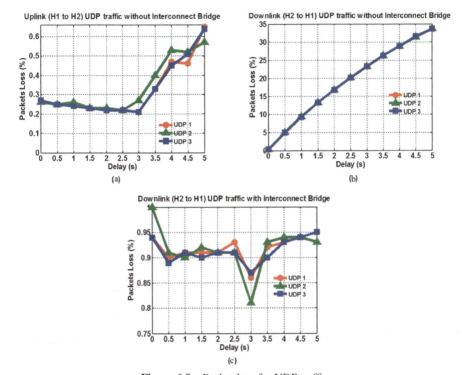

Figure 6.9 Packet loss for UDP traffic.

Figure 6.10(a) and (b) shows obtained results. Both figures show a staircase type of behavior. This is due to a combined effect of the periodic delay D_1 and an increasing TCP wait time caused by packet losses during D_1. During D_1, no ACK packets make it back to the sender until the X2 interface tunnel is activated. As D_1 is incremented, the delay in receiving ACK packets increases, and after a certain threshold, TCP extends its wait time. In Figure 6.10(b), for D_1 between 3 and 6 seconds, the TCP wait time is consistent because it is longer than the total delay time. Once the total delay time exceeds the TCP wait time, TCP reacts by further extending its wait time, which results in the increase between 6 and 7 seconds. This process is expected to continue generating the staircase-shaped plot, when D_1 keeps increasing beyond 8 s until a maximum threshold is received and the TCP connection is reset. Figure 6.10(b) shows that the total transmission time is significantly reduced when the X2 interface tunnel is used. Figure 6.10(c) and (d) reports results obtained when $D_2 = 0.25\ s$ and $D_2 = 0.5\ s$, respectively. Curves in the figures show that the

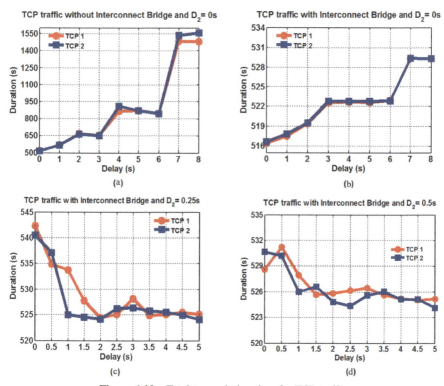

Figure 6.10   Total transmission time for TCP traffic.

duration of the data transfer is independent from D_1 even when there is a secondary delay before the X2 interface tunnel becomes available.

6.5 Extending the Design to Support Fixed WLAN Users

This section extends the proposed solution to support WLAN users from fixed network as well. This integration leads to a unified architecture, which allows an operator to support both cellular and fixed (WLAN) users through common infrastructure.

6.5.1 Network Architecture

We assume that the fixed network possesses a separate access network. This assumption is originated from a practical consideration that an operator may already have an existing fixed access network. The fixed network user traffic is aggregated in the core network. This design essentially implies that the architecture of the customer site and the access network remains identical with the architecture described in TR-101 [15].

The reference architecture from [15] is presented in Figure 6.11 for reader's convenience. We assume that the residential gateway (RG) works in switch mode, i.e., the IP address is assigned by the BNG, directly to the client device. We are using the *option g* described for the U interface and the *option c* for the V interface, as described in [15]. We further assume that IPv6 is used for addressing, which makes this solution scalable.

In the fixed network architecture, the broadband network gateway (BNG) is placed in the location analogous to the location of the S-GW in mobile networks. We envision the unified network in such a way that the traffic from the fixed access network merges with the traffic from the mobile backhaul network in the S-GW router. Since we are handling both mobile network and fixed network, we rename the S-GW router as unified gateway router (UG-Router). The control functions of the BNG are virtualized in cloud, which communicates with the SDN controller through northbound API. As we are using the same cloud entity to control both mobile network and fixed network, common functions like IP address assignment can be implemented only once, and thus, consumption of resources is optimized (Figure 6.12).

Using the same core infrastructure for both mobile network and fixed network gives us the advantage of statistical multiplexing. If required, traffic can be segregated in the core network using network virtualization techniques. A more compact and economic architecture is also possible, which increases the efficiency of the statistical multiplexing, by using a unified access/backhaul

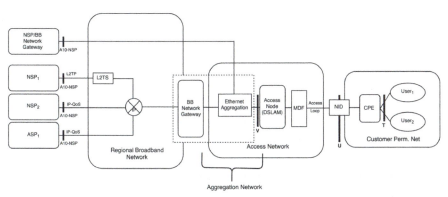

Figure 6.11 Reference architecture from TR-101 [15].

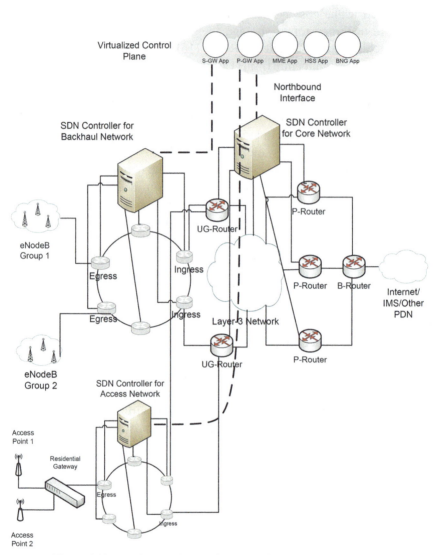

Figure 6.12   Unified architecture for supporting WLAN user mobility.

network. However, this scenario may not be always feasible, and therefore, the choice of implementation depends solely on the operators.

6.5.1.1 Uplink (UE → PDN) packet flow

The packet transmission in both directions is similar to what we proposed in the previous section. In open system authentication (OSA), when a user

(user equipment (UE)) initiates a session, it exchanges two pairs of messages with the WLAN access point (AP). The first pair is used for authentication, while the second pair indicates the association process. Depending on the security schemes deployed, additional messages may be exchanged between the UE and the WLAN AP. However, as soon as the association process is completed, the AP transmits a special message to the RG, which is forwarded by the RG to the access network, and the access network sends the packet to the SDN controller assigned for the access network. This special message does not have any hard and fast format. For example, gratuitous ARP can be used for this purpose, which is similar to the approach described in [6]. The main idea is to notify the SDN controller about the UE attachment to the AP. The SDN controller notes down the RG MAC address.

This MAC address acts as an identifier for the UE. As the egress switch gets the packet, it will choose an ingress switch based on any operator-defined criteria, e.g., shortest path and path congestion. From ingress switch, the packet is forwarded to the core network through the UG-Router. Packet routing in the core network is described in Section 6.2.3.

6.5.1.2 Downlink (PDN → UE) packet flow

A packet in downlink direction is forwarded to the UG-Router from the PDN/Internet according to the procedure mentioned in Section 2.3.3. The UG-Router forwards the packet to the ingress switch, through which it received the packet during uplink transmission. As the ingress switch receives the packet, it pushes the packet to the designated egress switch, suggested by the SDN controller. From the egress switch, the packet is forwarded to the intended RG interface, identified by the MAC address.

6.5.2 Handling Mobility

In typical WLAN-supported networks, we can observe two types of mobility—the layer 2 mobility and the layer 3 mobility. The layer 2 mobility, as the name indicates, is the agnostic of the layer 3 information. This kind of mobility can be achieved without any intervention from the core network. However, in our proposed architecture, IP address is assigned from a centralized point, and addresses are not attached to any particular BNG. Any address can belong to any part of the network. As a result, we can handle the mobility from layer 2 information. Similar to our proposition for the mobile networks, this solution does not deploy any tunnel in the user plane. We track the mobility by using

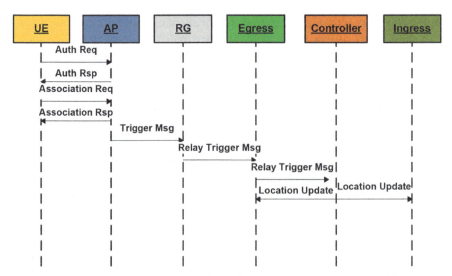

Figure 6.13   Mobility diagram of a unified network for downstream traffic.

RG MAC address as an identifier. User mobility may result in a change in the RG interface, egress interface, ingress interface, or BNG interface. Such a change is reflected in the mobility table of the SDN controller for the access network.

As the user moves between the APs (which do not necessarily belong to the same RG), it generates authentication and reassociation messages once the handover is completed. This association/reassociation process triggers a special message to the SDN controller, to update the RG MAC address, and other necessary information, e.g., ingress and egress interface addresses (Figure 6.13). The intermediate or transient packets that arrive the previous ingress/egress switch or UG router can be stored in the buffer and forwarded to the corresponding new entity by using a bidirectional tunnel, as described in RFC 5949 [16]. As mentioned in [16], this tunnel can be predictive or reactive and contributes significantly to reduce the packet loss, while a handover is taking place.

6.6 Conclusion

This chapter addresses several aspects of cellular networks. Firstly, by eliminating tunnels from the user plane, this method saves overhead in terms of packet size, and it reduces the signaling overhead in EPC. Secondly, the computation-intensive task of managing mobility is delegated to a number of controlling nodes in the network, and in this way, the method helps to avoid centralized operation. Thirdly, by means of pursuing the path of network virtualization, this method makes the network management and operation inherently flexible. In addition to all these, SDN technologies ensure a less expensive core network. Last but not least, the proposed solution encompasses a broader range of technologies, as it also integrates WLAN users from fixed network in the same unified architecture.

Ease of implementation has been given priority while designing the architecture. As a result, the proposed architecture requires a limited number of changes in the existing signaling standards.

References

[1] "Cisco Visual Networking Index: Global Mobile Data Traffic Forecast Update, 2014–2019, White paper," 2013.

[2] Varis, N., et al. (2011). A layer-2 approach for mobility and transport in the mobile backhaul, in *11th International Conference on ITS Telecommunications (ITST)*,

[3] Hawilo, H., et al. (2014). NFV: state of the art, challenges, and implementation in next generation mobile networks (vEPC). *IEEE Network*, 28.

[4] Gurusanthosh, P., et al. (2013). SDMA: A semi-distributed mobility anchoring in LTE networks, in *International Conference on Selected Topics in Mobile and Wireless Networking (MoWNeT)*.

[5] Matsushima, S. and Wakikawa, R. (2014). Stateless user-plane architecture for virtualized EPC (vEPC). *Work in Progress, draft-matsushima-stateless-uplane-vepc-03*.

[6] Banik, J., et al. (2015). A Software Defined Semi Distributed Mobility Management System Based on Layer 2 Backhaul Network, in *IEEE 82nd Vehicular Technology Conference (VTC Fall)*.

[7] Banik, J., Tacca, M., Fumagalli, A., Sarikaya, B., and Xue, L. (2015). Enabling Distributed Mobility Management: A Unified Wireless Network Architecture Based on Virtualized Core Network, in *2015 24th International Conference on Computer Communication and Networks (ICCCN)*.

[8] Yin, H., et al. (2012). SDNi: A message exchange protocol for software defined networks (SDNs) across multiple domains. *Internet Engineering Task Force, Internet Draft*.

[9] Yazici, V., et al. (2014). Controlling a software-defined network via distributed controllers. *arXiv preprint arXiv:1401.7651.*

[10] Liang, B. and Haas, Z. (2003). Predictive distance-based mobility management for multidimensional PCS networks. *IEEE/ACM Transactions on Networking*, 11(5).

[11] Camp, T., et al. (2002). A survey of mobility models for ad hoc network research. *Wireless Communications and Mobile Computing*, 2(5).

[12] Yap, K.-K., et al. (2010). Blueprint for introducing innovation into wireless mobile networks, in *Proceedings of the Second ACM SIGCOMM Workshop on Virtualized Infrastructure Systems and Architectures.*

[13] Jones, R., et al. (1996). NetPerf: A network performance benchmark. *Information Networks Division, Hewlett-Packard Company.*

[14] Avallone, S., et al. (2004). D-ITG distributed Internet traffic generator, in *Proceedings. First International Conference on the Quantitative Evaluation of Systems, 2004. QEST 2004.*

[15] Anschutz, T. (2011). Migration to Ethernet-Based Broadband Aggregation Broadband Forum TR-101. Issue 2.

[16] Yokota, H., et al. (2010). Fast handovers for proxy mobile IPv6. *IETF Draft (March 2009)*, 20.

7

Improving the Effectiveness of Data Transfers in Mobile Computing Using Lossless Compression Utilities

Armen Dzhagaryan[1], Aleksandar Milenković[1]
and Martin Burtscher[2]

[1]Department of Electrical and Computer Engineering, The University of Alabama in Huntsville, Huntsville, AL 35899, USA
[2]Department of Computer Science, Texas State University, San Marcos, TX 78666, USA

Abstract

The data traffic originating on mobile computing devices has been growing exponentially over the last several years. Lossless data compression can increase communication throughput, reduce latency, save energy, and increase available storage. However, compression introduces additional overhead that may exceed any gains due to transferring or storing fewer bytes. Compression utilities on mobile computing platforms differ in compression ratio, compression and decompression speeds, and energy requirements. When transferring data, we would like to have an agent to determine whether compressed transfers are beneficial, and if so, select the most beneficial compression utility. A first step toward designing such an agent is to obtain a good understanding of various parameters impacting data transfers. This chapter presents results from an experimental study that evaluates the effectiveness of several compression utilities on a modern smartphone in three typical usage scenarios that involve local and network transfers. We define several metrics that capture the effectiveness of uncompressed and compressed data transfers and provide practical guidelines for selecting the most effective utilities and configurations for each usage scenario. We also introduce an analytical framework for estimating the effectiveness of various compression utilities in transferring mobile data.[1]

Keywords: Mobile computing, Lossless Compression, Data compression, Performance Evaluation, Energy-aware systems, Frequency scaling, Performance, Energy Efficiency, Networks.

7.1 Introduction

Mobile computing devices such as smartphones, tablets, and e-readers have become the dominant platforms for consuming digital information. According to estimates for 2014 [1, 2], vendors shipped 1.2 billion smartphones, up 28.4% from the prior year, and 216 million tablets. Annual sales of smartphones exceeded those of feature phones for the first time in 2013 [3], totaling 1,807 million mobile devices [4]. The total number of mobile devices shipped in 2014 reached 1,839 million, with ~65% being smartphones. Global mobile data traffic continues to grow exponentially. A report from Cisco states that the global mobile data traffic grew 69% in 2014 relative to 2013, reaching 2.5 exabytes per month, which is over 30 times greater than the total Internet traffic in 2000 [5]. It is forecasted that the global mobile data traffic will grow nearly 10-fold from 2014 to 2019, reaching 24.3 exabytes per month. A report on global mobile economy by GSMA [6] states that the number of global SIM connections and the number of unique mobile users in 2014 were 7.3 billion and 3.6 billion, respectively. These two numbers are expected to reach 10 billion and 4.6 billion by 2020, respectively. The share of 3G/4G connections accounted for approximately 40% of the active connections by the end of 2014, and it is expected to reach 70% by 2020.

Data compression is critical in mobile data communication. It can help improve operating time, lower communication latencies, and make more effective use of available bandwidth and storage. The general goal of data compression is to reduce the number of bits needed to represent information. Data can be compressed in a lossless or lossy manner. Lossless compression means that the original data can be reproduced exactly by the decompressor. In contrast, lossy compression, which often results in much higher compression ratios, can only approximate the original data. This is typically acceptable if

---

[1]This material is based upon work supported in part by the National Science Foundation under Grants No. 1141022, 1205439, 1217231, 1217470, 1406304, and 1438963. Any opinions, findings, and conclusions or recommendations expressed in this material are those of the authors and do not necessarily reflect the views of the National Science Foundation.

the data are meant for human consumption such as audio and video. However, program code and input, medical data, e-mail, and other text generally do not tolerate lossy compression. We focus on lossless compression in this chapter.

Lossless data compression is currently being used to reduce the required bandwidth during file downloads and to speed up Web page loads in browsers. Google's Flywheel proxy [7], Google Chrome [8], Amazon Silk [9], and the mobile applications Onavo Extend [10] and Snappli [11] use proxy servers to provide HTTP compression for all pages during Web browsing. For file downloads, several Google services, such as Gmail and Drive, provide *zip* compression [12] of files and attachments [13]. Similarly, application stores such as Google Play and Apple's App Store use *zip* or *zip*-derived containers for application distribution. Several Linux distributions are also using common compression utilities such as *gzip*, *bzip2*, and *xz* for their software repositories. The importance of lossless compression in network data transfers has also been recognized in academia. This chapter extends our prior work [14] and complements earlier studies [15–19]. Unlike the prior studies, which were performed years ago when handheld computing devices were still in their infancy, we consider the most recent compression utilities including some with parallel implementations, our setup supports more accurate energy measurements, and we use a state-of-the-art smartphone with four processor cores and wireless as well as cellular communication interfaces.

The choice of the compression algorithm, the compression level, and the quality of the implementation affect the performance and energy consumption. Whereas the energy consumed for compression and decompression is not critical on desktop PCs and workstations, it can be a decisive factor in battery-powered mobile devices. Achieving a higher compression ratio requires more computation and, therefore, energy, but better compression reduces the number of bytes, thus saving energy when transmitting the data. Hence, we believe it is important to take a close look at both the performance and the energy efficiency of lossless compression algorithms on state-of-the-art mobile platforms. In particular, we want to provide answers to the following questions.

1. Can (de)compression reduce the latency and energy consumption of data transfers in mobile environments?
2. What common compression algorithms should be used?
3. Which configurations result in the best performance and energy efficiency?
4. Do parallel implementations of compression utilities outperform corresponding serial implementations?

5. How do the number of cores, processor frequency, and network through-put impact effectiveness of compression utilities?

In this chapter, we describe a comparative, measurement-based study of the most recent versions of several popular compression utilities, including *gzip*, *lzop*, *bzip2*, *xz*, *pigz* (a parallel implementation of *gzip*), and *pbzip2* (a parallel implementation of *bzip2*) on Google's Nexus 4 smartphone. For each utility, we analyze the effectiveness of all supported compression levels. We examine several performance metrics, including the compression ratio and the compression and decompression throughputs. Using our experimental setup for energy measurements, we study the amount of energy consumed by compression and decompression tasks and report the energy efficiency.

We evaluate the compression utilities in three typical usage scenarios. *LOCAL* involves compression and decompression tasks performed locally on the smartphone. *WLAN* and *CELL* involve compression tasks that stream data to and from a remote server over a secure communication channel. *WLAN* uses a wireless LAN interface, and *CELL* uses a mobile broadband network. Each experiment is conducted with three processor clock frequencies (1.512 GHz, 0.810 GHz, and 0.384 GHz) to examine the impact of frequency scaling on the performance and energy efficiency of data transfers.

The main findings from our experimental study are as follows:

- The throughput and energy efficiency of compression utilities vary widely across different utilities and compression levels, often spanning an order of magnitude. We identify combinations of utilities and compression levels that result in the best throughput and energy efficiency for typical usage scenarios.

- In the LOCAL experiments, *pigz* with compression level –1 and *lzop* –1 to –6 achieve the best compression throughput and energy efficiency. *lzop* achieves the best decompression throughput and efficiency.

- In the *WLAN* experiments, compressed uploads with *gzip* with a low compression level perform best, providing over twice more energy-efficient transfers than uncompressed uploads. For downloads, *pigz* with compression levels –6 to –9 achieves the best decompression through-put and energy efficiency, providing ~2.7 times more energy-efficient transfers than uncompressed downloads. We show that the effectiveness of individual compression utilities depends on the available network throughput.

- In the CELL experiments, compressed uploads with *bzip2* perform best, with up to ~2.7 times more energy-efficient transfers than uncompressed uploads. For decompression tasks, *xz* with the highest compression levels achieves the best decompression throughput and energy efficiency, providing up to 3 times more energy-efficient transfers than uncompressed downloads. Lowering the processor's clock frequency from 1.512 GHz to 0.384 GHz further improves the energy efficiency by up to 25% for compressed data downloads.

Based on the results of our experiments, we devised an analytical model for estimating the network throughput and energy efficiency of individual utilities in the WLAN and CELL scenarios. The model relies on parameters from the LOCAL experiments and parameters characterizing the network interface. This model is a first step toward developing an agent for selecting effective modalities in data transfers that originate on mobile computing platforms.

The rest of this chapter is organized as follows. Section 7.2 describes the operation of the six compression utilities we have studied. Section 7.3 presents the experimental setup, including the smartphone (Section 7.3.1), the measurement setup (Section 7.3.2), and the dataset used to obtain the results (Section 7.3.3). Section 7.4 explains the measured and derived metrics used (Section 7.4.1) as well as the experiments (Section 7.4.2). Section 7.5 discusses the results, including the compression ratio (Section 7.5.1), compression and decompression throughputs (Section 7.5.2), energy efficiency (Section 7.5.3), and overall insights (Section 7.5.4). Section 7.6 surveys related work. Section 7.7 summarizes our findings and draws conclusions.

7.2 Lossless Compression Utilities

Table 7.1 lists the six lossless compression utilities we have studied along with the supported range of compression levels. We chose the relatively fast *gzip* and the slower but better compressing *bzip2* because of their widespread use. *lzop* is included because of its high speed. *xz* is gaining ground and is known for

Table 7.1 Compression utilities

Utility	Compression Levels (Default Level)	Version	Notes
gzip	1 – 9 (6)	1.6	DEFLATE (Ziv-Lempel, Huffman)
lzop	1 – 9 (3)	1.03	LZO (Lempel-Ziv-Oberhumer)
bzip2	1 – 9 (9)	1.0.6	RLE+BWT+MTF+RLE+Huffman
xz	1 – 9 (6)	5.1.0a	LZMA2
pigz	1 – 9 (6)	2.3	Parallel implementation of *gzip*
pbzip2	1 – 9 (9)	1.1.6	Parallel implementation of *bzip2*

its high compression ratio, relatively slow compression and fast decompression. As many modern smartphones include multicore CPUs, we also included *pigz* and *pbzip2*, which are parallel versions of *gzip* and *bzip2*, respectively. All of these utilities operate at byte granularity and support a number of compression levels that allow the user to trade off speed for compression ratio. Lower levels favor speed, whereas higher levels offer higher compression ratios.

gzip [20] implements the deflate algorithm, which is a variant of the LZ77 algorithm [21]. It looks for repeating strings, i.e., sequences of bytes, within a 32-KB sliding window. The length of the strings is limited to 256 bytes. *gzip* uses two Huffman coders, one to compress the distances in the sliding window and another to compress the lengths of the strings as well as the individual bytes that were not part of any matched sequence. The algorithm finds duplicated strings using a chained hash table that is indexed with 3-byte strings. The selected compression level determines the maximum length of the hash chains and whether lazy evaluation should be used.

lzop [22] uses the LZO block-based compression algorithm that favors speed over compression ratio and requires little memory to operate. It splits each block of data into sequences of matches (a sliding dictionary) and non-matching literals, which are then compressed. LZO requires essentially no memory for decompression and only 64 KB for compression. The implementation on our test device supports only compression levels –1 to –6.

bzip2 [23] implements a variant of the block-sorting algorithm described by Burrows and Wheeler (BWT) [24]. It applies a reversible transformation to a block of inputs, uses sorting to group bytes with similar contexts together, and then compresses them with a Huffman coder. The selected compression level adjusts the block size between 100 KB and 900 KB (with compression levels –1 to –9).

xz is based on the Lempel–Ziv–Markov chain compression algorithm (LZMA) [25] developed for *7-Zip* [26]. It uses a large dictionary to achieve good compression ratios and employs a variant of LZ77 with special support for repeated match distances. The output is encoded with a range encoder, which uses a probability model for each bit (rather than whole bytes) to avoid mixing unrelated bits, i.e., to boost the compression ratio.

pigz is a parallel version of *gzip* for shared memory machines that is based on *pthreads* [27]. It breaks the input up into 128 KB chunks and concurrently compresses multiple chunks. The compressed data are outputted in their original order. Decompression operates mostly sequentially; however, separate threads are created for reading and writing.

pbzip2 is a multithreaded version of *bzip2* that is also based on *pthreads* [28]. It works by compressing multiple blocks of data simultaneously. The resulting blocks are then concatenated to form the final compressed file, which is compatible with *bzip2*. Decompression is also parallelized.

7.3 Experimental Setup

Section 7.3.1 describes the main characteristics of the smartphone used in our experiments as well as its software setup. Section 7.3.2 describes our measurement setup and the way the energy consumed by the compression and decompression tasks is calculated. Finally, Section 7.3.3 describes the dataset used.

7.3.1 Smartphone

We use Google's Nexus 4 smartphone [29] as the target platform. The Nexus 4 is powered by a Qualcomm Snapdragon S4 Pro (APQ8064) system on a chip that features a quad-core ARM Cortex A15 processor running at up to 1.512 GHz clock frequency and a powerful Adreno 320 graphics processor [30]. The smartphone has 2 GB of LPDDR2 RAM and 16 GB of built-in internal storage. It uses a 4.7-inch display and includes a 1.3-megapixel front-facing camera and an 8-megapixel rear-facing camera. It supports a range of connectivity options including WLAN 802.11n, Bluetooth 4.0, and several cellular network protocols such as GSM/EDGE/GPRS, 3G UMTS/HSPA+/DC-HSP+, and HSDPA+.

To prepare the smartphone for energy profiling, its plastic shield was removed to reveal connections on its motherboard and daughter boards as shown in Figure 7.1. We replaced the smartphone's battery with power connectors coming from a battery simulator. During measurements, connectors to the smartphone components such as the LCD display, touch screen, USB, and others can be removed to reduce the impact of these components on the consumed energy.

The smartphone's Android (Jelly Bean) operating system is upgraded to (a) add common (de)compression utilities not readily supported on Android and (b) to add utilities for managing performance and energy measurements. We flashed the smartphone with CyanogenMod version 10.2 [31], an open source operating system for smartphones and tablet computers based on official releases of Android that include third-party code. The included compression utilities are *gzip*, *bzip2*, *lzop*, and *xz*. The parallel versions of *gzip* and *bzip2*,

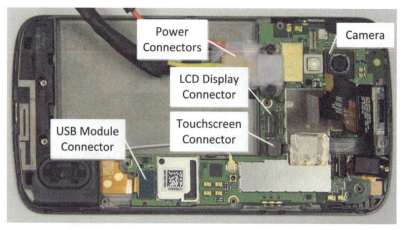

Figure 7.1   Nexus 4 prepared for energy measurements.

pigz and *pbzip2*, respectively, were compiled directly on the smartphone. In addition, the secure shell (*ssh*) can be configured to support uncompressed and compressed data transfers to/from a remote file server, accessed either over the smartphone's WLAN interface or its mobile broadband interface.

7.3.2 Measurement Setup

Our setup for measuring the energy consumed on the smartphone, shown in Figure 7.2, consists of an NI PXIe-4154 battery simulator [32], the smartphone, and a workstation. Figure 7.3 shows a block diagram of the setup including the main components and communication channels between them. The battery simulator, a specialized programmable power supply, resides inside the NI PXIe-1073 chassis [33], which is connected to an MXI-Express Interface card inside the workstation. The battery simulator is used (a) to power the smartphone through probes on channel 0 by providing 4.1 volts, thus bypassing the actual smartphone battery and (b) to measure the current drawn by the smartphone while running applications. The battery simulator is optimized for powering devices under test, including cellular handsets, smartphones, tablets, and other mobile devices. Its +6 V, ±3 A Channel 0 is designed to simulate a lithium-ion battery cell's transient speed, output resistance, and 2-quadrant operation (source/sink) [32]. Acting as a data acquisition system (DAQ), the battery simulator samples the current drawn on its channels with a configurable sampling frequency of up to 200,000 samples per second and a sensitivity of 1 μA. This means that for the processor on the Nexus 4 running

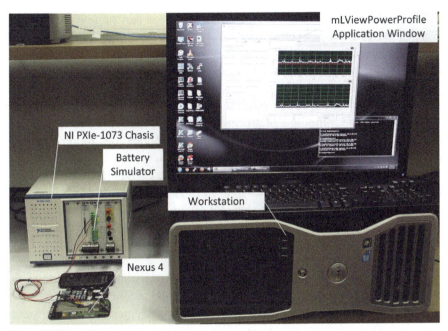

Figure 7.2 Hardware setup for energy profiling.

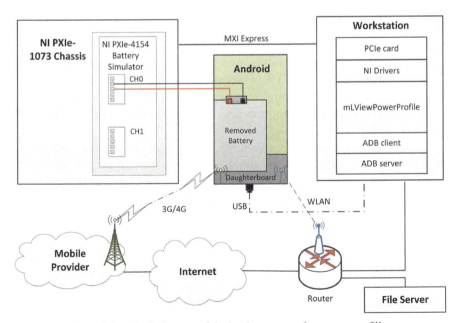

Figure 7.3 Block diagram of the hardware setup for energy profiling.

at its maximum clock frequency of 1.512 GHz, we can sample the current every 7,560 CPU clock cycles.

The workstation is a Dell T7500 Precision with an Intel Xeon processor, 12 GB of system memory, running in the Windows 7 Pro operating system. It runs *mLViewPowerProfile*, our custom software tool for automated capturing of power traces and evaluating the energy efficiency of applications running on mobile computing platforms. *mLViewPowerProfile* interfaces with (a) the smartphone to manage the activities and applications running on the smartphone that are being profiled and (b) the battery simulator to configure the channel and collect the current samples. The communication with the smartphone is carried out over the Android Debug Bridge (*adb*) [34]. *adb* is a client–server program that includes the following components: a client, which runs on the workstation; a server, which runs as a background process on the workstation; and a daemon, which runs on the smartphone. *adb* can connect to the smartphone over a USB or WLAN link.

Figure 7.4 shows the *mLViewPowerProfile*'s graphical user interface. A user configures the channels of the battery simulator. This involves setting the voltage and the current limits, the sampling frequency, the transient time, as well as software driver parameters that control fetching the current samples from the battery simulator. *mLViewPowerProfile* can average multiple samples from the battery simulator, which is controlled by a value set in the graphical user interface. We experimented with different sampling frequencies in the range of 10,000 samples/s to 200,000 samples/s and evaluated their impact on the energy calculations. We found the energy calculated using 20,000 samples/s to be within 1% of the energy calculated using the maximum sampling rate of 200,000 samples/s, so for our experiments, we choose to average ten samples, thus recording 20,000 samples per second in a user-specified file (*appsSamples.txt*).

To run a compression or a decompression task on the smartphone, a sequence of *adb* commands is launched from the workstation to be executed on the smartphone as shown in Figure 7.5. The third line executes one of the command scripts, which are prepared in advance and placed in a working directory of the smartphone. A sample command script with commands for invoking *gzip* with the default compression level is shown in Figure 7.6. The execution of a (de)compression task is typically preceded and followed by a five-second delay (head and tail delays) during which the smartphone is idle. The (de)compression task is wrapped by commands that take time stamps corresponding to the moments when the task is launched and completed. These times are used to determine the task execution time as well as to identify the

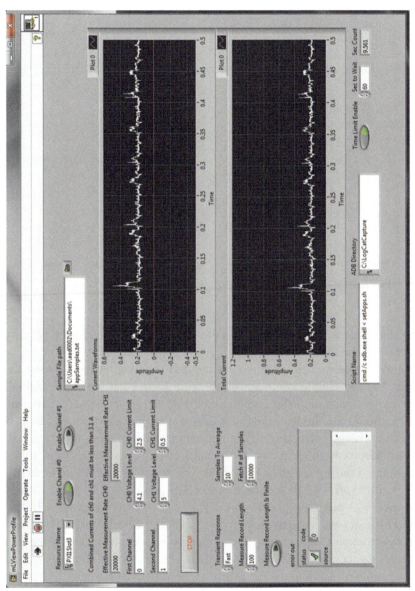

Figure 7.4 *mLViewPowerProfile* user interface.

```
1.  su                    # start as superuser
2.  cd /data/working      # move to working directory
3.  ./runGzip.sh          # start compression tasks
4.  exit                  # exit su session
5.  exit                  # exit adb
```

Figure 7.5   Sample *adb* script launched from the workstation to execute on the smartphone.

```
1.  #!/bin/bash
2.  path="/home/Ubuntu/datainput"     # input location
3.  file="totalInput"                 # filename
4.  filePath="$path/$file.tar"        # complete file path
5.  # run gzip compression task
6.  sleep 5                           # sleep for 5 seconds
7.  date %s.%N >> timestamps.txt      # starting timestamp
8.  gzip -fc6 $filePath >/dev/null    # compress input file
9.  date %s.%N >> timestamps.txt      # ending timestamp
10. sleep 5                           # sleep for 5 seconds
```

Figure 7.6   Command script *runGzip.sh* for running a compression task.

appropriate current samples logged on the workstation to calculate the energy consumed by the task.

Figure 7.7 shows the measured current drawn by the Nexus 4 during the execution of the sample command script from Figure 7.6. The head and tail delays are 5 s each, and the compression task takes roughly 17 s. The top graph in the figure shows the current drawn during the experiment as it is used in our energy calculations. The bottom graph shows the filtered signal, provided here only to enable easier visual inspection of the changes in the current drawn during program execution. The Nexus 4 with all unnecessary services turned off (LCD disconnected, and GPS and WLAN interfaces turned off) draws $\sim$11 mA ($I_{\text{IDLE}} = 11$ mA). The start of the compression task is marked by a step increase in the current drawn of $\sim$270 mA to 280 mA, and the current remains high during the compression and goes back to the idle level after the compression has terminated. The number of samples during the execution of a compression utility is $n = T.C \times SF$, where $T.C$ is the compression time for a given file and SF is the sampling frequency of the battery simulator (with respect to the number of recorded samples). The total energy consumed ($ET.C$) is calculated as shown in Equation (7.1), where V_{BS}

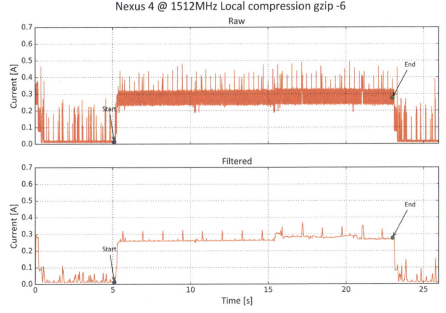

Figure 7.7 Current drawn by the Nexus 4 when executing the gzip utility.

is the supply voltage on the battery simulator (V_{BS} = 4.1 V) and each I_j is a current sample during compression.

$$ET.C = V_{BS} \cdot \frac{1}{SF} \cdot \sum_{j=1}^{n} I_j \qquad (7.1)$$

$$ET.C\,(0) = ET.C - I_{idle} \cdot V_{BS} \cdot T.C \qquad (7.2)$$

In addition to *ET.C*, we also calculate the overhead energy of the compression task alone, *ET.C(0)*, which excludes the energy needed to run the platform when idle. This overhead is calculated as shown in Equation (7.2). We similarly calculate the total energy and the overhead energy for decompression tasks using the decompression time *T.D* instead of the compression time *T.C*.

7.3.3 Datasets

In selecting the data to evaluate the effectiveness of the compression utilities, we compiled a set of diverse input files that are representative of mobile computing environments. The input file formats include text, an executable, an image, a file with comma-separated values from a wearable health monitor,

Table 7.2 Dataset

I	Type	Raw Size [KB]	Notes
1	text (txt)	15,711.6	Project Gutenberg works of Mark Twain
2	exec (so)	12,452.5	Open source web content engine library
3	image (bmp)	16,777.2	An image of Earth from space
4	table (csv)	9,988.9	Activity data from a health monitor
5	code (tar)	11,233.2	Perl 5.8.5 source code

and source code. Table 7.2 provides the input files, their types, the size in bytes, and a description. The files are merged into a single archive file (tar) that is used as an input for the compression utilities.

7.4 Metrics and Experiments

In Section 7.4.1, we describe the metrics used in the evaluation of the compression utilities, including the compression ratio, compression and decompression throughputs, and energy efficiency of compression and decompression tasks. Section 7.4.1 describes the type of experiments conducted.

7.4.1 Metrics

Table 7.3 summarizes the metrics used as well as their definitions.

Table 7.3 Metrics

Symbol	Description	Units	Definition
US	Uncompressed file size	MB	Measured
CS	Compressed file size	MB	Measured
CR	Compression ratio	–	US/CS
T.C [T.D]	Time to [de]compress	s	Measured
T.UUP [T.UDW]	Time to upload [download] the uncompressed file	s	Measured
ET.C [ET.D]	Total energy for [de]compression	J	Computed
ET.UUP [UDW]	Total energy for upload [download] of the uncompressed file	J	Computed
ET.C(0) [ET.D(0)]	Overhead energy for [de]compression	J	$ET.C - I_{IDLE} \times V_{BS} \times T.C$ [$ET.D - I_{IDLE} \times V_{BS} \times T.D$]
Th.C [Th.D]	[De]compression throughput	MB/s	US/T.C [US/T.D]
Th.UUP [Th.UDW]	Uncompressed upload [download] throughput	MB/s	US/T.UUP [US/T.UDW]
EE.C [EE.D]	[De]compression energy efficiency	MB/J	US/ET.C [US/ET.D]
EE.C(0) [EE.D(0)]	[De]compression overhead energy efficiency	MB/J	US/ET.C(0) [US/ET.D(0)]

Compression ratio. We use the compression ratio to evaluate the compression effectiveness of an individual utility and its levels of compression. The compression ratio (CR) is calculated as the size of the uncompressed input file (US) divided by the size of the compressed file (CS), CR=US/CS.

Performance. To evaluate the performance of individual compression utilities and compression levels, we measure the time to compress the raw input file (T.C) and the time to decompress (T.D) a compressed file generated by that utility with the selected compression level. Each of the compression and decompression tasks is repeated three times, and the average time is calculated. Instead of reporting the execution times directly, we report the compression and decompression throughputs (Th.C and Th.D) expressed in megabytes per second (Th.C=US/T.C and Th.D=US/T.D).

The throughput captures the efficiency of data transfers from the user's point of view—users produce and consume uncompressed data and care more about the time it takes to transfer data than about what approach is used internally to make the transfer fast. In addition, this metric is suitable for evaluating networked data transfers and comparing compressed and uncompressed transfers.

Energy efficiency. For each compression task with a selected compression level, we calculate the total energy for compression (ET.C) using the method described in Equation (7.1) as well as the total energy for decompression (ET.D). For each combination of a compression utility and a compression level, three measurements are conducted and the average energy is calculated. Instead of reporting the energy directly in joules, we report the energy efficiency (EE.C and EE.D) in megabytes per joule (EE.C=US/ET.C and EE.D=US/ET.D). To eliminate the effects of the idle current, we also consider the overhead energies ET.C(0) and ET.D(0) and the energy efficiency metrics EE.C(0) and EE.D(0) calculated using the method described in Equation (7.2).

7.4.2 Experiments

To evaluate the throughput and energy efficiency of compression and decompression tasks, we consider two typical usage scenarios as illustrated in Figure 7.8. The first experiment (LOCAL) involves measuring the time and energy of compression and decompression tasks performed locally on the smartphone. To eliminate latencies and energy overheads caused by

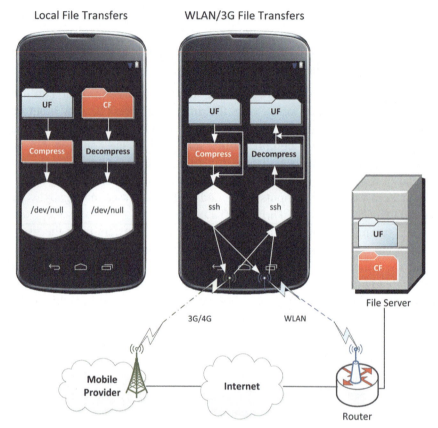

Figure 7.8   Data flow of the experiments (blue file icons refer to uncompressed files, and red file icons refer to compressed files).

writing files to the internal flash memory, the output of the compression and decompression tasks is redirected to the null device (/dev/null)—a special "file" that discards all data written to it.

The second group of experiments (WLAN and CELL) involves measuring the time and energy of compression and decompression tasks performed on the smartphone while transferring data to/from a remote server. For the compression tasks, the uncompressed input file (UF) is read from the local file system, compressed on the smartphone, and streamed to the remote server over a secure channel. The output files are redirected to the null device of the remote server. For the decompression tasks, the compressed files (CF) are retrieved from the temporary file system of the remote server through a secure channel and decompressed on the smartphone. The output files are redirected to the null device of the smartphone.

The communication between input, (de)compression, and output operations is carried out through pipes. The execution times include file transfer latencies as well as (de)compression times. Similarly, the energies are measured for completing the entire chain of tasks. WLAN and CELL correspond to typical file transfer tasks on smartphones: compressing and uploading files to a remote server, and downloading files from a remote server and decompressing them. In addition to the transfers that involve compression and decompression operations, we evaluate the time and energy needed to upload and download the uncompressed input file over a secure communication channel. In the WLAN experiments, the smartphone connects to the file server through the local router. In the CELL experiments, the file server is reached over the cellular network.

The Nexus 4's Snapdragon S4 Pro chip supports a total of 12 non-overclocked frequencies, ranging from the minimum 0.384 GHz to the maximum 1.512 GHz. To examine the impact of frequency scaling on energy efficiency, we conduct each experiment for three characteristic clock frequencies: the minimum (0.384 GHz), the middle (0.810 GHz), and the maximum (1.512 GHz). The processor clock frequency is set using a utility called *cpupower*. The *cpupower frequency-info* utility allows a user to inspect the current clock frequency setup on each of the four processor cores. The *cpupower frequency-set* utility allows a user to set the minimum, the maximum, and the current clock frequency, as well as a governor that specifies a power scheme for each processor core. The power schemes, such as *Performance*, *Powersave*, *Userspace*, *Ondemand*, and *Conservative* [35], dictate to the Linux kernel how to dynamically adjust the processor frequencies. For example, the *Performance* governor forces processor cores to constantly run at the maximum allowed frequency. In our experiments, we force all processor cores to run at a specific clock frequency, which remains fixed during an experimental run.

7.5 Results

This section discusses the results of our experimental analysis, including the compression ratio (Section 7.5.1), compression and decompression throughputs (Section 7.5.2), and energy efficiency as well as the overhead energy efficiency (Section 7.5.3). The energy efficiency and throughput findings are summarized in Section 7.5.4.

7.5.1 Compression Ratio

Figure 7.9 shows the compression ratio (CR) on the input dataset of all considered compression utilities and levels. *pigz*'s and *pbzip2*'s compression ratios are equivalent to those of *gzip* and *bzip2*, respectively. Generally, the compression ratio increases with higher compression levels. The best overall compression ratio is achieved by *xz*, ranging from 3.38 with –0 to 4.31 with –9, and by *bzip2/pbzip2*, ranging from 3.49 with –1 to 3.90 with –9. The lowest compression ratio is achieved by *lzop*, ranging from 2.07 with –1 through –6 to 2.62 with –9.

7.5.2 Compression and Decompression Throughputs

LOCAL. Figure 7.10 shows the compression and decompression throughputs for *LOCAL* at the three selected processor clock frequencies. The compression throughput varies widely across compression utilities and even across compression levels within a single utility. The higher compression levels result in lower throughputs because of the increased computational complexity, especially for *gzip*, *xz*, and *pigz*, where the throughput differences approach an order of magnitude. *pigz* with –1 achieves the highest compression throughput of ~36.5 MB/s at 1.512 GHz and ~33.3 MB/s at 0.810 GHz. It is followed by *lzop* with –1 to –6 that achieves ~30 MB/s at 1.512 GHz and ~19.5 MB/s at 0.810 GHz. At 0.384 GHz, the highest compression throughput of 8.2 MB/s is

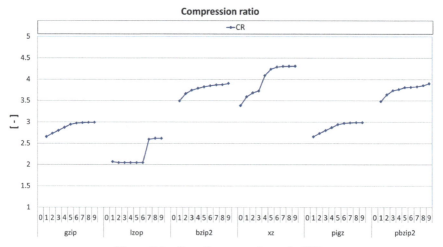

Figure 7.9 Overall compression ratio (CR).

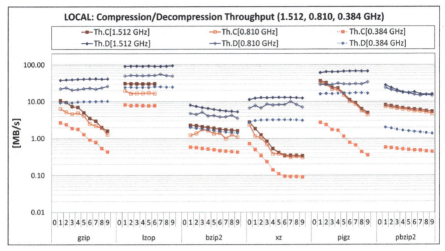

Figure 7.10 LOCAL: Compression and decompression throughput.

achieved by *lzop* –1, followed by *gzip* and *pigz* with ~2.7 MB/s. *xz* and *bzip2* achieve significantly lower compression throughputs when compared to other three compression utilities: 2.3–1.6 MB/s for *bzip2* and 2.8–0.4 MB/s for *xz* at 1.512 GHz; 1.7–1.0 MB/s for *bzip2* and 2.4–0.3 MB/s for *xz* at 0.810 GHz; and 0.6–0.4 MB/s for *bzip2* and 0.7–0.1 for *xz* at 0.384 GHz.

pigz fully utilizes the four processor cores to almost quadruple the throughput relative to *gzip* at higher clock frequencies. However, *pigz* does not offer any speedup relative to *gzip* at 0.384 GHz. Similarly to *pigz*, *pbzip2* achieves a linear compression throughput speedup relative to *bzip2* at 1.512 and 0.810 GHz and no speedup at 0.384 GHz.

The decompression throughputs are much higher than the compression throughputs. They remain relatively flat or slightly increase with an increase in compression level for *gzip*, *lzop*, *xz*, and *pigz* as the decompression computational complexity remains unchanged, but the input files get smaller. Notable exceptions are *bzip2* and *pbzip2*, where the decompression throughputs slightly decrease for higher compression levels due to increased computational complexity. For all three clock frequencies, the highest decompression throughput is achieved by *lzop* (~91–94 MB/s at 1.512 GHz, ~49–55 MB/s at 0.810 GHz, and ~24 at 0.384 GHz), followed by *pigz* (~67 MB/s at 1.512 GHz, ~33 MB/s at 0.810 GHz, and ~17 MB/s at 0.384 GHz), *gzip* (~38 MB/s at 1.512 GHz, ~24 MB/s at 0.810 GHz, and ~10 MB/s at 0.384 GHz), and *pbzip2* (~28–15 MB/s at 1.512 GHz and 0.810 GHz, and ∼2–1.3 MB/s at 0.384 GHz). *xz* and *bzip2* achieve significantly lower decompression throughputs compared to *pigz*, *gzip*, and *lzop*. *pbzip2* fully utilizes four processor cores to almost quadruple the throughput relative to *bzip2* for higher processor clock frequencies. *pigz*, which utilizes separate threads for reading and writing during decompression, achieves a speedup relative to *gzip* of ∼1.6, regardless of the clock frequency.

WLAN. Figure 7.11 shows the measured compression and decompression throughputs for the WLAN experiments at three selected processor clock frequencies. The dashed lines represent the effective WLAN throughput when the uncompressed input files are uploaded to the remote server (Th.UUP = 2.33 MB/s at 1.512 GHz, 2.01 MB/s at 0.810 GHz, and 1.42 MB/s at 0.384 GHz) and downloaded from the remote server (Th.UDW = 4.46 MB/s at 1.512 GHz, 4.27 MB/s at 0.810 GHz, and 1.99 MB/s at 0.384 GHz). The results indicate that the effective WLAN throughput for uncompressed uploads and downloads degrades as we lower the processor clock frequency, due to slower memory reads and writes.

The results from Figure 7.11 show that compressed data uploads outperform uncompressed uploads for the following combinations: *gzip* with –1 to –7, *lzop* with –1 to –6, *pigz* with –1 to –9, and *pbzip2* with –1 to –9 at 1.512 and 0.810 GHz. At the 0.384 GHz clock frequency, only *lzop* –1 to –6 and *gzip* –1 outperform uncompressed uploads. *bzip2* and *xz* effectively decrease

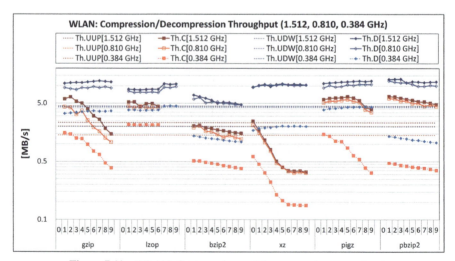

Figure 7.11   WLAN: Compression and decompression throughput.

the throughput relative to uncompressed uploads. This result is expected as the compression throughput of *bzip2* and *xz* in the LOCAL experiment falls below the uncompressed upload throughput over WLAN. With a processor clock frequency of 1.512 GHz and 0.810 GHz, the best performing utilities are *pbzip2* with −1 at 6.6 MB/s and 6.1 MB/s and *pigz* with −5 at 6.5 MB/s and 5.8 MB/s, respectively. At 0.384 GHz, the best performing utility is *lzop* with −1 to −6 achieving 2.2 MB/s, which is 1.5 times higher than the uncompressed upload throughput at the same frequency.

The compressed data downloads improve the effective throughput for all utilities and compression levels relative to the uncompressed downloads at 1.512 and 0.810 GHz. At 0.384 GHz, only *gzip*, *lzop*, and *pigz* improve the effective throughput relative to the uncompressed downloads. *pbzip2* with −1, *gzip* with −7 to −9, and *pigz* with −5 to −9 all reach ∼12 MB/s, which is an almost a 2.6-fold improvement of the effective download throughput at 1.512 GHz and 0.810 GHz relative to the uncompressed downloads. *pigz* with −5 to −9 and *lzop* with −7 to −9 reach ∼4.5 MB/s, which is an almost ∼2.3-fold improvement in the effective download throughput at 0.384 GHz clock. The decompression throughput of the individual utilities remains fairly constant when changing the compression level. *pbzip2* almost doubles the effective throughput relative to *bzip2* at 1.512 GHz and 0.810 GHz, but no speedup is observed at 0.384 GHz. *pigz* achieves a 1.2-fold improvement over *gzip*.

If we know the network uncompressed upload and download throughputs (Th.UUP^{WLAN}, Th.UDW^{WLAN}), the throughputs from the LOCAL experiments, and the compression ratios, we can estimate the lower and upper bounds for the compressed upload and download throughputs. Equation (7.3) gives boundaries for the effective WLAN compression throughput when using utility *i* with compression level *j*, Th.C$_{(i,j)}^{\text{WLAN}}$. Similarly, Equation (7.4) gives boundaries for the effective WLAN decompression throughput, Th.D$_{(i,j)}^{\text{WLAN}}$. The lower estimates (Th.C.LL$_{(i,j)}^{\text{WLAN}}$ and Th.D.LL$_{(i,j)}^{\text{WLAN}}$) assume that the (de)compression on the smartphone and the data transfer over the WLAN are not overlapped in time, whereas the upper estimates (Th.C.UL$_{(i,j)}^{\text{WLAN}}$ and Th.D.UL$_{(i,j)}^{\text{WLAN}}$) assume that they are fully overlapped and thus bounded by the network throughput and the compression ratio. The utilities with a low computational complexity achieve a (de)compression throughput close to the upper bound, whereas those with a high computational complexity achieve a (de)compression throughput close to the lower bound. Figure 7.12 illustrates the estimated bounds for WLAN compression and decompression performed

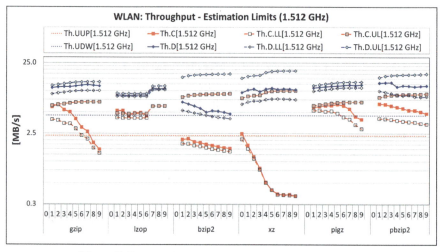

Figure 7.12   WLAN: Estimated throughputs at 1.512 GHz.

at 1.512 GHz. For example, the estimated lower and upper bounds for the compression throughput of *gzip* with –1 are 3.9 MB/s and 6.2 MB/s, and the measured compression throughput is 5.9 MB/s; in contrast, the estimated bounds for *bzip2* with –1 are 1.8 MB/s and 8.1 MB/s, and the measured compression throughput is 2.04 MB/s [2, 4].

$$\frac{1}{\frac{1}{CR_{(i,j)} \cdot Th.UUP^{WLAN}} + \frac{1}{Th.C_{(i,j)}^{LOCAL}}} \le Th.C_{(i,j)}^{WLAN} \le CR_{(i,j)} \cdot Th.UUP^{WLAN} \quad (7.3)$$

$$\frac{1}{\frac{1}{CR_{(i,j)} \cdot Th.UDW^{WLAN}} + \frac{1}{Th.D_{(i,j)}^{LOCAL}}} \le Th.D_{(i,j)}^{WLAN} \le CR_{(i,j)} \cdot Th.UDW^{WLAN}$$

$$(7.4)$$

CELL. Figure 7.13 shows the compression and decompression throughputs for data transfers over the cellular network interface (3G). The dashed lines represent the measured effective upload and download throughputs when transferring the uncompressed input file over the cellular network. The effective upload throughput Th.UUP^{CELL} ranges from 0.30 to 0.28 MB/s, and Th.UDW^{CELL} ranges from 0.44 to 0.56 MB/s depending on the processor clock frequency. Thus, the impact of the clock frequency on the effective uncompressed upload and download throughputs is fairly limited relative to the WLAN experiments.

The results show that the compressed uploads increase the effective network throughput for all utilities and compression levels, with the exception of *xz* with –3 to –9 at 0.384 GHz. The low network throughput favors utilities with higher compression ratios. Thus, unlike in the WLAN experiments,

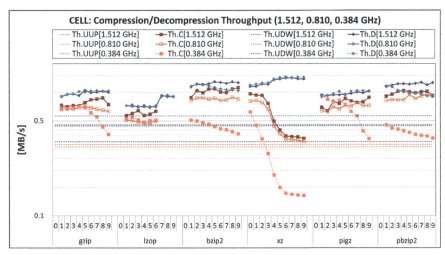

Figure 7.13 CELL: Compression and decompression throughput.

higher compression levels help improve the effective throughput. The highest compression throughputs of 1.11 MB/s, 0.94 MB/s, and 0.92 MB/s are achieved by *bzip2* with –9, *pbzip2* with –9, and *pigz* with –1 for 1.512 GHz, 0.810 GHz, and 0.384 GHz, respectively. Both *bzip2* and *pbzip2* perform well with all compression levels at 1.512 GHz and 0.810 GHz. Other notable combinations are *xz* with –0, which achieves 0.96 MB/s and 0.90 MB/s, and *gzip* with –6, which achieves 0.83 MB/s and 0.70 MB/s at 1.512 GHz and 0.810 GHz, respectively. The parallel utilities do not improve the overall throughput because their higher throughputs relative to the sequential utilities far exceed the network throughput, i.e., $\text{Th.C}^{\text{LOCAL}} \gg \text{Th.UUP}^{\text{CELL}}$.

With the low effective throughput for downloads offered by the cellular network—$\text{Th.UDW}^{\text{CELL}}$ ranges from 0.44 to 0.56 MB/s—all decompression utilities increase the available bandwidth ($\text{Th.D}^{\text{CELL}} > \text{Th.UDW}^{\text{CELL}}$) for all compression levels across all tested clock frequencies. The best performing utility is *xz*, which achieves ∼1.40–1.46 MB/s across all clock frequencies. *bzip2/pbzip2* closely follow *xz*, delivering ∼1.26 MB/s and ∼1.14 MB/s at 1.512 GHz and 0.384 GHz, respectively.

The lower and upper throughput boundaries shown in Equations (7.3) and (7.4) also hold for cellular data upload and download transfers, respectively. Thus, the effective compression throughput is limited by the network upload throughput and is always below $\text{CR} \times \text{Th.UUP}^{\text{CELL}}$. Although we face limited controllability of the experiments that utilize the cellular network interface due to network provider coverage, utilization of the network at the time the experiments are conducted, and unpredictable Internet latencies, our results are within the expected range defined by Equations (7.3) and (7.4).

7.5.3 Energy Efficiency

LOCAL. Figure 7.14 shows the energy efficiency of the *LOCAL* compression (EE.C) and decompression (EE.D) tasks as well as the energy efficiency when only the overhead energy is considered, EE.C(0) and EE.D(0) (where the

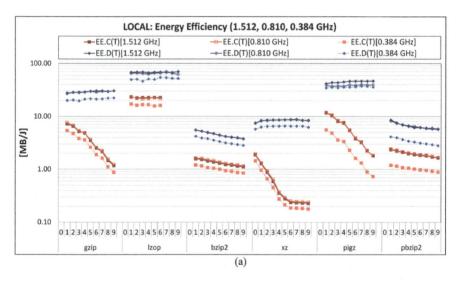

(a)

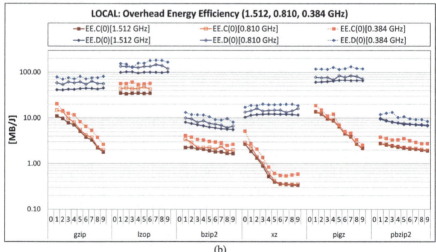

(b)

Figure 7.14   LOCAL: Energy efficiency and overhead energy efficiency.

idle current is assumed to be zero, $I_{IDLE} = 0$) for the three selected clock frequencies.

Expectedly, the energy efficiency for compression varies widely for different utilities and for different compression levels within each utility, especially for *gzip*, *pigz*, and *xz*. The most energy-efficient compression utility by far is *lzop* with compression levels −1 to −6, regardless of the clock frequency. At 1.512 GHz, it achieves an energy efficiency of ∼23.5 MB/J and an overhead energy efficiency of 35 MB/J. A distant second best is *pigz* with −1, which achieves ∼11.5 MB/J of energy efficiency and ∼13.5 MB/J of overhead efficiency at 1.512 GHz. Whereas the total energy efficiency favors faster clock rates, the overhead energy efficiency favors lower clock rates. Following the trends in compression throughputs, the higher compression levels for *gzip*, *pigz*, and *lzop* result in a dramatic decrease in energy efficiency. *pigz* and *pbzip2* are more energy efficient than their sequential counterparts when the total energy is considered because they reduce the compression time and thus the relative contribution of the idle energy for clock frequencies of 1.512 GHz and 0.810 GHz. They do not offer any energy savings when running at 0.384 GHz. *pbzip2* and *bzip2* exhibit low energy efficiencies as does *xz*, the least attractive choice when using high compression levels.

The energy efficiency for decompression (Figure 7.14) also varies widely for different utilities. The decompression energy efficiency is relatively stable for individual utilities—it increases slightly for higher compression levels for all utilities except *bzip2* and *pbzip2*. At 1.512 GHz, EE.D is ∼69 MB/J for *lzop*, ∼46 for *pigz*, ∼30 for *gzip*, and ∼8 MB/J for *xz*. The decompression energy efficiencies for 1.512 GHz and 0.810 GHz are close to each other. For a clock frequency of 0.384 GHz, the decompression energy efficiencies are reduced by 73–76% relative to those at the higher clock frequencies. At all frequencies, *lzop* emerges as the most energy-efficient choice in spite of its low compression ratio.

WLAN. Figure 7.15a shows the energy efficiency for compressed and uncompressed data transfers over the WLAN interface at the three processor clock frequencies. Figure 7.15b shows the overhead energy efficiency (EE(0)). From these graphs, one can easily identify cases when compressed data uploads offer higher energy efficiencies than uncompressed uploads (EE.C > EE.UUP) and when downloads with decompression offer higher energy efficiencies than uncompressed downloads (EE.D > EE.UDW).

At 1.512 GHz, the energy efficiency of the uncompressed data upload is EE.UUP = 1.42 MB/J and EE.UUP(0) = ∼2 MB/J. The energy efficiency trends are similar to those of the effective throughput. The energy efficiency

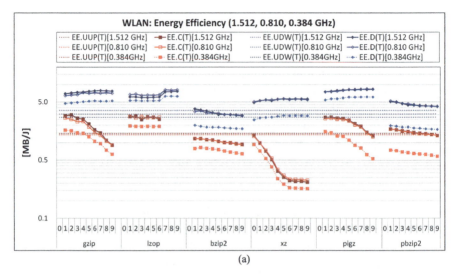

(a)

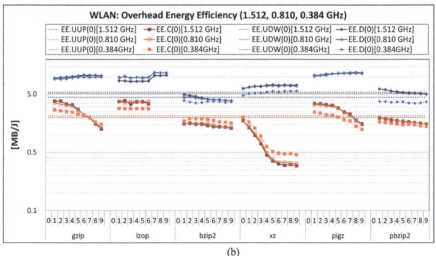

(b)

Figure 7.15   WLAN: Energy efficiency.

decreases for higher compression levels, especially for *gzip*, *xz*, and *pigz*. Only a small subset of utilities and compression levels improves the energy efficiency relative to the uncompressed upload. At 1.512 and 0.810 GHz, the subset includes *gzip* with −1 to −6, *lzop* with −1 to −6, and *pigz* with −1 to −8. At 0.384 GHz, the subset includes *gzip* with −1 to −3, *lzop* with −1 to −6, and *pigz* with −1. *xz*, *bzip2*, and *pbzip2* do not offer improvements over the uncompressed transfers for any compression level or clock frequency.

The most energy-efficient approach at 1.512 and 0.810 GHz is to use *gzip*, *pigz*, or *lzop* with the lowest compression level (–1 or –2). At 0.384 GHz, *lzop* is the most energy-efficient approach. For example, *gzip* with –2 achieves an energy efficiency of ~2.9 MB/J at 1.512 GHz, a more–than-twofold improvement over the uncompressed upload.

At the 1.512 GHz clock frequency, the energy efficiency of the uncompressed data download is EE.UDW = 3.1 MB/J, and EE.UDW(0) is 4.4 MB/J. At 0.810 and 0.384 GHz, EE.UDW and EE.UDW(0) increase by 16–20%. Decompression using *gzip*, *lzop*, or *pigz* exceeds the energy efficiency of the uncompressed download for both the total energy and the energy overhead. *bzip2* and *pbzip2* are less energy-efficient and barely outperform the uncompressed download with low compression levels. At 1.512 GHz, the highest energy efficiency of ~8.3 MB/J is achieved when using *pigz* with –6 to –9. This is 2.7 times better than the uncompressed download. Overall, the energy efficiency at 0.810 GHz is similar to the energy efficiency at 1.512 GHz. At 0.384 GHz, the energy efficiency decreases from 77% for *lzop* to 40% for *pbzip2* when compared to the energy efficiency at 1.512 GHz.

Similar to the estimation of the effective throughput in networked transfers of compressed data, we can also estimate the energy efficiency. By knowing the energy efficiency for uploads and downloads, the energy efficiencies in the local experiment, and the compression ratios, a lower and an upper bound for the compression and decompression energy efficiencies for utility i and compression level j, i.e., $\text{EE.C}_{(i,j)}{}^{WLAN}$ and $\text{EE.D}_{(i,j)}{}^{WLAN}$, can be computed as shown in Equations (7.5) and (7.6), respectively. The lower estimates $(\text{EE.C.LL}_{(i,j)}{}^{WLAN}$ and $\text{EE.D.LL}_{(i,j)}{}^{WLAN})$ assume that compression or decompression on the smartphone and the data transfer over the WLAN are not overlapped in time and the total energy needed is the energy to perform the compression and decompression plus the energy to perform the transfer of the compressed data. The upper estimate $(\text{EE.C.UL}_{(i,j)}{}^{WLAN}$ and $\text{EE.D.UL}_{(i,j)}{}^{WLAN})$ assumes complete overlap in time and is thus bounded by the energy efficiency of the network transfer and the compression ratio. The total energy, in this case, is the sum of the energy overhead (ET.C(0) and ET.D(0)) and the energy for the transfer of the compressed data.

$$\frac{1}{\frac{1}{CR_{(i,j)} \cdot EE.UUP^{WLAN}} + \frac{1}{EE.C_{(i,j)}^{LOCAL}}} \leq EE.C_{(i,j)}^{WLAN} \leq$$

$$\frac{1}{\frac{1}{CR_{(i,j)} \cdot EE.UUP^{WLAN}} + \frac{1}{EE.C(0)_{(i,j)}^{LOCAL}}} \tag{7.5}$$

$$\frac{1}{\frac{1}{CR_{(i,j)} \cdot EE.UDW^{WLAN}} + \frac{1}{EE.D_{(i,j)}^{LOCAL}}} \leq EE.D_{(i,j)}^{WLAN} \leq$$

$$\frac{1}{\frac{1}{CR_{(i,j)} \cdot EE.UDW^{WLAN}} + \frac{1}{EE.D(0)_{(i,j)}^{LOCAL}}} \tag{7.6}$$

Figure 7.16 illustrates the estimated bounds for WLAN data transfers performed at the 1.512 GHz clock frequency. Let us consider *gzip* −1 as an example: The estimates for the lower and upper bounds for the compression energy efficiency are 2.5 MB/J and 2.8 MB/J, and the measured energy efficiency is 2.8 MB/J. The measured energy efficiency for *bzip2* with −9 is 0.9 MB/J, and its estimated bounds are between 0.9 MB/J and 1.2 MB/J. Thus, utilities with a low computational complexity achieve efficiency close to the upper bound and those with a high computational complexity achieve efficiency close to the lower bound.

CELL. Figure 7.17 shows the energy efficiency of compression and decompression tasks, EE.C and EE.D, respectively, when transferring data over the cellular network at the three selected processor clock frequencies. We also show the energy efficiency that considers only the energy overhead, EE.C(0) and EE.D(0). In addition, the graphs show the energy efficiency for uncompressed upload (EE.UUP and EE.UUP(0)) and uncompressed download (EE.UDW and EE.UDW(0)).

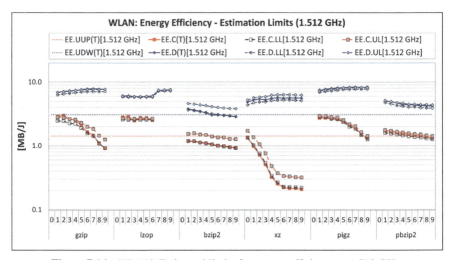

Figure 7.16   WLAN: Estimated limits for energy efficiency at 1.512 GHz.

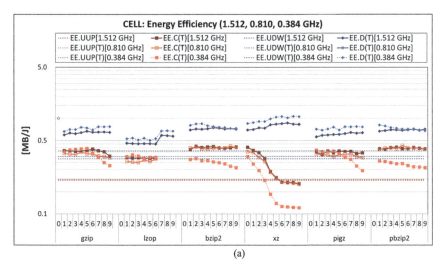

(a)

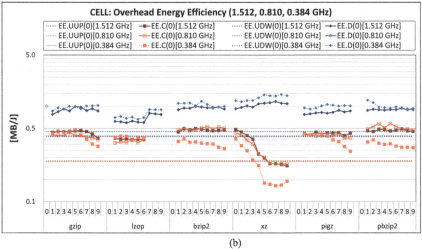

(b)

Figure 7.17 CELL: Energy efficiency.

The energy efficiency of the uncompressed upload and download is as follows: EE.UUP = 0.15 MB/J and EE.UDW = 0.3 MB/J at 1.512 and 0.810 GHz, and EE.UUP = 0.14 and EE.UDW = 0.36 MB/J at 0.384 GHz. Due to the relatively low energy efficiency of uncompressed data transfers, employing any combination of compression utility and compression level offers improvements in energy efficiency, except for *xz* with –6 to –9 at 1.512 GHz and 0.810 GHz, and except for *xz* with –4 to –9 at 0.384 GHz.

The higher computational complexity when increasing the compression level is compensated by the resulting higher compression ratio. Thus, the energy efficiency of *bzip2* and *pbzip2* stays almost constant. In fact, at 1.512 and 0.810 GHz, *bzip2* and *pbzip2* offer the highest energy efficiency when uploading data and reach 0.42–0.43 MB/J, a 2.8-fold improvement over the uncompressed transfer. At 0.384 GHz, a similar ~2.8-fold improvement over the uncompressed transfer is achieved by *gzip*, which offers the highest energy efficiency of 0.39 MB/J.

All decompression options offer an energy efficiency that exceeds the energy efficiency of uncompressed downloads from the remote server. The energy efficiency of uncompressed downloads, EE.UDW, is 0.28 MB/J, 0.30 MB/J, and 0.36 MB/J at 1.512 GHz, 0.810 GHz, and 0.384 GHz, respectively. Generally, downloading files that were compressed with higher compression levels increases the energy efficiency. At 1.512 GHz, the energy efficiency is ~0.65 for *gzip* and 0.7 MB/J for *bzip2/pbzip2*. At 1.512 GHz, the best energy efficiency of 0.84 MB/J is achieved by *xz* with –9. *xz* benefits from providing a superior compression ratio in conditions where communication energy dominates the overall energy cost. Interestingly, at 0.810 and 0.384 GHz, the overall energy efficiencies are higher than the energy efficiencies achieved at 1.512 GHz, with an energy efficiency of ~1.0 MB/J achieved by *gzip* and ~0.8 MB/J achieved by *bzip2/pbzip2* at both 0.810 and 0.384 GHz. With frequency scaling, the best energy efficiencies of 1.28 MB/J and 1.07 MB/J are achieved by *xz* –7 to –9 at 0.810 GHz and 0.384 GHz, respectively, a 1.48- and 1.27-fold improvement over the maximum achievable energy efficiency at 1.512 GHz. Overall, this is a ~3.8-fold improvement over the uncompressed data downloads at the maximum clock frequency.

The lower and upper energy efficiency boundaries shown in Equations (7.5) and (7.6) hold for cellular data upload and download transfers as well. By knowing the energy efficiency for uploads and downloads, the energy efficiency in the local experiment, and the compression ratios, a lower and an upper bound for the compression and decompression energy efficiencies for utility i and compression level j, i.e., EE.C$_{(i,j)}^{CELL}$ and EE.D$_{(i,j)}^{CELL}$, can be computed.

7.5.4 Putting It All Together

Tables 7.4 and 7.5 list the combinations of utilities and compression levels that achieve the best upload and download throughputs, respectively. We show the throughputs for compression and decompression tasks (Th.C and Th.D), for

Table 7.4 Upload: Utilities that achieve peak throughputs

Experiment	Compression		Raw UUP	Speedup
	Best Utility	Th.C [MB/s]	Th.UUP [MB/s]	Th.C/Th.UUP [–]
LOCAL				
1.512 GHz	pigz –1	36.50	–	–
0.810 GHz	pigz –1	33.25	–	–
0.384 GHz	lzop –1 to –6	8.00	–	–
WLAN				
1.512 GHz	pbzip2 –1	6.56	2.33	2.82
0.810 GHz	pbzip2 –1	6.05	2.01	3.01
0.384 GHz	lzop –1 to –6	2.15	1.42	1.51
CELL				
1.512 GHz	bzip2 –9	1.11	0.30	3.70
0.810 GHz	pbzip2 –5 to –9	0.94	0.27	3.48
0.384 GHz	pigz –1	0.92	0.28	3.28

Table 7.5 Download: Utilities that achieve peak throughputs

Experiment	Compression		Raw UUP	Speedup
	Best Utility	Th.C [MB/s]	Th.UUP [MB/s]	Th.C/Th.UUP [–]
LOCAL				
1.512 GHz	lzop –1 to –9	92.50	–	–
0.810 GHz	lzop –1 to –9	52.50	–	–
0.384 GHz	lzop –1 to –9	25.00	–	–
WLAN				
1.512 GHz	pbzip2 –1	13.10	4.46	2.94
0.810 GHz	pbzip2 –1	12.20	4.27	2.86
0.384 GHz	lzop –7 to –9	4.50	1.99	2.26
CELL				
1.512 GHz	xz –6 to –8	1.40	0.44	3.18
0.810 GHz	xz –7 to –9	1.42	0.45	3.15
0.384 GHz	xz –6 to –9	1.46	0.56	2.61

uncompressed file transfers (Th.UUP and Th.UDW), as well as the speedup that compressed transfers achieve over uncompressed transfers (Th.C/Th.UUP and Th.D/Th.UDW).

For the *LOCAL* compression tasks, we find that *pigz* with –1 achieves superior throughputs at 1.512 and 0.810 GHz. The best compression throughput at 0.384 GHz is achieved by *lzop* with –1 to –6.

When uploading data to a remote server over WLAN, the best utilities for sending compressed data are *pbzip2* with –1 at 1.512 and 0.810 GHz and *lzop* with –1 to –6 at 0.384 GHz. The compressed uploads increase the effective throughput relative to the uncompressed uploads by a factor of 2.82, 3.01, and 1.51 at 1.512 GHz, 0.810 GHz, and 0.384 GHz, respectively. When uploading data over CELL, *bzip2* with –9, *bzip2* with –5 to –9, and *pigz* with –1 increase the effective throughput by a factor of 3.70, 3.48, and 3.28 relative to the uncompressed transfer at 1.512 GHz, 0.810 GHz, and 0.384 GHz, respectively. For both WLAN and CELL transfers, the highest upload throughputs are achieved when using the maximum processor clock frequency.

When downloading data over WLAN, the highest download throughput is achieved by *pbzip2* with –1 at 1.512 and 0.810 GHz, and by *lzop* with –7 to –9 at 0.384 GHz. They increase the effective throughput by a factor of 2.94, 2.86, and 2.26 relative to the uncompressed download at each clock frequency. When downloading data over CELL, *xz* with –6 to –8 increases the throughput by a factor of 3.18, 3.15, and 2.61 relative to the uncompressed transfers at 1.512 GHz, 0.810 GHz, and 0.384 GHz, respectively. The download throughput of CELL transfers tends to benefit from frequency scaling—both compressed and uncompressed transfers achieve higher throughput at lower clock frequencies.

Tables 7.6 and 7.7 list the combinations of utilities and compression levels that achieve the best upload and download energy efficiencies, respectively. We show the energy efficiencies for compression and decompression tasks (EE.C and EE.D), for uncompressed file transfers (EE.UUP and EE.UDW), as well as the energy efficiency improvements that compressed transfers deliver over uncompressed transfers (EE.C/EE.UUP and EE.D/EE.UDW).

Table 7.6   Upload: Utilities that achieve peak energy efficiency

Experiment		Compression		Raw UUP	Improvement
		Best Utility	EE.C [MB/J]	EE.UUP [MB/J]	EE.C/ EE.UUP [–]
LOCAL					
$I_{idle} = 11$ mA	1.510 GHz	lzop –1 to –6	23.00	–	–
$I_{idle} = 11$ mA	0.810 GHz	lzop –1 to –6	23.00	–	–
$I_{idle} = 11$ mA	0.384 GHz	lzop –1 to –6	17.10	–	–
WLAN					
$I_{idle} = 11$ mA	1.510 GHz	gzip –1 to –2	2.90	1.40	2.07
$I_{idle} = 11$ mA	0.810 GHz	lzop –1	2.80	1.36	2.06
$I_{idle} = 11$ mA	0.384 GHz	lzop –1	1.94	1.45	1.34
CELL					
$I_{idle} = 11$ mA	1.510 GHz	bzip2 –9/xz –0	0.41	0.15	2.73
$I_{idle} = 11$ mA	0.810 GHz	bzip2/pbzip2	0.41	0.15	2.87
$I_{idle} = 11$ mA	0.384 GHz	gzip –5	0.39	0.14	2.78

Table 7.7 Download: Utilities that achieve peak energy efficiency

Experiment		Compression		Raw UUP	Improvement
		Best Utility	EE.C [MB/J]	EE.UUP [MB/J]	EE.C/ EE.UUP [–]
LOCAL					
I_{idle} = 11 mA	1.510 GHz	lzop –1 to –6	70.00	–	–
I_{idle} = 11 mA	0.810 GHz	lzop –1 to –6	67.00	–	–
I_{idle} = 11 mA	0.384 GHz	lzop –1 to –6	55.00	–	–
WLAN					
I_{idle} = 11 mA	1.510 GHz	pigz –6 to –9	8.30	3.10	2.68
I_{idle} = 11 mA	0.810 GHz	pigz –6 to –9	8.44	3.61	2.34
I_{idle} = 11 mA	0.384 GHz	lzop –7 to –9	6.28	2.74	2.29
CELL					
I_{idle} = 11 mA	1.510 GHz	xz –7	0.85	0.28	3.04
I_{idle} = 11 mA	0.810 GHz	xz –7	1.05	0.30	3.50
I_{idle} = 11 mA	0.384 GHz	xz –7	1.07	0.36	2.97

For the LOCAL compression tasks, we find that *lzop* with –1 to –6 achieves superior energy efficiency for all tested clock frequencies. When the total energy is considered, compression performed using *lzop* with –1 to –6 outperforms the next best utility by more than a factor of 2. For *LOCAL* decompression tasks, *lzop* with –1 to –9 achieves superior energy efficiency for all three clock frequencies.

When uploading data to a remote server over WLAN, the best utilities for sending compressed data are *gzip* with –1 to –2 at 1.512 GHz, and *lzop* with –1 at 0.810 and 0.384 GHz. They increase the effective energy efficiency by a factor of 2.07, 2.06, and 1.34 relative to the uncompressed uploads, respectively. When uploading data over CELL, *bzip2* with –9, *xz* with –0 at 1.512 and 0.810 GHz, and *gzip* with –5 at 0.384 GHz increase the effective energy efficiency by a factor of ∼2.7–2.8 relative to the uncompressed transfer. For both WLAN and CELL uploads, the highest energy efficiencies are achieved when using the maximum processor clock frequency.

When downloading data over WLAN, the highest download energy efficiency is achieved by *pigz* with –6 to –9 at 1.512 and 0.810 GHz, and by *lzop* with –7 to –9 at 0.384 GHz. They increase the effective energy efficiency by a factor of ∼2.7, 2.34, and 2.3 relative to the uncompressed downloads. When downloading data over CELL, *xz* with –7 increases the energy efficiency by a factor of 3.04, 4.27, and 2.97 relative to the uncompressed transfer at 1.512 GHz, 0.810 GHz, and 0.384 GHz, respectively. The highest download energy efficiencies are achieved at 0.810 GHz in the WLAN experiments and at

0.810 and 0.384 GHz in the CELL experiments. Selecting 0.810 or 0.384 GHz increases the effective energy efficiency by a factor of 1.25 relative to the best achievable energy efficiency at 1.512 GHz and by a factor of $\sim$3.5 relative to the uncompressed download at 1.512 GHz.

When transferring data over WLAN, the energy efficiency is a function of the chosen (de)compression utility, the compression level, the processor clock frequency, as well as the network conditions. In the measurements presented so far, we reported the throughputs and energy efficiency for unbounded network conditions while varying the clock frequency. For example, the uncompressed upload and download throughputs at 1.512 GHz are 2.33 MB/s and 4.46 MB/s, respectively. Since the effective throughput is a function of the processor clock frequency, the experiments with 0.810 and 0.384 GHz yielded lower uncompressed throughputs. In addition, the WLAN and CELL experiments confirmed that uncompressed transfers with lower processor clock frequencies typically maintain or improve energy efficiency in spite of lower effective throughputs. An interesting question is whether our findings would differ if we limited the effective WLAN throughput. To answer this question, we repeated the WLAN experiments under conditions where the WLAN uncompressed throughput is bounded to 0.5 MB/s, 1.5 MB/s, and 2.0 MB/s for upload data transfers and to 0.5 MB/s, 2.0 MB/s, and 4.0 MB/s for download data transfers. The effective WLAN throughput is controlled using the Linux *tc* (traffic control) utility.

Table 7.8 shows the combinations of utilities and compression levels that achieve the best throughput and the best energy efficiency for data uploads for different WLAN throughputs and processor clock frequencies. We show the effective throughput (Th.C) and energy efficiency (EE.C) for the best performing combinations. In addition, we report the throughput and energy efficiency of uncompressed uploads (Th.UUP and EE.UUP). The results indicate that combinations offering superior throughputs also offer poorer energy efficiency. For example, when the WLAN uncompressed throughput is set to 2.0 MB/s, the best effective throughput of 6.1 MB/s is achieved by *pbzip2* with –1. However, its energy efficiency is only $\sim$1.6 MB/J. In contrast, *lzop* with –1 at 0.810 GHz achieves a throughput of only 4.4 MB/s, but its energy efficiency is 2.8 MB/J. Similar findings can be observed for other network setups (1.5 MB/s and 0.5 MB/s). When the WLAN uncompressed throughput is set to 0.5 MB/s, the best effective throughput of 2.02 MB/s is achieved by *pbzip2* with –9, which has an energy efficiency of 0.99 MB/J. *gzip* with –2 at 0.384 GHz achieves a throughput of 1.38 MB/s, but its energy efficiency is 1.78 MB/J. Thus, these experiments confirm (a) that effective

Table 7.8 Upload: Effect of frequency scaling during fixed WLAN throughput

	Highest Th			Highest EE			Uncompressed	
Experiment	Best Utility	Th.C [MB/s]	EE.C [MB/J]	Best Utility	Th.C [MB/s]	EE.C [MB/J]	Th.UUP [MB/s]	EE.UUP [MB/J]
tc = 2.00 MB/s								
1.510 GHz	pbzip2 –1	**6.09**	1.57	gzip –1	5.63	2.78	2.16	1.38
0.810 GHz	pbzip2 –1	6.05	1.71	lzop –1	4.43	**2.80**	2.01	1.36
0.384 GHz	–	–	–	–	–	–	–	–
tc = 1.5 MB/s								
1.510 GHz	pbzip2 –2	**4.93**	1.50	gzip –1	3.72	3.66	1.40	1.09
0.810 GHz	pbzip2 –3	4.78	1.55	gzip –1	2.26	**2.50**	1.41	1.32
0.384 GHz	lzop –1	2.15	1.94	lzop –1	2.15	1.94	1.42	1.45
tc = 0.5 MB/s								
1.510 GHz	pbzip2 –9	**2.02**	0.99	gzip –2	1.43	1.30	0.53	0.55
0.810 GHz	pbzip2 –9	2.01	1.08	gzip –4	1.50	1.58	0.53	0.66
0.384 GHz	gzip –2	1.39	1.78	gzip –2	1.38	**1.78**	0.53	0.71

upload throughputs can be traded for better energy efficiency and (b) that in those conditions processor clock frequency changes can provide better energy efficiencies.

Table 7.9 shows combinations of utilities and compression levels that achieve the best throughput and the best energy efficiency for data downloads for different WLAN throughputs and processor clock frequencies. We show

Table 7.9 Download: Effect of frequency scaling during fixed WLAN throughput

	Highest Th			Highest EE			Uncompressed	
Experiment	Best Utility	Th.C [MB/s]	EE.C [MB/J]	Best Utility	Th.C [MB/s]	EE.C [MB/J]	Th.UUP [MB/s]	EE.UUP [MB/J]
tc = 4.00 MB/s								
1.510 GHz	pbzip2 –3	**12.48**	4.61	pigz –8	10.90	7.78	3.96	2.85
0.810 GHz	pbzip2 –4	11.85	4.64	pigz –9	10.34	**8.53**	3.90	3.14
0.384 GHz	–	–	–	–	–	–	–	–
tc = 2.0 MB/s								
1.510 GHz	xz –9	**8.04**	4.95	pigz –9	5.78	5.02	2.02	1.76
0.810 GHz	xz –7	8.03	5.26	pigz –9	5.79	5.83	2.01	2.19
0.384 GHz	lzop –8	4.51	**6.28**	lzop –8	4.51	**6.28**	1.99	2.74
tc = 0.5 MB/s								
1.510 GHz	xz –9	**2.34**	2.71	xz –7	2.33	2.72	0.55	0.81
0.810 GHz	xz –7	**2.34**	3.03	xz –7	**2.34**	3.03	0.55	0.89
0.384 GHz	xz –6	2.09	**3.10**	xz –6	2.09	**3.10**	0.55	0.93

the effective throughput (Th.D) and energy efficiency (EE.D) for the best performing combinations. In addition, we report the throughput and energy efficiency of uncompressed downloads (Th.UDW and EE.UDW). Again, the results indicate that combinations offering superior throughputs also offer poorer energy efficiency. For example, when the WLAN network throughput is set to 4.0 MB/s, the best effective throughput of 12.48 MB/s is achieved by *pbzip2* with –3. However, its energy efficiency is only ∼4.6 MB/J. In contrast, *pigz* with –9 at 0.810 GHz achieves a throughput of only 10.34 MB/s, but its energy efficiency is 8.53 MB/J. Similar findings hold for other network setups (2.0 MB/s and 0.5 MB/s). When the WLAN is set to 0.5 MB/s, the best effective throughput of 2.34 MB/s is achieved by *xz* with –9, which has an energy efficiency of 2.71 MB/J. Yet, *xz* with –6 at 0.384 GHz achieves a throughput of ∼2.1 MB/s, but its energy efficiency is 3.1 MB/J. As was the case for data uploads, these experiments confirm (a) that effective download throughputs can be traded for better energy efficiency and (b) that in those conditions processor clock frequency changes can provide better energy efficiencies.

7.6 Related Work

We are aware of two related studies that investigate data compression in the context of energy efficiency on embedded and mobile systems [18, 36]. Both studies examine the feasibility of using compression to reduce energy consumption and explore trade-offs between time, compression ratio, and energy.

The most closely related work to ours is a study by Barr and Asanović [18, 19]. It also investigates the energy efficiency of lossless data compression on a wireless mobile device. Their excellent publications include details that are beyond the scope of our work, such as the frequency with which different types of instructions are executed, the branch prediction accuracy, and the performance of the memory hierarchy. Their experimental setup has several advantages over ours. For example, their Skiff platform, which mimics an iPAQ mobile device, enabled them to separately measure the energy drawn by the CPU, the memory subsystem, peripherals, and the wireless interface. However, our test environment is superior in other aspects. Some of them are simply a result of about a decade of advances in technology. For instance, their now obsolete processor had a single core, a clock frequency of 233 MHz, and 32 MB of DRAM. The Skiff platform was further limited to 4 MB of nonvolatile flash memory. Thus, the root file system had to be mounted externally via an Ethernet port using NFS. In comparison, we use a Nexus 4 with four processor cores, which run at up to 1.512 GHz, and share 2 GB of RAM. Another advantage of our test bed is the sampling frequency of 200 kHz, which is about 5,000 times higher than theirs, presumably yielding more accurate measurements. Even when accounting for the difference in clock frequency, our hardware takes a sample every 7,560 CPU clock periods, whereas theirs sampled once per five million clocks.

There are also substantial software differences between Barr and Asanović's study and ours. Whereas several of their compression utilities are predecessors of the utilities we have evaluated, they only tested a few compression levels (we test all of them), and we include newer utilities such as *xz* and the parallel implementations *pigz* and *pbzip2*. Furthermore, their input data were limited to 1 MB of text and 1 MB of Web data. We cover a wider range of data types, and our files are over an order of magnitude larger. Due to their hardware's low sampling rate, they were forced to run the same compression or decompression task in an infinite loop to obtain sufficiently many samples. We are able to run our tests individually, that is, in a manner that is more representative of actual usage.

7.7 Conclusions

This chapter describes the experimental evaluation of recent implementations of common compression utilities on a Nexus 4 smartphone. We measure compression and decompression times, as well as the total and the overhead energy consumed by compression and decompression tasks, and we report metrics such as compression ratio, (de)compression throughput, and (de)compression energy efficiency across different compression levels. Our measurements mimic typical usage scenarios of mobile devices involving transfers of data over wireless and cellular networks.

Based on the results of our analysis, we provide practical guidelines for selecting the most energy-efficient utilities depending on the usage scenario and the selected clock frequency. For compression tasks, utilities that are fast and have low computational complexity, such as *lzop* with compression levels –1 to –6 and *pigz* with –1, are the most energy-efficient options in spite of their relatively low compression ratios. For decompression tasks, *lzop*, *gzip*, and *pigz* are good choices, as is *xz* when transferring data over a low-throughput wireless network. The selected compression level can have a significant impact. For example, at the highest tested clock frequency, the energy efficiency of compressed uploads over the wireless network using *gzip* with the default compression level –6 is 1.66 MB/J, whereas *gzip* –2 achieves 3 MB/J, an 81% improvement in energy efficiency.

We also demonstrate that lowering the processor's clock frequency can improve energy efficiency of data transfers over a cellular network. For instance, the energy efficiency of uncompressed data downloads is 0.28 MB/J at the maximum processor frequency of 1.512 GHz. The energy efficiency of compressed downloads using *gzip* with –6 is 0.65 MB/J when running at 1.512 GHz. However, the energy efficiency of *xz* with –7 is ∼1.07 MB/J at 0.810 and 0.384 GHz, an almost 1.65-fold improvement over *gzip* with –6 or a 3.82-fold improvement over uncompressed downloads.

Whereas the results of our analysis rely on a single type of smartphone, we expect that our recommendations for compression utilities that perform well in networked file transfers do not change significantly for other types of mobile platforms as shown elsewhere [16]. In addition, the analytical models we present in this chapter estimate the effective throughput and energy efficiency as a function of the following parameters: (a) network conditions, (b) platform performance and energy efficiency (reflected through local (de)compression throughputs and energy efficiencies), and (c) file characteristics (reflected through the compression utility). Using these analytical models, a user can easily evaluate the effectiveness of different file transfer modalities, provided that the parameters are known.

Our findings may guide energy optimizations of data transfers in mobile applications and encourage the development of data transfer frameworks that are conscientious of the mobile device's energy status, the user's activity level, and the benefits of frequency scaling. For example, a server could easily store multiple copies of the same file, compressed with different utilities and compression levels, to allow the mobile device to choose, based on its capabilities, currently available network bandwidth, energy status, user preferences, which version of a file to download, and at what clock frequency to perform the download and decompression. Based on the similar criteria, the mobile device could choose which format and configuration to use for uploading a file, and the server could then convert the file, if necessary, to the best download format(s). The next steps in the development of optimized data transfer methods will involve a detailed characterization of (de)compression utilities over varying file sizes, file types, network latencies, and network throughputs, as well as over varying performance and hardware parameters of the mobile devices.

References

[1] Gartner, Inc., "Gartner Says Smartphone Sales Surpassed One Billion Units in 2014," 2014. [Online]. Available: http://www.gartner.com/news room/id/2996817. [Accessed: 17-Apr-2016].

[2] Gartner, Inc., "Gartner Says Tablet Sales Continue to Be Slow in 2015," 2014. [Online]. Available: http://www.gartner.com/newsroom/id/ 2954317. [Accessed: 17-Apr-2016].

[3] Gartner, Inc., "Market Share Analysis: Mobile Phones, Worldwide, 4Q13 and 2013," 2014. [Online]. Available: http://www.gartner.com/newsroom/ id/2665715. [Accessed: 17-Apr-2016].

[4] Gartner, Inc., "Forecast: PCs, Ultramobiles, and Mobile Phones, World-wide, 2011–2018, 1Q14 Update," 2014. [Online]. Available: http:// www.gartner.com/newsroom/id/2692318. [Accessed: 17-Apr-2016].

[5] CISCO, "Cisco Visual Networking Index: Global Mobile Data Traffic Forecast Update 2014–2019 White Paper." 03-Feb-2015.

[6] "The Mobile Economy 2015." GSMA, 2015.

[7] Agababov, V., Buettner, M., Chudnovsky, V., Cogan, M., Greenstein, B., McDaniel, S., Piatek, M., Scott, C., Welsh, M., and Yin, B. (2015). Flywheel: Google's Data Compression Proxy for the Mobile Web. In *Proceedings of the 12th USENIX Conference on Networked Systems Design and Implementation*, Berkeley, CA, USA, pp. 367–380.

[8] Google, "Data Server – Google Chrome," 2014. [Online]. Available: https://developer.chrome.com/multidevice/data-compression. [Accessed: 17-Apr-2016].

[9] Amazon, "What Is Amazon Silk? – Amazon Silk," 2015. [Online]. Available: http://docs.aws.amazon.com/silk/latest/developerguide/. [Accessed: 17-Apr-2016].

[10] Onavo, "Onavo," *Onavo*, 2015. [Online]. Available: http://www.onavo. com. [Accessed: 17-Apr-2016].

[11] Snappli, "Snappli," 2014. [Online]. Available: http://snappli.com/. [Accessed: 17-Apr-2016].

[12] zlib, "zlib Home Site," 2015. [Online]. Available: http://www.zlib.net/. [Accessed: 17-Apr-2016].

[13] Google, "About Attachment Manager," 2014. [Online]. Available: http://www.google.com/support/enterprise/static/postini/docs/admin/en/ admin_msd/attach_overview.html. [Accessed: 17-Apr-2016].

[14] Dzhagaryan, A., Milenković, A., and Burtscher, M. (2015). Quantifying Benefits of Lossless Compression Utilities on Modern Smartphones.

In *24th International Conference on Computer Communication and Networks (ICCCN)*, Las Vegas, NV.

[15] Dzhagaryan, A., Milenković, A., and Burtscher, M. (2013). Energy efficiency of lossless data compression on a mobile device: An experimental evaluation. In *Performance Analysis of Systems and Software (ISPASS), 2013 IEEE International Symposium on*, Austin, TX, 2013, pp. 126–127.

[16] Milenković, A., Dzhagaryan, A., and Burtscher, M. (2013). Performance and Energy Consumption of Lossless Compression/Decompression Utilities on Mobile Computing Platforms. In *2013 IEEE 21st International Symposium on Modeling, Analysis Simulation of Computer and Telecommunication Systems (MASCOTS)*, San Francisco, CA, pp. 254–263.

[17] Dzhagaryan, A., and Milenković, A. (2015). On effectiveness of lossless compression in transferring mhealth data files. In 2015 *IEEE 17th International Conference on e-Health Networking, Applications and Services (Healthcom)*, Boston, MA.

[18] Barr, K., and Asanović, K. (2003). Energy aware lossless data compression. In *Proceedings of the 1st International Conference on Mobile Systems, Applications and Services (MobiSys'03)*, San Francisco, CA, pp. 231–244.

[19] Barr, K. C., and Asanović, K. (2006). Energy-aware lossless data compression. *ACM Trans. Comput. Syst.*, 24 (3), 250–291, Aug.

[20] "The gzip home page." [Online]. Available: http://www.gzip.org/. [Accessed: 17-Apr-2016].

[21] Ziv, J., and Lempel, A. (1977). A universal algorithm for sequential data compression. *IEEE Trans. Inf. Theory*, 23 (3), 337–343.

[22] Oberhumer, M. (2016). lzop file compressor (oberhumer.com Open-Source). [Online]. Available: http://www.lzop.org/. [Accessed: 17-Apr-2016].

[23] "bzip2: Home," 2015. [Online]. Available: http://www.bzip.org/. [Accessed: 17-Apr-2016].

[24] Burrows, M., and Wheeler, D. J. (1994). A block-sorting lossless data compression algorithm. Digital SRC, Report 124, May.

[25] "XZ Utils." [Online]. Available: http://tukaani.org/xz/. [Accessed: 17-Apr-2016].

[26] Pavlov, I. (2016). "7-Zip." [Online]. Available: http://www.7-zip.org/. [Accessed: 17-Apr-2016].

[27] "pigz – Parallel gzip," 2015. [Online]. Available: http://zlib.net/pigz/. [Accessed: 17-Apr-2016].

[28] Gilchrist, J. (2106). "Parallel BZIP2 (PBZIP2)." [Online]. Available: http://compression.ca/pbzip2/. [Accessed: 17-Apr-2016].

[29] Google, "Nexus – Google," 2014. [Online]. Available: http://www.google.com/intl/all/nexus. [Accessed: 17-Apr-2016].

[30] Qualcomm, "Adreno," *Qualcomm Developer Network*, 2015. [Online]. Available: https://developer.qualcomm.com/software/adreno-gpu-sdk/gpu. [Accessed: 17-Apr-2016].

[31] CyanogenMod, "CyanogenMod | Android Community Operating System," 2014. [Online]. Available: http://www.cyanogenmod.org/. [Accessed: 17-Apr-2016].

[32] NI, "NI PXIe-4154 – National Instruments." [Online]. Available: http://sine.ni.com/nips/cds/view/p/lang/en/nid/209090. [Accessed: 17-Apr-2016].

[33] NI, "NI PXIe-1073 – National Instruments," 2014. [Online]. Available: http://sine.ni.com/nips/cds/view/p/lang/en/nid/207401. [Accessed: 17-Apr-2016].

[34] "Android Debug Bridge | Android Developers," 2015. [Online]. Available: http://developer.android.com/tools/help/adb.html. [Accessed: 17-Apr-2016].

[35] Brodowski, D. (2016). CPU frequency and voltage scaling code in the Linux(TM) kernel. [Online]. Available: https://www.kernel.org/doc/Documentation/cpu-freq/governors.txt. [Accessed: 17-Apr-2016].

[36] Sadler, C. M., and Martonosi, M. (2006). Data compression algorithms for energy-constrained devices in delay tolerant networks. In *Proceedings of the 4th International Conference on Embedded Networked Sensor Systems*, New York, NY, pp. 265–278.

PART III

Spectrum

8

Scheduling-Inspired Spectrum Assignment Algorithms for Mesh Elastic Optical Networks

Mahmoud Fayez[3], Iyad Katib[1], George N. Rouskas[1,2] and Hossam M. Faheem[3,4]

[1]King Abdulaziz University, Jeddah, Saudi Arabia
[2]North Carolina State University, Raleigh, NC 27695, United States
[3]Fujitsu Technology Solution, Munich, Germany
[4]Ain Shams University, Khalifa El-Maamon St, Cairo, Cairo, Egypt

Abstract

Spectrum assignment has emerged as the key design and control problem in elastic optical networks. We have shown that spectrum assignment in networks of general topology is a special case of scheduling multiprocessor tasks on dedicated processors. Based on this insight, we develop and evaluate efficient and effective algorithms for mesh and chain networks that build upon list scheduling concepts.

Keywords: Elastic optical networks, Spectrum assignment, Multiprocessor scheduling theory.

8.1 Introduction

Optical networking technologies are crucial to the operation of the Internet and its ability to support critical and reliable communication services. In response to rapidly growing IP traffic demands, 40 and 100 Gbps line rates over long distances have been deployed, while there is a substantial research and development activity targeted to commercializing 400 and 1000 Gbps rates [1]. On the other hand, emerging applications, including IPTV, video-on-demand, and interdatacenter networking, have heterogeneous bandwidth

225

demand granularities that may change dynamically over time. Accordingly, it has been proposed that mixed line rate (MLR) networks [2] may be able to accommodate variable traffic demands. Nevertheless, optical networks operating on a fixed wavelength grid [3] allocate a full wavelength even to traffic demands that do not fill its entire capacity [4]. This inefficient utilization of spectral resources is expected to become an even more serious issue with the deployment of higher data rates [5, 6].

Elastic optical networks [7, 8] have the potential to overcome the fixed, coarse granularity of existing WDM technology and are expected to support flexible data rates, adapt dynamically to variable bandwidth demands by applications, and utilize the available spectrum more efficiently [6]. The enabling technology for such an agile network infrastructure is orthogonal frequency division multiplexing (OFDM), a modulation format that has been widely adopted in broadband wireless and copper-based communication systems, and is a promising candidate for high-speed (i.e., beyond 100 Gbps) optical transmission [9]. Other key technologies include distance-adaptive modulation, bandwidth-variable transponders, and flexible spectrum selective switches; for a recent survey of optical OFDM and related technologies, refer to [9].

OFDM, a multiple carrier modulation scheme, splits a data stream into a large number of substreams [10]. Each data substream is carried on a narrowband subchannel created by modulating a corresponding carrier with a conventional scheme such as quadrature amplitude modulation (QAM) or quadrature phase shift keying (QPSK). The modulated signals are further multiplexed by frequency division multiplexing to form what is referred to as multicarrier transmission. The composite signal is a broadband signal that is more immune to multipath fading (in wireless communications) and intersymbol interference. The main feature of OFDM is the orthogonality of subcarriers that allows data to travel in parallel, over subchannels constituted by these orthogonal subcarriers, in a tight frequency space without interference from each other. Consequently, OFDM has found many applications, including in ADSL and VDSL broadband access, power line communications, wireless LANs (IEEE 802.11 a/g/n), WiMAX, and terrestrial digital TV systems.

In recent years, OFDM has been the focus of extensive research efforts in optical transmission and networking, initially as a means to overcome physical impairments in optical communications [11, 12]. However, unlike wireless LANs or xDSL systems where OFDM is deployed as a transmission technology in a *single link*, in optical networks it is considered as the technology underlying the novel elastic network paradigm [6]. Consequently, in the quest for a truly agile, resource-efficient optical infrastructure, *network-wide spectrum management* arises as a key challenge; the routing and spectrum assignment (RSA) problem has emerged as an essential network design and control problem [13, 14].

In offline RSA, the input typically consists of a set of forecast traffic demands, and the objective is to assign a physical path and contiguous spectrum to each demand so as to minimize the total amount of allocated spectrum (either over the whole network or on any link). Several variants of the RSA problem have been studied in the literature that takes into account various design aspects including the reach versus modulation level (spectral efficiency) trade-off [15], traffic grooming [16], and restoration [17]. These problem variants are NP-hard, as RSA is a generalization of the well-known routing and wavelength assignment (RWA) problem [18]. Therefore, while most studies provide integer linear program (ILP) formulations for the RSA variant they address, they propose heuristic algorithms for solving medium-to-large problem instances. Such *ad hoc* solution approaches have two drawbacks. First, they do not provide insight into the structure of the optimal solution; hence, they cannot be easily adapted to other problem variants. Second, it is quite difficult to characterize the performance of heuristic algorithms; our recent work has demonstrated that heuristics for the related RWA problem produce solutions that are far from optimal even for problem instances of moderate sizes [19]. For a survey of spectrum management techniques in elastic optical networks, including a review of solution approaches to RSA problem variants, we refer to our recent survey [20].

When the path of each demand is provided as part of the input and is not subject to optimization, the RSA problem reduces to the spectrum assignment (SA) problem. The main contribution of this work is that we build upon well-understood scheduling theory techniques to develop efficient and effective algorithms for the SA problem in optical networks. We consider two types of networks: *mesh networks* of general topology, which are representative of commercial wide area networks, and *chain (linear) networks* consisting of an acyclic sequence of nodes connected in a linear chain. Our results are important because (1) the SA algorithms for mesh networks may be used to tackle the RSA problem by applying them in parallel to different routing configurations (which are independent of each other) and selecting the best solution; (2) the SA algorithms for chain networks may be used to analyze approximately large networks of general topology, e.g., by extending path-based decomposition approaches that we have developed for the case of wavelength assignment [21]; and (3) these algorithms may be applied to large-scale task scheduling problems in multiprocessor environments.

Following the introduction, we review our earlier work in Section 8.2 and provide insight into the properties and structure of the spectrum assignment problem as a special case of a general multiprocessor scheduling problem, in which a task must be executed by multiple machines simultaneously. In Section 8.3, we develop algorithms for multiprocessor scheduling, which can be used to solve the corresponding spectrum assignment problem; we present two algorithms—one for general topology (mesh) networks and a faster one for chain networks. We present numerical results to evaluate the performance of the algorithms in Section 8.4 and conclude the chapter in Section 8.5.

8.2 SA in Mesh Networks: A Special Case of Multiprocessor Scheduling

We consider the following general definition of the spectrum assignment (SA) problem in elastic optical networks:

- *SA Inputs*: (1) A graph $G = (\mathcal{V}, \mathcal{A})$, where $\mathcal{V}$ is the set of nodes and $\mathcal{A}$ is the set of arcs (directed edges); (2) a spectrum demand matrix $T = [t_{sd}]$, such that t_{sd} is the number of spectrum slots required to carry the traffic from source s to destination d; and (3) a fixed route r_{sd} from node s to node d.
- *SA Objective*: For each traffic demand, assign spectrum slots along all the arcs of its route such that the total required amount of spectrum used on any arc in the network is minimized.
- *RSA Constraints*: (1) Spectrum contiguity: each demand is assigned contiguous spectrum slots; (2) spectrum continuity: each demand uses the same spectrum slots along all arcs of its route; and (3) non-overlapping spectrum: demands that share an arc are assigned non-overlapping parts of the available spectrum.

Now, consider the multiprocessor scheduling problem $P|fix_j|C_{max}$, defined as [22]:

- $P|fix_j|C_{max}$ *Inputs*: A set of m identical processors, a set of n tasks, the processing time p_j of task j, and a set fix_j of processor sets that will execute each task j.
- $P|fix_j|C_{max}$ *Objective*: Schedule the tasks so as to minimize the makespan $C_{max} = \max_j C_j$ of the schedule, where C_j indicates the completion time of task j.

- $P|fix_j|C_{max}$ *Constraints*: (1) No preemption is allowed, (2) all the processors in the selected set must work on task j simultaneously, and (3) each processor may execute one task at most at any given time.

In earlier work [23], we have proved that the SA problem in mesh networks transforms to the $P|fix_j|C_{max}$ multiprocessor scheduling problem; however, the reverse is not true. In other words, SA is a special case of $P|fix_j|C_{max}$; hence, any algorithm for the latter problem may also solve the former. A formal proof of the transformation is available in [23]. In the transformation, the various entities of the elastic network domain transform to entities in the task scheduling domain as shown in Table 8.1. Specifically, each arc in the SA problem maps to a processor in the scheduling problem, each traffic demand to a task, the number of spectrum slots of a demand to the processing time of the corresponding task, and the maximum number of spectrum slots used on any link to the makespan of the schedule. Accordingly, minimizing the maximum spectrum allocation on any arc of the SA problem is equivalent to minimizing the makespan of the schedule in the corresponding problem $P|fix_j|C_{max}$. Furthermore, the spectrum contiguity constraint of SA is equivalent to the no preemption constraint of $P|fix_j|C_{max}$; the spectrum continuity constraint maps to the constraint that all required processors must execute a task simultaneously; and the non-overlapping spectrum constraint maps to the constraint that a processor works on at most one task at a time.

8.2.1 Illustrative Example

As an example, Figure 8.1(a) shows an instance of the SA problem on a mesh network with five directed links, $L1, L2, L3, L4$, and $L5$. There are six demands (shown as dotted lines) with the number of slots required by each demand shown next to the corresponding line. These demands are the inputs to the SA problem and are also listed in the left three columns of Table 8.2, which provides the transformation from traffic demands and corresponding paths in the elastic optical network domain to tasks and sets of processors in the task scheduling domain under the assumption that link $L1$ maps to processor

Table 8.1 Transformation of entities between the elastic network and task scheduling domains

Elastic Network Domain	Task Scheduling Domain
Arc	Processor
Traffic demand	Task
Spectrum slots of a demand	Processing time of a task
Path of a demand	Set of processors for a task
Highest assigned slot on an arc	Completion time of a processor

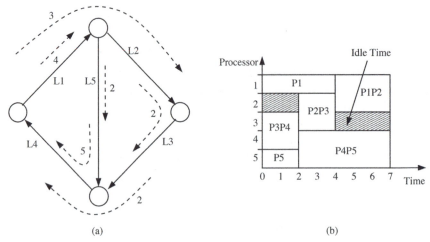

Figure 8.1 (a) Instance of the SA problem on a mesh network with five directed links (arcs). (b) Optimal schedule of the corresponding $P|fix_j|C_{max}$ problem.

Table 8.2 Mapping from the elastic network domain to the task scheduling domain

Elastic Network Domain			Task Scheduling Domain		
Demand #	Path	Spectrum Slots	Task #	Processors	Time Slots
1	L1	4	1	P1	4
2	L1, L2	3	2	P1,P2	3
3	L5	2	3	P5	2
4	L5, L4	5	4	P5, P4	5
5	L2, L3	2	5	P2, P3	2
6	L3, L4	2	6	P3, P4	2

$P1$, link $L2$ to processor $P2$, and so on. Figure 8.1(b) shows the optimal schedule for the $P|fix_j|C_{max}$ problem corresponding to this SA instance. As we can see, the demand of size 3 that follows the path $L1$–$L2$ is mapped to a task that is scheduled in the time interval $[4, 7]$ on the corresponding processors $P1$ and $P2$, similarly for the other demands. The schedule is optimal in that $C_{max} = 7$ is equal to the total processing time required for processors $P1$, $P4$, and $P5$. Also, the value of C_{max} is equal to the total number of spectrum slots required for links $L1$, $L4$, and $L5$.

8.2.2 Complexity Results

It has been shown [22] that the three-processor problem $P3|fix_j|C_{max}$ is strongly NP-hard for general processing times, but if the number of processors

m is fixed and all tasks have unit times, i.e., $Pm|fix_j, p_j = 1|C_{max}$, then the problem is solvable in polynomial time. Approximation algorithms and/or polynomial time approximation schemes (PTASs) have been developed for several versions of the problem [24].

By building upon this new perspective, it was shown in [23] that (1) $P3|fix_j|C_{max}$ transforms to the SA problem in a unidirectional ring with three links; hence, the latter is NP-hard, and (2) the SA problem is solvable in polynomial time on chain networks with at most three links, but is NP-hard on chains with four or more links. The latter result confirms the conclusion in [25] that the SA problem is harder than the wavelength assignment problem which can be solved in polynomial time on chains of any length. In [23], we also developed a set of list scheduling algorithms specifically designed for the SA problem in chains; the algorithms are both fast and effective, in that they produce solutions that are within 5% of the lower bound on average.

8.3 Scheduling Algorithms for Spectrum Assignment in Mesh Networks

We now present a new efficient scheduling algorithm for the $P|fix_j|C_{max}$ problem, which may also be used to solve the SA problem in mesh networks. The input to the algorithm is a list of n tasks, $j = 1, \ldots, n$, along with their corresponding processing times, p_j, and sets of processors, fix_j. Tasks in the list may be sorted in differently; in this work, we consider and compare the following two distinct orders:

- *Longest First (LF)*: Tasks appear in the list in decreasing order of their processing time p_j.
- *Widest First (WF)*: Tasks are listed in decreasing order of the size $|fix_j|$ of their processor set.

In either case, we assume that ties are broken arbitrarily. We also define two processor sets as *compatible* if they are disjointed, in which case the two processor sets can be scheduled simultaneously.

Figure 8.2 presents a pseudocode description of the scheduling algorithm for the $P|fix_j|C_{max}|$ problem. Depending on whether the tasks are listed in longest first or widest first order, we will refer to the scheduling algorithm as *SA-LF* or *SA-WF*, respectively[1].

---

[1]In this notation, we use the acronym "SA" to emphasize both that this is a *scheduling algorithm* for the $P|fix_j|C_{max}$ problem and the fact that this algorithm solves the *spectrum assignment* problem.

Scheduling Algorithm (SA-LF/WF) for $P|fix_j|C_{max}$
Input: A list L of n tasks on m processors, each task j having a processing time p_j and a set $fix_j \subseteq \{1, 2, \ldots, m\}$ of required processors
Output: A schedule of tasks, i.e., the time S_j when each task j starts execution on the multi-processor system

begin
1. Sort the tasks in list L based on longest-first or widest-first criteria
2. $L_p[1, \ldots, n] \leftarrow false$ //The list of in-progress tasks
3. $t \leftarrow 0$ //Scheduling instant
4. $F_p \leftarrow m$ //Counter of free processors
5. $Counter \leftarrow 0$ //Counter of finished tasks
6. $F[1, \ldots, m] \leftarrow true$ //The list of idle (free) processors
7. **while** $Counter \neq n$ **do**
8. $\quad j \leftarrow 0$
9. $\quad$ ScheduleTasks(L, L_p, F, t, j)
10. $\quad$ AdvanceTime($L_p, F, t, Counter$)
11. **end while**
12. **return** the task start times S_j
end

Procedure ScheduleTasks(L, L_p, F, t, j)
Operation: Schedules as many tasks from the input list L to start execution at time t, and moves these tasks from L to the list of in-progress tasks L_p

begin
1. **while** $j \neq n$ **and** $F_p > 0$ **do**
2. $\quad$ **if** $L_{p_j} = false$ **and** $F_{fix_j} = true$ **then**
3. $\quad\quad$ $S_j \leftarrow t$ // Task j starts execution at time t
4. $\quad\quad$ $L_{p_j} = true$
5. $\quad\quad$ $F_{fix_j} = false$
6. $\quad\quad$ $F_p \leftarrow F_p - count(fix_j)$
7. $\quad\quad$ ScheduleTasks($L, L_p, F, t, j + 1$)
8. $\quad\quad$ **break**
9. $\quad$ **endif**
10. $\quad$ $j \leftarrow j + 1$
11. **end while** //no more tasks may start at time t
end

Procedure AdvanceTime($L_p, F, t, Counter$)
Operation: Finds the first task or tasks to complete after time t, removes them from the list of in-progress tasks, and advances time to the time these tasks end

begin
    1. $j \leftarrow 0$
    2. $j_{min} \leftarrow -1$ //Index of earliest task to finish
    3. $t_{min} \leftarrow \infty$ //the default value to find minimum finish time
    4. **while** $j \neq n$ **do**
    5.   **if** $L_{p_j} = true$ **and** $S_j + p_j > t$ **and** $S_j + p_j < t_{min}$ **then**
    6. $j_{min} \leftarrow j$
    7. $t_{min} \leftarrow S_j + p_j$
    8.   **endif**
    8. $j \leftarrow j + 1$
    9. **end while**
    10. $F_{fix_{j_{min}}} \leftarrow true$ //Set processors to free again
    11. $F_p \leftarrow F_p + count(fix_{j_{min}})$
    12. $t \leftarrow t_{min}$ //Advance time
end

Figure 8.2 The scheduling algorithm SA-LF/WF for $P|fix_j|C_{max}$ and the corresponding spectrum assignment problem.

The algorithm maintains an array of m Booleans to keep track of the set of free processors, initialized to all processors, and a list of in-progress tasks initialized to the empty set. Initially, the tasks in the input list L are sorted by the longest first or widest first order. The algorithm then repeatedly calls two procedures, *ScheduleTasks()* and *AdvanceTime()*, until all the tasks in list L have been scheduled (i.e., until L becomes empty).

The *ScheduleTasks()* procedure takes the list L of unscheduled tasks, the list L_p of in-progress tasks, and the set F of free processors at time t as arguments. It then considers tasks in L one at a time in an attempt to schedule them starting at time t; note that a task j can be scheduled if all processors in fix_j are free at time t; i.e., it is pairwise compatible with all in-progress tasks. Every task that can be scheduled is marked as running in list L_p of in-progress tasks. This process continues until either the end of list L is reached or all processors become busy. At that point, procedure *AdvanceTime()* is called. This procedure finds the first in-progress task that ends and advances the time to the time t' this task ended. It then frees all involved processors for this task. Consequently, the procedure *ScheduleTasks()* is called again to schedule any remaining tasks starting at time t', and this process repeats until all tasks have been scheduled.

Each of the two procedures *ScheduleTasks()* and *AdvanceTime()* is called n times in the worst case, where n is the number of tasks. In worst-case scenarios, the *ScheduleTasks()* will consider all n tasks in list L, and for each task, it will check whether its processor set is a subset of the set F of free processors (line 2), and if so, it will remove these processors from the set (line 5); these operations take time $O(m)$ in the worst case, where m is the number of processors. Since all other operations are constant, the running time of this procedure is $O(nm)$. Procedure *AdvanceTime()* checks all in-progress tasks to identify the ones with minimum finish time and frees their processors; therefore, its worst-case running time is $O(m + n)$. Therefore, the overall running time complexity of the algorithm is $O(mn^2)$.

8.3.1 Scheduling Algorithm for Chain Networks

In the special case of chain networks, the corresponding scheduling problem is such that the m processors, each corresponding to a link of the chain, can be labeled linearly as $1, \ldots, m$. Furthermore, the processors required by each task are contiguous and may be represented as a range, a fact that has two implications. First, checking whether a task's processors are free at some time t can be performed in constant time, rather than in time $O(m)$ as in the general case of mesh networks. Second, assigning processors to a task naturally divides the previous free range of processors into (at most) two parts: one part consisting of free processors with labels smaller than that of the processor with the lowest label required by the task and one consisting of free processors with labels larger than that of the processor with the highest label required by the task. Therefore, it is possible to schedule tasks by recursively searching the (at most) two free ranges of processors created when a task is scheduled.

The recursive version of *ScheduleTasks()* for chain networks is shown in Figure 8.3. Since, as we mentioned above, the operations in Steps 2 and 5 now take constant time, the worst-case running time of the procedure for chain networks is $O(n)$. Therefore, the overall running time of the scheduling algorithm is $O(n^2)$ and is independent of the number m of processors (or links of the chain).

Procedure ScheduleTasks(L, L_p, F, t)
Operation: Schedules as many tasks from the input list L to start execution at time t, and moves these tasks from L to the list of in-progress tasks L_p

begin
begin
   1. **while** $j \neq n$ **and** $F_p > 0$ **do**
   2.   **if** $L_{p_j} = false$ **and** $F_{fix_j} = true$ **then**
   3. $S_j \leftarrow t$ // Task j starts execution at time t
   4. $L_{p_j} = true$
   5. $F_{fix_j} = false$
   6. $F_p \leftarrow F_p - count(fix_j)$
   8. $F_l \leftarrow$ new left range of free processors
   9. $F_r \leftarrow$ new right range of free processors
   7.   ScheduleTasks($L, L_p, F_l, t, j+1$)
   7.   ScheduleTasks($L, L_p, F_r, t, j+1$)
   8.   **break**
   9.   **endif**
  10. $j \leftarrow j+1$
  11. **end while** // no more tasks may start at time t
end

Figure 8.3 A specialized version of the *ScheduleTasks()* procedure for the $P|fix_j|C_{max}$ problem corresponding to a chain network.

8.4 Numerical Results

In this section, we present the results of simulation experiments we carried out to evaluate the performance of the scheduling algorithms for mesh and chain networks. We assume that the elastic optical network supports the following data rates (in Gbps): 10, 40, 100, 400, and 1000. For each problem instance, we generate random traffic rates between each pair of nodes based on one of the three distributions:

1. *Uniform*: Traffic demands may take any of the five discrete values in the set $\{10, 40, 100, 400, 1000\}$ with equal probability;
2. *Skewed low*: Traffic demands may take one of the five discrete values above with probabilities $0.30, 0.25, 0.20, 0.15$, and 0.10, respectively (i.e., the lower data rates have higher probability to be selected); or

3. *Skewed high*: Traffic demands may take one of the five discrete values above with probabilities $0.10, 0.15, 0.20, 0.25$, and 0.30, respectively (i.e., the higher data rates have higher probability to be selected).

In our experiments, we also used various other probability values the skewed low and high distributions, but the results are very similar to those shown below.

Once the traffic rates between every source–destination pair have been generated, we calculate the corresponding spectrum slots as follows. We assume that the slot width is 12.5 GHz and that there is a 16-QAM modulation format, such that demands of size 10, 40, 100, 400, and 1000 Gbps require 1, 1, 2, 8, and 20 slots, respectively, consistent with the values used in [26, Table 8.1]. We then transform the SA problem instance to the equivalent instance of $P|fix_j|C_{max}$ and run the scheduling algorithm described in the previous section to schedule the tasks.

In order to evaluate the performance of our algorithms, and since the optimal solution is not known due to the fact that $P|fix_j|C_{max}$ is NP-complete, we compute the lower bound as follows. Consider an instance of $P|fix_j|C_{max}$, and let T_k denote the set of tasks that require processor k, i.e., $T_k = \{j : k \in fix_j\}$. Clearly, all the tasks in T_k are pairwise incompatible; hence, they have to be executed sequentially. Let Π_k denote the sum of processing times of tasks that require processor k:

$$\Pi_k = \sum_{j \in T_k} p_j, \quad k = 1, \dots, m. \tag{8.1}$$

Then, a lower bound LB for the problem instance can be obtained as follows:

$$LB = \max_{k=1,\dots,m} \{\Pi_k\}. \tag{8.2}$$

We then compute the ratio of the makespan produced by the algorithm to this lower bound. It should be noted that the lower bound is not tight, as it ignores any gaps introduced by the scheduling of incompatible tasks in the optimal solution.

Each point in the figures shown in the remainder of this section is the average of 200 randomly generated problem instances for the specified parameters. We also estimated confidence intervals using the method of batch means, but they are so narrow that they are omitted. All the experiments were carried out using the resources of the high-performance computing cluster installed by Fujitsu at King Abdulaziz University, Jeddah, Saudi Arabia.

8.4.1 Mesh Networks

We have carried out simulation experiments on three network topologies[2]: (1) a 10-node, 32-link network; (2) the 32-node, 108-link network shown in Figure 8.4; and (3) a 75-node, 200-link topology. For each network topology, we generate random traffic demands between each source–destination pair[3] using each of the three distributions described earlier. Each demand is routed over the shortest path between its source and destination nodes.

The results for the 10-node network are shown in Figure 8.5. In four problem instances, the SA-LF algorithm found solutions with a makespan of about 10% higher than the lower bound, while for the remaining instances, the solutions constructed by the algorithm had a makespan equal to the lower bound, and hence, they are optimal. For the 32- and 75-node networks, the SA-LF produced optimal solutions for all three traffic distributions and all random problem instances we generated.

Figures 8.6 and 8.7 show the performance of the SA-LF and SA-WF algorithms, respectively, on complete mesh networks of varying sizes, in which traffic between each pair of nodes is routed over a randomly chosen path.

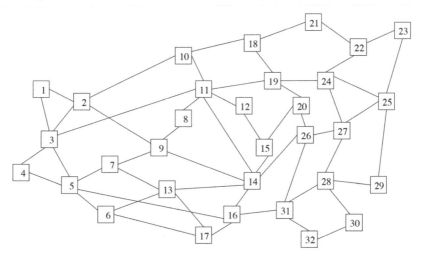

Figure 8.4 The 32-node, 108-link topology used in the experiments.

---

[2]The number of links of each network topology refers to directional links (arcs) since each direction of a link is considered independently for the purpose of spectrum assignment.

[3]The number of traffic demands for the 10-node, 32-node, and 75-node networks are 90, 992, and 5550, respectively.

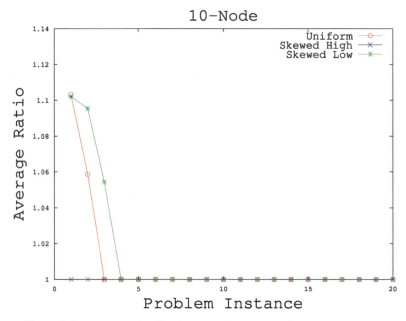

Figure 8.5   Average ratio of makespan to lower bound for 10-node network.

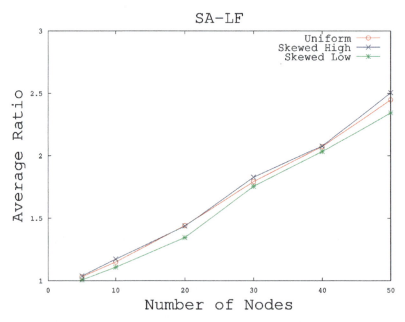

Figure 8.6   Average ratio of makespan to lower bound, complete mesh with SA-LF.

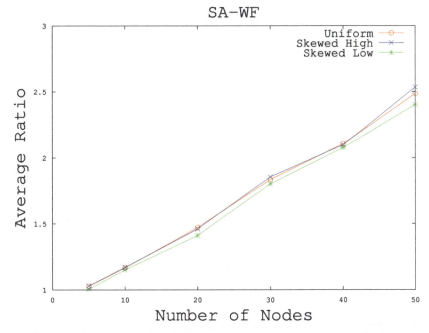

Figure 8.7 Average ratio of makespan to lower bound, complete mesh with SA-WF.

The performance of the two algorithms is similar and is not affected significantly by the traffic distribution (uniform, skewed high, or skewed low). The reason that the ratio of makespan to lower bound increases with the size of the network is due to the fact that, for a complete mesh, the optimal solution would be for each demand to take the single-link shortest path to its destination. However, since each demand is routed along a randomly selected path, and the lengths of these random paths increase with the size of the network, such a solution will move further from the lower bound (optimal) as the network size increases. The performance of the algorithms in the two figures simply reflects this observation.

8.4.2 Chain Networks

Figures 8.8 and 8.9 show results for chain networks of varying sizes and the SA-LF and SA-WF algorithms, respectively. We observe that the SA-LF algorithm performs better for all three traffic distributions and produces results that are within 5% of the lower bound. The SA-WF algorithm, which considers tasks for scheduling based on the number of processors they require,

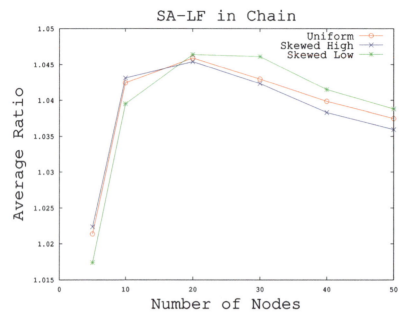

Figure 8.8 Average ratio of makespan to lower bound, chain networks with SA-LF.

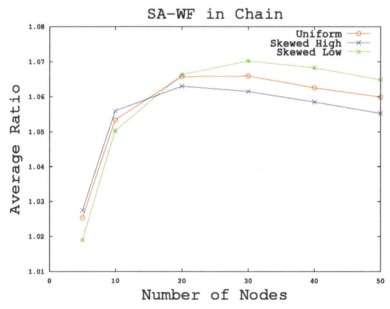

Figure 8.9 Average ratio of makespan to lower bound, chain networks with SA-WF.

may pair long tasks with short ones, thus creating gaps that result in a longer makespan.

8.4.3 Running Time Scalability

Let us now turn our attention to the scalability of the SA-LF and WA-LF algorithms in terms of running time. Figure 8.10 plots the running time of the two algorithms, in seconds, as a function of the number of tasks in the multiprocessor scheduling problem. Two plots are shown in the figure. For the curve labeled "general scheduling," the processors assigned to a particular task were randomly selected from the set of all processors without any restrictions; consequently, these problem instances correspond to the general version of the $P|fix_j|C_{max}$ problem. For the curve labeled "chain scheduling," on the other hand, processors were labeled from 1 to m. For each task, an integer k and start processor p, $1 \leq p \leq m - k$ were selected randomly, and the set of k sequentially labeled processors, $p, \ldots, p + k - 1$, were assigned to execute the task. These problem instances correspond to the special case of the $P|fix_j|C_{max}$ problem derived from chain networks.

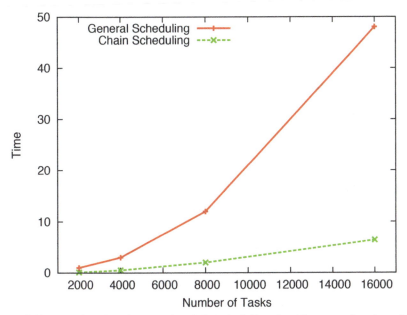

Figure 8.10   Running time, in seconds, of the scheduling algorithms as a function of the number of tasks.

As Figure 8.10 indicates, the scheduling algorithms scale to large instances of the two types of $P|fix_j|C_{max}$ problem. For the most general variant of the problem, instances of up to 16,000 tasks may be scheduled in less than one minute. Note that 16,000 tasks roughly correspond to the number of bidirectional traffic demands in a network with 125 nodes. For the special case of "chain scheduling," 16,000 tasks may be scheduled in less than seven seconds, due to the recursive nature of the corresponding algorithms. These results provide a strong indication that our algorithms may be used to tackle efficiently and effectively large instances of the spectrum assignment and corresponding task scheduling problems.

8.5 Concluding Remarks

We have developed scheduling algorithms that efficiently solve the spectrum assignment problem in mesh networks with good performance, under the assumption that the routing path of each traffic demand is fixed, i.e., it is part of the input to the problem and not subject to optimization. Our current research efforts are directed toward algorithms that jointly tackle the routing and spectrum assignment problems. More specifically, we are developing *parallel algorithms* for the RSA problem that applies the scheduling algorithms we presented in this chapter to a large number of reasonable routing configurations simultaneously.

Acknowledgments

This work was supported in part by the National Science Foundation under Grant CNS-1113191 and by the High-Performance Computing Project at King Abdulaziz University.

References

[1] Winzer, P. J. (2010). Beyond 100G Ethernet. *IEEE Commun. Mag.*, 48(7), 26–30.
[2] Nag, A. and Tornatore, M. (2009). Optical network design with mixed line rates. *Opt. Switch. Netw.*, 6(3), 227–237.
[3] ITU-T G.694.1. Spectral grids for WDM applications: DWDM frequency grid, February 2002.
[4] Shen, G. and Zukerman, M. (2012). Spectrum-efficient and agile CO-OFDM optical transport networks: architecture, design, and operation. *IEEE Commun. Mag.*, 50(5), 82–89.

[5] Jinno, M., Ohara, T., Sone, Y., Hirano, A., Ishida, O., and Tomizawa, M. (2011). Elastic and adaptive optical networks: possible adoption scenarios and future standardization aspects. *IEEE Commun. Mag.*, 49(10), 164–172.

[6] Jinno, M., Takara, H., and Kozicki, B. (2009). Dynamic optical mesh networks: Drivers, challenges and solutions for the future. In *Proceedings of 35th European Conference on Optical Communication (ECOC)*, page 7.7.4, September.

[7] Gerstel, O., Jinno, M., Lord, A., and J B Yoo, S. (2012). Elastic optical networking: a new dawn for the optical layer? *IEEE Commun. Mag.*, 50(2), s12–s20.

[8] Jinno, M., Takara, H., Kozicki, B., Tsukishima, Y., Yoshimatsu, T., Kobayashi, T., Miyamoto, Y., Yonenaga, K., Takada, A., Ishida, O., and Matsuoka, S. (2008). Demonstration of novel spectrum-efficient elastic optical path network with per-channel variable capacity of 40 Gb/s to over 400 Gb/s. In *Proceedings of 34th European Conference on Optical Communication (ECOC)*, page Th.3.F.6, September.

[9] Zhang, G., De Leenheer, M., Morea, A., and Mukherjee, B. (2013). A survey on OFDM-based elastic core optical networking. *IEEE Commun. Surv. Tutor.* 15(1), 65–87, First Quarter.

[10] Shieh, W. (2011). OFDM for flexible high-speed optical networks. *J. Lightwave Technol.*, 29(10), 1560–1577.

[11] Lowery, A. J. and Armstrong, J. (2006). Orthogonal-frequency-division multiplexing for dispersion compensation of long-haul optical systems. *Opt. Express*, 14(6), 2079–2084.

[12] Lowery, A. J., Du, L. B., and Armstrong, J. (2007). Performance of optical OFDM in ultralong-haul WDM lightwave systems. *J. Lightwave Technol.*, 25(1), 131–138.

[13] Klinkowski, M. and Walkowiak, K. (2011). Routing and spectrum assignment in spectrum sliced elastic optical path network. *IEEE Commun. Lett.*, 15(8), 884–886.

[14] Wang, Y., Cao, X., and Pan, Y. (2011). A study of the routing and spectrum allocation in spectrum-sliced elastic optical path networks. In *Proceedings of IEEE INFOCOM*, pages 1503–1511.

[15] Christodoulopoulos, K., Tomkos, I., and Varvarigos, E. A. (2011). Elastic bandwidth allocation in flexible OFDM-based optical networks. *J. Lightwave Technol.*, 29(9), 1354–1366.

[16] Zhang, Y., Zheng, X., Li, Q., Hua, N., Li, Y., and Zhang, H. (2011). Traffic grooming in spectrum-elastic optical path networks. In *Proceedings of*

Optical Fiber Communication Conference and the National Fiber Optic Engineers Conference (OFC/NFOEC), page OTuI1, March.

[17] Wei, Y., Shen, G., and You, Sh. (2012). Span restoration for CO-OFDM-based elastic optical networks under spectrum conversion. In *Proceedings of Asia Communications and Photonics Conference (ACP)*, page AF3E.7, Novomber.

[18] Rouskas, G. N. (2001). Routing and wavelength assignment in optical WDM networks. In *J. Proakis (Editor), Wiley Encyclopedia of Telecommunications*. John Wiley & Sons.

[19] Liu, Z. and Rouskas, G. N. (2012). A fast path-based ILP formulation for offline RWA in mesh optical networks. In *Proceedings of IEEE GLOBECOM 2012*, pages 2990–2995, December.

[20] Talebi, S., Alam, F., Katib, I., Khamis, M., Khalifah, R., and Rouskas, G. N. (2014). Spectrum management techniques for elastic optical networks: A survey. *Opt. Switch. Netw.*, 13, 34–48.

[21] Zhu, Y., Rouskas, G. N., and Perros, H. G. (2000). A path decomposition approach for computing blocking probabilities in wavelength routing networks. *IEEE/ACM Trans. Netw.*, 8(6), 747–762.

[22] Hoogeveen, J. A., Van de Velde, S. L., and Veltman, B. (1994). Complexity of scheduling multiprocessor tasks with prespecified processor allocations. *Discrete Appl. Math.*, 55, 259–272.

[23] Talebi, S., Bampis, E., Lucarelli, G., Katib, I., and Rouskas, G. N. (2014). Spectrum assignment in optical networks: A multiprocessor scheduling perspective. *J. Opt. Commun. Netw.*, 6(8), 754–763.

[24] Bampis, E. and Kononov, A. (2001). On the approximability of scheduling multiprocessor tasks with time dependent processing and processor requirements. In *Proceedings of the 15th International Parallel and Distributed Processing Symposium*, San Francisco.

[25] Shirazipourazad, S., Zhou, Ch., Derakhshandeh, Z., and Sen, A. (2013). On routing and spectrum allocation in spectrum-sliced optical networks. In *Proceedings of IEEE INFOCOM*, pages 385–389, April.

[26] Jinno, M., Kozicki, B., Takara, H., Watanabe, A., Sone, Y., Tanaka, T., and Hirano, A. (2010). Distance-adaptive spectrum resource allocation in spectrum-sliced elastic optical path network. *IEEE Commun. Mag.*, 48(8), 138–145.

9

Wideband Spectrum Sensing in Cognitive Radio Networks

Prosanta Paul[1], ChunSheng Xin[1], Min Song[2] and Yanxiao Zhao[3]

[1]ECE Department, Old Dominion University, 5115 Hampton Blvd, Norfolk, VA 23529, USA
[2]CS Department, Michigan Technological University, 1400 Townsend Dr, Houghton, MI 49931, USA
[3]ECE Department, South Dakota School of Mines & Technology, 501 E St Joseph St, Rapid City, SD 57701, USA

Keywords: Hypothesis Testing, Energy Detection, Matched Filter Detection, Cyclostationary Feature Detection, Wavelet basis function, Mother wavelet, Continuous WT, Discrete wavelet basis function, Scaling function, Daubechies wavelet, Signal Edges, Spectrum Sensing by WTMM, Wavelet Transform, Spectrum Sensing by WTMP and WTMS, Edge Detection through EMAMS.

Spectrum sensing plays a critical role in *cognitive radio networks* (CRNs). There are two main objectives for spectrum sensing. First, *secondary users* (SUs) perform spectrum sensing to identify the unused *primary user* (PU) frequency bands for opportunistic spectrum access for SU communications. Second, SUs need to detect the presence of PU signals on a frequency band so that the PU transmission is protected from the interference of SU communications.

In this chapter, we first briefly discuss the concept of spectrum sensing in the context of hypothesis testing. Then, we introduce the major spectrum sensing methods to detect PU signals on a single band, including energy detection, matched filter detection, and feature detection. These spectrum sensing methods typically have difficulty to be used for wideband spectrum sensing or subdivided band spectrum sensing, which aims to identify all idle sub-bands in a wide band. Hence, we further introduce wideband spectrum sensing methods, which are mostly based on wavelet transform, such as the *wavelet transform multiscale summation* (WTMS), *wavelet transform multiscale product* (WTMP), and *wavelet transform modulus maxima* (WTMM). The objective of wideband spectrum sensing or subdivided band spectrum sensing is to find the *edges* of the signals, i.e., the start and the end frequencies of each signal, on the frequency domain. Hence, wideband spectrum sensing is also called *edge detection*.

The existing wideband spectrum sensing methods often have difficulty to accurately detect both the presence of the signals and the locations of the signal edges, in particular for real-world signals. As such, as the last part of this chapter, we present a new wavelet-based wideband spectrum sensing algorithm, called the *exponentially moving averaged multiscale summation* (EMAMS), to effectively address this issue.

9.1 Hypothesis Testing

The hypothesis testing in spectrum sensing is a concept in statistics derived by Neyman and Pearson [1]. A two-state hypothesis is widely assumed in spectrum sensing for CRNs. The two states are *absence* and *presence* of PU signal in a frequency band which are denoted by $\mathcal{H}_0$ and $\mathcal{H}_1$, respectively. The two hypothesis states for a received signal $r(t)$ at an SU are written as:

$$r(t) = \begin{cases} n(t), & \text{if } \mathcal{H}_0 \\ g(t)s(t) + n(t), & \text{if } \mathcal{H}_1 \end{cases} \tag{9.1}$$

where $s(t)$ is the PU-transmitted signal, $g(t)$ is the channel gain due to path loss and lognormal shadow fading, and $n(t)$ is a zero mean *additive white Gaussian noise* (AWGN), i.e., $n(t) \sim \mathcal{N}(0, \sigma_n^2)$. The performance of spectrum sensing is primarily characterized by two metrics: the *false alarm* probability P_f and the *miss-detection* probability P_m. P_f is the probability that the SU incorrectly decides that the PU signal is present when it is actually not. P_m is the probability that the SU misses to detect a PU signal when it is actually present. Note that the signal detection probability P_d is the complement of miss-detection probability, i.e., $P_d = 1 - P_m$. P_d is the probability that the SU correctly detects the PU signal. The Neyman–Pearson criterion maximizes the PU signal detection probability, P_d, with a given false alarm probability.

The *likelihood ratio test* (LRT) for deciding the presence or absence of a PU signal is given as follows [1]:

$$\Lambda(r) = \frac{f_{(r|\mathcal{H}_1)}}{f_{(r|\mathcal{H}_0)}} \begin{array}{c} \mathcal{H}_0 \\ \gtrless \\ \mathcal{H}_1 \end{array} \eta, \tag{9.2}$$

where $f_{(r|\mathcal{H}_1)}$ and $f_{(r|\mathcal{H}_0)}$ are the probability density functions of $r(t)$ under the $\mathcal{H}_1$ and $\mathcal{H}_0$ hypotheses, respectively, and η is the PU's signal detection threshold, which is determined by a specified maximum false alarm probability constraint in the Neyman–Pearson hypothesis testing. A PU signal in a frequency band is determined to be present if the LRT in (9.2) is greater than the detection threshold η.

The Bayes criterion minimizes the average cost function in spectrum sensing as:

$$\bar{C} = \sum C_{ij} P_{(\mathcal{H}_i|\mathcal{H}_j)} P_{\text{prior}}, \qquad i,j = 0, 1, \tag{9.3}$$

where $\mathcal{C}_{ij}$ is the cost of choosing hypothesis $\mathcal{H}_i$ when $\mathcal{H}_j$ is true, $P_{(\mathcal{H}_i|\mathcal{H}_j)}$ is the conditional probability denoting the probability of $\mathcal{H}_i$ given $\mathcal{H}_j$, and P_{prior} is the prior probability of a hypothesis, which is assumed to be a known parameter in this test. The Bayes criterion decision rule is chosen such that the average cost $\bar{C}$ in (9.3) is minimized. Furthermore, in the Bayes hypothesis testing, the detection threshold, η, for LRT in (9.2) is given as:

$$\eta = \frac{P_{(\mathcal{H}_0)} \cdot (\mathcal{C}_{10} - \mathcal{C}_{00})}{P_{(\mathcal{H}_1)} \cdot (\mathcal{C}_{01} - \mathcal{C}_{11})} \tag{9.4}$$

As an example, we illustrate the corresponding parameters in spectrum sensing for IEEE 802.22 WRAN in Table 9.1 [2].

9.2 Single-Band Spectrum Sensing Methods

Next, we discuss three widely used spectrum sensing methods to detect a PU signal on a single band: energy detection, matched filter detection, and

Table 9.1 802.22 WRAN spectrum sensing parameters

Parameter	Analog TV	Digital TV	Wireless Microphone
Bandwidth	6 MHZ	6 MHZ	200 KHZ
Detection probability	0.9	0.9	0.9
False alarm probability	0.1	0.1	0.1
Channel detection time	≤2 second	≤2 second	≤2 second
Incumbent detection threshold	−94 dBm	−116 dBm	−106 dBm
SNR	1 dB	−21 dB	−12 dB

cyclostationary feature detection. We will also briefly discuss other methods at the end of this section.

9.2.1 Energy Detection

Energy detection is a widely used method for spectrum sensing as it does not need any knowledge of the PU signal and is also simple to be implemented. It measures the average energy of the received signal in a sensing period and compares the signal energy with a prespecified threshold. If the received signal energy is greater than the threshold, it is assumed that a PU is present in the band; otherwise, the SU perceives the band is idle [3–9]. In the beginning of each sensing period, an SU measures the average signal energy on a band. Let $r(t)$ be the received signal as given in (9.1) and N denote the total number of samples taken from $r(t)$ in one sensing period, i.e., $\mathbf{r} = (r[1], r[2], \ldots, r[N])^T$, and the corresponding sampled PU signal is $\mathbf{s} = (s[1], s[2], \ldots, s[N])^T$. The decision statistics of the energy detection is given as follows:

$$\lambda(r) = \frac{1}{N}[\mathbf{r} \cdot \mathbf{r}^T] \mathop{\lessgtr}_{\mathcal{H}_1}^{\mathcal{H}_0} \eta \tag{9.5}$$

The decision statistics $\lambda(r)$ is compared with a threshold η to determine the presence of PU signal. The SU makes a decision that a PU is present when $\lambda(r) \geq \eta$; otherwise, it declares no PU is present. The decision statistics $\lambda(r)$ has a chi-square distribution. However, if N is large enough (practically $N \geq 20$ is sufficient [10]), according to the central limit theorem, $\lambda(r)$ can be modeled as a Gaussian distribution as follows:

$$\lambda(r) \sim \begin{cases} \mathcal{N}(N\sigma_n^2, 2N\sigma_n^4), & \text{if } \mathcal{H}_0 \\ \mathcal{N}(N(\sigma_n^2 + \sigma_s^2), 2N(\sigma_n^2 + \sigma_s^2)^2), & \text{if } \mathcal{H}_1 \end{cases} \tag{9.6}$$

where σ_n^2 and σ_s^2 are the noise and PU signal variance, respectively. When $r(t)$ is sampled with the Nyquist rate, the total number of samples $N = 2\tau_s B$, where τ_s is the sensing duration of SU and B is the PU channel bandwidth. The *signal-to-noise ratio* (SNR) is defined as the ratio of the PU signal power to the noise power which is given as:

$$\text{SNR} = \frac{\sigma_s^2}{\sigma_n^2} = \frac{||s||^2}{N\sigma_n^2}, \tag{9.7}$$

where $||s||^2/N$ represents the average PU signal power. In this way, the PU signal detection probability, P_d, and the false alarm probability, P_f, for energy detector are given as [5, 6, 9, 10]:

$$P_d = P(\lambda(r) > \eta | \mathcal{H}_1) = Q\left(\frac{\eta - N(\sigma_n^2 + \sigma_s^2)}{\sqrt{2N(\sigma_n^2 + \sigma_n^2)^2)}}\right)$$

$$= Q\left(\frac{\eta - N\sigma_n^2(1 + \text{SNR})}{\sigma_n^2(1 + \text{SNR})\sqrt{2N}}\right), \qquad (9.8)$$

$$P_f = P(\lambda(r) > \eta | \mathcal{H}_0) = Q\left(\frac{\eta - N\sigma_n^2}{\sqrt{2N\sigma_n^4}}\right), \qquad (9.9)$$

where $Q(.)$ is the Q-function.

A miss detection of a PU signal results in interference to the PU caused by SU transmissions. Thus, a large P_d is always desirable in spectrum sensing. In addition, P_f should be kept as small as possible to prevent underutilization of the PU spectrum band. However, there is a trade-off between P_f and P_d. If we have the knowledge of the noise and PU signal power, it is possible to select the η value to have a good balance between P_f and P_d. The noise power is estimated as AWGN with zero mean and variance σ_n^2. Nevertheless, the estimation of the PU signal $s(t)$ is difficult since the channel characteristic $g(t)$ in (9.1) is related to the path loss and shadow fading of the PU signal $s(t)$. The path loss is dependent on the distance between the SU and the PU transmitters, which is uncertain since the SU is not aware of the PU transmitter location. In practice, the decision threshold η is derived from (9.9) for a desired P_f value and the estimated σ_n^2 as follows:

$$\eta = N\sigma_n^2 \left(\sqrt{\frac{2}{N}}Q^{-1}(P_f) + 1\right) \qquad (9.10)$$

Moreover, in order to achieve a certain P_d and P_f, the required number of samples, N, is calculated using (9.8) and (9.9) as follows:

$$N = 2[(1 + \text{SNR})Q^{-1}(P_d) - Q^{-1}(P_f)]^2 \text{SNR}^{-2} \qquad (9.11)$$

The energy detection method is relatively simple to be implemented. There have been many works on energy detection in past few decades. Usually, this detection method does not require any PU signal information, although a better detection performance can be achieved if an SU knows the PU's location information [11]. However, this method may incur a higher computational complexity and larger sensing delay, in particular for wideband sensing [10]. Its performance also degrades when the SNR at the SU receiver decreases. In addition, it is a blind technique in that it is unable to distinguish the background noise from the PU signal when the energy of these two is at the same level.

9.2.2 Matched Filter Detection

The matched filter detection method is a coherent detection technique where a known PU signal, $s(t)$, correlates with the unknown received signal, $r(t)$, at an SU. The matched filter method can improve detection performance if $s(t)$ is deterministic and known *a priori*. If an SU knows some PU signal information such as the operating bandwidth and frequency, pulse shaping, frame format, and type of modulation, then the matched filter method can be an excellent detection method since it can maximize the received SNR at the SU receiver [12]. Let $r(t)$ and $s(t)$ denote the SU-received signal and the PU-transmitted signal, respectively. The decision statistic is calculated as follows:

$$\lambda(r) = \frac{1}{N} \sum_{t=1}^{N} [r(t)s^*(t)] \underset{\mathcal{H}_1}{\overset{\mathcal{H}_0}{\lessgtr}} \eta, \qquad (9.12)$$

where the *finite impulse response* (FIR) of the matched filter is $s^*(t)$ which is the complex conjugate of the PU signal $s(t)$.

We assume the decision statistic $\lambda(r)$ is normally distributed for both hypotheses given as [10]:

$$\lambda(r) \sim \begin{cases} \mathcal{N}(0, N\varepsilon\sigma_n^2), & \text{if } \mathcal{H}_0 \\ \mathcal{N}(N\varepsilon, N\varepsilon\sigma_n^2), & \text{if } \mathcal{H}_1 \end{cases} \qquad (9.13)$$

where ε is the average energy of $s(t)$, i.e., $\varepsilon = ||s||^2/N$. Hence, P_d and P_f are given as:

$$P_d = P(\lambda(r) > \eta | \mathcal{H}_1) = Q\left(\frac{\eta - N\varepsilon}{\sqrt{N\varepsilon\sigma_n^2}}\right), \qquad (9.14)$$

$$P_f = P(\lambda(r) > \eta | \mathcal{H}_0) = Q\left(\frac{\eta}{\sqrt{N\varepsilon\sigma_n^2}}\right). \qquad (9.15)$$

The detection threshold for a given P_f in this method is derived as:

$$\eta = \sqrt{N\varepsilon\sigma_n^2}(Q^{-1}(P_f)). \qquad (9.16)$$

Accordingly, the number of samples, N, required to meet a certain P_d and P_f is given as follows:

$$N = [Q^{-1}(P_d) - Q^{-1}(P_f)]^2 \text{SNR}^{-1} \qquad (9.17)$$

This method requires a comparatively shorter sensing duration than the energy detection method. It can also distinguish the difference between the noise and a PU signal. That is, it is not a blind detection technique like energy detection. The matched filter method is the optimal detector that can maximize the SNR in the presence of noise when the PU signal, $s(t)$, is known to SU *a priori*. However, the matched filter detector is not a suitable choice for wideband sensing since it requires the PU signal information. Moreover, the implementation complexity and the power consumption are higher compared with the energy detection method. Another disadvantage of this method is that if the SU-received signal SNR decreases for a specified maximum P_f, the analog-to-digital processing is required to analyze more samples, which is, however, a major cause of synchronization mismatch between the PU transmitter and the SU receiver [3].

9.2.3 Cyclostationary Feature Detection

The cyclostationary feature detection is a spectrum sensing technique which exploits some cyclostationary features of the received signal. PU signals are often cyclostationary. More specifically, most transmitted signals are modulated by sinusoidal carriers and have some cyclostationary features such as line coding, symbol periods, and cyclic prefixes, to provide redundancy [13]. This sensing method exploits such redundancy property to build the cyclic correlation function instead of analyzing the *power spectral density* (PSD) of the received signal. On the other hand, noise is a non-cyclostationary process. The autocorrelation function of $n(t)$ is given as:

$$R_n(t, \tau) = E[n(t)n(t - \tau)] = R_n(\tau), \qquad (9.18)$$

where $E[\cdot]$ is the expectation operator. The PU-transmitted signal, $s(t)$, exhibits some inherent periodicity both in its mean and in its autocorrelation function, which helps to design a unique feature pattern recognition filter. More specifically, $s(t)$ can be characterized as a *wide sense stationary* (WSS) process with its mean and autocorrelation showing periodicity. Mathematically, the mean and autocorrelation function of $s(t)$ are given as:

$$\mu_s = E[s(t)], \qquad (9.19)$$

$$R_s(t_1, t_2) = E[s(t_1)s^*(t_2)]. \qquad (9.20)$$

In a WSS process, for all time samples $s(t_1), s(t_2), \ldots, s(t_N)$, (9.19) and (9.20), are given as:

$$\mu_s(t) = \mu_s(t + T), \qquad (9.21)$$

$$R_s(t_1, t_2) = R_s(t_1 + T, t_2 + T), \qquad (9.22)$$

where T is the time period of signal $s(t)$. The *cyclic autocorrelation function* (CAF) of a WSS process with a cyclic frequency $\beta \neq 0$ is defined as [4, 10]:

$$R_s^\beta(\tau) = E[s(t)s^*(t+\tau)e^{-2\pi\beta t}], \qquad (9.23)$$

where τ is the sampling interval in signal $s(t)$. The *cyclic spectrum density* (CSD) is the Fourier series expansion of CAF which is denoted as [10]:

$$S_s^\beta(f) = \sum_{-\infty}^{+\infty} R_s^\beta(\tau)e^{-2\pi f\tau}. \qquad (9.24)$$

When the cyclic frequency β and the fundamental frequency of $s(t)$ are the same, the CSD function reaches its maximum value under the $\mathcal{H}_1$ hypothesis. On the other hand, the CSD function does not show any such peak under the $\mathcal{H}_0$ hypothesis since noise is not a cyclostationary process. Unlike energy detection, the feature detection technique can distinguish noise from the PU signal. Moreover, this detection method can be used to detect weak signals at the low-SNR region where the energy detection and the matched filter detection perform poorly.

The feature detection method can achieve better performance than the energy detection and the matched filter methods. However, this method can be applicable to only a few types of PUs, due to the fundamental assumption that the periodicity, T, of signal $s(t)$ is known *a priori* to an SU. For example, the TV band periodicity property can be known to SUs. Therefore, the feature detection is a good choice for TV signal detection since it has better performance. Nevertheless, the periodicity of many PU signals may be unknown to an SU. In this case, this detection scheme is not applicable [3, 4]. In addition, this technique faces challenges for wideband spectrum sensing, since the cyclic features vary over different PU frequency bands. The SU receiver complexity and the processing power consumption of this method are also higher compared with the energy detection and matched filter detection.

9.2.4 Other Methods

Aside from the spectrum sensing methods introduced above, there are some other techniques including the radio identification (RI)-based sensing, multiple threshold-based sensing, eigenvalue-based detectors, moment-based detectors, Hough transformation (HT), time–frequency estimation, and covariance-based detection [7, 14–21]. In the radio identification-based sensing techniques, the SU extracts some features of an unknown received signal, such as the PU type, range of bandwidth, center frequency, transmission range, standard deviation of the instantaneous frequency, and spectral correlation density, by employing the radial basis function (RBF) neural network [16]. Spectrum sensing based on combining multiple energy thresholds was discussed in [7], where the energy detection, matched filter detection, and cyclostationary detection techniques are used simultaneously in one sensing framework. An SU first measures the receive signal SNR and then decides which sensing technique is to be employed. If the SU has the PU information such as the modulated signal type, frequency range, and pulse shaping format, it directly performs the matched filter detection. On the other hand, if the SU does not have the PU information *a priori*, it uses the double SNR threshold method to choose between energy detection and cyclostationary detection. The SU receiver complexity and sensing time of this approach are very high. The eigenvalue-based spectrum sensing methods were discussed in [17, 19]. The maximum–minimum eigenvalue-based spectrum sensing method can achieve a high detection probability without requiring the PU signal information *a priori* and the noise power information. A blind moment-based spectrum sensing method was discussed in [18] where the PU signal power and noise variance are unknown. This technique estimates these unknown parameters from the PU signal constellation and then utilizes the eigenvalue-based detection method. Any type of radio signal that has periodic patterns such as lines, circles, and ellipses can be captured using the Hough transformation which is widely used for detecting radar pulses. The randomized Hough transformation is one type of Hough transformation discussed in [21] to identify radar pulses in the IEEE 802.11 frequency band. The statistical covariance-based spectrum sensing method was employed in [22] to detect digital television signals.

Generally, if SUs have no knowledge about the PU signal and its spectrum usage, energy detection is utilized for spectrum sensing. If SUs have some prior information about the PU signal, more sophisticated spectrum sensing methods discussed above can be exploited to obtain better performance. However, most of the existing methods are designed for spectrum sensing on one licensed band and are not suitable for wideband spectrum sensing or spectrum sensing in a subdivided licensed band. In wideband spectrum sensing, the objective is to detect all idle sub-bands in a broad range of spectrum that contains many licensed bands, with some active and some inactive. One may apply the spectrum sensing methods discussed in this section onto each licensed band independently. Nevertheless, this results in large overhead for spectrum sensing considering the potentially large number of licensed bands. For spectrum sensing in a subdivided licensed band, the licensed band is arbitrarily subdivided by the PU so that one sub-band is used by one PU communication session. In this case, the SU may not have the knowledge of the spectrum subdivision and hence cannot even apply the single-band spectrum sensing method such as energy detection on each subdivided band. For example, the energy detection has difficulty to select the threshold, as it depends on the number of active sub-bands, which is unknown to the SUs [23]. In the next section, we introduce the wavelet transform-based spectrum sensing and present one effective algorithm for wideband sensing or subdivided band spectrum sensing.

9.3 Wideband or Subdivided Band Spectrum Sensing

The wideband spectrum sensing and subdivided band spectrum sensing are often desirable in CRNs. First, with wideband spectrum sensing, the SUs can aggregate the bandwidth of multiple idle bands to achieve higher throughput and lower delay. Second, with subdivided band spectrum sensing, the SUs can still utilize the idle sub-bands in a licensed band even when the PU is using the licensed band as long as not all sub-bands are used by the PU communication sessions. This can dramatically reduce the band switching of SUs to avoid communication disruption. In principle, the wideband spectrum sensing is equivalent to the subdivided band spectrum sensing, where one band in the former case is equivalent to a sub-band in the latter case. Here, we allow that the subdivision is arbitrary, i.e., the bandwidth of different sub-bands can be different. Therefore, in this section, *we refer to wideband spectrum sensing and subdivided band spectrum sensing interchangeably.*

To detect active sub-bands in a licensed band, with the start and end frequencies of each sub-band unknown to the SUs, usually the received signal in the time domain is transformed into the frequency domain to get the *power spectral density* (PSD) of the signal, e.g., through the *fast Fourier transform* (FFT). The PSD function in the frequency domain gives the information of the start and end frequencies of each signal, which are called the *edges of a signal.* However, the PSD of sub-band signals can be highly irregular and unpredictable [23]. Figure 9.1 illustrates the PSD of a received signal along the frequency domain at a specific time. It can be seen that there are three signals on this band. Note that there may be multiple signals on several consecutive sub-bands, so that those signals may appear like one single signal. The objective of spectrum sensing for a subdivided band is to detect the start and the end frequencies of each signal, called *edge detection.* The PSD shape of a real-world signal is significantly different from a regular cosine or rectangular form that has been assumed in most existing studies of edge detection

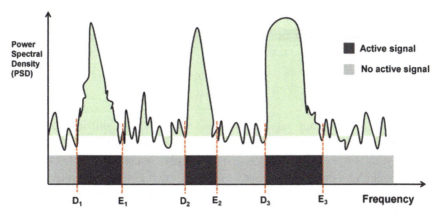

Figure 9.1   Three active signals in a licensed band. Edges are the start and end frequencies of each active signal represented by *D* and *E*, respectively.

[24–26]. Therefore, the precise localization of the edges of a signal is a great challenge in designing an edge detection scheme.

Wavelet transform (WT) is an important tool for wideband spectrum sensing, i.e., detecting signal edges in the PSD function. WT can effectively separate the *detail* and the *approximation* of a signal [27–29]. The detail is the high-frequency component of a signal, while the approximation is the low-frequency component. The edge features are hidden in the detail or high-frequency component of the WT-decomposed PSD function. The WT decomposition has a good edge localization property at a low scale, but the performance degrades when the scale increases. This is because the edges of a signal are dislocated on the frequency axis at higher WT scales. On the other hand, there may be many false edges due to random noise at lower scales [30, 31]. In the next section, we discuss the WT concept and the related WT-based spectrum sensing methods for detecting the PU signal edges.

9.3.1 Wavelet Transform (WT)

WT is a linear operation which decomposes a signal into components that appear at different scales [29]. It decomposes a function $f(t)$ to another function $W_f(s, b)$ by applying a certain wavelet function through a convolution with $f(t)$. s and b are called the *scaling* and *shifting* parameters, respectively, of WT. Specifically, $W_f(s, b)$ indicates the strength of a wavelet scaled by s and shifted by b at time t. The wavelet extends for only a short period. So its effect is limited to the area immediately surrounding t. WT gives the strength information of function $f(t)$ in the frequency domain at a given time [30]. We can form the dilated filter or *wavelet basis function* $\Psi_{sb}(t)$ with continuous parameters s and b as shown below:

$$\Psi_{sb}(t) = \frac{1}{\sqrt{s}} \Psi\left(\frac{t-b}{s}\right), \tag{9.25}$$

where $\Psi(t)$ is the mother wavelet or wavelet kernel. The shifting parameter b gives the position of the wavelet, while the scaling parameter s governs its frequency. If $|s| \ll 1$, the wavelet $\Psi_{sb}(t)$ is a highly concentrated or "shrunken" version of $\Psi(t)$ where the frequency content is mostly in the high-frequency range. On the other hand, if $|s| \gg 1$, the wavelet $\Psi_{sb}(t)$ is a highly "spread-out" version of $\Psi(t)$ which has mostly low frequencies (see [28, 29] for more details).

The function $\Psi(t)$ is called a wavelet if its average is zero and has the orthogonal property as shown below:

$$\int_{-\infty}^{\infty} \Psi(t)dt = 0, \tag{9.26}$$

$$\langle \Psi_l(t), \Psi_m(t) \rangle = \delta_{l,m} = \begin{cases} 1, & \text{if } l = m \\ 0, & \text{if } l \neq m \end{cases}, \tag{9.27}$$

where l and m are integers which represent two dilated versions $\Psi_l(t)$ and $\Psi_m(t)$ of wavelet function $\Psi(t)$ and $\delta_{l,m}$ is the Kronecker delta. Moreover, the mother wavelet $\Psi(t)$ is basically the impulse response of a continuous-time band-pass filter which satisfies the vanishing moment criterion that describes how a function decays toward infinity. The wavelet function $\Psi(t)$ has v vanishing moments if

$$\int_{-\infty}^{\infty} x^h \Psi(t)dt = 0 \text{ for } 0 \leq h \leq v. \tag{9.28}$$

A faster decay of $\Psi(t)$ implies better time localization of a signal $f(t)$. WT is based on the convolution of the signal $f(t)$ with the scaling or diluted filter in (9.25). The continuous WT of $f(t)$ is defined as:

$$W_f(s,b) = \langle f(t), \Psi_{sb}(t) \rangle = \int_{-\infty}^{+\infty} f(t)\Psi_{sb}(t)dt. \tag{9.29}$$

The signal $f(t)$ can be recovered back from its wavelet coefficients as follows:

$$f(t) = \frac{1}{K_\Psi} \int_{-\infty}^{+\infty} \int_0^{+\infty} \frac{1}{s^2} W_f(s,b)\Psi_{sb}(t)dsdb, \tag{9.30}$$

where $\mathcal{K}_\Psi$ is the wavelet coefficient.

The mother wavelet $\Psi(t)$ must satisfy the *admissibility* condition as follows:

$$\mathcal{K}_\Psi = \int_0^{+\infty} \frac{|\Psi(w)|^2}{w} dw < \infty, \qquad (9.31)$$

where $\Psi(w)$ is the Fourier transform of $\Psi(t)$.

In practice, the continuous WT is often difficult to be utilized, and the *discrete wavelet transform* (DWT) is often used. For DWT, wavelet parameters s and b need to be discretized. Generally, they are sampled and mapped to (s_o, b_o) to derive the *discrete wavelet basis function* as follows [32]:

$$\Psi_{jk}(t) = s_o^{-j/2} \Psi(s_o^{-j} t - k b_o), \quad j, k \in Z, \qquad (9.32)$$

where the scaling step $s_o > 1$ and the shifting step $b_o \neq 0$. s_o and b_o are positive real numbers. The corresponding scaling and shifting parameters $s = s_o^j$ and $b = k b_o s_o^j$. We can see that the shifting parameter b depends on the selected scaling rate. Like the continuous wavelet basis function, if $j \gg 1$, the discrete wavelet basis function $\Psi_{jk}(t)$ spreads out; conversely, it is highly shrunken if $j \ll 1$. That is, the DWT of function $f(t)$ with a large j decomposes the big picture or low-frequency component of $f(t)$, while with a small j, the decomposed function contains the *detail* or high-frequency component of $f(t)$.

The discrete wavelet coefficient is defined as the inner product of $\Psi_{jk}(t)$ and $f(t)$ as follows:

$$\mathcal{T}_{jk} = \langle \Psi_{jk}(t), f(t) \rangle = s_o^{-j/2} \int f(t) \Psi(s_o^{-j} t - k b_o) dt. \qquad (9.33)$$

Note that $\Psi_{jk}(t)$ satisfies the *admissibility* condition as (9.31). Hence, function $f(t)$ is reconstructed as:

$$f(t) = \sum_j \sum_k \mathcal{T}_{jk} \Psi_{jk}(t). \qquad (9.34)$$

DWT captures only the detail of a signal at each scale. Hence, it requires infinite number of wavelet scales in order to perfectly reconstruct $f(t)$ using (9.34). Moreover, when we increase the scale j in (9.33) by the factor of 2 in the time domain, its bandwidth in the frequency domain is reduced by half. In other words, in every successive wavelet stretch, we cover only half of the signal spectrum. To cover the whole spectrum of a signal, a complement of the wavelet basis function is introduced which is called the *scaling function*

$\Phi(t)$ [31]. Similar to the wavelet basis function, the scaling function $\Phi(t)$ also satisfies the orthonormality property as follows:

$$\int_{-\infty}^{\infty} \Phi(t)dt = 0, \tag{9.35}$$

$$\langle \Phi_l(t), \Phi_m(t) \rangle = \delta_{l,m} = \begin{cases} 1, & \text{if } l = m \\ 0, & \text{if } l \neq m \end{cases}. \tag{9.36}$$

We define a set of scaling functions as follows by scaling and shifting a unique function $\Phi(t)$ with $s_o = 2$ and $b_o = 1$

$$\Phi_{jk}(t) = 2^{-j/2}\Phi(2^{-j}t - k), \quad j, k, \in Z. \tag{9.37}$$

Note that the Fourier transform of the scaling function in (9.37) gives us a low-pass filter. That is, by applying this scaling function to a signal at different scales and shifting parameters, we achieve the low pass or obtain the *approximation* of that signal at different resolutions. Because of the low-pass nature of the scaling function, it is sometimes referred to as the averaging filter. The scaling coefficient or approximation of function $f(t)$ is shown below [27]:

$$\mathcal{A}_{jk} = \langle \Phi_{jk}(t), f(t) \rangle = 2^{-j/2} \int f(t)\Phi(2^{-j}t - k)dt. \tag{9.38}$$

Here, $\mathcal{A}_{jk}$ is a linear operator which approximates signal $f(t)$ at scale 2^j and being shifted by k. During this signal approximation at a scale 2^j, some information about $f(t)$ is lost since the approximation is a low-pass filtering process. When the scale increases, more details of function $f(t)$ are removed by the $\mathcal{A}_{jk}$ operator. More specifically, the information content of the approximation component of a signal at scale 2^{j+1} is smaller than the approximation component of the signal at scale 2^j. When the scale approaches zero, the approximation component $\mathcal{A}_{jk}$ converges to the original signal $f(t)$. Conversely, the approximation contains less and less signal information as the scale increases.

Function $f(t)$ can be recovered using both the detail component and the approximation component together within a finite number of scales, unlike using either only the detail or approximation component which needs an infinite number of scales as in (9.34). The representation of $f(t)$ is derived by utilizing the scaling and wavelet basis function as follows [31]:

$$f(t) = \sum_{k=-\infty}^{+\infty} \mathcal{A}_{Jk} 2^{-j/2} \Phi(2^{-J}t - k) + \sum_{j=1}^{J} \sum_{k=-\infty}^{+\infty} \mathcal{T}_{jk} 2^{-j/2} \Psi(2^{-j}t - k) \tag{9.39}$$

The first term in (9.39) is the approximation of signal $f(t)$ computed at scale J using the scaling function, and the second term is the summation of details up to scale J which is obtained using the wavelet basis function. The difference of information of the approximation between scale 2^{j+1} and 2^j is captured by WT of that signal at scale 2^{j+1}.

The following mother wavelet and mother scaling function, $\Psi(t)$ and $\Phi(t)$, can be used [27, 33]. Note that they satisfy the requirements specified in (9.26) and (9.35).

$$\Phi(t) = \sum_{k=-\infty}^{\infty} h_\phi(k)\sqrt{2}\Phi(2t - k), \qquad (9.40)$$

$$\Psi(t) = \sum_{k=-\infty}^{\infty} h_\psi(k)\sqrt{2}\Phi(2t - k). \qquad (9.41)$$

The $h_\phi(k)$ and $h_\psi(k)$ are the scaling and wavelet vectors at translation k, respectively. For instance, the discrete time domain representation of the scaling vector or low-pass filter for the Daubechies wavelet with the 2nd vanishing moment is given as:

$$h_\phi(k) = \frac{\sqrt{2} + \sqrt{6}}{2}\delta(k) + \frac{3\sqrt{2} + \sqrt{6}}{8}\delta(k - 1) + \frac{3\sqrt{2} + \sqrt{6}}{8}\delta(k - 2)$$
$$+ \frac{\sqrt{2} - \sqrt{6}}{8}\delta(k - 3). \qquad (9.42)$$

Note that the minimum $h_\phi(k)$ filter length is twice as the vanishing moment. The corresponding wavelet vector or high-frequency filter $h_\psi(k)$ is just a mirror of $h_\phi(k)$. The high-frequency filter $h_\psi(k)$ is given as:

$$h_\psi(k) = (-1)^k h_\phi(1 - k). \qquad (9.43)$$

9.3.2 Signal Edge Detection Using DWT

In this section, we discuss different methods of detecting signal edges using DWT. In the literature, there are some edge detection techniques including the *wavelet transform-based multiscale product* (WTMP), *wavelet transform-based multiscale sum* (WTMS), and *wavelet transform-based multiscale modulus maxima* (WTMM) [24, 25, 30, 34]. Before discussing those signal edge detection techniques, we first review an important property of WT: detecting the sharp transition in a smooth signal.

For signal edge detection, let us consider a one-dimensional signal $f(t)$. The DWT of $f(t)$ at scale 2^j is simplified as follows:

$$W_j f = f * \Psi_j(t), \tag{9.44}$$

where $*$ denotes the convolution operator. Here, we use only the scaling property of the wavelet basis function in order to compute $W_j f$ in (9.44). In other words, the function $W_j f$ is the DWT of signal $f(t)$ at scale 2^j and at position t. Accordingly, the wavelet basis function $\Psi_j(t)$ is formed by only scaling the mother wavelet (ignoring the shifting parameter) as follows:

$$\Psi_j(t) = 2^{-j/2} \Psi\left(\frac{t}{2^j}\right). \tag{9.45}$$

When the scale 2^j decreases, i.e., j decreases, the support of $\Psi_j(t)$ decreases. So $W_j f$ is sensitive to finer details of signal $f(t)$. The scale parameter is in dyadic sequence $(2^j)_{j\in z}$. So (9.45) is also called *dyadic wavelet*, and (9.44) is *dyadic wavelet transform*.

The DWT can find the sharp transition point of signal $f(t)$. The zero crossing of $W_j f$ is the location of the sharp transition point of signal $f(t)$ when the wavelet $\Psi(t)$ is the second derivative of a low-pass filter function $\mathcal{L}(t)$ [29]. Mathematically, if

$$\Psi(t) = \frac{d^2 \mathcal{L}(t)}{dt^2}, \tag{9.46}$$

then

$$\ddot{W}_j f = f * \Psi_j(t) = (2^{-j/2})^2 \frac{d^2}{dt^2}(f * \mathcal{L}_j)(t), \tag{9.47}$$

where $\mathcal{L}_j(t)$ is the dialation of $\mathcal{L}(t)$. The convolution operation of $\mathcal{L}_j(t)$ with signal $f(t)$ results in a smoother version of $f(t)$, i.e., it removes high-frequency components from $f(t)$ without modifying its shape. Therefore, (9.47) indicates that the DWT of signal $f(t)$ is proportional to the second derivative of that signal smoothed by function $\mathcal{L}(t)$. This WT property is utilized to detect the signal edges of the PU signals in a wide band.

9.3.3 Wideband Spectrum Sensing Using DWT

DWT has been widely used for wideband spectrum sensing in CRNs. In a wide band, there are usually heterogeneous wireless devices that exploit various wireless technologies in order to transmit signals at different frequency bands.

Assume that there are totally L PU frequency bands in a W Hz wide band in the frequency range $[w_0, w_L]$. At first, the SU-received signal in the time domain is transformed into the frequency domain using the Fourier transform to get the PSD. Let $w_0, w_1, w_2, \ldots, w_L$ denote the frequency boundaries of the L PU frequency bands. The bandwidth of the *lth* frequency band is $W_l = w_l - w_{l-1}$. Let $S(w)$ be the Fourier transform of the received signal $r(t)$. The PSD function $S(w)$ contains all PU signals in the wide band. The wideband spectrum sensing problem is essential to find the edges of each PU signal in the PSD function $S(w)$. The edges in $S(w)$ are the locations of frequency boundaries of the PU bands, i.e., an edge is one of the frequency boundaries $w_0, w_1, w_2, \ldots, w_L$.

9.3.3.1 Spectrum sensing by WTMM

The *wavelet multiscale modulus maxima* (WTMM) [25] is the local maxima of the first derivative of the DWT of the PSD function $S(w)$, when the wavelet in (9.45) is approximated as a first derivative of a smoothing function $\mathcal{L}_j(w)$ such that $\Psi_j(w) = d\mathcal{L}_j(w)/dw$. Note that both the wavelet basis function $\Psi_j(w)$ and the smoothing function $\mathcal{L}_j(w)$ are given in the frequency domain in order to apply onto the PSD function $S(w)$. The edges or sharp transitions of $S(w)$ are often visible in the shapes of its derivatives. The first derivative of WT using $\mathcal{L}_j(w)$ is given as:

$$\dot{W}_j S = S * \Psi_j(w) = (2^{-j/2})\frac{d}{dw}(S * \mathcal{L}_j)(w). \tag{9.48}$$

Therefore, the maximum peak of $\dot{W}_j S$ is at the locations with a sharp transition, i.e., at the edges of $S(w)$. Furthermore, the zero crossing at sharp transitions of $S(w)$ can also be found by deriving $\ddot{W}_j S$ using (9.47).

The edges of the PSD function $S(w)$ are given as follows:

$$\mathbf{w} = \arg\max{}_w\{|\dot{W}_j S|\} \tag{9.49}$$

or

$$\mathbf{w} = \arg\text{zeros}_w\{|\ddot{W}_j S|\}, \tag{9.50}$$

where (9.49) finds the indices of $S(w)$, i.e., w_l, at each local maxima of function $|\dot{W}_j S|$, vector $\mathbf{w}$ includes the indices of all local maxima of $|\dot{W}_j S|$, i.e., all edges, and (9.50) finds those $S(w)$ indices when function $|\ddot{W}_j S|$ is zero. As a result, both equations are used to find edges of $S(w)$.

For a PSD function that has a shape like a rectangular pulse and has a high SNR, the WTMM algorithm can detect signal edges very well. However, WTMM has a major drawback. As seen in Section 9.5, the PSD function of a real-world signal is highly irregular and has many noise impulses. When applying WTMM onto such real-world signals, a large number of spurious signal edges caused by the noise impulses are incorrectly detected as signal edges.

9.3.3.2 Spectrum sensing by WTMP and WTMS

Besides WTMM, WTMP is a widely used edge detection algorithm [30]. It multiplies the adjacent scales to detect the edges in a PSD function. The WTMP function is defined as the correlation between two adjacent wavelet transform scales as follows:

$$P_j^S(w) = W_j S \cdot W_{j+1} S. \tag{9.51}$$

The subscript j in (9.51) means that the correlation is computed between two adjacent scales 2^j and 2^{j+1}. The edges of PSD $S(w)$ are peaks of WT decomposition $W_j S$. Those peaks usually propagate across the scales. Hence, the product of adjacent scales enhances the edge feature. Moreover, when the noise is Gaussian, the DWT average local maxima at scale 2^{j+1} is half of that at scale 2^j. That is, when the scale increases, the Gaussian noise decreases. Therefore, multiplying the adjacent WT scales eventually dilutes the noise. As mentioned earlier, when the scale increases, the support of wavelet basis function W_j increases, which indicates that the decomposed signal $W_j S$ becomes smoother rapidly when the scale increases.

Using (9.51), we compute the edges of the PSD function $S(w)$ as follows:

$$w = \arg\max{}_w\{|P_j^3 w|\} \tag{9.52}$$

Again, the objective here is to find the indices of all local peak points of the scale product function $|P_j^S(w)|$, which are the edges of signals. WTMP has difficulty to detect a signal if the signal bandwidth is relatively narrow. Furthermore, when the WT scale increases, the edges of the PSD function are more visible as the noise is reduced. However, the edge locations also shift on the frequency domain. Therefore, WTMP often cannot accurately detect the start and the end frequencies of a signal.

WTMS is an alternative way of extracting the edge information by summing up the WT scales. Multiscale summation is given as follows:

$$U_J^S(w) = \sum_{j=1}^{J} W_j S. \tag{9.53}$$

The WTMS-based edge detection method is discussed in [24] for wideband spectrum sensing. It detects the sharp variation points, called *singularities,* of a synthetic cosine shape PSD function in a wide band to classify the sub-bands as occupied or vacant. However, directly summing up the transformed details of WT also has a problem, as it accumulates noise from the low scale to the high scale, i.e., the summation of the WT scales contains a high level of noise mixed with the signal edge information. Moreover, the PSD edges, **w**, are shifted in the frequency domain, which leads to dislocated edges of signals.

The authors in [8] studied a *discrete wavelet packet transform* (DWPT)-based energy detection algorithm under uncertain noise. The technique estimates the received signal noise level by performing DWPT and then uses energy detection over the wide band to detect signals. The problem of detecting sub-band signal edges using multiscale DWPT still remains in this approach. The WT-based wideband spectrum sensing method using the Hilbert transform was proposed in [25]. A rectangular shaped PSD function with additive Gaussian noise is examined for edge detection. Moreover, in [26], the authors proposed an edge detection-based spectrum sensing technique by using the phase-field segmentation and claimed that the sensing performance can be improved by properly setting the phase-field segmentation parameters.

The existing techniques detect the edges of synthetic PSD functions with regular shapes, such as cosine or rectangle, by using WT. However, in practice, the real-world signals have highly irregular shape in the frequency domain. Moreover, the real-world noise is also unpredictable. The performance of edge detection for real-world signals using the existing techniques introduced earlier in this section is often not satisfactory.

Specifically, these edge detectors are prone to miss the narrow bandwidth signals and/or dislocate the edges. In the next section, we present a robust and accurate edge detection algorithm based on DWT.

9.4 Exponentially Moving Averaged Multiscale Summation (EMAMS)

In this section, we present a DWT-based spectrum sensing method introduced in [23], called the *exponentially moving averaged multiscale summation* (EMAMS), to detect the active sub-band signals within a licensed or a wide band. This approach decomposes the PSD function into multiple DWT scales and extracts the edges, by assigning exponentially changing weights to different DWT scales. As mentioned earlier, the noise gets suppressed when the DWT scale increases, which is beneficial to detect the existence of signals. However, we lose the accuracy of the signal edge locations at higher scales. Hence, there is a trade-off between suppressing noise to detect the existence of signals and accurately finding the signal edge locations. The EMAMS method can effectively address this issue and successfully detect the edges of signals.

The motivation of EMAMS is to combine both the higher scales and the lower scales with different weights, such that the noise is suppressed by assigning larger weights to higher scales, while the signal edge information is preserved by incorporating lower scales. The edges are detected in two stages. In the first stage, we compute the approximation component using the scaling basis function and compare it with a threshold to detect the existence of signals. In the second stage, the DWT results are compared with another threshold. Finally, the spurious edges are discarded through a bandwidth threshold.

Without loss of generality, we consider signal edge detection in one licensed band that is subdivided into N_c sub-bands by PUs such as shown in Figure 9.1. SUs do not need to have the subdivision knowledge of the licensed band. Specifically, SUs do not need to know N_c and the frequency boundaries of the sub-bands. Let $F[n]$ ($1 \leq n \leq M$) denote the PSD of the received signals on the licensed band at a specific time. Here, we consider F as a discrete function. That is, the PSD of the licensed band is sampled with a frequency resolution ∇, with totally M number of samples. $F[1]$ is the PSD at the first sample, and $F[M]$ is the PSD at the last sample. A continuous PSD function over the licensed band can also be used, but it usually requires significantly more computation time.

In order to detect the signal edges, it is highly desirable that the wavelet basis function has the singularity property. We use the Daubechies wavelet basis function with the 4th vanishing moment which has a good singularity property. The dyadic dilation of Daubechies wavelet basis function $\Psi_j[n]$ in the discrete frequency domain is given as:

$$\Psi_j[n] = 2^{-j/2}\Psi[2^{-j}n]. \qquad (9.54)$$

Similarly, the scaling function $\Phi_j[n]$ is given as:

$$\Phi_j[n] = 2^{-j/2}\Phi[2^{-j}n]. \qquad (9.55)$$

Accordingly, the scaling and wavelet basis functions are formulated in the context of discrete frequency domain as below:

$$\Phi[n] = \sum_{k=-\infty}^{\infty} h_\phi[k]\sqrt{2}\Phi[2n - k], \qquad (9.56)$$

$$\Psi[n] = \sum_{k=-\infty}^{\infty} h_\psi[k]\sqrt{2}\Phi[2n-k]. \tag{9.57}$$

The detail and approximation of $F[n]$ at dyadic scale 2^j are defined as:

$$W_j F = F * \Psi_j[n] = 2^{-j/2}(F * \Psi[2^{-j}n]), \tag{9.58}$$

and

$$\mathcal{A}_j F = F * \Phi_j[n] = 2^{-j/2}(F * \Phi[2^{-j}n]), \tag{9.59}$$

respectively.

9.4.1 Edge Detection through EMAMS

Next, we first analyze the multiresolution approximation component $\mathcal{A}_j F$ of the PSD function $F[n]$ at multiple scales using (9.59) to find the existence of the signals on sub-bands. As mentioned earlier, by convolving the scaling basis function with a signal at different scales, we get different smoothed versions of that signal without changing its shape. The smoothing filter in this process just removes the high-frequency component from signal $F[n]$. In the next subsection, we will discuss how to set thresholds to detect signals from the approximation component $\mathcal{A}_j F$ at different scales. After detecting the presence of a signal in $F[n]$, we then locate the edges of that signal.

The EMAMS algorithm utilizes an exponentially moving average of details at different scales. As discussed earlier, on the detail component at the higher scale, the noise is low; hence, it is easier to detect the signal edges. Nevertheless, the location of the edges is also shifted in the frequency domain at the higher scale. In other words, with the detail at the higher scale, we can detect the existence of signals, but cannot find the accurate start and end frequencies of the signals, as the edges of the signals are shifted. On the other hand, on the detail at the lower scale, the noise is high; hence, it is often difficult to detect the existence of the signal. Nevertheless, once the existence of a signal is detected, then the edges of the signal can be often accurately detected, as the locations of the signal edges are not shifted on the detail component at the lower scale.

Therefore, the EMAMS approach intelligently combines the detail components at both lower and higher scales, so that the signal edge information is preserved, while the existence of signal is also accurately detected. The exponentially moving averaged multiscale summation for the DWT of function $F[n]$ is given as:

$$S_{j+1}[n] = (1-a)S_j[n] + aW_{j+1}F[n], \tag{9.60}$$

where a is a weighting factor. Equation (9.60) represents the EMAMS results after $j + 1$ scales. Specifically, the EMAMS results are as follows:

$$S_0[n] = 0,$$
$$S_1[n] = aW_1F,$$
$$S_2[n] = a(1 - a)W_1F[n] + aW_2F,$$
$$\vdots$$
$$S_j[n] = a(1 - a)^{j-1}W_1F + a(1 - a)^{j-2}W_2F + \cdots + aW_jF.$$

Here, W_1F, $W_2F, \ldots, W_jF$ are high-frequency components of the PSD function decomposed by DWT. If the weighting factor a is large, the higher-scale detail weighs more than the lower-scale detail. To further remove noise and the false signal edges, adaptive thresholds are used to detect signal edges, which are adaptable to the noise variance and signal bandwidth, to be discussed in the next subsection.

9.4.2 Adaptive Thresholds

At first, we can obtain a power threshold from the approximation component to be used for the edge detection algorithm as follows:

$$T_P^A = \frac{\max(A_jF)}{\sqrt{2}}, \tag{9.61}$$

where T_p^A is the 3 dB cutoff point of the approximation component A_jF. Alternatively, EMAMS can use one power threshold based on the DWT detail component instead of (9.61). We assume that the noise is AWGN with zero mean and variance σ^2. Let N_0 denote the single-sided power spectral density of AWGN and B denote the total bandwidth of the wide band in the unit of Hz. The noise variance is then given as $\sigma^2 = \frac{N_0B}{2}$. The power threshold for the EMAMS detector is calculated as $T_P^W = \frac{k\sqrt{(\sigma^2)}}{\max(S_{j+1}[n])}$ where K is a constant with a typical value 20.

In general, there is a trade-off to set the value for T_P^W. If T_P^W is too high, some signals may be missed. However, with a lower T_P^W, some noise impulses may appear as signals. As a matter of fact, we have conducted extensive study for spectrum measurement data and found that many noise impulses have even higher PSD magnitude than signals. Thus, those noise impulses would be treated as signals if T_P^W is not high. Fortunately, the bandwidth of the noise impulses is very narrow, typically in the range of a few kHz. This is much lower than the bandwidth of a normal signal. Therefore, a bandwidth threshold T_B is used in EMAMS to distinguish between signals and noise impulses. Let λ_B denote the minimum bandwidth for any signal. We compute the bandwidth threshold T_B as follows:

$$T_B = \left\lceil \frac{\lambda_B}{\nabla} \right\rceil, \tag{9.62}$$

where ∇ is the frequency resolution of the discrete PSD function $F[n]$. For example, if $F[n]$ is the values sampled every 30 Hz on the continuous PSD of a signal, then $\nabla = 30$.

Algorithm 9.1 EMAMS edge detection

Input: $F[n]$ $(1 \leq n \leq M)$, j, T_P^W, T_B
Output: K, $\alpha(k)$, $\beta(k)$ $(1 \leq k \leq K)$

  1:    $K = 0$
  2:    Compute S_j from $F[n]$ using (9.60)
  3:    **for** $n = 1$ to M **do**
  4:       **if** $S_j[n] \geq T_P^W$ **then**
  5:          $G[n] \leftarrow 1$
  6:       **else**
  7:          $G[n] \leftarrow 0$
  8:       **end if**
  9:    **end for**
10:    // Add two auxiliary values to facilitate edge detection
11:    Let $G[0] = 0$ and $G[M + 1] = 0$
12:    $n = 1$
13:    **while** $n \leq M$ **do**
14:       **if** $G[n] = 1$ and $G[n-1] = 0$ **then**
15:          // Find a segment of consecutive 1's in G
16:          Find m such that $n \leq m \leq M$, $G[n]$, $G[n+1]$, ..., $G[m]$ are all equal
              to 1, and $G[m+1] = 0$
17:          **if** $m - n + 1 \geq T_B$ **then**
18:             $K = K + 1$
19:             $\alpha(K) = n$
20:             $\beta(K) = m$
21:             $n = m + 1$ // continue to scan from $G[m+1]$
22:          **end if**
23:       **end if**
24:       $n = n + 1$
25:    **end while**

9.4.3 EMAMS Algorithm

Algorithm 9.1 describes how to detect the start and end frequencies of signals using the EMAMS method. The input is the PSD function $F[n]$, the number of scales j, and the power and bandwidth thresholds T_P^W and T_B. The output is the number of signals K, and the start and end frequency indices $\alpha(k)$ and $\beta(k)$ for each signal k. At first, it computes the multiscale summation by the exponentially moving average, S_j. Then, the algorithm uses the power threshold T_P^W to detect the existence of signals at each frequency, which is recorded in vector $G[n]$. Next, the bandwidth threshold is utilized to filter out the spurious noise impulses, and the start and end frequencies of real signals are recorded in $\alpha(k)$ and $\beta(k)$.

EMAMS is a reliable and robust signal edge detection algorithm which is able to preserve the start and end frequency edge structures of the PSD function even at higher DWT scales. Hence, the performance of signal detection improves. Compared with WTMP, EMAMS effectively avoids the edge shifting issue and more accurately detects the active sub-bands in a licensed band.

9.5 Performance Evaluation of EMAMS

In this section, we evaluate the performance of EMAMS. We compare the performance of EMAMS with the widely used WTMP edge detection algorithm [30]. Two performance metrics, the number of false alarms and true detections, are used. We assume the single-sided noise PSD N_0 is –90 dBm. The threshold constant k is 20, and the weight constant a is 0.5 in EMAMS. The measured PSD signal frequency resolution ∇ is 217 kHz. The λ_B is set as 868 kHz.

In the first experiment, we measure the PSD of the signals at the 2.4 GHz ISM band (2.4 GHz to 2.5 GHz) using a spectrum analyzer in a university building. Totally, 6000 time samples are measured with a 1-second interval. Figure 9.2 shows the measured PSD, which indicates the spectrum usage in the Wi-fi band that varies in both time and frequency domains. The total measured bandwidth B is 100 MHz, from 2.4 GHz to 2.5 GHz. Figure 9.3 illustrates the measured PSD at a specific time, i.e., one time sample of the measured PSD. It can be seen that there are four signals and those signals do not have the ideal sinusoidal or rectangular shapes. In fact, the PSD function usually has large fluctuations. There are also noise impulses that have high PSD magnitude but narrow bandwidth.

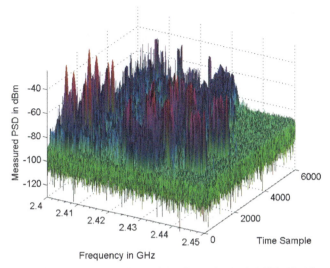

Figure 9.2   The measured PSD of the signals in the 2.4 GHz ISM band.

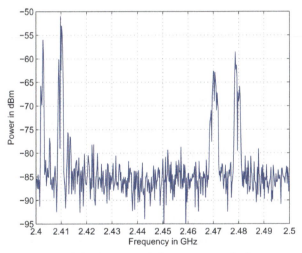

Figure 9.3   The measured PSD at a certain time.

Next, we decompose the PSD sample in Figure 9.3 to get the WT detail components at multiple scales using (9.58), as illustrated in Figure 9.4. We have shown the detail components of the first 4 scales from decomposition. In the first-scale decomposition, the edge features of signals are not clear since there is a high level of noise. When the scale increases, the detail

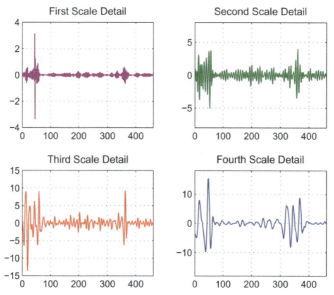

Figure 9.4   The detail component at multiple scales resulted from the decomposition of the PSD sample in Figure 9.3 using (9.58).

component of signals becomes apparent and the noise power remarkably decreases. However, it is difficult to accurately detect the edges of signals from the detail component at higher-scale decomposition since the edges shift in the frequency axis. In fact, in the 4th-scale detail component, it seems that there are 5 signals due to edge shifting.

Now, we evaluate the EMAMS algorithm and compare it with the widely used WTMP edge detection algorithm. Figure 9.5 depicts the signal edge detection results using 5 scales, by the EMAMS and WTMP approaches, respectively, for the PSD sample in Figure 9.3. There are four signals in this sample PSD function. Both EMAMS and WTMP have detected these four signals, and the edges of signals are shown by vertical solid lines. However, WTMP does not give the accurate locations for the signal edges. Figure 9.6 is the zoomed-in version of Figure 9.5 in the 2.465–2.485 GHz frequency range. There are two signals in this frequency range. However, WTMP fails to extract the correct edge locations of signals. Specifically, the right edge detected by WTMP for the first signal in Figure 9.6(b) is not at the correct location. It actually sits on the left of the true right edge. In addition, WTMP found three edges for the second signal which makes it difficult to determine the correct left and right edges, as any of them could be a false edge.

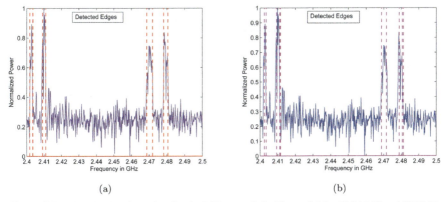

Figure 9.5 Signal edge detection for the PSD sample in Figure 9.3 by EMAMS and WTMP: (a) signal edges detected by EMAMS after the 5th-scale DWT decomposition and (b) signal edges detected by WTMP after the 5th-scale decomposition.

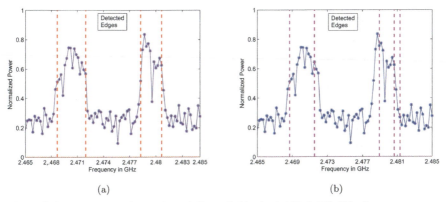

Figure 9.6 The zoomed-in version of Figure 9.5 in the 2.465–2.485 GHz frequency range: (a) signal edges detected by EMAMS and (b) signal edges detected by WTMP.

Next, we examine the number of true detections and false alarms. The number of true detections denotes the number of signals that have been correctly detected, while the number of false alarms denotes the number of "false" signals that have been detected, but are actually not true signals. Figure 9.7 illustrates the number of signals of 30 PSD samples, i.e., the measured PSD at 30 time points, and the corresponding number of detected signals by EMAMS and WTMP, respectively. To present the results better, we have sorted the sample indices at the ascending order of the number of occupied signals. We can see that the EMAMS edge detection algorithm outperforms

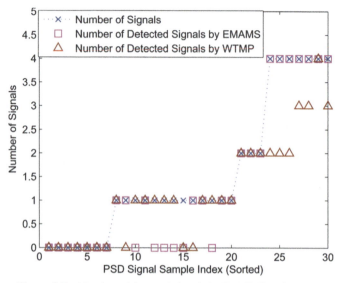

Figure 9.7   Number of detected signals by EMAMS and WTMP.

the WTMP algorithm. EMAMS misses 6 signals, while WTMP misses 12 signals.

Figure 9.8 plots the number of false alarms for the same PSD samples in Figure 9.7. The figure shows that with regard to false alarm, EMAMS significantly outperforms WTMP. There are totally 57 false alarms with WTMP, while there are only 6 false alarms with EMAMS. Figures 9.9 and 9.10 plot the number of true detections and false alarms for a different set of 30 PSD samples measured at a different date. For this PSD sample set, EMAMS misses 7 signals, while WTMP misses 14 signals. With regard to false alarms, there are totally 43 false alarms with WTMP, while EMAMS has only 5 false alarms.

9.6 Summary

In this chapter, we first discuss the concept of spectrum sensing and then introduce multiple methods for spectrum sensing of a signal on a single band. Next, we discuss spectrum sensing on a wide band or a subdivided licensed band to identify all idle sub-bands, which offers greater potential for SUs to exploit spectrum. Then, we discuss the major techniques for wideband spectrum sensing, the wavelet transform based spectrum sensing methods. The wideband spectrum sensing is reduced to detecting edges of signals on

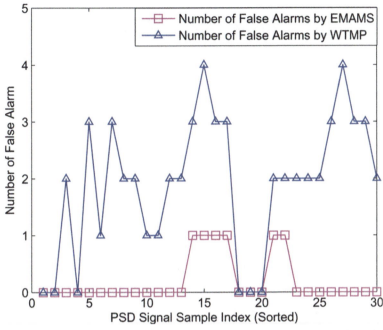

Figure 9.8 Number of false detections for the same PSD samples as in Figure 9.7.

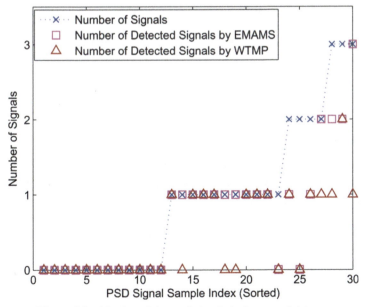

Figure 9.9 Number of detected signals in the second dataset.

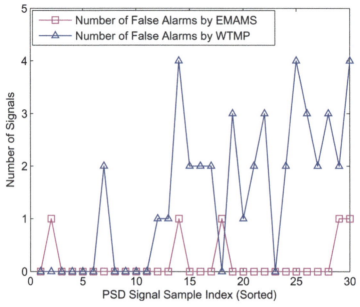

Figure 9.10   Number of false detections in the second dataset.

the frequency domains, i.e., finding the start and end frequencies of each signal. Hence, the wavelet transform-based wideband spectrum sensing is also called edge detection. At last, we present a new edge detection algorithm based on exponentially moving averaged multiscale summation, called EMAMS, to both detect the existence of signals and accurately locate the edges of the signals. The simulation results indicate that EMAMS outperforms the existing wavelet transform-based edge detection algorithms.

References

[1] Neyman, J. and Pearson, E. S. (1933). "On the problem of the most efficient tests of statistical hypotheses," *Phil. Trans. R. Soc. London A: Math. Phys. Eng. Sci.*, 231(694–706), 289–337.

[2] Qiu, R. C., Hu, Z., Li, H., and Wicks, M. C. (2012). *Cognitive Radio Communications and Networking (Principle and Practice)*. John Wiley and Sons Publication.

[3] Jun, M., Ye, L. G., and Hwang, J. B. (2009). "Signal processing in cognitive radio," *Proc. of the IEEE*, 97(5), 805–823.

[4] Yucek, T. and Arslan, H. (2009). "A survey of spectrum sensing algorithms for cognitive radio applications," *IEEE Commun. Surveys Tuts.*, 11(1), 116–130.

[5] Zhao, Y., Paul, P., Xin, C., and Song, M. (2014). "Performance analysis of spectrum sensing with mobile SUs in cognitive radio networks," in *Proc. IEEE International Conference on Communications (ICC)*, Jun.

[6] Zhao, Y., Song, M., and Xin, C. (2011). "A weighted cooperative spectrum sensing framework for infrastructure-based cognitive radio networks," *Comput. commun.*, 34(12), 1510–1517.

[7] Ejaz, W., ul Hasan, N., Lee, S., and Kim, H. S. (2013). "I3S: Intelligent spectrum sensing scheme for cognitive radio networks," *EURASIP J. Wirel. Commun. Netw.*, 2013(1), 1–10.

[8] Bing Zhang, S. and Jing Qin, J. (2010). "Energy detection algorithm based on wavelet packet transform under uncertain noise for spectrum sensing," in *Proc. 6th International Conference on Wireless Communications Networking and Mobile Computing (WiCOM)*, Sep.

[9] Zhao, Y., Pradhan, J., Huang, J., Luo, Y., and Pu, L. (2015). "Joint energy-and-bandwidth spectrum sensing with GNU radio and USRP," *ACM SIGAPP Appl. Comput. Rev.*, 14(4), 40–49.

[10] Quan, Z., Cui, S., Poor, H. V., and Sayed, A. H. (2008). "Collaborative wideband sensing for cognitive radios," *IEEE Signal Proc. Mag.*, 25(6), 60–73.

[11] Saad, W., Han, Z., Debbah, M., Hjorungnes, A., and Basar, T. (2009). "Coalitional games for distributed collaborative spectrum sensing in cognitive radio networks," in *Proc. IEEE INFOCOM*, Apr.

[12] Kay, S. M. (2013). *Fundamentals of Statistical Signal Processing: Practical Algorithm Development*. Pearson Education, vol. 3.

[13] Gardner, W. and Spooner, C. (1992). "Signal interception: performance advantages of cyclic-feature detectors," *IEEE Trans. Commun.*, 40(1), 149–159.

[14] Gandetto, M. and Regazzoni, C. (2007). "Spectrum sensing: A distributed approach for cognitive terminals," *IEEE J. Sel. Area. Commun.*, 25(3), 546–557.

[15] Gandetto, M., Guainazzo, M., Pantisano, F., and Regazzoni, C. S. (2004). "A mode identification system for a reconfigurable terminal using wigner distribution and non-parametric classifiers," in *Proc. IEEE Global Telecommunications Conference (GLOBECOM)*, Nov.

[16] Gandetto, M., Guainazzo, M., and Regazzoni, C. S. (2004). "Use of time-frequency analysis and neural networks for mode identification in

a wireless software-defined radio approach," *EURASIP J. Appl. Signal Proc.*, 2004, 1778–1790.

[17] Zeng, Y. and Liang, Y.-C. (2009). "Eigenvalue-based spectrum sensing algorithms for cognitive radio," *IEEE Trans. Commun.*, 57(6), 1784–1793.

[18] Kortun, A., Ratnarajah, T., Sellathurai, M., Zhong, C., and Papadias, C. B. (2011). "On the performance of eigenvalue-based cooperative spectrum sensing for cognitive radio," *IEEE J. Sel. Top. Signal Process.*, 5(1), 49–55.

[19] Kortun, A., Ratnarajah, T., Sellathurai, M., Liang, Y.-C., and Zeng, Y. (2014). "On the eigenvalue-based spectrum sensing and secondary user throughput," *IEEE Trans. Veh. Technol.*, 63(3), 1480–1486.

[20] Pillay, N. and Xu, H. (2013). "Eigenvalue-based spectrum sensing using the exact distribution of the maximum eigenvalue of a wishart matrix," *IET Signal Process.*, 7(9), 833–842.

[21] Challapali, K., Mangold, S., and Zhong, Z. (2004). "Spectrum agile radio: Detecting spectrum opportunities," in *Proc. International Symposium on Advanced Radio Technologies*, Mar.

[22] Zeng, Y. and Liang, Y.-C. (2007). "Covariance based signal detections for cognitive radio," in *Proc. 2nd IEEE International Symposium on New Frontiers in Dynamic Spectrum Access Networks (DySPAN)*, Apr.

[23] Paul, P., Xin, C., Song, M., and Zhao, Y. (2015). "Spectrum sensing for a subdivided band in cognitive radio networks," in *Proc. International Conference on Computer Communications and Networks (ICCCN)*, Aug.

[24] El-Khamy, S. E., El-Mahallawy, M., and Youssef, E. (2013). "Improved wideband spectrum sensing techniques using wavelet-based edge detection for cognitive radio," in *Proc. International Conference on Computing, Networking and Communications (ICNC)*, Jan.

[25] Jeng, S.-S., Chen, J.-M., Lin, H.-Z., and Tsung, C.-W. (2011). "Wavelet-based spectrum sensing for cognitive radios using hilbert transform," *World Acad. Sci. Eng. Technol.*, 596–600.

[26] Eslami, M. and Sadough, S.-S. (2010). "Wideband spectrum sensing for cognitive radio via phase-field segmentation," in *Proc. 6th IEEE Conference on Wireless Advanced (WiAD)*, Jun.

[27] Mallat, S. G. (1989). "A theory for multiresolution signal decomposition: the wavelet representation," *IEEE Trans. Pattern Anal. Mach. Intell.*, 11(7), 674–693.

[28] Daubechies, I. (1990). "The wavelet transform, time-frequency localization and signal analysis," *IEEE Trans. Inf. Theory*, 36(5), 961–1005.

[29] Mallat, S. (1991). "Zero-crossings of a wavelet transform," *IEEE Trans. Inf. Theory*, 37(4), 1019–1033.

[30] Zhang, L. and Bao, P. (2002). "Edge detection by scale multiplication in wavelet domain," *Pattern Recogn. Lett.*, 23(14), 1771–1784.

[31] Akansu, A. N., Serdijn, W. A., and Selesnick, I. W. (2010). "Emerging applications of wavelets: A review," *Phys. commun.*, 3(1), 1–18.

[32] Daubechies, I. (1988). "Orthonormal bases of compactly supported wavelets," *Commun. Pure Appl. Math.*, 41(7), 909–996.

[33] Mallat, S. (1999). *A wavelet tour of signal processing*. Academic press.

[34] Tian, Z., and Giannakis, G. B. (2006). "A wavelet approach to wideband spectrum sensing for cognitive radios," in *Proc. 1st International Conference on Cognitive Radio Oriented Wireless Networks and Communications*, Jun.

PART IV

Pervasive Computing/Sensor Networks/IoT

10

Assessing Performance of Smart Grid Applications Using Co-simulation

**Paul Moulema[1], Wei Yu[1], Sriharsha Mallapuram[1],
David Griffith[2] and Nada Golmie[2]**

[1]Department of Computer and Information Science, Towson University,
Towson, Maryland 21252, USA
[2]The Communications Technology Laboratory (CTL), National Institute
of Standards and Technology, Gaithersburg, Maryland 20899, USA

Abstract

The smart grid is a complex system that comprises components from both the
power grid and communication networks. To understand the behavior of such
a complex system, co-simulation is a viable tool to capture the interaction
and the reciprocal effects between a communication network and a physical
power grid. In this chapter, we systematically review the existing efforts of co-
simulation and propose a framework to explore co-simulation scenarios. Using
the demand response and energy price as examples of smart grid applications,
and operating the communication network under various conditions (e.g.,
normal operation, degraded performance, and security threats), we evaluate
the performance of smart grid applications by leveraging the co-simulation
platform. In addition, we conduct a co-simulation study to investigate the
impact of wireless network implementations, including the wireless mesh
architecture based on Ad hoc On-Demand Distance Vector (AODV) protocol, and the Worldwide Interoperability for Microwave Access (WiMAX)
architecture, on the performance of smart grid applications.

Keywords: Smart Grid, Co-simulation, Power Grid, Communication
network, Simulation, Demand response, Energy market, Framework,
GridLAB-D, FNCS, NS-3, Substation, Dynamic pricing, Security, Cyber-
attacks, Scenarios, GridMat.

10.1 Introduction

One of the major goals of the smart grid is to provide efficient and reliable energy service to consumers by integrating modern communication and networking technologies and renewable energy resources into the power grid [1]. Because the physical power grid and communication networks are essential components of the smart grid, it is critical to develop system-level modeling and simulation tools to study the interaction and the reciprocal effects between the power grid and the communication network.

A single, unified co-simulation framework that integrates the power grid and communication network is therefore a viable tool for the research and development of the smart grid. As defined by Norling et al. [2], *"co-simulation is a methodology for simulating individual components of a larger system concurrently by different simulation tools which mutually exchange information in a collaborative manner."* Co-simulation can and will provide cost-effective means to evaluate and test smart grid technologies before deployment in the field [3, 4]. Broadly speaking, unlike independent simulators that treat the electric grid and the communication network as isolated environments, co-simulation can capture the interaction and reciprocal effects between the two as an integrated smart grid system [5–8]. Particularly, co-simulation can be used to evaluate the impact of the performance, failure, and security of communication networks on smart grid operations. In this way, we can determine the communication system requirements to support smart grid applications and evaluate the behavior of these applications when the network is under various conditions such as congestion, failure, and/or cyber-attacks.

In this chapter, we first review existing efforts on smart grid co-simulation. We then describe the power grid and communication network models and present a framework to explore co-simulation scenarios. In our framework, we consider two orthogonal dimensions: one represents smart grid applications (demand response, dynamic market pricing, etc.), and the other depicts operation conditions of the communication network (normal operation, network under attack via false data injection and/or denial-of-service, network quality-of-service degradation, etc.).

Based on this framework, we then develop six scenarios. Using the capabilities and features of the Fenix framework for Network Co-Simulation (FNCS), a High Performance Computing (HPC) simulation platform[1] [9, 10],

---

[1]Certain commercial equipment, instruments, or materials are identified in this chapter in order to specify the experimental procedure adequately. Such identification is not intended to imply recommendation or endorsement by the National Institute of Standards and Technology,

we have conducted co-simulations to evaluate the performance of smart grid applications for individual scenarios. Our results show that the intertwined implementation of demand response and the dynamic energy market can effectively smooth power generation, maintain a balance between generation and demand, and reduce high peak power generation. All of these factors can consequently lower power losses and generation costs. In addition, Quality-of-Service (QoS) and security threats on communication networks are determining factors for the efficiency of smart grid applications. We have additionally extended our work by designing two wireless network models: wireless mesh network based on the Ad-hoc On-Demand Distance Vector (AODV) protocol and Worldwide Interoperability for Microwave Access (WiMAX) network. Applying these designs to our co-simulation framework, we have investigated their impact on the performance of demand response and energy market in the smart grid.

Notice that this chapter is an extension of our prior work in [11]. The remainder of this chapter is organized as follows: We provide the background and review the state-of-the-art of smart grid co-simulation in Section 10.2. In Section 10.3, we describe our simulation model and scenarios. In Section 10.4, we present a performance evaluation of the developed scenarios using the co-simulation platform. In Section 10.5, we extend our framework to investigate the impact of wireless network implementations on the performance of smart grid application using the co-simulation platform. Finally, we conclude the chapter in Section 10.6.

10.2 Background and Related Work

In this section, we provide the background and conduct a literature review of the existing efforts on co-simulations for the smart grid. There have been a number of research efforts on building a co-simulation environment for the smart grid. Generally speaking, the existing work in this area can be classified into two categories [12]: (i) the extension of a single simulator to allow co-simulation, and (ii) integration of a power grid simulator and a communication network simulator into a unified co-simulation framework.

In the first category, existing research efforts such as [5, 13] developed co-simulation frameworks either by extending power distribution features to enable the communication network or by integrating a power distribution system module within a communication network simulator. For example, in

---

nor is it intended to imply that the materials or equipment identified are necessarily the best available for the purpose.

Matlab/Simulink [5], a library named TrueTime [14, 15] was used to enable a communication network. Mets et al. [13] proposed a modular simulation environment to integrate a power distribution module within OMNET++. Zhang et al. [16] developed a Matlab-OMNET++ co-simulation framework (called COSMO) to integrate Matlab libraries within OMNET++ to enhance the simulation of power grids. These co-simulation frameworks have the advantage of bypassing the hurdle of synchronization. Nonetheless, they are not scalable as they are confined to simulating simple scenarios. Simulation results could become unreliable, inaccurate, and unpredictable as the system grows in complexity and size [12].

In the second category, the primary challenge is to synchronize two completely different systems: a continuous time-based simulator (e.g., power grid) and a discrete event-based simulator (e.g., network simulator) [12]. To this end, a considerable number of research efforts have been devoted to develop co-simulation tools that integrate two independent simulators into a single uniform simulation framework [9, 10, 17–26]. For example, Hopkinson et al. [21] developed a framework, namely the Electric Power and Communication Synchronizing Simulator (EPOCHS), which is capable of integrating simulator software: PSCAD/EMTDC for electromagnetic transient simulation, PSLF electromechanical transient simulation, and NS-2. Zhu et al. [26] proposed an integration of MATLAB Simulink [5] and OPNET [27–29] in a co-simulation framework. Davis et al. [18] proposed a co-simulation of a power system and a communication network by designing a SCADA (supervisory control and data acquisition) cyber security test bed to assess the vulnerability of smart grid communication network to cyber-attacks. Anderson et al. [17] proposed a real-time simulation for smart grid control and communications design, which is named as GridSim. Kohtamaki et al. [22] implemented a Platform for Simulation, Implementation, and Modeling (PiccSIM)—a Matlab/Simulink and NS-2 Platform for integrating the design of communications and control. Lin et al. [30] proposed co-simulation framework, called GECO, using a global event-driven mechanism. The developed framework allows the evaluation of wide area measurement and control schemes of the smart grid.

In addition, PNNL (Pacific Northwest National Laboratory) developed the Fenix Framework for Network Co-Simulation (FNCS) for the smart grid [9, 10, 19]. Generally speaking, FNCS is a High Performance Computing (HPC) simulation platform, which integrates GridLAB-D [31], MATPOWER [32] (simulators for power distribution systems), and NS-3 [6] (a simulator for communication networks). Liberatore et al. [25] proposed a co-simulator called PowerNet that combines the NS-2 network simulator [33] with a power grid simulator based on Modelica to evaluate the impact of networked control on smart grid scenarios. Georg et al. [20] proposed a generic hybrid simulator based on the IEEE standard 1516. Their developed modeling scheme of mapping network and power system components is based on the concept of a substation data processing unit. Nutaro et al. [34] used a hybrid modeling and simulation technique to develop a co-simulation framework. To validate the effectiveness of their developed co-simulation framework, they conducted a simulation based on a wide-area cooperative automatic load-control scenario.

Another prominent domain of co-simulation research has been addressing synchronization issues [35–37]. For example, Kounev et al. [36] proposed a scheduler to synchronize the simulators (e.g., MATLAB and OMNeT++) and test the effect of the power delivery stability through the co-simulation. Ezeme [35] proposed a synchronization scheme that tends to mitigate the accumulated synchronization errors related to fixed-time-step, while enhancing the effectiveness of co-simulation platform by designing a scheme to enable efficient data exchange. Lin et al. [37] proposed a co-simulation platform, in which the synchronization mechanism can be tunable based on the time scale requirements from measurement and control schemes. They then conducted a case study of an agent-based remote backup relay system based on the designed platform.

Beyond the categorization of co-simulation frameworks, there have been a number of research efforts devoted to expanding the realm of smart grid co-simulation solutions and developing techniques to conduct co-simulations in an effective and efficient manner [38–43]. For example, Li et al. [42] provided an overview of co-simulation frameworks and different simulation techniques for smart grid communications with guidelines on how to select from them for a given application. Lévesque et al. [40] developed a multi-simulation framework based on the IEC 61850 power grid model and the realistic traffic model. Chouikhi et al. [38] described the wireless sensor networks and provided a review of smart grid challenges in developing co-simulation platforms. Garau et al. [39] developed a co-simulation tool that combines OpenDSS and NS-2 network simulator [33] to examine the impact of communication delays and failures on distribution management systems. Lévesque et al. [41] used a co-simulation framework that integrates OMNET+ and OpenDSS to evaluate the effectiveness of reactive control algorithms for electric vehicles connected to the smart grid. Tariq et al. [43] studied scenarios such as demand response using a co-simulation platform that combines NS3 network simulator [6] and PowerWorld power system simulator [44]. Garau et al. [39] proposed an information and communication technology (ICT) device model in a co-simulation platform and tested the efficiency of the control algorithms and the effects of losing the control signals.

10.3 Co-simulation Models and Scenarios

In our co-simulation study, we use the FNCS developed by PNNL because it integrates GridLAB-D [31], MATPOWER [32], and NS-3 [6] and therefore fits our needs to study the interaction and the reciprocal effects between cyber components and power grid applications. More details on the FNCS can be found in [9, 10, 19]. In the following, we first describe the power grid and communication network models that we developed in our co-simulation environment. We then present the scenarios used for our co-simulations. Finally, we discuss so me other issues related to our study.

10.3.1 Power Grid and Communication Network Models

As shown in Figure 10.1, the power grid that we choose comprises a substation and a residential load. The substation consists of a three-phase swing bus with a nominal voltage of 7200 V and a power rating of 4500 kW (i.e., 1500 kW per phase). A meter between the substation transformer and the load measures the total load and senses the energy demand, enabling the substation to adjust the power supply accordingly. As the energy supplier, the substation sets the maximum power capacity available in the energy market and sets the energy reference price based on the time of the day and the current energy demand. The residential load is made up of 300 houses connected to the power line through triplex meters. Each individual house is equipped with a Heating, Ventilation, and Air Conditioning (HVAC) system with a controller.

As shown in Figure 10.2, the communication network that we use consists of 300 nodes representing smart meters installed in individual houses. Smart meters are organized and grouped into clusters of 20 nodes that form local area networks. An edge network device in each cluster routes the data to a data aggregator point through a point-to-point communication link. The data communication is based on the UDP (User Datagram Protocol) transport protocol. For the point-to-point link, its bandwidth is 4 Mbs and transmission delay is 2 ms. For the local area network, its bandwidth is 10 Gbs and transmission delay is 3 ms. We choose UDP that is a connectionless protocol because it incurs a lower transmission delay in comparison with TCP (Transmission Control Protocol).

10.3.2 Co-simulation Scenarios

To design the scenarios, we define a two-dimensions co-simulation space as illustrated in Figure 10.3. The first dimension shown on the x-axis represents different smart grid applications. Here, x_1 and x_2 are two

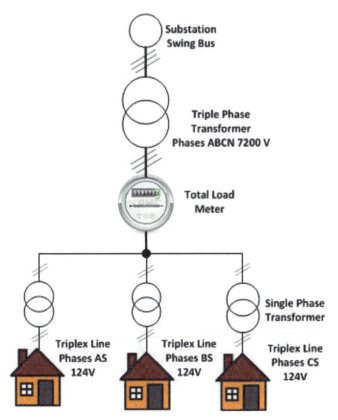

Figure 10.1 Power grid simulation model.

independent applications: demand/response and market/dynamic pricing. The second dimension shown on the y-axis represents three independent operation conditions or parameters, including y_1: normal operation, y_2: network attacks that involve false data injection and denial-of-service, and y_3: network performance/quality of service. It is worth noting that our developed framework is generic and can be expanded to include other smart grid applications and networks in different states. In the following, we will detail these dimensions. To support the communications for a large and complex system like the smart grid, wireless networks could provide easy and cost-effect deployment due to their ubiquitous nature. Therefore, in addition to conducting co-simulation based on wired networks in Section 10.4, we also extend our co-simulation work to evaluate two representative wireless network: mesh network based on AODV, and WiMAX network in Section 10.5.

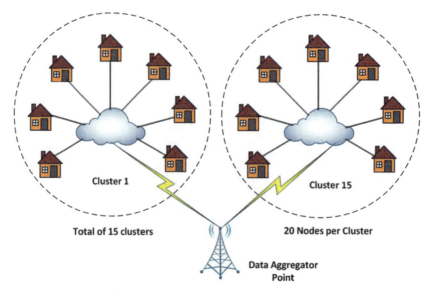

Figure 10.2   Communication network model.

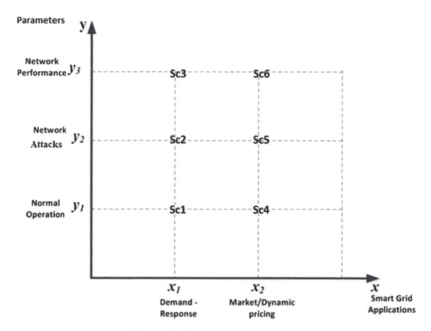

Figure 10.3   Co-simulation space.

10.3.2.1 Smart Grid applications

Through co-simulation, numerous smart grid applications that require a two-way communication infrastructure can be studied. In this chapter, we consider two representative applications: demand/response and market/dynamic pricing, listed below.

- *Demand/Response:* The goal here is to preserve a balance between power generation and load by adjusting energy demand dynamically. In the traditional power grid, customers cannot participate in the balancing process. Then, power generation was designed to address power imbalances and mitigate peak system conditions without taking endusers' dynamic load into consideration. With smart grid technologies, end users become active participants in the power-balancing process, adjusting demand, and therefore balancing generation, dynamically.

- *Market/Dynamic pricing:* Dynamic energy pricing is an effective regulator of energy demand/response. A control signal (e.g., a reference bid price) is sent to consumers, who can then adjust their energy use accordingly, intelligently managing their energy consumption versus cost. The goal is to maintain a mutually beneficial state, balancing the energy provider's profit with customer savings. The three typical methods of pricing are: Price based on Time of Use (ToU), Critical Peak Price (CPP), and Real-Time Price (RTP).

10.3.2.2 Operation conditions

To test the impact on system performance, we have configured the following network conditions:

- *Normal Operation:* This co-simulation environment represents optimal settings for the system, which maintain the grid operation in a faultless operational mode.

- *Security Threats:* Cyber-attacks include false data injection and denial-of-service. Particularly, in a false data injection attack, an adversary exploits system vulnerabilities and manipulates data collected from the grid with a goal of disrupting system operations. As an example, an adversary can change the current bid price in the energy market and thus cause the energy clearing price to be unexpectedly higher at low peak hours or very low at peak hours. In the same way, an adversary can send false high energy demands to the substation, causing the substation to unnecessarily supply more energy than the current demand.

- *Network Performance:* It involves network settings such as data rate, throughput, and delay, which have an impact on the overall operation and performance of the power grid. During co-simulation, we have tuned these network characteristics and settings and evaluate their impact on smart grid applications (e.g., demand/response and energy market).

10.3.2.3 Co-simulation scenarios

We now describe our co-simulation scenarios. We have run different scenarios using smart grid co-simulation environment by combining the two orthogonal dimensions X and Y that have been previously discussed. For example, scenario X_1Y_2 simulates demand response and load control applications under network-based attacks. The scenarios are listed as follows:

- *Scenario 1—x_1y_1. Demand/Response: Normal Operation:* In this scenario, co-simulation is performed with optimal settings. The goal is to observe the grid in a faultless operation environment and verify: (i) the effectiveness of bidirectional communications and interactions between the power domain and the cyber domain; (ii) the effectiveness of demand/response and power balance in a faultless environment; and (iii) the smoothing of power generation and load adjustment during peak periods. To this end, we compare the variation of total load with and without the demand/response feature enabled and collect output metrics, which include the total load and the statistics of HVAC systems.
- *Scenario 2—x_1y_2. Demand/Response: Security Attacks:* The goal of this scenario is to assess the overall performance and effectiveness of the demand/response feature described in Scenario 1: X_1Y_1, but under security attacks. Then, we can not only inject false data into the system, but also create a denial-of-service through an overflow of network traffic. Under these conditions, we can observe system behavior and collect performance data.
- *Scenario 3—x_1y_3. Demand/Response and Network Performance:* This scenario aims to demonstrate the effectiveness of balancing power within the constraints of the communication network settings. This can allow us to evaluate the impact of network performance on demand/response applications. To validate this, we can set the data rate lower than the normal setting and increase the minimal transmission delay of the point-to-point link between the smart meters of the local network and the gateway.
- *Scenario 4—x_2y_1. Energy Market and Normal Operation:* This scenario consists of evaluating economic benefits and the effectiveness of the

double auction energy market and real-time pricing in a faultless environment. The workflow of the energy market is shown in Figure 10.4 and can be described as follows: For every 5-minute period, energy suppliers bid the maximum power capacity and corresponding real-time price that they can provide to the market. Simultaneously, end users bid the energy amount that they desire along with the price, which they are willing to pay. When a period ends, the bidding process stops. Then, the market clearing price and market clearing quantity are determined.

- *Scenario 5—x_2y_2. Energy Market and Security Attacks:* This scenario aims to demonstrate the effectiveness of energy market applications when under security attacks. By injecting false data and creating a denial-of-service, we can evaluate the impact on the network under an attack by considering the market clearing price, market clearing quantity, total energy used, and HVAC statistics.
- *Scenario 6—x_2y_3. Energy Market and Network Performance:* By tuning the network link data rate sufficiently low so as to create congestion, we can assess the effectiveness of the energy market and the impact of the network quality-of-service on the market in terms of market clearing price, market clearing quantity, and energy billing.

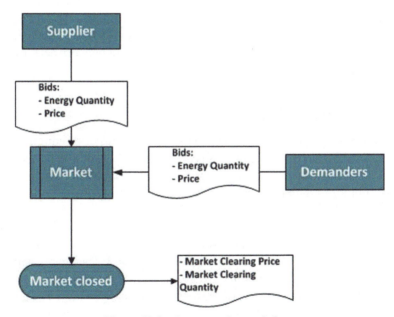

Figure 10.4 Energy market work flow.

10.3.3 Discussion

As noted previously, our developed framework is a generic one and can be extended in the various ways. For example, the smart grid application dimension can be expanded to include more applications, such as the integration of distributed energy resources, integrated electric vehicles, etc. In the dimension of operation conditions, we can encompass a variety of cyber and physical threats such as blackhole attack, wormhole attack, botnet attack, load-distribution attacks, false data injection attacks, etc. We could also introduce additional dimensions to not only implement defensive mechanisms against cyber-attacks on Integrity, Availability, and Confidentiality (IAC), failures, but also consider different network protocols and architectures.

There are other alternatives to carry out co-simulations. For example, GridMat [45, 46] is a co-simulation tool that serves as a middleware between a power grid such as GridLAB-D and Matlab. Generally speaking, GridMat leverages the capabilities of GridLAB-D to model and simulate the power grid system and provides a Matlab interface to control the power grid. Control algorithms such as load control, controlling distributed energy resources could be implemented and studied using GridMat. In addition, it is possible to run an offline co-simulation [47]. For example, we can first run the simulation in NS-3 that consists of a number of nodes as houses/loads in the power grid. We then record the results in terms of network performance such as packet delivery ratio, end-to-end delay, etc. The results from NS-3 can then feed into the Matlab/Simulink simulator to evaluate the impact of network performance on power grid.

10.4 Performance Evaluation

Our simulation was performed on an INTEL Core 3 workstation equipped with 8 GB of DRAM. The installation of FNCS required a 64-bit Linux operating system and consists of prerequisite libraries (e.g., the ZeroMQ message library, the MPI library, and updated versions of gcc compilers), a Fenix module that encompasses the FNCS simulator API, a GridLAB-D simulator, and an NS-3 simulator.

The settings for the co-simulation are shown in Table 10.1 and outputs for co-simulations include the following: (i) *Total Load:* The aggregated load demand for the grid at a given time, (ii) *Market Clearing Price:* Energy price where demand equals to supply; (iii) *Market Clearing Quantity:* The amount of energy sold when the market closes; and (iv) *HVAC Population Statistics:* The number of HVAC "ON" indicated at a given time. It is worth noting

Table 10.1 Co-simulation settings

Object	Settings
Substation	- Nominal voltage: 7200 V
	- Power rating: 4500 kW, 1500 kW/phase
Residential Load	300 houses
Double Auction	- Maximum bid Price: $3.78
	- Maximum capacity bid quantity: 1100 kW
	- Initial price: $0.042676
	- Appliances in the market: HVAC
	- Market clearing: every 5 minutes
Length of Simulation	1 day: 07–21 00:00 AM to 07–22 00:00 AM
Billing	Hourly
Communication Network	- Nodes: 300
	- Transport protocol: UDP
	- Data rate: 10 Mbps
	- Delay: 3 ms

that in our power grid simulation topology, every house is equipped with a HVAC system that can either be turned OFF or turned ON based on the current energy price. To collect the number of HVAC system that are running (ON), we used a GridLAB-D data collection tool called recorder to count and store the total number of HVAC that are ON in every time interval. Concerning the injection false data injection, our co-simulation platform that consists of GridLAB-D and NS3 does not have the ability of injecting data in runtime. To simulate the false data injection attack, we generated false data offline and then stored the data in a GridLAB-D object called scheduler that changed the data input every 30 minutes or based on a predetermined interval. In the following, we present the results of our simulations. Notice that the normal scenario represents the reference, showing the normal behavior of the grid in an ideal environment with optimal settings. To clearly visualize the impact of the individual scenario, we will show the result for both individual scenario and normal scenario. Because there is no randomness in these simulations, error bars are not shown in any figure.

10.4.1 Demand Response

Normal Operation Mode: Figure 10.5(a) and (b) represent the variation of the total load over 24 hours in the smart grid, where demand/response is enabled and not enabled, respectively. In the case where demand/response is enabled, the total load with demand/response is 1600 kW. In the case where

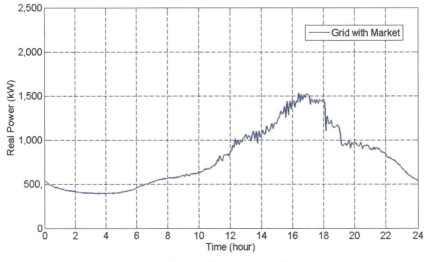

(a) Total Load with Market Enabled

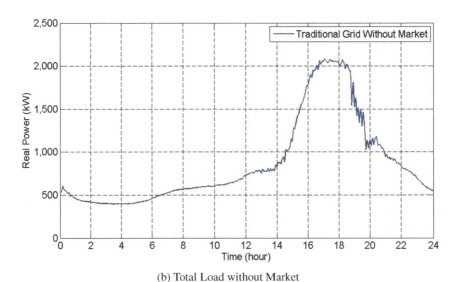

(b) Total Load without Market

Figure 10.5   Demand/Response normal operation.

demand/response is not enabled, the total load is 2290 kW. As we can see, for the same number of users and for the same energy demand, demand/response through market incitement can reduce the total power generation required to meet the demand and consequently reduce power losses. Effects of the market are more noticeable during peak periods. In the morning periods, where power demand is typically lower, the two curves are similar. From the afternoon to early evening, as the energy demand increases, the difference between the two curves is clearly visible. The total load in the grid system when demand/response is not enabled grows linearly until a peak condition occurs whereas we observe smooth variations on the system when demand/response is enabled. This demonstrates that demand/response is capable of smoothing power generation and mitigating peak system conditions. Thus, these conditions contribute to reducing power generation costs and power losses. During a low peak, users increase their energy use as the power is affordable whereas users reduce their energy use at high peak times as the price is relatively high. Notice that Figure 10.6 shows the normal reference bid price in 24 hours.

False Data Injection: High Capacity Bid (2500 kW) Figure 10.7(a) illustrates the variation of the total load over 24 hours for both normal capacity bid of 1100 kW and false capacity bid of 2500 kW, respectively. By injecting a false, large maximum capacity bid, there is more energy available in the market, causing the price to drop. As such, users can afford consuming more power even during peak periods without affecting their monthly bill. From 3 pm to 6 pm, a period of typically higher consumption, there is a total

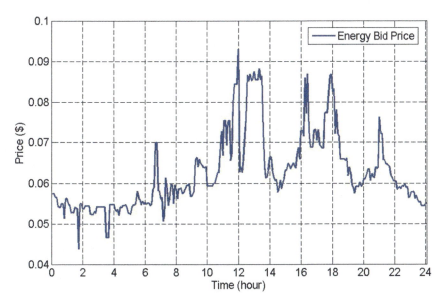

Figure 10.6 Market reference bid price.

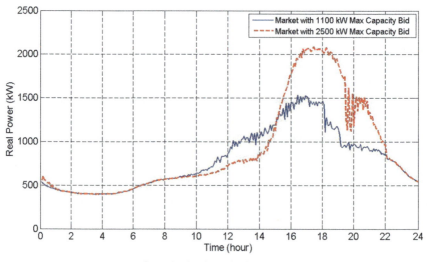

(a) Total Load with Capacity Bid at 2500 kW

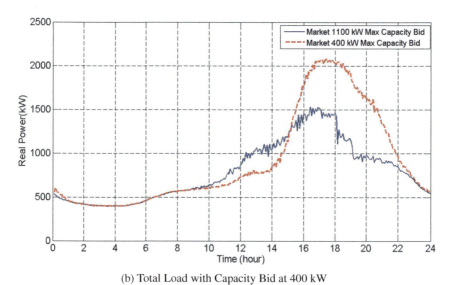

(b) Total Load with Capacity Bid at 400 kW

Figure 10.7   Demand/Response with false capacity bid injection attacks.

load increase of 588.612 kW. From 8 pm to 9 pm, a low peak period, the total load increases by 491.645 kW. Hence, to achieve a fair market and a win–win situation between energy providers and end users, the maximum capacity bid or the amount of energy the supplier provides to the market should be set optimally. On one hand, a small capacity bid indicates scarcity, and will generate substantial profit for energy suppliers at a higher cost for the customers. Conversely, an excessive maximum capacity bid will drive up energy usage, increasing power generation costs and power losses, resulting in an economic gain for customers and a financial loss for the supplier.

False Data Injection—Low Capacity Bid (400 kW): Figure 10.7(b) shows the variation of total load for a normal capacity bid of 1100 kW and a low capacity bid of 400 kW, respectively. By injecting a false, small maximum capacity bid, we provide less power to the energy market relative to the average demand. As expected, energy contention is high and consequently, the real-time price is high regardless of the time of use. In these circumstances, attempts by endusers to adjust their energy use will have minimal impact and the market operation is less efficient. In addition, we observe that since the market was very congested during peak period, endusers tend to increase their energy usage after peak periods as the energy becomes more affordable. Because of energy contention, we can also observe that a very high real-time price generates substantial profit for the supplier while increasing energy costs for end users. Nonetheless, the financial gain of the supplier will be offset by the power losses due to high power generation costs.

False Data Injection: High Bid Price: Figure 10.8 represents the variation of the total load when a false high bid price is injected from 12:30 pm to 4 pm. In contrast to the normal operation, we observe the intense fluctuations

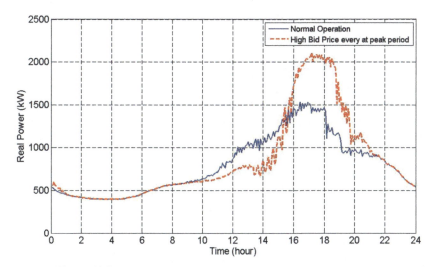

Figure 10.8 Demand/Response with high bid price injection attack.

of the load as customers attempt to reduce their energy usage from 12:30 pm to 4:00 pm, which is the time when the false high bid price is injected. Their energy usage then increases from 7 pm to 10 pm when the price has come down to normal. This shows the effectiveness of the power balance process as endusers intentionally reduce their power consumption when the price is high and increase it when the market is more affordable. Nonetheless, this scenario still creates disturbances, which are noticeable by the presence of high peaks.

Network Performance and Network Congestion: Figure 10.13(a) illustrates the variation of total load and market clearing price within the constraints of the communication network settings. The network is configured with the intent to create a poor performing network. The maximum data rate was set to 1 Mbps and the transmission delay set to 10 ms. In this configuration, the packet delivery ratio is very low as more packets are dropped. Both poor network performance and network congestion serve to disarticulate the market operation. The lack of a real-time information exchange between substations (suppliers) and endusers leads to a dysfunctional market, making the impact of demand response application to be not very noticeable. This makes a behavior similar to that of the traditional grid, which does not have demand response capability.

10.4.2 Energy Market: Market Clearing Price

Reference Bid Price: Figure 10.5(c) shows the variation of the energy reference bid price from the energy supplier over 24 hours. As we can see, the energy price is the highest during the peak hours from 12 pm to 1:30 pm and from 4 pm to 6 pm. During this period, price varies from $0.05 to $3.78 at peak time. Conversely, the energy price is relatively low, less than $0.08 late at night and early in the morning.

Market Clearing Price for False High Capacity Bids: Figure 10.9(a) represents the variation of the market clearing price for a high capacity bid. A false maximum capacity bid of 2500 kW creates the opposite effect, in which the energy price is always affordable regardless of the time of use. As shown in the figure, the energy price with 2500 kW capacity bid is almost the same for low consumption and peak periods and remains below $0.05. In this scenario, the gain of end users coincides with the significant loss of the energy suppliers. Manipulating and forging capacity bids can have tremendous impact on the market.

Market Clearing Price for False Low Capacity Bids: Figure 10.9(b) illustrates the variation of the market clearing price for a maximum capacity

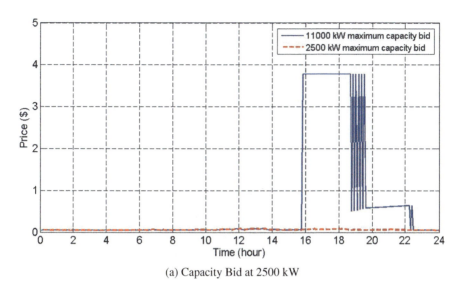

(a) Capacity Bid at 2500 kW

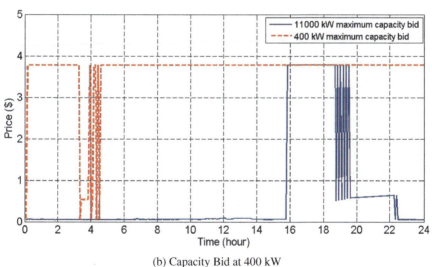

(b) Capacity Bid at 400 kW

Figure 10.9 Market clearing price with false capacity bid injection attacks.

bid set at 400 kW. During normal operation (1100 kW maximum bid), the price stays below $0.08 and reaches the maximum only during the peak hours between 4 pm and 7 pm. It is worth noting that during this high price period, users intentionally reduce their energy use in order to reduce energy costs.

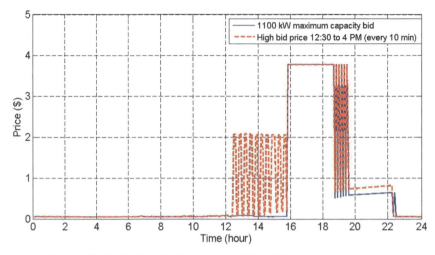

Figure 10.10   Market clearing price with high bid price injection attack.

With a false maximum capacity bid of 400 kW, which is far below the average demand, contention is high on the small amount of energy available in the market, causing the clearing price to reach the maximum of 3.78 even during the period of low consumption between 3 am and 4 am. This scenario gives a large profit for the supplier, but significant losses for end users.

Market Clearing Price for False and unstable High Price Bids: Figure 10.10 shows the variation of the market clearing price when a high bid price is injected in the system every 10 minutes from 12:30 pm to 4 pm. This causes the price to fluctuate continuously. As shown in the figure, the fluctuation and instability of bidding price reflect directly on the market clearing price.

Market Clearing Price with Network Congestion: Figure 10.13(b) illustrates the variation of the market clearing price with a poor performing or congested network. The result is a disarticulated energy market. The lack of a real-time information exchange between substations (suppliers) and end users leads to a dysfunctional market. The clearing price is similar to the reference clearing price (Figure 10.14) as the substation barely gets any bid or feedback from end users.

10.4.3 Energy Market: Market Clearing Quantity

Market Clearing Quantity with a False High Maximum Capacity Bid: Figure 10.11(a) represents the variation of market clearing quantity over 24 hours for a false maximum capacity bid of 2500 kW. Here, the clearing market quantity

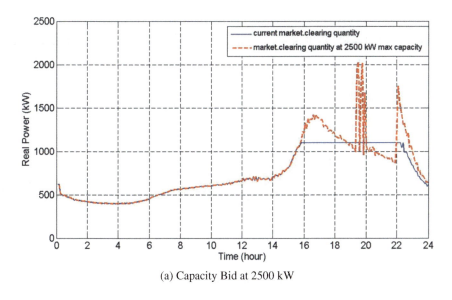

(a) Capacity Bid at 2500 kW

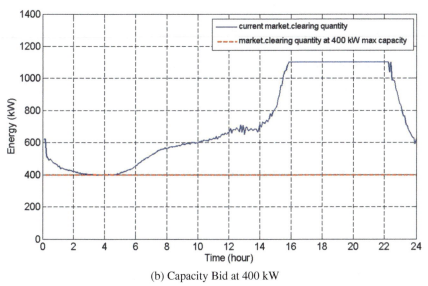

(b) Capacity Bid at 400 kW

Figure 10.11 Market clearing quantity with false capacity bid injection attacks.

increases during low and high peak hours as end users take advantage of the affordable energy rates.

Market Clearing Quantity with a False Low Maximum Capacity Bid: Figure 10.11(b) represents the variation of market clearing quantity over

24 hours for a false maximum capacity bid of 400 kW. With a false maximum capacity bid of 400 kW that is less than 1100 kW of the normal operation bid, and below basic end users' demand, the energy market is unable to maintain the energy clear quantity below the 400 kW limit. As a consequence, a too small capacity bid less than basic end users' energy demand has no significant effect on end users' energy usage and the total load could inevitably exceed the substation maximum capacity bid.

Market Clearing Quantity at False High Bid Prices: Figure 10.12 illustrates the variation of the clearing market quantity when a false high bid price is injected every 10 minutes between 12:30 pm and 4 pm. Users respond to the fluctuating bid price by adjusting their power demand every time when the price changes. This situation can be very disturbing for end users. The more unpredictable and unstable the market is, the more difficult it is for end users to efficiently adjust their energy use habits.

Economic Impact of Dynamic Markets: Table 10.1 represents the comparison of one-hour energy bill for a single house for different simulation scenarios. While the house used approximately the same amount of energy, a reduced maximum bid capacity incurs the most severe economic impact with a 189.24 percent increase or a financial loss of $144.107 when compared to the normal scenario. In addition, confirming our previous finding, a large maximum bid capacity is the most profitable for the end user with an 82 percent bill reduction or a gain of $62.468.

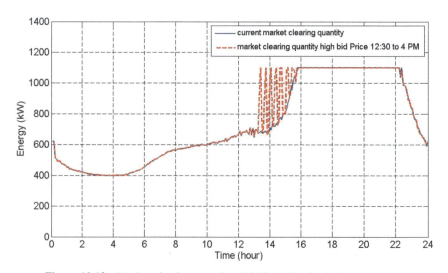

Figure 10.12 Market clearing quantity with high bid price injection attack.

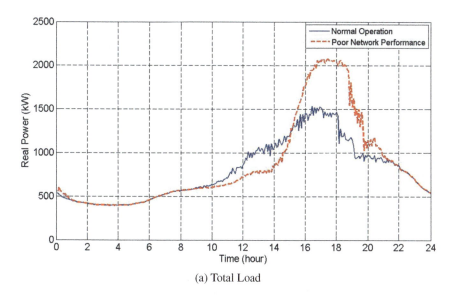

(a) Total Load

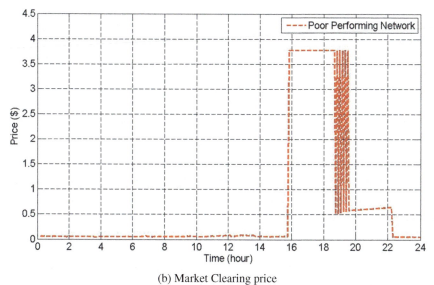

(b) Market Clearing price

Figure 10.13 Poor network performance.

10.4.4 HVAC Population Statistics

Maximum Capacity Bid—2500 kW: Figure 10.15(a) illustrates the variation of the number of HVAC systems running over time for a false maximum capacity

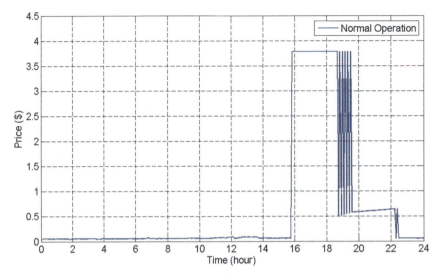

Figure 10.14   Market clearing price normal operation.

Table 10.2   Economy impact for 1 hour (single house)

Scenario	Energy Usage (kW)	Bill Amount ($)	Gain/Loss ($)
Normal Operation	56.833	76.150	N/A
Max capacity bid at 400 kW	56.847	220.257	−144.107
Max capacity bid at 2,500 kW	56.792	13.681	62.468
High bid price 12:30–4 PM	56.833	86.453	−10.302
Poor network Performance	56.833	76.035	0.114

bid of 2500 kW. Because this is greater than the 1100 kW for normal operation, there is more energy available and the market price drops. This motivates end users to use more energy and to keep their HVAC system running at low temperature setting points.

Maximum Capacity Bid—400 kW: Figure 10.15(b) illustrates the variation of the number of HVAC systems that are ON in a 24 hour period when a false maximum capacity bid of 400 kW is injected to the grid. This is less than the average demand, meaning that the contention on biddable energy is high, which causes the real-time energy price to significantly increase. Consequently, as shown in the figure, the number of HVAC systems running decreases during peak times (4 pm–7 pm). After 7 pm, as the demand and the real-time price are falling, more HVAC systems resume their operation and thus the number of HVAC systems increases.

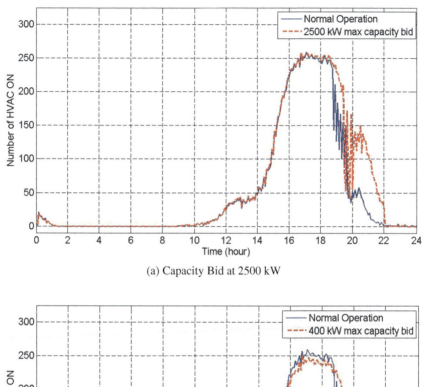

(a) Capacity Bid at 2500 kW

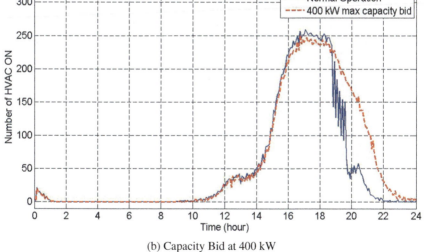

(b) Capacity Bid at 400 kW

Figure 10.15   HVAC population statistics with false capacity bid injection attacks.

High Bid Price: Figure 10.16 represents the variation of the number of HVAC systems running over time when a false high bid price is injected between 2 pm and 6 pm. As shown in the figure, a high bid price during

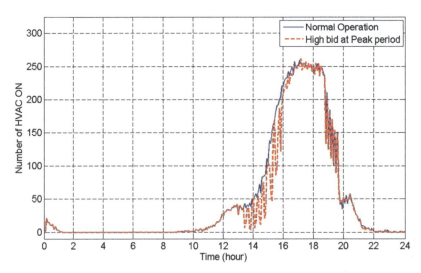

Figure 10.16   HVAC population statistics with high bid price injection attack.

that specific time frame causes a lot of fluctuations in the HVAC population. Many HVAC systems are flip-flopping, going from ON to OFF in order to adjust their energy use and to contain the energy bill within a reasonable range. We observe that after 4 pm, the market returns to normal operation and the HVAC population follows the normal trend.

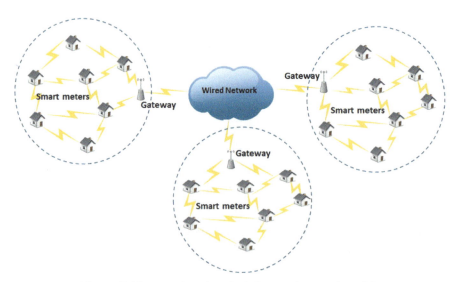

Figure 10.17   An AODV-based wireless mesh network model.

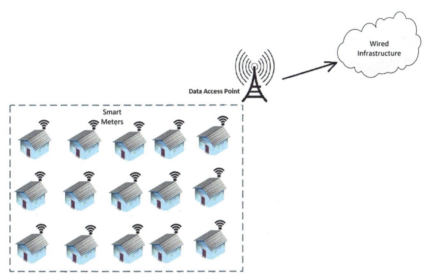

Figure 10.18 WiMAX network model.

10.5 Extension

Recall that we present a framework that consists of two dimensions to design simulation scenario in Section 10.3 and we present the results of co-simulation based on wired network in Section 10.4. We now extend the framework to consider various wireless networks protocols: wireless mesh network infrastructure based on Ad hoc On-Demand Distance Vector (AODV) protocol and WiMAX network infrastructure based on IEEE 802.16d-2004 WiMAX layer 2 protocol. The objective is to take advantage of the performance impact of wireless network protocols on smart grid applications, such as demand/response and market/dynamic pricing. In this chapter, we select AODV and WiMAX because they are two known network protocols and network technologies that can be used to support smart grid communications. AODV can be used to establish a mesh network topology that provides a resilient, flexible, and cost-effective network whereas WiMAX can be used to extend the area of network coverage for smart grid components. As an ongoing work, we are designing and implementing simulation scenarios based on other wireless techniques. Notice, for the sake of comparison, we also include the wired network, which is the similar to the one used in Section 10.4. In this case, smart meters are organized into clusters of 50 nodes and the cluster head is connected to the substation through a point-to-point wired communication link. In the following, we first briefly introduce wireless mesh network and WiMAX network and then show evaluation results.

10.5.1 Wireless Network Models

Wireless communication technologies show great potential in supporting smart grid applications due to intrinsic capabilities such as cost-effectiveness, efficiency, and scalability. In our experiment setting, the wireless mesh network model consists of multiple mesh clusters and each cluster consists of 50 nodes, where each node represents a smart meter. The Data Aggregator Point (DAP) in each cluster is connected to the wired network infrastructure through a dedicated point-to-point connection link. In our simulation, we consider the wireless mesh network based on Ad hoc On-Demand Distance Vector (AODV) protocol. Generally speaking, AODV is a distance vector routing protocol designed for mobile ad hoc networks. AODV can be classified as a reactive routing protocol, as a route to a designated destination is not pre-established during the protocol initialization stage and stored in the routing table. Instead, a route to the destination node is created upon request of the source node and the created route is closed once it ceases to be active. In our prior work [48], we carried out a simulation study using NS-3 and showed that AODV could maintain good performance in terms of packet delivery ratio, throughput, and end-to-end delay as the number of nodes increases. In the similar way, the WiMAX network model consists of a single cluster of nodes and a WiMAX base station that represents the DAP. A point-to-point connection allows bidirectional data exchange between the DAP and the wired communication infrastructure.

10.5.2 Evaluation Results

Using aforementioned smart grid applications, we conducted simulations and evaluated the impact of wireless network models versus the wired model on the performance of the smart grid. For each network model (i.e., wired, mesh/AODV, and WiMAX), we have conducted simulations on two grid topologies: 50 nodes and 300 nodes. The wired network is built on the Ethernet technology with a data transmission rate of 10 Gbps and a 2 milliseconds transmission delay. Specific to the mesh network, we have used a grid topology of 5×10 for each cluster. A topology of 50 nodes yields one single cluster, whereas a topology of 300 nodes yields six clusters. The distance between mesh nodes is set to 170 m and clusters are 1000 meters distant from each other. Concerning the WiMAX network, we have deployed nodes in an area of 1650×850 m^2 the equivalent of 20 city block areas.

In a point-to-multipoint setting, we have defined three types of nodes: (i) *Subscriber Station:* represents the residential smart meter that transmits the data through uplink to the base station; (ii) *Base Station:* represents the DAP that forwards data in both ways between the substation center and smart meters. The base station is deployed outside the city block area with the coordinates of 2000 m and 1500 m for the x and y axis, respectively. We selected the modulation and coding technique 16QAM 1/2 based on our prior work [48]; (iii) *AMI Head End or Substation Center Node:* responsible for collecting data from smart meters via the communication network. For each scenario, we run 9 hour simulations (from 11:00 am to 8:00 pm) for different network sizes and different parameters associated with data rates and transmission delays and observe how the network performance will affect the total load demand, the clearing energy price, and quantity. For example, for 300 nodes, we set three different tuples (4 Mbps, 5 ms), (100 Mbps, 2 ms), and (0.1 Mbps, 5 ms). The time interval of 11:00 am to 8:00 pm has been specifically chosen because it represents the time when the energy demand is high and when consequently the network contention is present. We evaluate the network performance by comparing changes to the clearing energy quantity, the clearing energy price, and total load demand (defined in Section 10.4).

Energy Market Clearing Quantity: Figure 10.19(a) illustrates the variation of the energy market clearing quantity from 11 am to 4 pm for the wired network and the wireless mesh network based on AODV when the total number of nodes is set to 300, the transmission rate of the point-to-point link is 100 Mbps and the transmission delay is 2 ms. As we can see from the figure, based on our network settings, more energy could be sold by the utility when a wireless mesh network is used. From 11 am to 4 pm on the month of July (summer day), the load demand is very high. Consequently, the network traffic is very intense as users are sending more load demand to the substation. In this contentious network condition, the multi-hop connectivity of the mesh network could incur less packet loss and less congestion, leading to a better throughput. As such, more users' demand could effectively reach the substation and consequently lead to an increase in energy clearing quantity. Conversely, the wired network becomes more congested as smart meters attempt to send load demands to the substation over a shared communication link. The subsequent increase in transmission delay and packet losses could cause the substation to become partially "blind" to the real energy demand, leading to a lesser energy clearing quantity. To summarize, a better network, with a high packet delivery ratio and less congestion, provides a more accurate picture of the real energy demand. Figure 10.20(a) compares the impact of the three network models on the market clearing quantity for 50 nodes. As we can see from the figure, based on our network setting, the two wireless

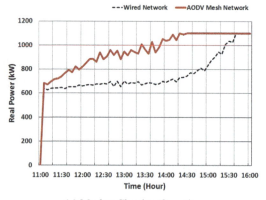

(a) Market Clearing Quantity

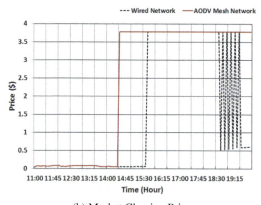

(b) Market Clearing Price

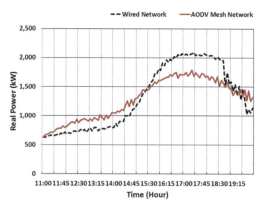

(c) Total Load Demand

Figure 10.19 Wired network vs. AODV wireless mesh network (300 nodes).

technologies outperform the wired network, and the wireless mesh network based on AODV performs slightly better than the WiMAX wireless network.

Energy Market Clearing Price: Figure 10.19(b) represents the variation of the energy market clearing price from 11 am to 8 pm for the wired network and the wireless mesh network based on AODV when the total number of nodes is set to 300, the transmission rate of the point-to-point communication link is 100 Mbps, and the transmission delay is 2 ms. As we can see in the figure, the energy clearing price for the wireless mesh network is overall greater than the price when a wired network is used during peak periods. Since the wireless mesh network is more efficient and robust with less congestion and high packet delivery ratio, users' demand could effectively reach the substation. This would lead to a higher energy market quantity. As more users compete for the limited amount of energy available in the market, the price consequently rises. Figure 10.20(b) represents a comparative view of the impact of three network models on the market clearing price with a network size of 50 nodes. Similar to the market clearing quantity, wireless mesh network-based AODV and WiMAX have nearly the same impact. Overall, a better communication network will allow the substation or the utility to have a better view of the market and apply a fair price consistently with the time of use. As seen in the figure, a poor performing network will likely lead to the inaccurate state of the market, leading to more fluctuation in the market price and financial losses.

Total Load Demand: Figure 10.19(c) represents the variation of the total load demand from 11 am to 8 pm for the wired network and the wireless mesh network based on AODV when the total number of nodes is set to 300, the transmission rate of the point-to-point communication link is 100 Mbps, and the transmission delay is 2 ms. During the peak periods from 3 pm to 7 pm, the total load demand for the wired network is greater than that of the wireless mesh network. This demonstrates the impact of network performance on the effectiveness of demand response application. Recall that the objective of demand response application is to preserve the balance between power generation and load demand and to avoid the high peak and excess power generation during the peak period. This is achieved through the price incentive that allows users to adjust the power consumption. Since the wireless mesh network performs better than the wired network in our network setting, load demands from users more effectively reach the substation. As the total load demand increases during peak periods, the substation will raise the energy price as shown in Figure 10.19(b). In response to the high energy price, users will turn off the HVAC systems, leading to a lower overall total load demand. Conversely, for the wired network during the bandwidth contention period,

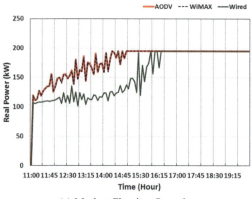

(a) Market Clearing Quantity

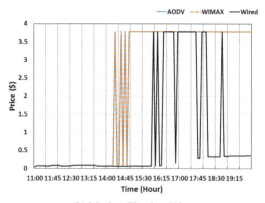

(b) Market Clearing Price

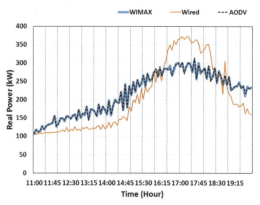

(c) Total Load Demand

Figure 10.20   Wired network vs. AODV wireless mesh network vs. WiMAX network (50 nodes).

more packets will be dropped, and the substation will be partially blind to the real load demand. As a consequence, the price is low and the amount of consumption increases. Figure 10.20(c) shows the impact of wireless networks on the total load demand. From the figure, we observe that a robust and efficient network could make the demand response application more effective, reducing the total power generation and subsequently power losses and generation cost.

10.6 Conclusion

In this chapter, we addressed the performance issue of the smart grid applications utilizing co-simulation. We systematically reviewed the existing efforts of co-simulation and designed a framework to perform co-simulation evaluations on diverse configurations. To understand the interaction and the reciprocal effects between the communication network and power grid applications in the smart grid, we investigated the performance of demand/response and dynamic market pricing under different network conditions (e.g., normal operation, degraded performance, and security threats). Leveraging a known co-simulation platform, we implemented the configuration scenarios and conducted a performance evaluation of the selected smart grid applications. We then extended our evaluation to implement additional network architectures and wireless protocols, comparing their performance impact on the effectiveness of smart grid applications.

References

[1] U.S. Department of Energy, Office of Energy Delivery and Energy Reliability. *Smart Grid*. http://energy.gov/oe/services/technology-devel opment/smart-grid/
[2] Norling, K., Broman, D., Fritzson, P., Siemers, A., and Fritzson, D. (2007). Secure distributed co-simulation over wide area networks. In *Proceedings of the 48th Conference on Simulation and Modelling (SIMS)*.
[3] Karnouskos, S. and De Holanda, T. N. (2009). Simulation of a smart grid city with software agents. In *Proceedings of IEEE Third UKSim European Symposium on Computer Modeling and Simulation (EMS)*.
[4] Pipattanasomporn, M., Feroze, H., and Rahman, S. (2009). Multi-agent systems in a distributed smart grid: Design and implementation. In *Proceedings of IEEE/PES Power Systems Conference and Exposition (PSCE)*.

[5] Mathworks. *Simulink, Simulation and Model Design.* http://www.math works.com/discovery/power-system-simulation-and-optimization.html

[6] Network Simulator V3 (NS-3). http://www.nsnam.org/overview/what-is-ns-3/

[7] OMNET++. http://www.omnetpp.org

[8] OpenDSS. http://smartgrid.epri.com/SimulationTool.asp

[9] Fuller, J. C., Ciraci, S., Daily, J. A., Fisher, A. R., and Hauer, M. (2013). Communication simulations for power system applications. In *Proceedings of IEEE 2013 Workshop on Modeling and Simulation of Cyber-Physical Energy Systems (MSCPES).*

[10] Pacific Northwest National Laborratory (PNNL). *Network and Data Management for Grid Analytics.* http://availabletechnologies.pnnl.gov/docs/NetworkDataMngmnt.pdf

[11] Moulema, P., Yu, W., Griffith, D., and Golmie, N. (2015). Performance evaluation of smart grid applications using co-simulation. In *Proc. of IEEE International Conference on Computer Communication and Networks (ICCCN).*

[12] Li, W., Zhang, X., and Li, H. (2014). Co-simulation platforms for co-design of networked control systems: An overview. *Control Eng. Pract.,* 23, 44–56.

[13] Mets, K., Verschueren, T., Develder, C., Vandoorn, T. L., and Vandevelde, L. (2011). Integrated simulation of power and communication networks for smart grid applications. In *Proceedings of 2011 IEEE 16th International Workshop on Computer Aided Modeling and Design of Communication Links and Networks (CAMAD).*

[14] Cervin, A., Ohlin, M., Henriksson, D., et al. (2007). Simulation of networked control systems using truetime. In *Proceedings of 3rd International Workshop on Networked Control Systems: Tolerant to Faults.*

[15] Henriksson, D., Cervin, A., and Årzén, K.-E. (2002). Truetime: Simulation of control loops under shared computer resources. In *Proceedings of the 15th IFAC World Congress on Automatic Control.*

[16] Zhang, Z., Lu, Z., Chen, Q., Yan, X., and Zheng, L.-R. (2010). Cosmo: Co-simulation with matlab and omnet++ for indoor wireless networks. In *Proceedings of IEEE Global Telecommunications Conference (GLOBECOM).*

[17] D. Anderson, H. Zhao, and D. Venkatasubramanian, Bakke. *A Virtual Smart Grid.* http://magazine.ieee-pes.org/january-february-2012/a-virtual-smart-grid/, February 2012.

[18] Davis, C., Tate, J., Okhravi, H., Grier, C., Overbye, T., and Nicol, D. (2006). Scada cyber security testbed development. In *Proceedings of the 38th North American Power Symposium (NAPS)*.

[19] Fuller, S. and Wynne. Middleware for the next generation power grid.

[20] Georg, H., Wietfeld, C., Muller, S. C., and Rehtanz, C. (2012). A hla based simulator architecture for co-simulating ICT based power system control and protection systems. In *Proceedings of 2012 IEEE Third International Conference on Smart Grid Communications (SmartGridComm)*.

[21] Hopkinson, K., Wang, X., Giovanini, R., Thorp, J., Birman, K., and Coury, D. (2006). Epochs: a platform for agent-based electric power and communication simulation built from commercial off-the-shelf components. *IEEE Trans. Power Syst.*, 21(2), 548–558.

[22] Kohtamaki, T., Pohjola, M., Brand, J., and Eriksson, L. M. (2009). Picc-sim toolchain-design, simulation and automatic implementation of wireless networked control systems. In *Proceedings of IEEE International Conference on Networking, Sensing and Control (ICNSC)*.

[23] Li, W. and Monti, A. (2010). Integrated simulation with vtb and opnet for networked control and protection in power systems. In *Proceedings of the 2010 Conference on Grand Challenges in Modeling & Simulation*.

[24] Li, W., Monti, A., Luo, M., and Dougal, R. A. (2011). Vpnet: A co-simulation framework for analyzing communication channel effects on power systems. In *Proceedings of 2011 IEEE Electric Ship Technologies Symposium (ESTS)*.

[25] Liberatore, V. and Al-Hammouri, A. (2011). Smart grid communication and co-simulation. In *Proceedings of IEEE 2011 Energytech Conference*.

[26] Zhu, K., Chenine, M., and Nordstrom, L. (2011). Ict architecture impact on wide area monitoring and control systems' reliability. *IEEE Trans. Power Del.*, 26(4), 2801–2808.

[27] Opnet Technologies. http://www.opnet.com

[28] Panchadcharam, S., Ni, Q., Taylor, G., Irving, M., Gershinsky, G., Lewin-Eytan, L., and Shagin, K. (2011). Evaluation of throughput and latency performance for medium voltage and low voltage communication infrastructures. In *Proceedings of 2011 46th International Conference Universities' Power Engineering (UPEC)*.

[29] Panchadcharam, S., Taylor, G., Ni, Q., Pisica, I., and Fateri, S. (2012). Performance evaluation of smart metering infrastructure using simulation tool. In *Proceedings of 2012 47th International Conference on Universities' Power Engineering (UPEC)*.

[30] Lin, H., Veda, S. S., Shukla, S. S., Mili, L., and Thorp, J. (2012). Geco: Global event-driven co-simulation framework for interconnected power system and communication network. *IEEE Trans. Smart Grid*, 3(3), 1444–1456.

[31] Pacific Northwest National Laboratory (PNNL). *GridLABD*. http://www. gridlabd.org

[32] Power Systems Engineering Research Center (PSERC). *MATPOWER*. http://www.pserc.cornell.edu/matpower/

[33] Issariyakul, T. and Hossain, E. (2011). *Introduction to network simulator NS2*. Springer Science & Business Media.

[34] Nutaro, J., Kuruganti, P. T., Miller, L., Mullen, S., and Shankar, M. (2007). Integrated hybrid-simulation of electric power and communications systems. In *Proceedings of IEEE 2007 Power Engineering Society General Meeting*.

[35] Ezeme, M. O. (2015). *A Multi-domain Co-simulator for Smart Grid: Modeling Interactions in Power, Control and Communications – PhD Thesis*. PhD thesis, https://tspace.library.utoronto.ca/bitstream/1807/ 70286/2/Ezeme_Mellitus_O_201511_MAS_thesis.pdf/.

[36] Kounev, V., Tipper, D., Levesque, M., Grainger, B. M., McDermott, T., and Reed, G. F. (2015). A microgrid co-simulation framework. In *Proceedings of 2015 IEEE International Workshop on Modeling and Simulation of Cyber-Physical Energy Systems (MSCPES)*.

[37] Lin, H., Sambamoorthy, S., Shukla, S., Thorp, J., and Mili, L. (2011). Power system and communication network co-simulation for smart grid applications. In *Proceedings of IEEE International Conference of PES Innovative Smart Grid Technologies (ISGT)*.

[38] Chouikhi, S., El Korbi, I., Ghamri-Doudane, Y., and Saidane, L. A. (2013). A comparison of wireless sensor networks co-simulation platforms for smart grid applications. *International Journal of Digital Information and Wireless Communications (IJDIWC)*, 3(1), 87–96.

[39] Garau, M., Celli, G., Ghiani, E., Soma, G. G., Pilo, F., and Corti, S. (2015). Ict reliability modelling in co-simulation of smart distribution networks. In *Proceedings of 2015 IEEE 1st International Forum on Research and Technologies for Society and Industry Leveraging a Better Tomorrow (RTSI)*.

[40] Lévesque, M., Béchet, C., Suignard, E., Maier, M., Picault, A., and Joós, G. (2014). From co-toward multi-simulation of smart grids based on hla and fmi standards. *arXiv preprint arXiv:1412.5571*.

[41] Lévesque, M., Xu, D. Q., Joós, G., and Maier, M. (2012). Communications and power distribution network co-simulation for multidisciplinary smart grid experimentations. In *Proceedings of the 45th Annual Simulation Symposium.*

[42] Li, W., Ferdowsi, M., Stevic, M., Monti, A., and Ponci, F. (2014). Cosimulation for smart grid communications. *IEEE Trans. Ind. Informat.*, 10(4), 2374–2384.

[43] Tariq, M. U., Swenson, B. P., Narasimhan, A. P., Grijalva, S., Riley, G. F., and Wolf, M. (2014). Cyber-physical co-simulation of smart grid applications using ns-3. In *Proceedings of the ACM 2014 Workshop on NS-3.*

[44] PowerWorld Simulator Overview. http://www.powerworld.com/products/simulator/overview

[45] Faruque, A., Abdullah, M., and Ahourai, F. (2014). Gridmat: Matlab toolbox for gridlab-d to analyze grid impact and validate residential microgrid level energy management algorithms. In *Proceedings of the 5th IEEE PES Innovative Smart Grid Technologies Conference (ISGT).*

[46] University Of California, Irvine. Electrical Engineering and Computer Science Department. *GriMat.* http://www.sourceforge.net/projects/gridmat

[47] Mallapuram, S., Moulema, P., and Yu, W. (2015). On a simulation study for reliable and secured smart grid communications. In *SPIE Defense+Security*, pages 945808–945808. International Society for Optics and Photonics.

[48] Moulema, P., Yu, W., Xu, G., Griffith, D., Golmie, N., Lu, C., and Su, D. (2013). On simulation study of mesh-based protocols for smart grid communication networks. In *Proceedings of the ACM 2013 Research in Adaptive and Convergent Systems (RACS).*

11

Tight Bounds on Localized Sensor Self-Deployment for Focused Coverage

Gokarna Sharma[1] and Hari Krishnan[2]

[1]Department of Computer Science, Kent State University, Kent, OH 44242, USA
[2]School of Electrical Engineering and Computer Science, Louisiana State University, Baton Rouge, LA 70803, USA

Abstract

We consider the self-deployment problem in mobile sensor networks with the objective of providing focused coverage for a point of interest (POI) such that the maximum area around it is covered by sensors without sensing holes. We present a local greedy algorithm, called TTGREEDY that solves this problem in at most $R + 2 \cdot (n - 1)$ time steps, where R is the distance to the farthest sensor position from the POI in the initial configuration of sensors and n is the number of sensors. This is a significant improvement over the best previously known $\mathcal{O}(D)$ time step algorithm of Blázovics and Lukovszki, where D is the sum of the distances to the sensors from the POI in their initial configuration. The main idea behind our algorithm is to synchronously drive mobile sensors along a locally computed triangle tessellation avoiding collisions of sensors. We also show that there are initial configurations of n sensors in this problem where at least $R + \frac{(n-1)}{2}$ time steps are needed by any greedy algorithm. Several simulations under different parameter settings confirm these findings. To the best of our knowledge, these results provide the first tight runtime (within a small constant factor) solution to this problem.

Keywords: Mobile sensor networks, Focused coverage, Self-deployment, Point of interest, Local algorithms, Runtime.

319

11.1 Introduction

Mobile sensor networks (MSNs) are typically formed from the collection of sensor nodes which have locomotion capabilities in addition to communication, sensing, and computation capabilities. MSNs are gaining attention in recent years since they provide low-cost decentralized solutions to many problems (compared to their centralized solutions) and handle very well the dynamism that arises in many practical scenarios. Furthermore, the solutions they provide are scalable with the size of the network and tolerant to unpredictable sensor failures. Several problems are solved in the literature using MSNs, such as barrier coverage, wireless coverage control and connectivity, environmental control, and target tracking (see [1–3]). These problems are solved by moving sensors independently such that the objective is fulfilled with no or very minimum communication with only the neighboring sensors; these kinds of problems that move sensors this way are generally called *self-deployment* problems.

In this chapter, we consider the self-deployment problem which deals with an important research issue of autonomous coverage formation.[1] The motivation is that there are several applications where sensors are designated to monitor concerned events around a given strategic site, e.g., the area discovery for the survivors around the epicenter of a disaster, monitoring the pollution impact on the soil/air in the vicinity of a chemical plant, and guarding a strategic site for intruder detection. The given strategic site is the point that needs coverage focus and is generally called *point of interest* (POI). This type of coverage of a POI is called *focused coverage* or simply F-*coverage* [5, 6], where F denotes the POI.

We focus on a local synchronous algorithm for the optimal F-coverage problem. The optimal F-coverage problem has maximized coverage radius with no holes as we define later in Section 11.3. The local algorithms are preferred as they cope up with unpredictable sensor failures, dynamical topological change, and network scale that may be potentially large. By "local," we mean that each sensor makes self-deployment decision independently using only its k-hop neighborhood information for some small integer $k \geq 1$. The algorithm is called *strictly local* when $k = 1$. The most important question here is *whether the sensors are able to provide optimal F-coverage through local self-deployment algorithms and how long it takes to do so.* The first

---

[1]A preliminary version of the results discussed in this chapter appeared in the 24th International Conference on Computer Communication and Networks (ICCCN), Las Vegas, USA, IEEE, August 3–5, 2015 [4].

part of the question is answered by Li et al. [5] giving several strictly local algorithms in an asynchronous setting; for the second part of the question, they only gave the proof that their algorithms take finite time. Recently, Blázovics and Lukovszki [6] gave an algorithm assuming $k = 2$ in a synchronous setting and proved that it provides optimal F-coverage in $\mathcal{O}(D)$ time steps answering both parts of the question, where D is the sum of the distances of the sensors from the POI in the initial configuration of sensors. However, the runtime of their algorithm is still significantly large and not suitable in real-world scenarios where fast area discovery is critical for quick response, for example, in disaster-hit areas. In this chapter, we give a new local algorithm to provide optimal F-coverage in the same setting of Blázovics and Lukovszki ($k = 2$) that has significantly better runtime, answering both parts of the question. Particularly, we have made the following three contributions.

- We present and analyze a local self-deployment algorithm, called TTGREEDY that provides the optimal F-coverage around the POI in $R + 2 \cdot (n - 1)$ time steps starting from any initial configuration of n sensors, where R is the distance from the position of the farthest sensor to the POI in the initial configuration.
- We also prove that there are initial configurations of n sensors where any local greedy algorithm needs at least $R + \frac{(n-1)}{2}$ time steps to obtain optimal F-coverage. To the best of our knowledge, these results provide the first solution to the F-coverage problem that is worst-case tight within a small constant factor.
- We compare TTGREEDY with previous algorithms due to [5, 6] through several simulations under different settings of number of sensors and network sizes. The simulation results confirm our theoretical findings.

TTGREEDY significantly improves the $\mathcal{O}(D)$ runtime of the best previously known algorithm of Blázovics and Lukovszki [6]. Moreover, the lower bound on runtime of any local algorithm for F-coverage was not given in the literature and this is the first time a lower bound is given for this problem for a class of greedy algorithms. An algorithm for this problem is called *greedy* if it tries to minimize the hop distance of sensors to the POI by 1 in every time step.

The main idea behind our algorithm TTGREEDY is that in each time step, sensors move synchronously toward the POI along a locally computed triangle tessellation unless they block each other (in fact, in each time step, they move one hop closer to the POI). When sensors block each other so that minimizing the distance to the POI is not possible, they try to rotate to the neighboring vertex from the current vertex without increasing/decreasing the distance to the POI. If the sensors block each other such that even rotation is not possible, they stay idle in the current vertex at that time step.

11.1.1 Chapter Organization

We proceed as follows. We discuss related work in Section 11.2. We present model and preliminaries in Section 11.3. We then present our algorithm in Section 11.4. In Section 11.5, we prove a runtime lower bound and analyze our algorithm in Section 11.6. We present experimental results in Section 11.7 and conclude the chapter in Section 11.8 with a short discussion.

11.2 Related Work

Several problems, such as barrier coverage, patrolling, wireless coverage control, and target tracking, were solved in the literature through sensor self-deployment algorithms [1–3, 7–10]. Sensor self-deployment algorithms are also studied for coverage formation over a region of interest (ROI) [3, 8]. The vector-based (or virtual force-based) approaches are studied in, e.g., [7, 9, 10]. Cortes et al. [2] studied Voronoi diagram-based self-deployment approach for the coverage of the ROI. Bartolini et al. [11] presented a local algorithm on a hexagonal grid map in which the sensors simultaneously use snap and spread activities in order to cover the given area. Yang et al. [12] presented a load-balanced sensor self-deployment algorithm that partitions the plane into a two-dimensional mesh and treats nodes as load.

However, the ROI coverage problem does not deal with the particular coverage focus that is needed for the F-coverage problem. Li et al. [5] were the first to introduce the F-coverage problem and solve it on an equilateral triangle tessellation (TT) graph layout (details in Section 11.3). However, the collision of the sensors during the self-deployment process was allowed, i.e., more than one sensor can occupy the same TT vertex at the same time. They only proved that their algorithms provide optimal F-coverage in finite time in an asynchronous setting. Recently, Blázovics and Lukovszki [6] showed that it can be performed faster avoiding collisions by giving a local algorithm that has runtime $\mathcal{O}(D)$ time steps. To get better understanding of the runtime, suppose that n sensors are deployed initially so that the distance from each of them to the POI F is R. Then, our algorithm needs at most $R + 2 \cdot (n - 1)$ time steps, whereas the algorithm of Blázovics and Lukovszki [6] needs $\mathcal{O}(R \cdot n)$ time steps. This says that our algorithm is linear in R and n and asymptotically better than the algorithm of Blázovics and Lukovszki [6].

Recently, Li et al. [13] presented carrier-based and Wu et al. [14] studied movement-assisted self-deployment approaches to provide coverage in MSNs with different objectives.

In the area of *robot gathering* problems, many algorithms were studied for the robots to gather in a single (not predefined) point in the Euclidean plane, e.g. [15–18], under various settings. Recently, runtime bounds of some of the gathering algorithms were studied in [17–19]. However, all these chapters assumed that the robots are point robots, they can share positions with other robots, and collisions are allowed. Recently, some chapters studied the gathering problem for the so-called *fat* robots that cannot share positions with other robots due to their extent [20–22]. The authors [22] give a runtime of $\mathcal{O}(nR)$ time steps for the discrete version of the problem over a two-dimensional grid, where R is the Manhattan distance from the farthest robot to the gathering point, and no runtime is given for [20, 21]. Moreover, the gathering problem of [22] is different than the F-coverage problem: (i) It tries to minimize the radius of the area around the gathering point by putting the robots as close as possible to each other, and (ii) it is studied only on grid.

11.3 Model and Preliminaries

We consider a set $\mathcal{S} := \{s_1, \ldots, s_n\}$ of n mobile sensors in the Euclidean plane. Sensors are assumed to have a common coordinate system and they all know the location of the POI F. Moreover, sensors are assumed to know their own spatial coordinates; this can be done by attaching GPS devices or any effective localization algorithm. Without loss of generality, F is assumed to be at the origin of the coordinate system.

A triangle tessellation (TT) is a planar graph composed of congruent equilateral triangles as shown in Figure 11.1. Given a common orientation, say the north, and edge length l_e, sensors in $\mathcal{S}$ are able to locally compute a unique TT in the Euclidean plane containing F as a vertex. Denote this TT by G in which we will work. Two vertices are neighbors (i.e., adjacent to each other) in G if there is an edge between them. We denote by $\mathrm{dist}(p, q)$ the *hop distance* between two nodes p and q in G, i.e., $\mathrm{dist}(p, q)$ is the minimum number of edges one needs to traverse in order to get from p to q and vice versa.

With the knowledge of its own location, each sensor can determine whether it is located at a vertex in G, which vertex, and its adjacent vertices. Initially, the sensors in $\mathcal{S}$ are assumed to be in the different vertices of G; the technique of Bartolini et al. [11] can also be used to achieve this. The assumption of sensors

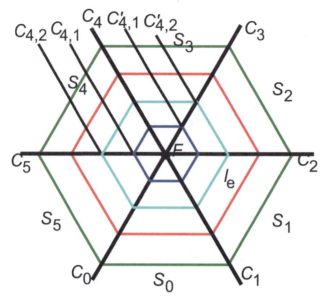

Figure 11.1 An illustration of G with POI F, sectors, sector lines, and hexagons $\mathcal{H}_i$ (colored) of each layer $i \geq 1$.

in the different vertices of G initially guarantees that collisions are avoided in the initial configuration of robots. The edge length l_e is set to $r_s\sqrt{3}$ because it automatically maintains network connectivity when the communication radius of each sensor $r_c \geq r_s\sqrt{3}$ and it minimizes the sensing range overlapping [7, 23, 24] (i.e., maximizes the coverage area), where r_s is the sensing radius of each $s_i \in \mathcal{S}$.

We assume that each $s_i \in \mathcal{S}$ has $r_c \leq 2\sqrt{3}r_s = 2l_e$, that is, s_i can communicate with s_j if and only if they are within 2 hops away in G; note that immediate neighbors in G are one hop away. They are *oblivious*—sensors do not remember decisions performed in previous time steps—except the information about F. They are *anonymous*, i.e., they cannot be distinguished from each other.

Sensors operate in synchronous *Look–Compute–Move* (LCM) steps: During each step t, all sensors act synchronously at the same time and travel an edge of G (i.e., one hop on G). We call the positions $p_{s_1}(t), \ldots, p_{s_n}(t)$ of the sensors at the beginning of step t the *configuration* at time t. When the context is clear, we sometimes use s_i and $p_{s_i}(t)$ interchangeably. We call the configuration at step 0 the *initial configuration*. A *path* taken by a sensor s_i is the concatenation of the edges in G that s_i traverses while going from some source vertex p to a destination vertex q. A vertex in G is called *free* if there is no sensor at that vertex.

We measure the quality of the algorithm by counting the number of synchronous steps until the optimal F-coverage around F is achieved and sensors stop moving. During each LCM step, first all sensors take synchronously the snapshot of the positions of the other sensors they see (*Look*); then, execute the algorithm, the same for all sensors, using the snapshot as an input (*Compute*); and move toward the computed destination, if any (*Move*); the *Move* is only performed when the computed destination position (which is one of its neighbor nodes) is different than the current position. When a sensor executes the algorithm, it makes decision on either staying idle or moving to one of its neighbor nodes. With respect to the synchronous LCM steps, there is a common notion of time which is assumed to be discrete and *all* sensors are active at each time step.

In F-coverage, the distance from the POI F to uncovered areas is important; therefore, the *optimal* F-coverage has maximized coverage radius and no sensing holes. The *coverage radius* is the radius of the maximum sensing hole-free disk enclosed by sensors centered at F; note that the disk is *sensing hole-free* if there is no point inside the disk that is not covered by the sensing range of at least one sensor. If the number of sensors in S is unlimited, sensors can be deployed densely and the coverage achieved becomes near close to a circular shape; in this case, coverage radius is called *circular* radius. However, since both S and r_s are finite, the coverage radius here is a discrete variant of the circular radius which we refer to as *polygonal* radius. This polygonal radius can also be measured by *layer distance* which represents the number of successive complete convex polygons adjacently surrounding the POI F. More precisely, we consider a discrete set of convex polygons $P_i, i = 0, 1, 2, \ldots$, that are composed of sensors centered at POI F and having a diameter of $i \cdot d$ for some constant d. The coverage radius is the maximum value χ such that P_χ is completely in the coverage region. As we consider G, the polygons defining the layers are hexagons centered at F. There are $6i$ vertices with equal hop distance i to F in G for any $i \geq 0$. These $6i$ nodes constitute an *i-layer* hexagon, denoted by $\mathcal{H}_i$. Each $\mathcal{H}_i, i \geq 0$, is concentric to F. The total number $v(i)$ of vertices enclosed by $\mathcal{H}_i$ (inclusive), for any $i \geq 0$, is $v(i) = 1 + \sum_{q=1}^{i} 6q = 1 + 3i(i + 1)$.

For simplicity in the analysis, we assume that $n = |v(i)|$. Therefore, the vertices up to layer i form the set of *target positions*. The target positions can be partitioned into $i + 1$ layers $T_0, \ldots, T_i$ such that T_0 contains F and T_j contains all the target positions that are in $\mathcal{H}_j$. For T_0, only target position

$F \in T_0$ is called *occupied* if a sensor is at vertex F. For $T_j, 1 \leq j \leq i, x \in T_j$ is called occupied if all target positions on layers $i' < j$ are occupied and there is a sensor at vertex x. If a target position is not occupied, it is called *free*. We call layer j occupied if all target positions in T_j are occupied. An occupied target position on any layer j can be free again, but once a layer is occupied, it cannot become free anymore.

We call the vertices located at hexagon corners *corner vertices* and the others *edge vertices*. Corner vertices are denoted as $C_0, \ldots, C_5$ in counter-clockwise direction dividing the entire graph G into 6 sectors and they form mutual angle of $\pi/3$ at F. The sector toward south (between C_0 and C_1) is called "Sector 0" and denoted by S_0; the other sectors are named $S_1, \ldots, S_5$, respectively. The path from F to any corner vertex is called *sector line*, which separates nodes and edges that are common to two adjacent sectors. When the context is clear, we sometimes denote sector line segment by simply $C_{(l+1) \bmod 6}, 0 \leq l \leq 5$, i.e., $C_{(l+1) \bmod 6}$ separates sectors S_l and $S_{(l+1) \bmod 6}$. However, the edges and vertices that are in the sector lines are counted in both the adjacent sectors divided by that line. Moreover, the line segment that is in the sector S_4 parallel to C_4 at hop distance 1 from the nodes in C_4 is denoted as $C_{4,1}$. This definition can be extended to the line segment parallel to C_4 in the sector S_3 which we denote by $C'_{4,1}$. $C_{4,2}$ and $C'_{4,2}$ can also be similarly defined in the right and left of $C_{4,1}$ and $C'_{4,1}$, respectively (see Figure 11.1).

11.4 The Algorithm

The pseudocode of our algorithm, TTGREEDY, is given in Algorithm 11.1. TTGREEDY performs self-deployment in each time step based on two operations: GREEDY and ROTATION, which are similar to the greedy and rotation operations of [5, 22]. The GREEDY operation performed by a sensor s_i at any time step t decreases its distance to F by exactly one hop (Lines 2, 3 of Algorithm 11.1). The ROTATION operation performed by a sensor s_i at any time step t moves that sensor to its adjacent vertex in G in the same layer (Lines 4, 5 of Algorithm 11.1) in the counterclockwise direction. Similar to the rotation movement of [5, 22], the ROTATION operation is restricted to counterclockwise direction only so as to avoid collisions among rotating sensors. If both the GREEDY and ROTATION operations are not possible at any time step t, then s_i stays idle at the current vertex in that time step (Line 6 of Algorithm 11.1). If several sensors simultaneously try to move to the same vertex of G, TTGREEDY avoids collisions through priorities (Lines 7–13 of Algorithm 11.1), which we define

---

Algorithm 11.1 TTGREEDY algorithm for $s_i \in S$ at any time step $t > 0$

Input: A set $S := \{s_1, \ldots, s_n\}$ of n sensors in the distinct
vertices of G.
Output: A maximum sensing hole-free disk of sensors centered
at POI $F \in G$.

1. **Operations:**
2.   **If** there is a free vertex $x \in G$ that minimizes s_i's distance
       to F by 1 hop **then**
3.   Perform GREEDY to x;
4.   **Else If** there is a free vertex $x \in G$ next to s_i in the
       counterclockwise direction that does not increase/decrease
       s_i's distance to F **then**
5.   Perform ROTATION to x;
6.   **Else** Stay idle at the current vertex;

7. **Collision Avoidance:**
8.   **If** s_i is trying to perform GREEDY and s_j is trying to
       perform ROTATION to the same vertex x **then**
9.   Give higher priority to s_i;
10.   **If** s_i is trying to perform ROTATION and s_j is trying to
       perform GREEDY to the same vertex x **then**
11.   Give higher priority to s_j;
12.   **If** both s_i and s_j are trying to perform GREEDY to the same
       vertex x **then**
13.   Choose one between s_i and s_j arbitrarily;

---

adapting the technique of [22]. Particularly, two sensors that are targeting the
same position are doing GREEDY, then one among them is chosen arbitrarily;
if one is doing GREEDY and another is doing ROTATION, then the sensor that is
doing GREEDY is chosen. We provide a lower bound proof in the next section
and analyze this algorithm in Section 11.6.

11.5 The Lower Bound

We prove a runtime lower bound for a class of greedy algorithms. A *greedy*
algorithm tries to minimize the hop distance of sensors to the POI by 1 in every
time step to guarantee progress toward the solution. Consider a configuration
where each $s_i \in S$ is positioned on the vertices of G such that $\text{dist}(s_i, F) = R$.
This can be achieved by positioning $n := v(i)$ sensors, $i \geq 1$, in the vertices
of $\mathcal{H}_R$ in the sectors S_1 and S_2. Denote this configuration by $\mathcal{CONF}_{start}$ (an
example is in the left of Figure 11.2). Moreover, $n := v(i)$ guarantees that in

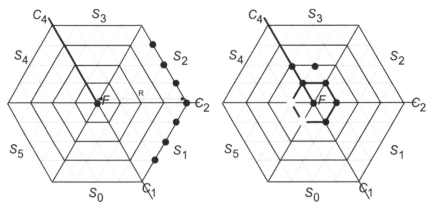

Figure 11.2   (left) An illustration of $CONF_{start}$ with 7 sensors at hop distance R from F; (right) A possible configuration $CONF_R$ after R time steps.

the optimal configuration, there will be sensors in all the layers of G up to i (F inclusive). Denote this configuration by $CONF_{final}$. In this section, we will prove the following result.

Theorem 11.5.1   *Given $CONF_{start}$, any greedy algorithm takes at least $R + \frac{(n-1)}{2}$ time steps to achieve optimal F-coverage around F.*

We prove Theorem 11.5.1 by proving two different lemmas below, namely Lemmas 11.5.2 and 11.5.7. We first prove in Lemma 11.5.2 the properties of the configuration (which is not final) that the sensors will reach in first R time steps. We then prove in Lemma 11.5.7 that starting from the configuration reached in R steps, any greedy algorithm needs at least $\frac{(n-1)}{2}$ time steps to reach $CONF_{final}$. We proceed with the following observation:

Observation 1 Given $CONF_{start}$ and any time step t, while executing a greedy algorithm, no sensor crosses the sector line C_1 to reach nodes in sector S_0.

Observation 1 immediately follows as GREEDY operations of a greedy algorithm never drive sensors inside a sector out of its two-sector boundary lines and sensors perform ROTATION operations only in counterclockwise direction.

We now prove that given $CONF_{start}$, no sensor crosses the sector line C_4 in first R time steps of any greedy algorithm.

Lemma 11.5.2   *Given $CONF_{start}$, no sensor crosses the sector line C_4 to reach nodes of sector S_4 in first R time steps of any greedy algorithm.*

Proof. Recall that each sensor s_i is exactly R hop away from F in $\mathcal{CONF}_{start}$. Moreover, in $\mathcal{CONF}_{start}$, the distance from each sensor s_i to the nodes in sector line C_4 is at least R. Therefore, even if all the sensors that are inside sector S_2 are able to move (doing GREEDY or ROTATION operation) in every time step, they are only be able to reach the nodes in sector line C_4. (Note that in every time step, a GREEDY operation decreases the distance to POI by exactly one hop and a ROTATION operation keeps the same distance to POI.) This lemma holds even when we consider the sensors that are in sector S_1. This is because if some of the sensors in S_1 are able to move at some time step, then the sensors in S_2 cannot perform these operations, and from Observation 1, no sensor crosses the line C_1.

Denote the resulting configuration of sensors after executing any greedy algorithm for first R time steps by $\mathcal{CONF}_R$. We now aim to show that starting from $\mathcal{CONF}_R$, we need at least $\frac{(n-1)}{2}$ time steps to achieve optimal F-coverage. As sensors do not cross sector line C_4, all the vertices in sectors S_4, S_5, and S_6 are unoccupied in $\mathcal{CONF}_R$. Denote by $\mathcal{U}(i)$ the unoccupied vertices in layer $i \geq 1$ in S_4, S_5, and S_6 in $\mathcal{CONF}_R$. We have that $\mathcal{U}(i) = 3i - 1$ for every $i \geq 1$, i.e., $\mathcal{U}(1) = 2, \mathcal{U}(2) = 5$. For example, the unoccupied vertices in layer 1 in a possible $\mathcal{CONF}_R$ are shown in white circles in the right of Figure 11.2. For simplicity in the analysis, we assume that all the i layers (including the POI F and layer i) around the POI F in sectors S_1, S_2, and S_3 are already occupied (i.e., covered) by sensors in $\mathcal{CONF}_R$ and the remaining sensors are either in the vertices of the line C_4 or concentrated in the vertices nearby C_4 in sector S_3. This allows us to focus only on determining the minimum number of steps needed to occupy the vertices in S_4, S_5, and S_6 starting from $\mathcal{CONF}_R$.

Observation 2 All the unoccupied vertices in S_4, S_5, and S_6 will be occupied by the sensors advancing to those sectors crossing the sector line C_4.

Observation 2 follows from Observation 1 as vertices in S_4, S_5, and S_6 will only be occupied through ROTATION and GREEDY operations from sensors, and sensors perform ROTATION operations only in counterclockwise direction.

Lemma 11.5.3 *Given* $\mathcal{CONF}_R$, *it takes at least* $\mathcal{U}(1)+1$ *time steps to occupy* $\mathcal{U}(1)$ *unoccupied vertices in layer* 1.

Proof. Starting from $\mathcal{CONF}_{start}$, we have that the only node F that is in T_0 is occupied in exactly R steps. This is because all the sensors are R distance away from F in $\mathcal{CONF}_{start}$ and one sensor always greedily progresses toward F in each time step. The sensor in F will not perform any operation after R. Moreover, there is no sensor in sectors S_4, S_5, and S_6 in $\mathcal{CONF}_R$ (Lemma 11.5.2). Therefore, in T_1, we have at least $\mathcal{U}(1) = 3 * 1 - 1 = 2$ vertices that need to be occupied (white nodes in the right of Figure 11.2). We prove that at least $\mathcal{U}(1) + 1 = 3$ steps after R are necessary for any greedy algorithm to cover them.

At time step $R+1$, each sensor on the sector line C_4 performs the ROTATION operation; if there are scenarios where some sensors in C_4 do not need to perform a ROTATION operation to the nodes in S_4, then either they have the chance to do a GREEDY operation in the vertices in C_4 or they are in S_3 and they can only reach to the vertices in C_4 after performing a operation at this time step. Moreover, no new sensor enters layer 1 due to the ROTATION operations in the time step $R + 1$ (i.e., only the sensor that is already in layer 1 moves from one position of layer 1 to other position in layer 1), and hence, $\mathcal{U}(1)$ does not decrease at that time step. A possible resulting configuration at time step $R + 1$ is depicted in Figure 11.3a (sensors in nodes g and f moved to a and b, respectively) for the example of the right of Figure 11.2 starting from the configuration $\mathcal{CONF}_R$.

At time step $R + 2$, the sensors outside layer 1 in the line segments $C'_{4,1}$ and $C_{4,1}$ compete to perform GREEDY operation to the vertices of C_4 ($C'_{4,1}$ and $C_{4,1}$ are defined in Section 11.3). Let the sensors in line $C'_{4,1}$ perform a GREEDY operation each to the vertices of C_4. Then, the sensors in line $C_{4,1}$ must perform ROTATION operations to line $C_{4,2}$. The sensor at b in layer 1 will perform a ROTATION operation to h as no GREEDY operation is possible for it. At this time step, $\mathcal{U}(1)$ decreases by exactly one as a new sensor enters to layer 1 at vertex f. However, no sensor can occupy b at $R + 2$ to avoid collision as the sensor that was at b moves to h at the same time step. Figure 11.3b shows an example execution at the time step $R + 2$ starting from $\mathcal{CONF}_R$ of the right of Figure 11.2 (the sensor at c performs a GREEDY operation to f, and sensors at a and b perform a ROTATION operation each to u and h, respectively). Therefore, the vertex b can only be occupied at the time step $R + 3$ (see Figure 11.3c).

We now show that any other combination of sensor operations in these time steps and placement of sensors in S_3 result in at least $\mathcal{U}(1) + 1$ time steps to occupy $\mathcal{U}(1)$ unoccupied nodes using any greedy algorithm. The sensor that occupies vertex h is the sensor that was at vertex f at the time step R and it needs at least 2 time steps to reach h after R in any combination. When it reaches h at $R + 2$, no sensor can be at b, and at the earliest, a sensor can arrive at b at time step $R + 3$. The lemma follows.

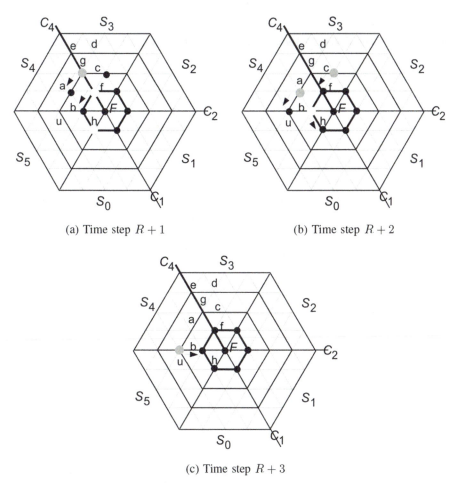

(a) Time step $R + 1$

(b) Time step $R + 2$

(c) Time step $R + 3$

Figure 11.3 Sensor movement for 3 time steps after R starting from $\mathcal{CONF}_R$ for the example of the right of Figure 11.2.

Let $\mathcal{CONF}_{layer,i}$ be the configuration at the time step in which all $\mathcal{U}(i)$ vertices up to T_i from F are occupied executing any greedy algorithm. We prove the following lemma when T_1 is occupied.

Lemma 11.5.4 *If $\mathcal{CONF}_{layer,1}$ is reached in exactly $\mathcal{U}(1) + 1$ time steps, then there is no sensor in the area of S_4 in the left side of the line segment $C_{4,1}$.*

Proof. The ROTATION operations from sensors at C_4 at time step $R + 1$ take them to line $C_{4,1}$. At time step $R + 2$, the sensors in line $C_{4,1}$ (outside $\mathcal{H}_1$)

may need to perform again a ROTATION operation each to reach $C_{4,2}$ if there are sensors at the line $C'_{4,1}$ and they win the competition between the sensors at $C_{4,1}$ to perform GREEDY operations to the positions in C_4 (the target positions at line C_4 were empty at the end of the time step $R + 1$ due to ROTATION operations). After sensors reach the vertices of $C_{4,2}$ at $R+2$, they will definitely perform GREEDY operations to the positions of the line $C_{4,1}$ in time step $R+3$, since GREEDY operations are preferred compared to ROTATION operations by the greedy algorithm. That is, sensors at C_4 cannot reach $C_{4,1}$ in time step $R + 3$ as they need to perform ROTATION operations, whereas the sensors at $C_{4,2}$ are eligible to perform GREEDY operations. The lemma follows.

We now extend the above two lemmas, namely Lemmas 11.5.3 and 11.5.4, for T_1 to all $T_{i>1}$. We first have the following observation.

Observation 3 After all the unoccupied vertices $\mathcal{U}(i)$ in layer $T_{i \geq 1}$ are occupied, the sensors up to layer T_i inclusive no more perform any GREEDY or ROTATION operation in the future time steps.

This observation immediately follows as ROTATION operations are no more possible at T_i after it is occupied, and GREEDY operations are not allowed as they increase their distance to F. We now prove the following two lemmas. They generalize the proof ideas of Lemmas 11.5.3 and 11.5.4.

Lemma 11.5.5 *After the configuration $\mathcal{CONF}_{layer,i-1}$ is reached at some time step t, there is no sensor in the area of S_4 in the left side of the line $C_{4,i}$ at time step $t + \mathcal{U}(i) + 1$.*

Proof. We have from Lemma 11.5.4 that when all $\mathcal{U}(1)$ unoccupied nodes in layer 1 at $\mathcal{CONF}_R$ are occupied in exactly $\mathcal{U}(1) + 1$ time steps after R, there is no sensor in the left side of the line $C_{4,1}$. This works as the base case for this lemma. For any layer $i \geq 2$, in the first time step after $\mathcal{CONF}_{layer,i-1}$, all the nodes in $C_{4,i-1}$ perform ROTATION operations. In the next time step, the sensor at the layer i vertex in the line $C_{4,i}$ performs a ROTATION operation; all other sensors in the line $C_{4,i}$ that are in the vertices outside layer i perform a GREEDY operation back to vertices in the line $C_{4,i-1}$. This is because the sensors in the vertices of $C_{4,i}$ that are outside layer i and the sensors in the line $C_{4,i-2}$ that are outside layer $i - 1$ compete for the positions in $C_{4,i-1}$ to GREEDY or ROTATION into. The sensors from $C_{4,i}$ are able to move back to the positions in $C_{4,i-1}$ as they can do GREEDY operations compared to the sensors in the line $C_{4,i-2}$ which can only do ROTATION operations at that time step. This scenario of the first two steps repeats until all the unoccupied vertices $\mathcal{U}(i)$ are occupied. Therefore, when $\mathcal{U}(i)$ is odd, no sensor will be at the left side of the line $C_{4,i-1}$, and when $\mathcal{U}(i)$ is even, no sensor will be at the left side of the line $C_{4,i}$ for at least $\mathcal{U}(i) + 1$ time steps after the time step in which $\mathcal{CONF}_{layer,i-1}$ is reached.

Lemma 11.5.6 *After the configuration $\mathcal{CONF}_{layer,i-1}$ is reached at some time step t, any greedy algorithm takes at least next $\mathcal{U}(i) + 1$ time steps to occupy all the $\mathcal{U}(i)$ unoccupied vertices in layer $T_{i\geq 1}$.*

Proof. We first focus on $i > 1$. We have from Lemma 11.5.5 that there is no sensor in the area of S_4 in the left side of the line $C_{4,i}$ for at least $\mathcal{U}(i) + 1$ time steps after the configuration $\mathcal{CONF}_{layer,i-1}$ is reached at some time step t. The $\mathcal{U}(i)$ unoccupied nodes in layer i are occupied by the ROTATION operations of the sensors that reach to the layer $i-1$ vertex in the line $C_{4,i-1}$. When $\mathcal{CONF}_{layer,i-1}$ is reached at some time step t, there are still $\mathcal{U}(i) - (i-1)$ unoccupied vertices in layer i. As one new sensor is entered in the layer i in every two time steps, we need $2 \cdot (\mathcal{U}(i) - (i-1))$ time steps to occupy all the vertices in layer i. As the time steps we need are $\mathcal{U}(1) + 1$ for layer 1, we can simply say that at least $2 \cdot (\mathcal{U}(i) - (i-1)) \geq \mathcal{U}(i) + 1$ steps are needed for any layer $i \geq 1$ starting from $\mathcal{CONF}_{layer,i-1}$ that is reached at some time step $t \geq R$.

Lemma 11.5.7 *Starting from the configuration $\mathcal{CONF}_R$, any greedy algorithm needs at least $\frac{(n-1)}{2}$ time steps to reach the final configuration $\mathcal{CONF}_{final}$.*

Proof. We have from Lemmas 11.5.3 and 11.5.6 that unoccupied nodes $\mathcal{U}(i)$ of layer $i \geq 1$ are occupied in at least $\mathcal{U}(i) + 1$ time steps after the unoccupied nodes of layer $i - 1$ are occupied. The only node F in layer 0 is occupied in exactly R time steps. Therefore, the number of time steps until all the unoccupied nodes $\mathcal{U}(i)$ in layer T_1 up to T_i are occupied is at least

$$1 + \sum_{q=1}^{q=i} \mathcal{U}(q) = 1 + \sum_{q=1}^{q=i} (3q - 1)$$
$$= 1 + \frac{3i(i+1)}{2} - 1$$
$$= \frac{(n-1)}{2},$$

since $3i(i+1) = n - 1$.

We are now ready to prove the main result of this section.

Proof of Theorem 11.5.1. Combining the results of Lemmas 11.5.2 and 11.5.7, we have the proof of Theorem 11.5.1.

11.6 Analysis of the TTGREEDY Algorithm

We start with some definitions and observations which are adapted from [22, 25]. We say that two sensors u and v are in *conflict* if they target the same vertex $x \in G$ in any time step. We say that a sensor $s \in T_i$ is in the *innermost* hexagonal layer if s cannot move to another hexagonal layer $T_j, j < i$, in any future time steps. Furthermore, we say that a sensor is *live* if it has not reached the innermost hexagonal layer. We have the following two observations.

Observation 4 In Algorithm 11.1, (i) once a layer is occupied, it cannot become free anymore; (ii) the number of occupied target positions in T_j is monotonically increasing; and (iii) the ith layer (i.e., the outermost layer) is occupied if and only if the sensors reach optimal F-coverage configuration.

Observation 5 Each sensor s_a that is not in the innermost layer can move toward F in every time step if it is not in conflict with another sensor s_b and s_b wins the conflict. Otherwise, s_a either stays idle in its current vertex or rotates on the next vertex in its current layer in the counterclockwise direction (if the next vertex is free), without increasing its distance to F.

Theorem 11.6.1 *Given any arbitrary initial configuration of n sensors in G, Algorithm 11.1 eventually leads to an optimal F-coverage.*

Proof. We prove this theorem borrowing some ideas from Cord-Landwehr et al. [22], and the proof follows the similar structure of Cord-Landwehr et al. [22, Theorem 11.6.1]. Consider the innermost hexagonal layer $\mathcal{H}_i$ around F in G that has less sensors than in the optimal F-coverage. Let $\mathcal{H}_{next} > \mathcal{H}_i$ denote the next layer that contains at least one sensor at some time step t. If $\mathcal{H}_{next} > \mathcal{H}_{i+1}$, at least one sensor at layer $\mathcal{H}_{next}$ performs a GREEDY operation at time step $t + 1$. If $\mathcal{H}_{next} = \mathcal{H}_{i+1}$, the sensors at layer $\mathcal{H}_{next}$ will perform ROTATION operations until one of them finds a free position at layer $\mathcal{H}_i$ to perform a GREEDY operation. Note that there must be a free position somewhere at layer $\mathcal{H}_i$ as it is assumed that it has less sensors than in the optimal F-coverage. One of the sensors that is performing a ROTATION operation will be able to see this free position and move to it with a finite time step as the sensors are rotating around the POI in counterclockwise direction. The theorem follows.

We now show that the optimal configuration of Theorem 11.6.1 can be reached in at most $R + 2 \cdot (n - 1)$ time steps.

Definition 1 [25] In Algorithm 1, the path that a sensor s_a is supposed to take starting at a certain time step at a certain vertex $v \in G$ is the path that s_a would take if it was the only live sensor in G. The path that s_a is supposed to take with respect to a given set of sensors S is the path that s_a would take if the sensors in S were the only live sensors in G.

The path that s_a would take (or supposed to take) may not be *unique* even if it was the only live sensor in G and the path uniqueness is not required as Algorithm 11.1 is greedy.

Lemma 11.6.2 *For some sensor s_a and an arbitrary set of live sensors S, suppose that $e_1, \ldots, e_m$ is one of the paths composed of edges that s_a is supposed to take in G with respect to S starting at time step t. Then, at time step $t + i' - 1, 1 \le i' \le m$, a sensor will be moving on edge $e_{i'}$.*

Proof. If no other sensor is moving on the edge $e_{i'}, 1 \le i' \le m$, starting at time step t, then s_i is moving on it and the lemma immediately follows. If s_i is not moving on that path, some other live sensor from S must be moving on $e_{i'}$ winning the conflict over s_i or some other third sensor.

Lemma 11.6.3 *Suppose that a sensor s_a is the only live sensor in G at time step t at layer T_j. Let s_a reach a layer $T_{i'} < T_j$ on its way to its innermost layer following one of the paths that it is supposed to take without performing any ROTATION operation. Then, s_a must have reached the layer $T_{i'}$ in time step $t + (j - i')$.*

Proof. Note that in TTGREEDY, any path that a sensor would take (and supposed to take) will decrease its distance to the POI by 1 in every time step until it reaches a layer from which further GREEDY operations are not possible without at least a ROTATION operation. Moreover, TTGREEDY prefers GREEDY operations over ROTATION operations, whenever possible. Therefore, as s_i is the only live sensor at time step t at layer j, it must have reached $i' < j$ in time step $t + (j - i')$.

The following definition will be useful later in proving the runtime of our TTGREEDY algorithm.

Definition 2 The sensors destined to provide optimal coverage around F are called *dense around* F if there exists an integer $l > 0$ such that for all $1 \leq \xi \leq l$ the number of sensors in each layer T_ξ is at least 1, and there is no sensor in any layer $\ell > l$.

We can prove the following lemma for Algorithm 11.1.

Lemma 11.6.4 *Given any arbitrary starting configuration of sensors and the POI F in G, the sensors will be dense around F after at most R time steps of Algorithm 11.1.*

Proof.   Consider some sensor s_a during the first R time steps. Since Algorithm 11.1 is greedy, s_a tries to perform a GREEDY operation in every time step. If s_a did not arrive to F during that time, let d be its hop distance to F at time step R. Since s_a did not arrive after R time steps, it must have been stayed idle or performed a ROTATION operation at least once during the R time steps. Now, using a simple induction argument, it follows that from the time step after the last time it stayed idle or performed a ROTATION operation until time step R, there is always a sensor $s_b \neq s_a$ that is at least two time steps closer to F than r_a.

We are now ready to prove the main result.

Theorem 11.6.5 *Given any arbitrary starting configuration of sensors and the POI F in G, Algorithm 11.1 needs at most $R + 2 \cdot (n - 1)$ time steps to achieve the optimal F-coverage.*

Proof.   Consider a sensor s_a that is initially the closest to F and which defeats all other sensors in conflicts. Note that there exists such a sensor when there is at least one sensor in S. Then, this sensor arrives at F by the time step R. Moreover, we have from Lemma 11.6.4 that after R time steps, all other sensors will be dense around F. Therefore, it remains to show that after the time step R, the layer $T_{j \geq 1}$ will be occupied in next at most $12 \cdot j$ time steps of Algorithm 1 after the layer T_{j-1} is occupied. As only node F in layer 0 is occupied in the time step R, we first show that all the target positions in T_1 are occupied in at most next 12 steps and argue through an inductive hypothesis for the layers $T_{j \geq 2}$. T_1 will be occupied in 12 steps as there are only 6 target positions in T_1, and at least one new sensor will enter T_1 in every second time step starting from any dense around configuration (Lemma 11.6.4). It is easy to see that the dense around configuration remains dense around even after one new sensor enters T_1. Now, we can argue inductively for $T_{j \geq 2}$ since the sensors in T_{j-1} do not perform any GREEDY or ROTATION operation in the future (Observation 4) and it is easy to see that sensors in $\mathcal{S}$ outside T_{j-1} satisfy dense around property after every time a sensor enters T_{j-1}. Therefore, summing the total time steps to occupy layers up to T_i (the outermost layer) inclusive, we have that the total runtime of Algorithm 11.1 is at most

$$
\begin{aligned}
R + \sum_{q=1}^{i} 12q &= R + 12 \cdot \frac{i(i+1)}{2} \\
&= R + 2 \cdot (n-1),
\end{aligned}
$$

since $1 + 3i(i+1) = n$.

11.7 Experiments

We implemented TTGREEDY and compared its performance with two previous algorithms: GRG due to Li et al. [5] and mGRG due to Blázovics and Lukovszki [6], in a synchronous setting (where the speed and the distance traveled are the same for each sensor in each time step). We followed the simulation approach as that of Blázovics and Lukovszki [6] for the better comparison of our results. In all the simulations, sensors are initially placed uniformly at random at distinct vertices in the TT graph G. The POI was considered to be the center vertex of G. Note that the algorithm mGRG [6] and our algorithm TTGREEDY use the 2-hop neighborhood information for sensor self-deployment, whereas the algorithm GRG [5] uses only the 1-hop neighborhood information. The simulations are done in Java, and the results are plotted in GNUplot.

We conducted two kinds of experiments. In the first kind of experiments, we kept the *dropping area* of sensors constant (30×30 square) and varied the number of sensors dropped in that area from 37 (sensors to fully cover 3 layers around the POI) to 331 (sensors to fully cover 10 layers around the POI). We then computed the total number of time steps to cover the POI, total number of moves, average mileage, average progress, and average mileage over progress. Here, *mileage* for a sensor is the number of hops it traveled to reach to its final position from its position in the initial configuration. Moreover, *progress* for a sensor is the minimum number of hops it needs to travel to reach to its final position from its position in the initial configuration; this is computed by comparing the positions of each sensor in the initial configuration and final configuration reached by Algorithm 11.1. In the second kind of experiments, we fixed the number of sensors to cover only 5 layers around the POI (91 sensors) and varied the size of the dropping area from 25×25 square to 44×44 square. We then computed again the total number of time steps to cover the POI, total number of moves, average mileage, average progress, and average mileage over progress. We plot the average of 40 simulations in all the results.

The simulation results for the first kind of experiments are given in Figure 11.4. It is clearly evident from the top left of Figure 11.4 that TTGREEDY needs roughly the half number of time steps compared to that of both GRG and mGRG. Note that both GRG did not terminate in some of our simulations but TTGREEDY achieved termination at all simulations. The top right of Figure 11.4 shows the total number of moves of the algorithms during self-deployment. TTGREEDY performs less number of moves than both GRG and mGRG. mGRG performs worse compared to both TTGREEDY and GRG, as sensors keep moving at all times in mGRG during self-deployment. In GRG and TTGREEDY, sensors do not need to move at all times, and hence, the number of moves is minimized. The middle left of Figure 11.4 shows the average mileage per sensor, which is the average number of hops sensors traveled during self-deployment. The algorithm mGRG has high average mileage per sensor, since the sensors move at all times and all sensors make almost the same number of moves. As sensors do not move all the times in both GRG and TTGREEDY, the average mileage per sensor in them is not as high as mGRG. Moreover, TTGREEDY does not have priority rule as such in GRG which allows more parallel movements. The middle right of Figure 11.4 shows the average progress per sensor, which is the average of the minimum number of hops sensors need to travel to reach to their final positions during self-deployment. Similar to average mileage per sensor, the algorithm mGRG has high average progress per sensor, since the sensors move at all times and all sensors make almost the same number of moves.

As mileage is related to energy consumption, we compared average mileage per sensor to average progress per sensor and obtain the average mileage over progress ratio. This gives an idea about the costs of the zigzag movements performed by the algorithms [5]. If mileage over progress ratio of an algorithm is higher, then the algorithm performs many zigzag movements during self-deployment and sensors consume more energy. The results for this performance metric are in the bottom plot of Figure 11.4, which shows that TTGREEDY is better in terms of energy consumption as well compared to GRG and mGRG.

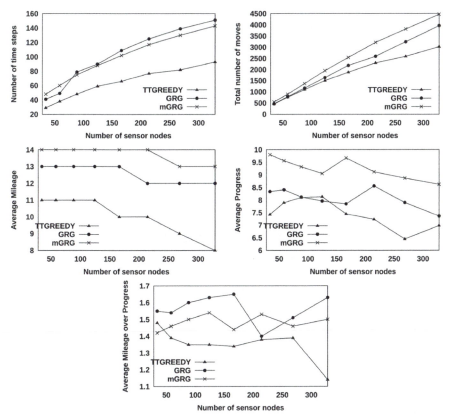

Figure 11.4 Results when the dropping area is fixed to 30×30 square, and the sensors in the dropping area vary from 37 to 331.

The simulation results for the second kind of experiments are given in Figure 11.5. The top left of Figure 11.5 shows that TTGREEDY converges to optimal configuration significantly faster compared to both GRG and mGRG. The top right of Figure 11.5 shows that TTGREEDY also performs better in total number of moves compared to both GRG and mGRG. This is because TTGREEDY controls the sensor movements better than GRG and mGRG. The middle left, the middle right, and the bottom plots of Figure 11.5 cumulatively show that TTGREEDY consumes less energy compared to both GRG and mGRG during self-deployment. Moreover, the performance of all three algorithms is monotonically varying with the dropping area size in all performance metrics due to the fixed number of sensors.

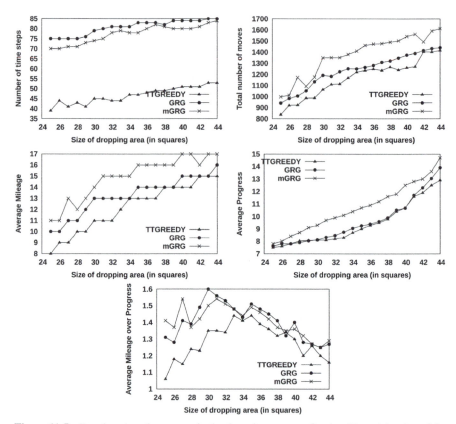

Figure 11.5   Results when the sensors in the dropping area are fixed to 91, and the size of the dropping area varies from 25×25 square to 44×44 square.

To summarize our findings, TTGREEDY provided significantly faster run-time in the self-deployment process, required less number of moves per node in average, and consumes less energy compared to the existing algorithms in all our simulations. This makes TTGREEDY particularly suitable for fast sensor self-deployment in energy-constrained scenarios.

11.8 Conclusions

We have studied the self-deployment problem in MSNs under a synchronous setting. The objective of this problem is to provide the sensing hole-free coverage of a POI maximizing the coverage radius around it. We have given a local greedy self-deployment algorithm that solves the problem avoiding sensor collisions with the runtime that is significantly better than that of the best previously known algorithm of Blázovics and Lukovszki [6]. We have also studied the tightness of the algorithm and showed that it is in fact time-optimal within a very small constant factor. The simulation results under different parameter settings confirmed our theoretical findings. Our contribution is appealing as fast self-deployment is crucial in many real-world scenarios. In the future, it will be interesting to extend our synchronous algorithm to semi-synchronous/asynchronous settings, and also, it will be interesting to make our algorithm tolerant to sensor failures.

References

[1] Saipulla, A., Liu, B., Xing, G., Fu, X., and Wang, J. (2010). Barrier coverage with sensors of limited mobility. In *MobiHoc*, pp. 201–210.

[2] Cortes, J., Martinez, S., Karatas, T., and Bullo, F. (2004). Coverage control for mobile sensing networks. *IEEE Trans. Robot. Automat.*, 20 (2), 243–255.

[3] Brass, P. (2007). Bounds on coverage and target detection capabilities for models of networks of mobile sensors. *ACM Trans. Sen. Netw.*, 3 (2).

[4] Sharma, G., and Krishnan, H. (2015). Tight bounds on localized sensor self-deployment for focused coverage. In *ICCCN*.

[5] Li, X., Frey, H., Santoro, N., and Stojmenovic, I. (2011). Strictly localized sensor self-deployment for optimal focused coverage. *IEEE Trans. Mob. Comput.*, 10 (11), 1520–1533.

[6] Blzovics, L., and Lukovszki, T. (2013). Fast localized sensor self-deployment for focused coverage. In *ALGOSENSORS*, pp. 83–94.

[7] Ma, M., and Yang, Y. (2005). Adaptive triangular deployment algorithm for unattended mobile sensor networks. In *DCOSS*, pp. 20–34.

[8] Nayak, A., and Stojmenovic, I. (2010). *Wireless Sensor and Actuator Networks: Algorithms and Protocols for Scalable Coordination and Data Communication.* Wiley.

[9] Howard, A., Matariæ, M. J., and Sukhatme, G. S. (2002). An incremental self-deployment algorithm for mobile sensor networks. *Auton. Robots*, 13 (2), 113–126.

[10] Wang, G., Cao, G., and La Porta, T. F. (2006). Movement-assisted sensor deployment. *IEEE Trans. Mob. Comput.*, 5 (6), 640–652.

[11] Bartolini, N., Calamoneri, T., Fusco, E. G., Massini, A., and Silvestri, S. (2008). Snap and spread: A self-deployment algorithm for mobile sensor networks. In *DCOSS*, pp. 451–456.

[12] Yang, S., Liz, M., and Wu, J. (2007). Scan-based movement-assisted sensor deployment methods in wireless sensor networks. *IEEE Trans. Parallel Distrib. Syst.*, 18 (8), 1108–1121.

[13] Li, X., Fletcher, G., Nayak, A., and Stojmenovic, I. (2013). Randomized carrier-based sensor relocation in wireless sensor and robot networks. *Ad Hoc Networks*, 11 (7), 1951–1962.

[14] Wu, J., and Yang, S. (2007). Optimal movement-assisted sensor deployment and its extensions in wireless sensor networks. *Simulat. Modell. Practi. Theor.*, 15 (4), 383–399.

[15] Ando, H., Oasa, Y., Suzuki, I., and Yamashita, M. (1999). Distributed memoryless point convergence algorithm for mobile robots with limited visibility. *IEEE Trans. Robot. Automat.*, 15 (5), 818–828.

[16] Izumi, T., Izumi, T., Kamei, S., and Ooshita, F. (2009). Randomized gathering of mobile robots with local-multiplicity detection. In *SSS*, pp. 384–398.

[17] Cohen, R., and Peleg, D. (2005). Convergence properties of the gravitational algorithm in asynchronous robot systems. *SIAM J. Comput.*, 34 (6), 1516–1528.

[18] Degener, B., Kempkes, B., and auf der Heide, F. M. (2010). A local $\mathcal{O}(n^2)$ gathering algorithm. In *SPAA*, pp. 217–223.

[19] Degener, B., Kempkes, B., Langner, T., Meyer auf der Heide, F., Pietrzyk, P., and Wattenhofer, R. (2011). A tight runtime bound for synchronous gathering of autonomous robots with limited visibility. In *SPAA*, pp. 139–148.

[20] Agathangelou, C., Georgiou, C., and Mavronicolas, M. (2013). A distributed algorithm for gathering many fat mobile robots in the plane. In *PODC*, pp. 250–259.

[21] Czyzowicz, J., Gasieniec, L., and Pelc, A. (2009). Gathering few fat mobile robots in the plane. *Theor. Comput. Sci.*, 410 (6–7), 481–499.

[22] Cord-Landwehr, A., Degener, B., Fischer, M., Hüllmann, M., Kempkes, B., Klaas, A., Kling, P., Kurras, S., Märtens, M., Meyer auf der Heide, F., Raupach, C., Swierkot, K., Warner, D., Weddemann, C., and Wonisch, D. (2011). Collisionless gathering of robots with an extent. In *SOFSEM*, pp. 178–189.

[23] Bai, X., Kumar, S., Xuan, D., Yun, Z., and Lai, T. H. (2006). Deploying wireless sensors to achieve both coverage and connectivity. In *MobiHoc*, pp. 131–142.

[24] Zhang, H., and Hou, J. C. (2005). Maintaining sensing coverage and connectivity in large sensor networks. *Ad Hoc & Sensor Wireless Netw.*, 1 (1–2).

[25] Ben-Aroya, I., Eilam, T., and Schuster, A. (1995). Greedy hot-potato routing on the two-dimensional mesh. *Distrib. Comput.*, 9 (1), 3–19.

12

Toward Resident Behavior Prediction in Wireless Sensor Network-Based Smart Homes

Christopher Osiegbu[1], Seifemichael B. Amsalu[1], Fatemeh Afghah[2], Daniel Limbrick[1] and Abdollah Homaifar[1]

[1]ECE Department, North Carolina A&T State University, Greensboro, NC 27411, USA
[2]ECS Department, Northern Arizona University, Flagstaff, AZ 86011, USA

Abstract

The smart home has gained widespread attention due to its flexible integration into everyday life. This next-generation green home system transparently unifies various home appliances, smart sensors, and wireless communication technologies. It can integrate diversified physical sensed information and control various consumer home devices, with the support of active sensor networks having both sensor and actuator components. Although smart homes are gaining popularity due to their energy saving and better living benefits, there is no standardized design for smart homes. In this chapter, we take one step toward modeling the resident behavior in smart home equipped with noninvasive sensors. We put forward a concept by designing and implementing a smart home system that can classify and predict the state of home. In this system, various sensors collect the home information over the course of a few months. These historical data are utilized as an input in a supervised machine learning technique to predict the current state of the home. This enables us to recognize any potential abnormal conditions and notify authorities in case of emergencies. A possible application of this model would be in senior care facilities to assure the resident safety with minimal human supervision.

Keywords: Smart home, DigiMesh, Supervised machine learning, Wireless sensor networks, Senior care smart facility, Support vector machine.

12.1 Introduction

Smart home is the term commonly used to describe a home with devices that are capable of communicating with each other and can be controlled remotely by the user [1, 2]. The primary objective of a smart home is to enhance comfort, energy saving, and security for the house residents. The smart home is a great asset when it comes to senior care. In 2010, the population of senior citizens (65 and older) globally was 470 million, and by 2025, that number is projected to be 820 million [3]. This triggers an effect in all aspects of our society—business, health care, policy, and technology. As our aging population continues to grow, so does the demand for both in-home care and residential care facilities. This is a niche for the smart home. According to a 2012 survey across the USA by Genworth, assisted living, semi-private, and private nursing homes cost up to $32,568, $65,160, and $73,800 annually, respectively. The projected rise in the elderly population will lead to greater demand, and therefore, the cost can only increase. In addition to cost saving, the smart home promotes independent living and social interaction.

The US department of energy stated that in 2012, the average person spent $3,052 on energy [4]. This cost can be reduced by a smart home equipped with motion, glare, and luminosity sensors, constantly providing the actuating system with data leading to situational awareness. The smart home can provide endless security/safety solutions, depending on the requirements of the user. Common solutions tend to emphasize on intruder alert and gas detection.

Still, the long-term adoption and commercialization of the smart home are hindered by unanswered questions in terms of reliability, implementation cost, standardization, and security. Smart homes need to be reliable so as not to fail when users become reliant on its services. In a home that relays real-time patient information to a hospital, system failure can lead to misdiagnosis. Standardization of smart homes will lead to reduction in the cost of implementation. Being that data mining is integral to the learning process of smart homes, data security is also very important. Data gathering is the foundation for every smart home's successful implementation, but even with reliable data gathering, how best do we utilize this data? Several machine learning-based algorithms have been recently proposed were residents' routines were studied and used to predict the future states in smart homes [5–8]. As claimed, the reliability and autonomy of the smart home is improved using these approaches. The current solutions are suited for permanent residents and will be effective after the system has been in place for a period of time. However, they do not cater to commercial buildings with frequently changing residents.

Our proposed design models the activities which can occur in a particular space not taking the routine or schedule of a particular user into context. For example, in a bedroom sleeping, reading, grooming, toileting and entertainment are classes which the smart home will recognize as activities in that space. Mapping these activities to an actuating system will produce smart reactions. In this work, a wireless sensor network (WSN)-based smart home was implemented. We installed a WSN in a one-bedroom apartment to gather data, considering benefits of WSNs such as accuracy, scalability, cost, and the possibility of upgrading for future applications [9–12]. A secure communication is assumed among the sensors, and investigation of the authentication algorithms and potential threats is out of the scope of this work. The gathered data were used to train a classification model that enabled the determination of the real-time state of the home by classifying real-time WSN data and serve as input to a prediction model that prompts the retraining of the system, allowing learning of new environment. The contribution of this work is to predict the state of home with a fairly good level of accuracy without connecting any sensors to the residents or implementing any cameras that may disturb the resident or compromise their privacy. Furthermore, the minimal level of human intervention makes this design an appropriate model to be used in senior care facility to assure the resident safety.

12.2 Related Work

In this section, a brief overview of the current literature on the implementation of smart homes is provided. In [13], the performance of different topologies of WSN with varying node scales and loads was studied for smart home environment. A small node scale with light loads where acknowledgment (ACK) was disabled suited the star topology. When the load and nodes were large, mesh topology was more efficient.

Byun et al. [14] proposed a ZigBee-based intelligent self-adjusting sensor (ZiSAS) design, which can autonomously reconfigure middleware, network topology, sensor density, and sensing rate based on the environmental situation. In essence, the sensors' self-adjustment property saves power as well as bandwidth when the information being sensed is repetitive.

Different machine learning algorithms have been utilized in smart home applications. Roy et al. [5] utilized information theory and proposed a framework for asymptotic equipartition property to predict routes of inhabitants. Route prediction was utilized to automate device control and proactively reserve resources for example, electrical energy or scarce wireless bandwidth for mobile multimedia applications along the inhabitant's most probable locations and routes.

In [6], two prediction algorithms, namely dynamic Bayesian network artificial neural networks (DBN-ANN) and dynamic Bayesian network reinforcement learning (DBN-R) based on the deep learning framework, were used to predict various activities in a home. Using home activity datasets, these algorithms were compared with existing methods such as nonlinear SVM and k-means, in terms of prediction accuracy of newly activated sensors. Sequence prediction via enhanced episode discovery (SPEED) was introduced in [7]. SPEED is a variant of the sequence prediction algorithm. It works with the episodes of smart home events that have been extracted based on the on–off states of home appliances. An episode was defined as a set of sequential user activities that periodically occur in the home. The extracted episodes were processed and arranged in a finite-order Markov model. A method based on the prediction by partial matching (PPM) algorithm was applied to predict the next activity from the previous history.

Youngblood et al. [8] introduced a data-driven approach for building a hierarchical model of inhabitant activities and learning decision policies in smart environments. They designed the ProPHeT decision-learning algorithm that learns a strategy for controlling a smart environment based on sensor observation, power line control, and the generated hierarchical model. The performance of the algorithm was evaluated using data collected from a smart home and office environments.

Most of the related works in this section studied the human behavior in a smart home and used the result to predict activities in the home. The predictions were mostly based on frequency; hence, these models are suitable for permanent residents with strict schedules. However, these models do not present reliable solutions for homes with spontaneous residents or frequently changing occupants such as hotel guests. To that effect, we investigated the smart home learning problem from a different perspective. This perspective is described in detail in Section 12.3.

12.3 System Design

The smart home comes with the promise of better living, security, and energy efficiency, but due to lack of standardization, there is not a global generic design of the smart home. Smart home setup may be as simple as switching on the lights when a sensor senses presence or as advanced as utilizing user identity, environmental conditions, and past preferences, as parameters in making control decisions. A typical smart home architecture is shown in Figure 12.1.

The architecture consists of sensors, a communication system, actuators, user interface, and a data processing and decision module. In this work, we focused on the sensors, communication system, and the data processing and decision module. We designed and implemented a WSN-based smart home, where the collected data were used to classify the state of the home. This model contains two data sources: a historical database and a WSN continuously collecting information in real time as shown in Figure 12.2.

In Figure 12.2, the data from the historical database are used to train classification and prediction models using support vector machine (SVM). SVM is a method used for learning separating functions in pattern recognition tasks. The main idea of SVM is to separate the training data into two classes with a hyperplane that maximizes the margin between them [15, 16]. In this system, the real-time classified state is always compared to the home's predicted state. If the two states match, the output is sent to an actuating system. If they do not match, the real-time state will be sent to the actuating system, terming the differential an anomaly. When the WSN at $t = \tau$ continuously

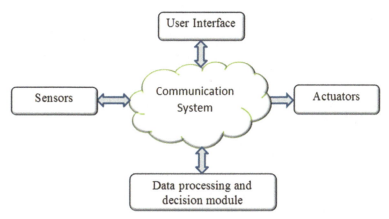

Figure 12.1 Smart home architecture.

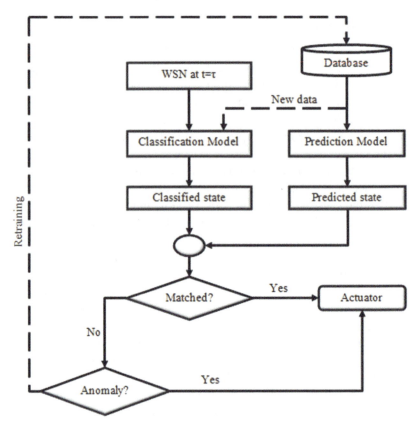

Figure 12.2   Smart home system flowchart.

produces dataset that cannot be classified by the classification model, the dataset cannot be termed an anomaly; hence, it is sent back to the database for retraining of the classification model (the broken lines). For example, if the WSN from a room produces data at 3:15 PM, the context of the data prompts the class "sleeping" and the predicted state is "resident away." The smart home will implement actions for a sleeping resident such as locking the doors and switching off the TV set. However, if a room has been re-purposed and it continuously (days) produces data with high temperatures like that of a steam room. This can no longer be considered an anomaly. Therefore, the data will be sent back to the database for retraining, so this occurrence can be classified as one of the states of the home. In subsequent sections, we look at the building blocks of our system in detail.

12.4 Test Bed

In this section, we provide a detailed description of the hardware devices and the communication protocol utilized in our WSN-based smart home test bed. We required sensor nodes and a central sink node. For this, we utilized the Waspmote and Meshlium, respectively. These devices were made by Libelium [17].

1) Waspmote: The Waspmote is furnished with an Xbee radio transceiver, a microprocessor, an electronic circuit, and a battery. The Waspmote is a programmable board which when interfaced with a sensor board and sensors can collect data from the environment and either transmit it to a central node or to other Waspmotes. The Waspmote does not have an operating system. It has a bootloader through which programs are loaded through USB. The programming is done via the Waspmote pro integrated development environment (IDE). Its sensed data can be transmitted using a variety of protocols ZigBee, 802.15.4, DigiMesh, Bluetooth, Wi-fi, and GPRS at frequencies 2.4 GHz, 868 MHz, and 900 MHz. The Waspmote is shown in Figure 12.3a.

2) Meshlium: Meshlium is a Linux router which contains 6 different radio interfaces: Wi-fi 2.4 GHz, Wi-fi 5 GHz, 3G/GPRS, Bluetooth, and RF (ZigBee, DigiMesh, and 802.15.4) communication protocols. The Waspmote equipped with the appropriate transceiver module can communicate with the Meshlium via the aforementioned protocols. A picture of the Meshlium is shown in Figure 12.3b.

After all the nodes were placed in the apartment, the received signal strength indication (RSSI) was used to ensure a 10 dB margin in every link between the mean RSSI and the receiver sensitivity. This protected the

(a)   (b)

Figure 12.3   (a) Waspmote and (b) Meshlium.

link from unwanted, unpredictable fading effects, interference, and moving obstacles.

3) Wireless sensor network setup: Nodes in WSNs need to be logically arranged so as to bring out the best performance. In our research, we required a central node to store all data, so the star topology was best suited. However, in larger-scale implementations, the tree topology will be better suited because of its scalability and ease of troubleshooting.

WSN can communicate using several communication protocols, namely ZigBee, Bluetooth, Wi-fi, and GPRS. The selection of a protocol depends on the task to be accomplished. The radio frequency (RF) protocols suited our purpose due to their low data rate and power consumption. In our research, we used DigiMesh at 2.4 GHz; DigiMesh is a proprietary RF protocol developed by Digi which is similar to ZigBee and is built on 802.15.4. DigiMesh offers only one node type; therefore, there are no parent–child relationships. All nodes can be configured to sleep when there are no events; this is ideal for energy conservation. DigiMesh offers different RF data rate options of 900 MHz (10, 125, 150 Kbps) and 2.4 GHz (250 Kbps). DigiMesh uses one layer of addressing which is the 64-bit MAC address. Its simplified addressing method improves network setup and troubleshooting. DigiMesh combats potential interference from other devices working in the same frequency and also the interference among the sensors by using direct-sequence spread spectrum (DSSS) at 2.4 GHz and can send a payload amounting to 256 bytes which yields an improved throughput. To improve the energy efficiency of the sensor nodes in the proposed smart home, sleep timers were embedded into node source codes in order to save energy. The status of the batteries was also sent along with the sensed information in the case of power replenishment.

12.4.1 Data Gathering

In this section, we will detail our data collection mechanism. The WSN was installed in a one-bedroom apartment shown in Figure 12.4. The dots in the figure indicate the areas where sensor nodes were placed. The apartment was occupied by a volunteer who lived there full time. The test bed was setup with sensors designed to capture presence, door breach, pressure, and environmental conditions such as air composition and light intensity.

In Figure 12.4, there are 21 sensor nodes and a Meshlium. The one-bedroom apartment was equipped with 13 PIR, 1 temperature, 1 luminosity,

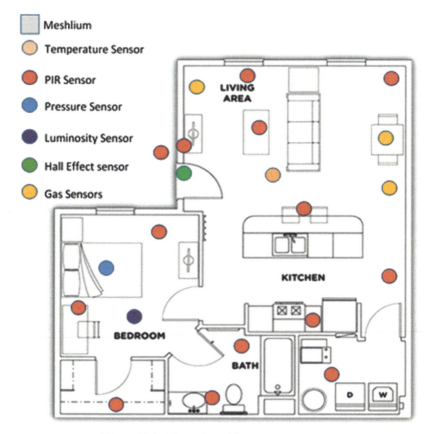

Figure 12.4 Floor plan with sensor arrangement.

1 pressure, 1 of each gas (oxygen, carbon monoxide, carbon dioxide, and humidity), and 1 Hall effect sensor. All nodes transmitting at the same time can present an internal interference problem. Therefore, the radio transceivers were configured using XCTU software to implement a clear channel assessment protocol. This protocol senses the energy of the channel before it transmits. The amount of time specified to scan each channel affected the detected energy therefore, selecting the minimum energy value was insufficient in detecting channel energy so a higher value was used. Usually, the energy value for a free channel is around –84 dBm and –37 dBm for an occupied one [18]. We configured the clear channel assessment threshold value for a free channel to be –44 dBm.

12.5 Software

12.5.1 Data Classification

In this section, we will discuss data classification. When faced with a large set of data, patterns in the data need to be recognized in order to classify the data. There are three methods through which this can be done: supervised, semi-supervised, and unsupervised. Supervised classification (e.g., discriminant analysis) is a classification method in which the input pattern is identified as a member of a predefined class. Unsupervised classification is a classification method in which the pattern is assigned to a hitherto unknown class [19]. Semi-supervised learning (SSL) is halfway between supervised and unsupervised classifications, and in addition to unlabeled data, the algorithm is provided with some supervision information but not necessarily for all examples. Hence, new patterns are discovered.

With unsupervised learning it is possible to learn larger and more complex models than with supervised learning. However, unsupervised learning was not suited to our research at this stage because we knew the possible outcomes from the installed sensors; therefore, supervised learning was utilized in this work. Supervised learning can be applied using machine learning techniques such as naive Bayes, decision trees, support vector machines (SVMs), artificial neural networks, and hidden Markov models [20]. SVM was utilized in this research due to its high performance in many applications such as prediction, handwritten recognition, face detection, text categorization, and speech recognition. A comprehensive survey on the applications of the SVMs in pattern recognition is presented in [19]. In most of these applications, SVM outperforms other machine learning techniques in generalizing the pattern recognition problem.

We collected approximately eight weeks of data. Our labeling exercise captured seven states in the home. When labeling, we considered safety to be integral to our design, so the gas levels were taken into consideration when defining every state except arrival and departure. Table 12.1 describes the states and the parameters which make up the states.

1. Safe Sleeping (SS): The resident was on the bed sleeping, and the gas levels were within acceptable levels.
2. Safe Away (SA): The resident was away from the home, and gas levels were within adequate levels.
3. Safe Home Activity (SH): The resident was active in the home, and the gas levels were adequate.

4. Alarm (AL): The gas levels exceeded recommended indoor levels but were not up to emergency levels, though further exposure may have been dangerous.
5. Emergency (EMR): The gas levels have reached emergency levels.
6. Departure (DP): This state indicated the departure of the resident.
7. Arrival (AR): This state indicated the arrival of the resident.

The rules guiding our labeling are found in Table 12.1.

From Table 12.1, in order to classify the home state as SH, the O_2 level will have to be greater than 20% but less than 25%, the CO_2 level will have to be equal to or between 0.3% and 0.1%, the CO level will have to be equal to or between 0 and 9 ppm, the relative humidity will have to be equal to or between 30 and 60%, the PIR living room (PIRL) will have to be equal to 1, the luminosity (LUM) will have to be equal to or between 2 and 60, and finally Hall effect is equal to zero. The other states can be understood by following the explanation of the SH state.

12.5.2 Support Vector Machines

SVMs are supervised learning models with associated learning algorithms that analyze the data and recognize patterns, used for classification and regression analysis. It is unlike classical pattern recognition algorithms that work in an L1 or L2 norm and minimize the absolute value of an error or of an error square, the SVM performs structural risk minimization (SRM) that builds a model with a minimized VC bounds [16].

Table 12.1 Sensor data labeling rules

Sensor	SS	SH	SA	AR	DP	AL	EMR
				States			
O_2 (%)	>20 & <25	>20 & <25	>20 & <25	–	–	>8 & ≤17	≤8
CO_2 (%)	≥0.3 & ≤0.1	≥0.3 & ≤1000	≥0.3 & <0.1	–	–	>0.1 & <0.15	≥15
CO (ppm)	≥0 & ≤9	≥0 & ≤9	≥0 & ≤9	–	–	≥9 & ≤15	>15
%RH	≥30 & ≤60	≥30 & ≤60	≥30 & ≤60	–	–	≤20 & ≤60	≤10 & ≥70
PIR living room	–	1	1	1	0	1	1
PIR bedroom	1	–	0	–	0	–	–
PIR outside	–	–	0	1	1	–	–
Luminosity	≥2 & ≤600	≥2 & ≤60	–	–		–	–
Free fall	–	–	–	–		–	1
Pressure (Pa)	>7	–	–	–		–	–
Hall effect	0	0	0	1	1	<2	<2

Since the smart home problem is a multiclass classification problem (safe sleeping, safe home activity, safe away, alarm, emergency, departure, and arrival), the SVM cannot be applied directly. There are different methods to apply SVM for multiclass classification. The most common ones are one versus one and one versus the rest [21].

The one-versus-one method trains a classifier for each possible pair of classes. For M classes, this results in $(M - 1)$ M/2 binary classifiers. When it tries to classify a test pattern, it evaluates all the binary classifiers and classifies according to which of the classes gets the highest number of votes. The one versus the rest gets M-class classifiers, and it constructs a set of M binary classifiers, each trained to separate one class from the rest and combine them by performing the multiclass classification according to the maximal output. In this work, one-versus-one multiclass SVM classification method is used to estimate the state of the smart home. Figure 12.5 shows the pattern recognition algorithm.

In Figure 12.5, the processed features of the sensor data at t = τ are presented to the multiclass SVM classifier. In the classification stage, the

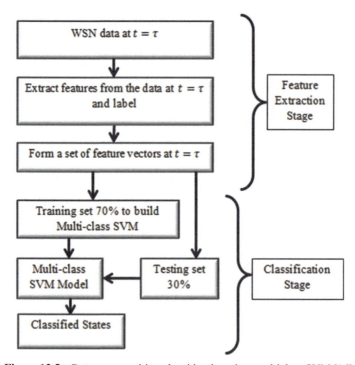

Figure 12.5   Pattern recognition algorithm based on multiclass SVM [16].

dataset (the data points with their corresponding labels) is divided into 70% training set and 30% testing set. The training set is used to build the multiclass SVM classifier, and the testing set is used to calculate the classifier performance.

12.5.3 Prediction

The next phase of our research involved prediction of future states in the home. At this stage, because the data were already classified, what we were predicting was the next state in the home at a particular time and not individual sensor readings. For example, a week or three hours from a given time, what will be the state of the home? The EMR and AL states were anomalies. We normalized the data by switching those labels to the previous label before their occurrence. For example, if the state of the home was SS before AL occurred, we returned the label to SS.

SVM was utilized for regression, extracting, and labeling sequential home states from the historical data. The time duration of a sequence varied from weeks to hours, and eventually, a 2.5-hour window was used. The data from the home were preprocessed, so each dataset was ten minutes apart; therefore, a 2.5-hour window will consist of fifteen determined states. In order to determine the state at time $t = \tau$, the previous 14 states were used. The historical states of the home were converted into time series data, and features were extracted every 15 states (2.5 hours) and labeled.

$$\phi(t = \tau) = [D(\tau - \lambda), D[\tau - 2\lambda] \ldots \ldots \ldots D(\tau - N\lambda)]$$

where λ is 10 minutes, N is 14 states in this experiment and $\phi(t)$ is a feature vector.

For a given sequence, SVM provides a label;
$L(t = \tau) = D(\tau)$.

Seventy percent of the data were randomly selected and used to build the multiclass SVM and 30% of the dataset was used to test the predictor. The future state of L(t) was determined using the sequence of past states $\phi(t)$ as feature vector for the SVM. For instance, given five home states and a classification model capturing the state of the home every ten minutes, for two and a half hours, the feature vector will be [111222112231444] capturing fifteen states. Using historical sequence of data, what will be the next state if the past states were [11122211223144]? This is the question answered by the SVM prediction model, as it will predict state 4.

12.6 Results

12.6.1 Classification

The data analysis results for the proposed smart home test bed are presented in this section. The total dataset contained 8496 data points with corresponding labels. Seventy percent of the data points, that is, 5666, were randomly selected from the dataset and used to train the model. The performance of the model was evaluated using 30% of the remaining dataset for testing. The testing result is presented in the confusion matrix (Table 12.2), where each column represents the number of instances in a class and each row represents the number of instances in an actual class. The testing phase had an accuracy of 99.72%.

12.6.2 Prediction

In the prediction model, one-versus-one multiclass SVM is used to predict the states at time $t = \tau$ with the RBF kernel. The confusion matrix given in Table 12.3 shows the performance of the prediction model which had 96.25% accuracy when tested with 30% of the data. And also, Figure 12.6 shows the actual and predicted classes over 7 days, that is, for every ten minutes a total

Table 12.2   The confusion matrix after testing the classification model with 30% of the dataset

Actual States $t = \tau$	Classified States at $t = \tau$						
	SS	SA	SH	DP	AR	AL	EMR
SS	963	0	0	0	0	0	0
SA	0	888	0	0	0	0	0
SH	4	0	932	0	0	0	0
DP	4	0	0	16	0	0	0
AR	0	0	0	0	21	0	0
AL	0	0	0	0	0	1	0
EMR	0	0	0	0	0		1

Table 12.3   The confusion matrix after testing the prediction model with 30% of the dataset

Actual States at $t = \tau$	Predicted States at $t = \tau$				
	SS	SA	SH	DP	AR
SS	945	1	18	0	0
SA	5	863	1	0	19
SH	22	0	914	0	1
DP	1	0	19	0	0
AR	1	17	1	0	2

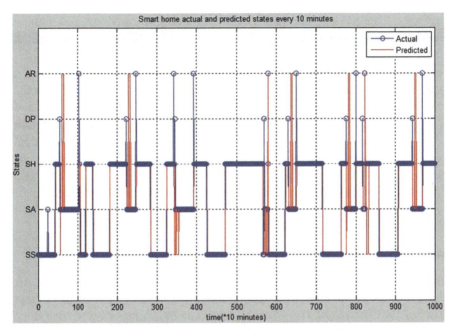

Figure 12.6   Actual and predicted classes over 7 days.

of about 1008 data points. The Statistics Toolbox of MATLAB was used to implement the multiclass SVM.

The performance of the model shows that the AR and DP states were not accurately predicted. This was in line with our objective because we wanted the home to predict states within the home. AR and DP are more in line with schedules. Our smart home design predicted domineering activities which happened within the home; hence, our system prediction performance was satisfactory.

12.7 Conclusion

In this chapter, a system model was designed consisting of a data gathering WSN, a classification and prediction model built with SVM. Using pattern recognition, we were able to classify the state of the home achieving an accuracy of 98.6%. Utilizing this information, we predicted the state of the home and compared our prediction to real-time WSN data with satisfactory results. The proposed model can be utilized in applications with larger number of states, while greater accuracy may be achieved using other machine learning methods. This system is capable of being fully autonomous (no human interface), as the historical state of the home can be utilized as a vital cog in the decision-making module when future situations arise. Our model provides an advantage over previously designed implementations being that it has a generic design which can be applied to smart homes and commercial buildings such as senior homes and hotels. It does not only classify real-time data, but also predict the future home state. This can be beneficial in anomaly detection and smart home autonomy. In order to identify and predict the individual behavior of the resident, the use of biometric sensors or RFID devices is necessary that it can be further investigated. The current model can also be extended to address the reliability and autonomy issues using techniques proposed in previous works [5–8].

Acknowledgement

The third and last authors would like to acknowledge the support from Air Force Research Laboratory for sponsoring this research under the agreement number FA8750-15-2-0116. The US Government is authorized to reproduce and distribute reprints for governmental purposes notwithstanding any copyright notation thereon. The views and conclusions contained herein are those of the authors and should not be interpreted as necessarily representing the official policies or endorsements, either expressed or implied, of Air Force Research Laboratory, or the US Government.

References

[1] Stefanov, D., Bien, Z., and Bang, W.-C. (2004). The smart house for older persons and persons with physical disabilities: structure, technology arrangements, and perspectives. *IEEE Trans. Neural Systems Rehabilitat. Eng.*, 12(2), 228–250.

[2] Abedi, A., Afghah, F., and Razi, A. (2016). Resource Management in Cyber Physical Systems. In Zeadally, S., and Jabeur. N. (eds.) *Cyber-Physical System Design with Sensor Networking Technologies*, IET, ISBN-13: 978-1849198240, April. 28 p.

[3] Hicks, J. (2014). Forbes, [Online]. Available: http://www.forbes.com/sites/jenniferhicks/2013/06/24/robots-and¬smart-homes-for-the-silver-economy/ [Accessed 17 October 2014].

[4] Smart Home automation, "Solution Center," Smart Home, [Online]. Available: http://www.smarthome.com/sc-save-energy-the-smarthome-way. [Accessed 17 February 2015].

[5] Roy, N., Das, S., and Basu, K. (2007). A predictive framework for location-aware resource management in smart homes. *IEEE Trans. Mobile Comput.*, Vol. 6, No. 11, pp. 1270–1283.

[6] Choi, S., Kim, E., and Oh, S. (2013). Human behavior prediction for smart homes using deep learning. In RO-MAN, Gyeongju.

[7] Alam, M., Reaz, M., and Ali, M. (2011). SPEED: An inhabitant activity prediction algorithm for smart homes. *IEEE Trans. Syst. Man Cybernet. Part A: Syst. Humans*, 42(4), 985–990.

[8] Youngblood, G., and Cook, D. (2007). Data Mining for Hierarchical Model Creation. *Systems, Man, and Cybernetics, Part C: Applications and Reviews*, 37(4), 561–572.

[9] Razi, A., Afghah, F., and Abedi, A. (2011). Binary source estimation using a two-tiered wireless sensor network. *IEEE Commun. Lett.*, 15(4), 449–451, Apr.

[10] Abedi, A., Razi, A., and Afghah, F. (2011). Smart battery-free wireless sensor networks for structural health monitoring. *International Workshop on Structural Health Monitoring*, Stanford University, CA, Sep. 2011.

[11] Akyildiz, I. and Vuran, M. (2010). *Wireless sensor networks, West Sussex*: John Wiley & Sons Ltd.

[12] Razi, A., Afghah, F., and Abedi, A. (2013). Power optimized dstbc assisted DMF relaying in wireless sensor networks with redundant super nodes. *IEEE Trans. Wireless Commun.* 12(2), 635–645, Feb.

[13] Zhang, C., and Luo, W. (2013) *Topology performance analysis of zigbee network in the smart home environment*. College of Electrical and Information Engineering Guangxi University of Science and technology, Guangxi Zhuang.

[14] Byun, J., Boungju, J., Noh, J., Kim, Y., and Park, S. (2012). An intelligent self-adjusting sensor for smart home services based on zigbee communications. *IEEE Trans. Consumer Electronics*, 58(3), 794–802.

[15] Aarohi, V., Paunwala, C., and Paunwala, M. (2014). Statistical analysis of various kernel parameters on SVM based multimodal fusion. *India Conference (INDICON)*, Pune, 2014.

[16] Amsalu, S., Homaifar, A., Afghah, F., Ramyar, S., and Kurt, A. (2015). Driver Behavior Modeling near Intersections Using Support Vector Machines based on Statistical Feature Extraction. In *IEEE Intelligent Vehicles Symposium*.

[17] Libelium, "Event sensor board 2.0," Libelium, 2014.

[18] Libelium, "Networking guide," Libelium, 2014.

[19] Lee, S. and Verri, A. (2002). "SVM," Springer-Verlag Berlin Heidelberg, Lncs 2388, pp. 213–236.

[20] Stenudd, S. (2010). "Using machine learning in the adaptive control of a smart environment," Utigivare, Vuorimiehentie.

[21] Jain, A., Duin, R., and Mao, J. (2000). *Statistical Pattern Recognition: A Review. IEEE Trans. Pattern Analysis and Machine Intell.* 22(1), 4–37.

13

Mobile Node Scheduling in MANETs for Resource Assignment: From Hospital Assignment to Energy Charging

Peng Liu[1], Biao Xu[1], Zhen Jiang[2] and Jie Wu[3]

[1] Institute of Computer Application Technology,
Hangzhou Dianzi University, Hangzhou, China
[2] Department of Computer Science, West Chester University,
West Chester, PA 1938, USA
[3] Department of Computer and Information Sciences,
Temple University, Philadelphia, PA 19122, USA

Abstract

In the large-scale MANETs, mobile nodes could act as resource providers, relays, or requestors in the network. Since the storage space, the channel capacity, and the energy supply are often limited, it is always worth considering the assignment between requestors and resources. We model two scenarios in this chapter, namely hospital assignment and energy charging. In the first scenario, the mobile node carries the patient to a hospital. Patients with different conditions may compete for the same hospital. In the battery-powered MANETs, mobile nodes need constantly charging to remain active. The problem lies in that energy-critical nodes may not find an available charger in short distance and a temporary stationary node being assigned to a mobile charger.

In both cases, our goal is to optimize QoS (in terms of latency) in assigning requestors to resources. We extend the scheme to consider resource reservation along the timescale by estimating from the past records in history. As a result, the occupancy is balanced in order to reduce the risk of critical nodes being delayed. Then, we develop a local waiting queue to keep serious nodes waiting locally, when it costs more to reach another available resource. Simulation results show the substantial improvement of our approach in average delay and number of failure of assignment.

Keywords: Ambulance service, Resource allocation, Bipartite matching, Wireless ad hoc network, Energy charging.

13.1 Introduction

Indicated by Li [1], preconsultation is one of the most important facts of delay in emergency treatment in addition to transportation time of ambulances. After that, an emergent patient would be assigned a sickbed and a doctor for next step treatment. In China and India [2, 3], when there are not enough sickbeds, some life-critical patients would die, which also leads to patient–doctor dispute [4]. Traditional local greedy patient-hospital assignment methods, which assign a patient to the available hospital with shortest delay in a first-come-first-serve (FCFS) manner. The problem is that a life-critical patient may encounter non-vacancy of sickbeds in nearby hospitals, which have been occupied by non-critical patients, forcing an ambulance choose a further hospital and causing deadly delay [5]. This is either because non-critical patients have smaller delay, or due to their earlier appearances before life-critical patients. In an outburst of emergent patients [6], the amount could be far more beyond the hospital accommodation. Therefore, we must balance the average delay of each patient and reserve some space for life-critical patients in advance. Also, the chargers in the mobile ad hoc networks have limited resources. Each one of them can only support a fixed number of mobile nodes. When there exist multiple requests in the system, careless assignment such as greedy-based methods would bring extra delay and void assignment which is severe to energy-critical nodes. To remain the network working in a green and lasting manner, the assignment between the nodes and chargers must be carefully considered.

In this chapter, we first propose a hospital assignment for emergent patients in a big city, denoted by HAEP, to minimize the average delay for sending a patient to a hospital. Then, apply the algorithm to the scenario of energy charging in MANETs. An ambulance can obtain a reserved sickbed in a certain hospital within limited distance without continuously waiting for preconsultation. In this way, many life-critical situations can be treated in the limited time. We take advantage of the recent technical advances in wireless networks and vehicular networks [7, 8] to collect the real-time information from hospitals and patients and to solve the above optimization problem in a way that is derived from the Hungarian algorithm [9]. In our system, there are three kinds of patients, namely life-critical, serious, and cared patients. Two kinds of hospitals, i.e., premium and primitive, are considered, where primitive hospitals can only treat non-critical patients. Furthermore, the occupied bed is not pre-emptable, regardless life-critical or not, due to patient-doctor dispute. To avoid the situation of lack of bed, a number of preserved beds are set aside for life-critical patients, who are diagnosed by the ambulance. In our chapter, the preserved beds are calculated based on history records. We propose our solution on the extension of maximum weight bipartite matching [10], a problem to assign n resource to n resource requestors. We assign n patients to m hospitals with capacity C at each hospital and provide the preserving sickbeds and waiting queue along the timescale. Our contribution is threefold:

1. We propose a new assignment to balance the bed requirement of life-critical and non-critical patients with the purpose to reduce the delay of treatment for life-critical patient as well as the average delay for all patients.
2. We extend the solution from Hungarian algorithm, with the capacity and consideration of reservation along the timescale and then the application on cases of inadequate resources.
3. We simulate a real scenario of patient assignment in Shanghai [11]. The experimental results derived from real trace data show our substantial improvement in delay and failure of assignment in terms of their impact on efficient treatment of life-critical patients.
4. We further apply the scheme to an energy charging problem in MANETs where average cost for searching chargers is minimized.

13.2 Target Problem and Related Works

In this section, we discuss the problem in existing system in the ambulance services. Greedy algorithms are often used in resource allocation problems. When the constraints determine a polynomial and the objective is linear, the greedy procedure results in an optimal solution [12]. However, most current research work ignores the fact that all the requests take place along the timescale which is not a one-time optimization problem or resource assignment. M. Xu studied another interesting phenomenon that service delay for non-emergent patients will be significantly affected due to the arrival of emergent patients [13]. They conducted a retrospective study in real trace of a large hospital in Hong Kong and estimated waiting time and length of stay for individual non-emergent patients as a function of the presence of emergent patients and other related factors. The study convinces us that the competition of critical and non-critical patients must be carefully addressed.

As demonstrated in Figure 13.1, there are two hospitals, in which a is a premium hospital and b is a primitive hospital. There are also two patients in which 1 is a non-critical patient and 2 is a life-critical patient. We use *number:time* in the patient ellipse to show the patient number and his appearance time. The weight in the arrow denotes the delay. Assuming available capacity in each hospital is 1, requests are submitted from patients 1 and 2 at time 0 as shown in Figure 13.1a. They will compete for a, and according to the FCFS greedy algorithm, 1 will be assigned to a since it has the smallest delay, leaving 2 no place to go. As the resource is enough for both patients, the best solution is to assign 1 to b and 2 to a. In Figure 13.1b, when two patients raise requests one after another, FCFS results 1 occupying a although the delay of 1 is greater than that of 2, and 2 has no hospital to go. The desired assignment is indicated in Figure 13.1b.

We assume hospitals have fixed locations and patients will appear anywhere to call ambulance so that localizing resources [14, 15] cannot be used here to solve the problem. Through route planning or traffic control,

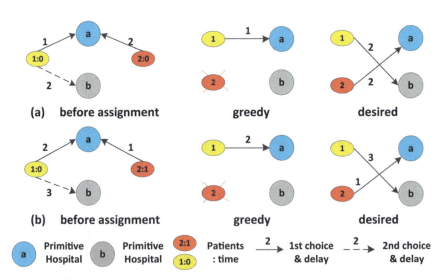

Figure 13.1 A demonstration of patients and hospitals' assignment.

we assume there is no extra delay in taking patient to hospitals [16–18], so that the delay can be calculated based on the distance and preconsultation for any hospitals. Also, the resource allocation problem inside the hospitals is addressed in many literatures, such as operation room planning [19], admission arrangement [20], improving the use of computed tomography facilities [21], and surgery scheduling [22], which enables us to study the historic records and model capability of available beds. We also consider the ability of different hospitals to treat different kinds of patients. Hospitals can be categorized into different classes according to their ability to do medical care. Therefore, the problem becomes the one with multiple kinds of resources (each has a different capacity) and multiple kinds of requesters, which include the consideration of future reservation along the timescale.

In this chapter, we solve the optimized resource allocation problem using the bipartite matching in [9], first to balance the requirements among critical and non-critical patients in terms of the average delay, then to develop a preservation-based method on top of the bipartite matching along the timescale for life-critical patients, and finally to introduce the local waiting queue for serious patients. The number of available beds is estimated using possibility observation and history statistics. The waiting queue is built up by leveraging local waiting time against reassignment cost. The proposed method will greatly balance the need of both life-critical and non-critical patients along the timescale.

13.3 System Model

In our system, patients, hospitals and the service center are three main components. Ambulance can send the request to the service center via wireless ad hoc networks or cellular networks. Hospitals count vacancy and estimate the capacity growth with the advanced technology of [7, 8]. All the ambulances are equipped with on-vehicle communication devices and can be guided by the service center to transport patients to the target hospital.

Table 13.1 summarizes all of the notations used in this chapter, which will be explained in the following. There are three kinds of patients: life-critical, serious, and cared. Denote $x_i \in C_x \subseteq X$, as the ith critical patient, $x_j \in S_x \subseteq X$, as jth serious patient, and $x_k \in N_x \subseteq X$, as kth cared patients. There are also two kinds of hospitals; i.e., H_i^P means ith premium hospitals, and H_j^p means jth primitive hospitals. As we consider the timescale, the in and out of patients from hospital y could be monitored, managed, and predicted; thus, the capacity of sickbeds at $t > 0$, denoted as C_y^t, is calculated

Table 13.1   Notations

X	Patient set $X = C_x \cup S_x \cup N_x = \{1,2,3,\ldots\}$
$\|X\|$	total number of patients ($\in X$) in schedule
Y	hospitals $Y = H^P \cup H^p = \{a,b,c,\ldots\}$
$\|Y\|$	number of inpatient beds
C_y^X	available (also called bed capacity) of $y \in Y$ where $X \in \{C_x, S_x, N_x\}$
$R(x,y)$	cost (≤ 0) for $x \in X$ to reach $y \in Y$ in terms of elapsed time where "$-$" indicates an initial/unreachable status
$m(x,y)$	bipartite matching between $x \in X$ and $y \in Y$ where 1 denotes a saturated assignment, 0 denotes a possible assignment, and "$-$" indicates not reachable currently
$L(v)$	labeling function of Hungarian algorithm [9], $v \in X \cup Y$
$L'(v)$	previous record of L for any given $v \in X \cup Y$
α	the difference between $L(v)$ and $L'(v)$ each time
S	patient set in the current consideration of allocation, $\subseteq X$
$N(S)$	hospitals ($\subseteq Y$) that are reachable by patients $\in S$, or arriving), i.e., $\{j \mid \exists m(i,j) = 0 \text{ or } 1\}$
ⓐ	common beds available at a hospital, that could be assigned to all types of patients, say $y \subseteq Y$, that have bed(s), denoted by ⓐy, i.e., $\{y \mid ⓐy = C_y - \sum_{x \in X} m(x,y)\}$
Ⓡ	Availability of preserved beds for critical patients at a hospital, e.g., Ⓡ$y = R_i$
Ⓦ	Availability of waiting queue for serious patients at a hospital, e.g., Ⓦ$y = w_k$
E_u^*	an alternating tree [9] derived from m, with the root u, simply called E-tree

based on the prediction of patient-leaving amount and allocation at round t. Especially, when $t = 0$, $C_y^0 = C_y$. In our method, we also predict the amount of critical patients in the future time slice. Since critical patients can appear for many reasons, e.g., a burst of a serious disease and a severe traffic accident, although there are some mathematic tools [23, 24] to estimate one kind of situation, it is hard to estimate superimposed situations. However, there are two key facts that we could use. First is that there will be regular days and peak days regarding the burst rate of critical patients due to epidemic seasons, holidays, or bad weather. Second, no matter what the accident or disease is, there is always a trend rather than sporadic ups and downs. Therefore, we can use the last three data to predict the future possibility. Our method can be described in two steps, i.e., cost matrix buildup and hospital assignment.

The goal of this chapter is to assign each proposed patient $\in X, X = C_x \cup S_x \cup N_x$ to a proper hospital $\in H, H = H^P \cup H^p$, so that the total delay (from the time a patient is picked up by an ambulance, waiting at hospitals, until he gets treatment in a hospital). By giving a critical patient more weight in the delay cost, our method can balance the requirements between the life-critical and non-critical patients. The impact on patient treatment is also measured by the failure of assignment of critical patients. Existing research utilizing bipartite matching often formalizes similar problems as maximum weight matching [10], while this problem is a minimum weight matching. By setting the delay cost table in [9] to its contrariety, such as $R(x, y) = -cost(x, y)$, our minimal delay problem could be implemented as maximum weight matching. Inherited from the bipartite matching, we use matrix m to indicate whether there is an assignment from patient x to hospital y, e.g., $m(x, y) = 0, 1$. Then, we formalize the problem as follows:

$$\max \sum_{x \in X} \sum_{y \in Y} \sum_{0 \leq t < \infty} R(x, y) m(x, y)$$

$$\text{s.t.} \quad \text{every } m(x, y) = 0, 1, \text{ or "}-\text{"} \tag{i}$$

$$\sum_{y \in Y} m(x, y) = 1 \quad \text{for every } x \in X \tag{ii}$$

$$\sum_{x \in X} m^t(x, y) \leq C_y^t \quad \text{for every } y \in Y \text{ and any } t \tag{iii}$$

"$-$" indicates an initial or unreachable status. Constraint (i) ensures the sickbed assignment as a bipartite matching. Constraint (ii) guarantees such an assignment without double assignment. Constraint (iii) asserts the use of resources under the capacity constraint. The problem cannot be solved by minimizing average delay in each patient level, but to minimize the average delay for all patients, especially between the non-critical and serious.

When a hospital could have multiple capacities and more than one type of patients, the problem becomes complicated, since the current optimal does not mean global optimal. As shown in Figure 13.2, different patients have different views of capacity of a premium hospital. In our method, we first consider the extension of capacity to bipartite matching and then the optimization across adjacent time periods. Based on the estimated rate of critical patients that will enter the hospital in the near future, we set up the preservation in our algorithm HAEP (indicated as R in Figure 13.2) to optimize the allocation across time periods. Furthermore, for patients in serious condition, we arrange a special waiting room to have them stay rather than ask them to go (indicated as w

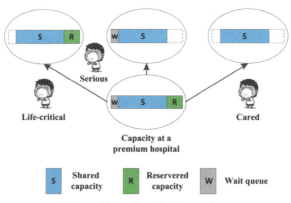

Figure 13.2   Capacity illustration.

in Figure 13.2, capacity is based on the number of patients leaving the hospital and time cost to another nearest hospital).

13.4 Method to Solve Multidimension Hospital Assignment

13.4.1 Cost Matrix Buildup

The basic requirement of the bipartite matching is the buildup of the cost matrix. In our scenario, the main cost is the response time from the ambulance submitting a request until the patient successfully checks in at a hospital. It is composed of two parts: One is the transportation time from where the patient is to the location of the destination hospital. The other is the waiting time at the destination hospital, which is necessary. To simplify the model, we define Euclidean distance between them as the metric of transportation time. In practice, the transportation time can be controlled [16–18] so that this will not affect the proposed method.

We show an example in Shanghai as indicated in Figure 13.3 and Table 13.2. To quickly simplify the algorithm, we use grids to achieve the same goal. We first divide the whole region into small grids. The size of a grids is closely related to the organization of the city, such as by living area and by postcode. With grid, we can easily find out the hospital sequence for any grid according to the distance.

After applying the grid participation as seen in Figure 13.3, we get Figure 13.4. Then, we can easily get the cost matrix as shown in the Equations (13.1) and (13.2). As indicated here, for each time slot, we can

Figure 13.3 A demonstration of patients and hospitals assignment.

Table 13.2 A simple example of patients requests

Patient	Requested Assignment Time	Situation	If No Vacancy
U1	0	Cared	no wait
U2	0	Cared	no wait
U3	1	Serious	wait
U4	1	Life-critical	no wait

get one cost matrix where "_" indicates an impossible reach from patient to hospital. The initial capacity (C_a, C_b, C_c, C_d) at each hospital is $(1, 0, 2, 1)$.

$$CostMatrix(slot0) = \begin{array}{c} \\ U1 \\ U2 \end{array} \begin{array}{cccc} a & b & c & d \\ \left(\begin{array}{cccc} 1 & 3 & 2 & 4 \\ 0 & 2 & 3 & 5 \end{array} \right) \end{array} \quad (13.1)$$

$$CostMatrix(slot1) = \begin{array}{c} \\ U3 \\ U4 \end{array} \begin{array}{cccc} a & b & c & d \\ \left(\begin{array}{cccc} 2 & 0 & 5 & 3 \\ 1 & - & - & 6 \end{array} \right) \end{array} \quad (13.2)$$

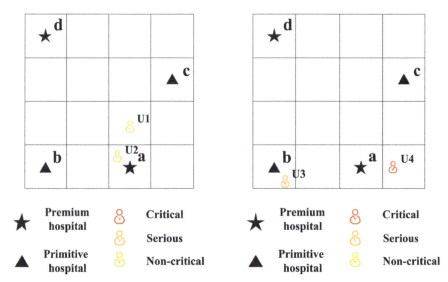

Figure 13.4   Example illustration.

For the FCFS schedule, at slot 0, patient $U1$ will be allocated to hospital a, and patient $U2$ will be allocated to hospital c. At slot 1, the capacity of each hospital (C_a, C_b, C_c, C_d) changes from $(1, 0, 2, 1)$ to $(0, 0, 1, 1)$. Therefore, conflict appears in which two emergent patients cannot be assigned to close hospitals. More terribly, patient $U4$ will have no hospital in which to stay, since the last eligible hospital d is assigned to patient $U3$ since he is earlier than patient $U4$.

13.4.2 Hospital Assignment

For the regular matching algorithm, at each time slot, there is an augment path found so that $U1$ and $U2$, and $U3$ and $U4$ will switch assignments to get the optimization. Finally, after the switch, patient $U4$ could find a hospital with time cost 6. It is local optimal; however, in the view of slot 0 and slot 1 together, the assignment is not optimal.

In our algorithm, assume that there is a reservation of 1 bed for critical patients at hospital a at slot 0. Therefore, the capacity of (C_a, C_b, C_c, C_d) that the two patients see is $(0, 0, 2, 1)$. After applying a matching algorithm, the result of slot 0 is $U1$ to c and $U2$ to c.

At slot 1, consider the patient leaving rate at hospital b: It allows a waiting queue of 1 room; therefore, the capacity of (C_a, C_b, C_c, C_d) that the serious and critical patients see is $(0, 1, 0, 0)$ and $(1, -, -, 0)$, so the matching would be as shown in Figure 13.5. The total cost of our algorithm is $6 + waitingtime(lessthan1)$, while that of the regular matching is 13, and the FCFS has no answer.

Non-critical patients can switch with critical patients, but with limitations. In the experiment, we set the cost of life-critical patient three times as the same distance at the non-critical patient. The value could be adjusted when applying the framework to practice. The assignment problem is actually a specific capacity to a different requester. As shown in Figure 13.2, each category of patients can use a different portion of the entire capacity of a hospital, which is designed to meet the requirement of each patient.

Reserved capacity is an evolutional parameter, which can only accommodate critical patients since there are less applicable hospitals, and critical patients obviously cannot wait. Overprediction will cause waste, and less than enough will cause inefficiency. Wait-allowed capacity is a small list which is related to the hospital-leaving rate. For serious patients, the earlier the treatment, the better it is; therefore, it is good to have them have some basic treatment and wait in a nearby hospital when it will cost additional time to

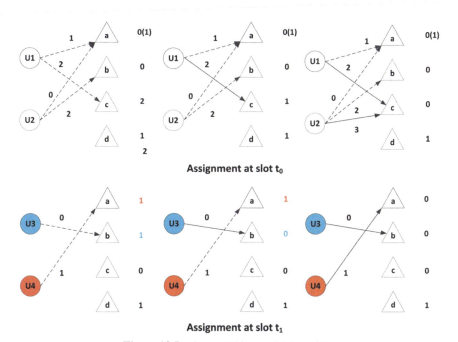

Figure 13.5 Our matching at slot t_0 and t_1.

reach a further hospital. The capacity of a queue in a waiting room is based on the number of patients leaving the hospital and the time cost to another nearest hospital. A non-critical patient is not allowed to wait and will always be sent to the nearest hospital with capacity.

The schedule is expected to apply whenever there is a new request. However, to avoid local optimization, the method needs more requests together to perform the schedule. A time slice δ is adopted with trade-off between waiting time and global cost. If the δ is too large, it will incur additional waiting time. If it is too small, the local optimization will incur global cost. We have the following definition:

Definition 1. Any $x \in X$ that has not seized the reservation is called unsaturated, and it has $m(x, y) \neq 1$ for every $y \in Y$. Any $y \in Y$ still available for allocation is called unsaturated, and it has $\sum_{x \in X} m(x, y) < C_y$.

Definition 2. Any $y \in Y$ still available for allocation is called available and ⓐy or Ⓡy or Ⓦ$y > 0$ according to the category of $v \in S$ where $m(v, y) = 0$.

Our algorithm is shown as Algorithm 13.1. The first phase is to initialize cost matrix R, matching matrix m, and label L(v). The second phase is to check whether the matching could stop and converge otherwise to build a new augmenting tree with root x. The remaining phases are similar to the

---

Algorithm 13.1 Hospital assignment based on Hungarian algorithm with capacity and reservation

---

Require: X, Y, R, and C_y for each $y \in Y$
Ensure: a bipartite matching in Y for each $x \in X$

1. *Initialization* (i.e., R(i,j), m(i,j), and L(v) with Equation (13.3)).
2. *Completion check*:
   For any unsaturated (Definition 1) patient $x \in X$, prepare S, T and E_x^* to start the matching process in the following; otherwise, successfully end the entire process.
3. *Label update for new opportunities for x to match with y in m*:
   IF $N(S) \neq T$ *GOTO* phase 4 *ELSE*
   Update reaching opportunities of patients by (first α with Equation (13.3), and then L with Equation (13.4), in order to reset m with Equation (13.5).
4. Table construction (E_x^*, m, T and S):
   Find any $y \in N(S) \backslash T$; *IF* y is available (Definition 2), an alternating path from x (root of the tree E_x^*) to y exists, apply augment matching along this path as the Hungarian algorithm, decrease capacity of y by 1, *GOTO* phase 2. *ELSE* find all z that are matched with y, update E_x^* with (z, y) respectively, $S = S \cup \{z\}$, $T = T \cup \{y\}$, *GOTO* phase 3.

Hungarian algorithm, except we use Definition 2 to implement multidimensional matching, and we alter the table construction phase to apply on multiple capacities.

For critical patients, first use reserved capacity. For any $C_y > 1$, multiple reservations are allowed on hospital y. Any existing reservation will be added into our records (E-tree and S), for later reservation shuffle with the augment path. This phase of hospital matching will continue until all patients have been checked under the capacity constraint. For serious and cared patients, they can only see the capacity at premium hospitals after removing critical preservation. For serious patients, they will be allowed to stay at a waiting list before getting a bed in a hospital which is denoted as $\widehat{w}y$. All the capacity checks imply the above strategy. The entire process will stop at phase 2 when every patient finds its target hospital or all the beds are allocated (the method will stop when the sum of the cost of allocated patients and unallocated patients is minimal). Equation (13.3) is used to calculate α for the greedy progress along the timescale of patient arrivals.

$$\alpha = \min_{x \in S, y \in Y \setminus T} \{L(x) + L(y) - R(x, y)\} \tag{13.3}$$

$$L(v) = \begin{cases} L'(v) - \alpha, & \text{if } v \in S \\ L'(v) + \alpha, & \text{if } v \in Y \text{ has been considered} \\ & \text{before for } S, \text{ i.e., } \{y \in Y \mid \\ & \exists m(x, y) = 1 \text{ where } x \in X\} \\ L'(v), & \text{otherwise} \end{cases} \tag{13.4}$$

$$m(x, y) = \begin{cases} 0 & L(x) + L(y) = R(x, y) \\ \text{``\_''} & \text{otherwise} \end{cases} \tag{13.5}$$

When there are more patients than available resources, the original Hungarian algorithm cannot give a solution since the result is obviously not a perfect matching. When all the resources (say y) are allocated, there are still requesters that remain unsaturated (say x). Therefore, in our algorithm, we add a virtual resource hospital beyond the real ones. By setting the cost to be a very big number for each patient to reach along with infinite capacities, our algorithm will converge and assure an optimization for all allocated y. As indicated in Equation (13.6), there are two hospitals and two patients; we add virtual hospital c and a cost of 99 time units for each patient. Assume the initial availability at each hospital a and b is (1, 0) without the virtual hospital, the matching process would be as shown in Figure 13.6. Without the virtual hospital c, the final assignment from the extended Hungarian

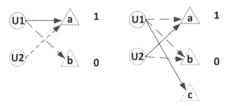

Figure 13.6   Illustration of virtual hospital.

algorithm would be as the left with figure in Figure 13.6 with a cost of 2 ($U1$ to a), while our algorithm will stop as the right figure in Figure 13.6 with a cost of 1 ($U2$ to a).

$$CostMatrix = \begin{matrix} & a & b & c \\ U1 \\ U2 \end{matrix} \begin{pmatrix} 2 & 3 & 99 \\ 1 & 2 & 99 \end{pmatrix} \tag{13.6}$$

Theorem 1 *The bipartite matching achieved with* Algorithm 13.1 *is optimal on total transportation time in R when each R_k is accurate.*

Proof. Algorithm 13.1 is derived from the Hungarian algorithm [10]; first, with multiple capacity, we can still guarantee optimality since in phase 4 every possible available capacity is considered and the multiple z assigned with y will be added to the augmenting tree properly. According to Def. 2, if the estimation of R_k is accurate, the total cost will be minimized. As we add a virtual node, the algorithm will finally stop at phase 2, no matter whether the resource is adequate or not. Therefore, we have this statement proven.

13.4.3 Parameter Formulation

The waiting capacity of a hospital, i.e., w_k, is very important to serious patients, since they could make sure that they would have beds after waiting and get some basic treatment while waiting. It takes extra traffic time (may be greater than the delay cost of waiting) to get to the next hospital. The parameter w_k is related to the hospital-leaving rate and the time to the furthest hospital. We adopt Poisson distribution [19] and find λ using the statistics. Since the average inpatient hours d_k is in Poisson distribution [14] as $P(k, \lambda) = \frac{\lambda^k e^{-k}}{k!}$ and normally most hospitals are running in a high occupancy, we could get w_k as below in Equation (13.7):

$$\sum_{i=w}^{\infty} P\left(i, \frac{C_k}{d_k} \times Min\{T_k\}\right) > 0.8 \tag{13.7}$$

where $Min\{T_k\}$ denotes the distance (represented by time) of the nearest hospital to hospital y_k. C_k and d_k denote the capacity and average inpatient time of y_k, respectively. Namely, we hope the total possibility of leaving w_k patients in $Min\{T_k\}$ time is larger than 80%. The larger the $Min\{T_k\}$ is, the larger the w_k is. It is normal that patients would want to wait at the closer hospital, as opposed to going to the further one.

The reservation number of beds R_k at hospital y_k would greatly affect the performance of the framework. Cared patients can only use rest capacity excluding R, and critical patients will first use R and then the rest of the capacity. In our chapter, we suggest a method that calculates R_k in a less complex way. Since the reservation depends on the distribution of inpatient possibility, a concrete method could be adopted according to a different realistic model. Here, we assume that the number of critical patients in a region is always smaller than the capacity of a premium hospital.

Most diseases have their unique distribution and disciplinarian. For any single illness, ARIMA [23] can be used to estimate the happening rate, or Markov process [24]. However, the inpatient requirements are also from incidents such as a car crash, fire, and alcohol poisoning. These incidents vary, but will remain stable in a period of time. Illness and incidents, altogether, will make orderliness along the timescale, but will be smooth in a divided period of time. The goal is to try to match the reservation beds with the critical situations along a month scale.

The idea is simple, and if we observe a waste of additional reservation beds, we cut the budget. If we continually encounter insufficient reservations, we increase the budget. The method is not as accurate as the ARIMA or Markov process, but it requires less training and history data, as well as a great reduction of computational complexity, while the accuracy is very acceptable. We record the last three datasets as history to estimate the future. Accuracy could be improved with a higher sample rate and more history data. For example, let r_i denote the ith amount of inpatient critical patients, then the estimation of next time slot's reservation could be calculated as shown in Equation (13.8)

$$R_{i+1} = r_i + \frac{\alpha \times (r_i - r_{i-1}) + \beta \times (r_{i-1} - r_{i-2})}{2} + \gamma \qquad (13.8)$$

Here, α, β, and γ are all constant coefficients where $\alpha + \beta = 1$. α and β are the weights of the history data, where they are set to 0.6 and 0.4 in our experiment. γ is a compensatory factor, which is set to 0 in our experiment.

13.5 Experimental Evaluation and Scenario Overview

We evaluate our algorithm using simulations, but the data are derived from real statistics [20]. Here is the simulation parameter setup: a 4×4 grid (adjacent grid's distance is 5 min) with 4 hospitals (2 premium and 2 primitive), each having 120 beds. The average in-hospital time is 24 h (shrink pro rata according to 10 days, on average, in practice [11]). We use 15 min as a schedule time slot. We use Poisson distribution function to generate x patients including critical, serious, and non-critical every 15 min and distribute them to 16 grids randomly. Generate dataset and record the cost matrix. The patient number along the timescale is shown in Figure 13.7. Here, we use unit number rather than real number of patients. Therefore, one unit of patients could be hundreds of patients in a real scenario.

The first competitor is a local heuristic method denoted by $FCFS$ [12]. Whenever there is a new request, it will be assigned to the nearest appropriate hospital, i.e., a critical patient will be sent to a nearest premium hospital, while a non-critical patient will be sent to any nearest hospital with vacancy. The second competitor is a basic bipartite matching method [10] denoted by $K - M$ which divides time into slots and uses bipartite matching to achieve global optimization. The third method is $FCFS$ with preservation, denoted as $FCFS - res$. The fourth method is our method which uses a critical safe reservation on top of $K - M$, denoted by $HAEP$. Apply $FCFS$, $K - M$, and our algorithm $HAEP$ onto the same dataset, run simulations for 24 h and repeat 10 times. If any of the algorithms could not allocate a critical patient, increase failure of assignment by 1 and add the maximum cost of the grid. Every 15 min, generate patients and try to assign them in 4 hospitals. To lever the different emergency situations, we set the coefficients as 1:1:3 and 2:1:3 for critical, serious, and non-critical patients, respectively, as regular and peak days.

Figure 13.7 shows the dynamic of patient number in each round. The difference is that the number of critical patients rises one time in Figure 13.7b against Figure 13.7a. All the dynamics abide by the λ of the Poisson distribution.

From Figure 13.8, we could compare the round waiting time for 4 methods. The costs vary since there are different patient requests in each round. $FCFS$ will cost more than $K - M$, $FCFS - res$, and $HAEP$. $K - M$ will be better than $FCFS$, but not as good as $FCFS - res$ and $HAEP$. They will give more benefit to serious and critical patients so that some non-critical patients will be affected. $HAEP$ is obviously the best.

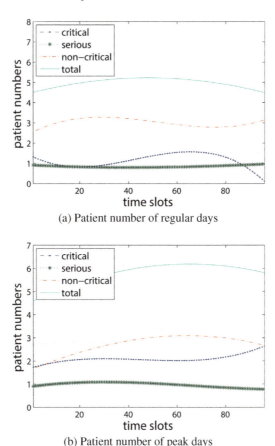

(a) Patient number of regular days

(b) Patient number of peak days

Figure 13.7 Patient number varies along time.

We repeat the test 10 times so that we get more visual facts. The total cost of four methods is shown in Figure 13.9. $HAEP$, denoted by a light blue line, is about 25% less than $K - M$, which implies that we both have a good total cost and better care for critical patients. $FCFS - res$ also has preservation, so the total cost is a little bit lower than $K - M$. The more the critical patients, the better the performance of our method $HAEP$.

As the distribution of patients in each hospital, shown in Figure 13.10, $FCFS$ has nearly average distribution due to the fact that it is only based on distance. $FCFS - res$ and $HAEP$ consider the requirements from critical patients so that premium hospitals a and d get more patients in. $HAEP$ uses the reservation so that some non-critical patients will go further, and primitive

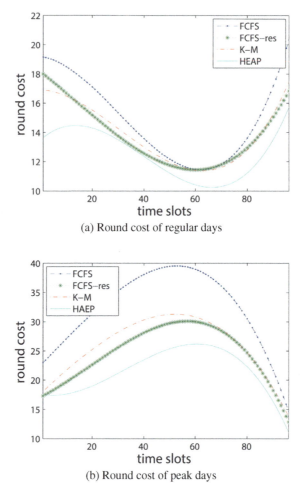

(a) Round cost of regular days

(b) Round cost of peak days

Figure 13.8   Cost for each round.

hospital will get more patients. However, the distribution also relies on the patients' distribution in grids.

When there are not enough resources (beds), there are always some patients that cannot be assigned to a hospital. We compare the total mismatch times of four algorithms, as shown in Figure 13.11. As for $HAEP$, critical patients are the first to consider; therefore, some non-critical patients maybe become mismatched. That is why the improvement is not very obvious. However, if we look into the details, $HAEP$ will have more critical patients scheduled than the other two methods.

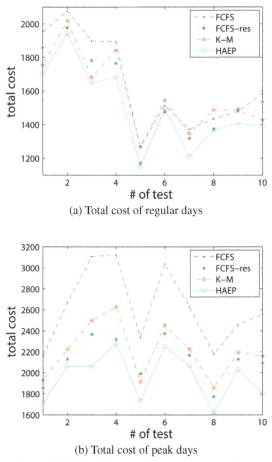

(a) Total cost of regular days

(b) Total cost of peak days

Figure 13.9 Total cost in different number of test.

We also consider the failure of assignment of critical and serious patients. This is because the incoming of patients is based on the possibility that there could be times that the total resource cannot meet the requirement. Under this situation, we compare the total number of failure of assignment of four methods as shown in Figure 13.12. $HAEP$ definitely gets the minimum failure of assignment.

We also study the impact on our method using different reservation percentages. From Figure 13.13, we illustrate the unsuccessful assignment and total cost change according to preservation adjustment. The experiment is based on one extremely completive test instance of 96 rounds. We name

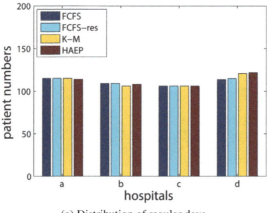

(a) Distribution of regular days

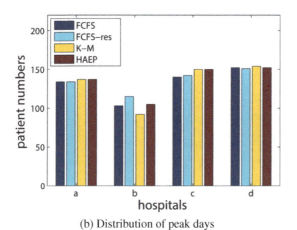

(b) Distribution of peak days

Figure 13.10 Distribution of patients to hospitals.

our choice as 100% and decrease/increase accordingly. The unsuccessful assignments of critical patients decrease with the increase in preservation beds, but the unsuccessful assignments of all patients are increasing in Figure 13.13a. Meanwhile, the total cost increases very fast when the preservation is added beyond our value, as shown in Figure 13.13b. It is obviously a trade-off between the total cost and benefit of critical patients to find the best point of preservation size.

Our observations are summarized as follows: (1) From Figure 13.9, our results show that the *HAEP* always outperforms *FCFS*, $FCFS - res$, and $K - M$, no matter in regular days or peak days. During regular days, the critical

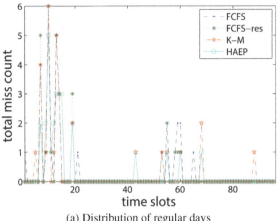

(a) Distribution of regular days

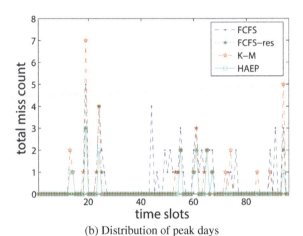

(b) Distribution of peak days

Figure 13.11 Distribution of patients to hospitals.

patients are not many, so that the competition is not serious. *FCFS* has the worst performance, while *FCFS − res* and *K − M* are quite similar. During the peak days, consider the ratio: *HAEP* is 30% better than *FCFS* and 16% better than *K − M* and 10% better than *FCFS − res*. *FCFS − res* is rated as the second best which means that critical patients have been well taken care of and comprise a good portion of the total performance. (2) From Figure 13.12, regarding the death rate, our algorithm *HAEP* has a very low failure of assignment over the other three algorithms. The *FCFS* is the worst algorithm, since it does not consider the future situation of critical patients at all.

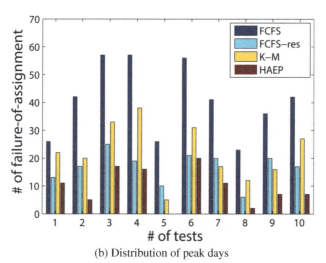

(a) Distribution of regular days

(b) Distribution of peak days

Figure 13.12   Distribution of patients to hospitals.

$FCFS - res$ considers the requirement of critical patients, so it has a good performance as well. (3) Figure 13.10 shows the distribution of assigned patients. Since only successfully assigned patients will be calculated, the figure shows that *HAEP* has the best utilization of hospital beds, while *FCFS* has the worst. $K - M$ comes as the second, since it has the local optimal for each round.

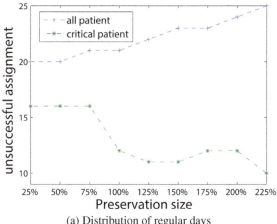

(a) Distribution of regular days

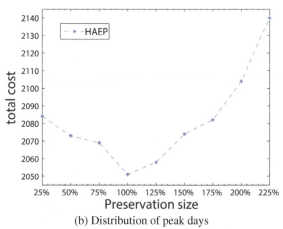

(b) Distribution of peak days

Figure 13.13 Evaluation of different preservation size.

13.6 Charger Assignment in MANETs

In mobile ad hoc networks, most mobile nodes such as smartphones, sensor nodes, and UAVs are powered by batteries. To ensure their proper working conditions, constant charging is necessary. Wireless charging is a promising technology [25], where a power cord is not needed anymore between the charger and the receiver, so that it is very convenient and easy to use for mobile nodes. The technology has many advantages such as no requirement of line of sight, supporting multiple mobile devices, and no additional alignment needed. The main challenge is how to satisfy all user's requirements while keeping the cost low.

Due to the decrease in fast energy transfer efficiency [26], a wireless charge can only cover a small range of area. Thus, it is costly to deploy ample chargers in the whole area. Charging sequence arrangement [27], which schedules a mobile node to a serial of chargers at the specified time and location, is also not feasible, since it is impossible to gather all nodes' trajectory information not mentioning that many of which are stochastic. Wireless charging vehicle [28, 29] will not work in our scenario because the trajectory planning needs the location information for each node which is not trivial in MANETs. In most cases, a mobile node needs to move to the charger itself when it feels necessary. However, since each node has its own planned trajectory and energy consumption rate, it is not possible to schedule a node to a charger in advance. From the viewpoint of the node itself, the desired situation is that a nearest charger will be ready when it raises a request of energy replenishment. Therefore, additional movement could be minimized. Unlike the above studies, in this chapter, the wireless chargers are already installed in the area with respect to efficiency and cost. The main goal is to schedule a mobile node to a charger as close as possible when it posts an energy replenishment request so that the average delay (moving duration) and cost (moving distance) can be minimized.

13.6.1 Charger Assignment Problem in MANETs

Similar to the hospital assignment problem, there are resources and requestors, and we need to make the best match between them. Since the residual energy of each mobile node tends to be quite different when they post a request, we can also group them into three categories which are critical, serious, and mild, respectively. The energy depletes very quickly in some devices so that they need to find a charger as soon as possible. We call them critical nodes in that case. Some nodes would like to remain their residual energy above a threshold for the sake of safety. They will raise charging request more frequently. But, actually, they can bear some time before being charged. We call these nodes mild nodes. All other nodes between critical and mild could be regarded as serious. They can hold for some time but not too long. Otherwise, they will also shut down due to flat energy. Regarding chargers, we consider they are identical in functioning so that the background is different from the one in the scenario of hospital assignment. Below, we will surmise the feature of the scenario.

13.6.1.1 Capacity of chargers

Apparently, considering the energy storage and radiation capability, each charger will have a charging capacity, denoted by C_y. We assume that each energy receiver possesses same portion of resources so that C_y can be explained as the number of mobile nodes that the charger y can support at a time. For simplicity, C_y is a positive integer.

13.6.1.2 Effective charging distance

Theoretically, a charger can transfer energy to a receiver through wireless means within some range. However, the energy transfer efficiency drops quickly when the distance increases [25]. Therefore, to ensure the reception power at a receiver is larger than a threshold so that it is large enough to charge the node, mobile nodes requiring charging must come into the effective charging distance of a charger denoted by D. Because the size of the D will not affect the result of the assignment, we set the D as 0.

13.6.1.3 Mobility of chargers

In the first scenario, all hospitals are stationary, while in this scenario, the chargers can be small in size so that they are possible to be mounted on public transporters such as trains, buses, and ferries. Therefore, the mobility of chargers must be considered. However, the mobility is relative which does not mean a charger will move toward a receiver in purpose. Let us consider a case that a user is waiting in a subway at the platform and the train is coming. We should not assign the user to the charger in the platform, but to the one in the coach. We group chargers into two groups: One is moving with the mobile nodes (denoted by H^M), and the other is not (denoted by H^S). Obviously, when a user is in the coach, it should not be assigned with the charger located in the platform. Then, we can model the scenario in the similar form with the previous one. Here, we adopt a simple model where we do not consider the dynamic distance changing between the charger and the node. Otherwise, the speed and direction should be involved. The dynamic changing of distance will only affect the weight, and it will be considered in the future work.

13.6.1.4 Charging duration

The algorithm will end in the first scenario when a patient is successfully assigned to a hospital. In this scenario, due to node mobility, such as a passenger riding in a subway, a node may only stay with a charger for a short of time. During that short time duration, the job of charging may not be finished. Therefore, a node may be scheduled to multiple chargers until it is satisfied with the residual energy. We divide the time into T equal time slots indexed by $1, 2, \ldots, T$, and each time slot is denoted by U_t. Note that the battery capacity and energy consumption rates of different nodes may significantly vary, due to different hardware configuration and behaviors, and we set a charger duration of U_x for each node x where U_x is an integer multiple of time slot U_t. Also, for some nodes, the charging duration could be shorter than U_t.

13.6.1.5 Appearance of charging request

A charging request could appear anywhere and anytime. It could be critical, serious, and mild as well, just like the patient in a big city. The appearance pattern is similar so that we model the appearance of charging request with Poisson distribution.

13.6.1.6 Local waiting queue

When the delay to a further charger is greater than the expected delay of waiting locally, a node will be assigned to the waiting queue of the local charger.

$$\sum_{i=w}^{\infty} P\left(i, \frac{C_k}{d_k} \times Min\{T_k\}\right) > 0.9 \tag{13.9}$$

where $Min\{T_k\}$ denotes the distance (represented by time) of the nearest charger to the charger y_k. C_k and d_k denote the capacity and average charging time of y_k, respectively. Namely, we hope the total possibility of leaving w_k nodes in $Min\{T_k\}$ time is larger than 90%. Otherwise, it is better for the node going to a further charger. We use parameter 90% which is stricter than that in hospital assignment. The larger the $Min\{T_k\}$ is, the larger the w_k is. It is normal that nodes would want to wait at the closer charger, as opposed to going to the further one.

13.6.1.7 Reservation

The same contradiction exists in this scenario. For example, when a critical node posts a charging request, the nearest charger is all occupied. And the residual energy will run out before it reaches the further charger. Therefore, we need to hold some resources as the reservation for critical nodes. The calculation of the reservation amount can also use Equation (13.8).

13.6.2 Bipartite Matching-Based Algorithm

From the above discussions, we will formulate the charger assignment problem as follows:

$$\max \sum_{x \in X} \sum_{y \in Y} \sum_{0 \leq t < \infty} R(x, y) m(x, y)$$

$$\text{s.t.} \quad \text{every } m(x, y) = 0, 1, \text{ or } \text{``}-\text{''} \qquad \text{(i)}$$

$$\sum_{y \in Y} m(x, y) = 1 \quad \text{for every } x \in X \qquad \text{(ii)}$$

$$\sum_{x \in X} m^t(x, y) \leq C_y^t \quad \text{for every } y \in Y \text{ and any } t \qquad \text{(iii)}$$

where x denotes the mobile nodes, y denotes the chargers, and "$-$" indicates an initial or unreachable status. We use constraint (i) to ensure the charger assignment as a bipartite matching. Constraint (ii) does not allow a mobile node being charged by more than one charger at the same time. Constraint (iii) ensures that the total number of charging mobile nodes does not exceed the limitation. It is not surprise that two different scenarios could be modeled as a same problem. Therefore, we can also use the Algorithm 13.1 derived from the first scenario to solve this problem. Here, we will skip the steps of describing since it has been illustrated thoroughly in the first part of this chapter.

13.6.3 Results Analysis and Discussion

The scenario and parameter are different in two systems, but the key problem is the same, i.e., the optimization of the resources. When each requestor can only be assigned to one resource and has different weights for each resource, we could consider using the bipartite matching to solve it. In the hospital assignment, the patient is the requestor and the hospital is the resource, while in charger assignment, the mobile node is the requestor and the charger is the resource. If the total number of requestors is not too big and central scheduling is available, the algorithm is compatible with respect to cost and efficiency. Resource assignment is a common problem in communication and network applications, and in this chapter, we propose solution extended from the traditional bipartite matching which considers the mobility and time involved of the system. It is an online algorithm with less proactive knowledge requirement, and the search space could be limited to subarea so as to bring down the time and space complexity.

13.7 Conclusions

In this chapter, we consider the assignment problem between requesters and resources in MANETs. The resource can act either as a data storage or as a provider of energy. Two scenarios are discussed, i.e., hospital assignment and energy charging. In both scenarios, there are multiple types of resources and requesters which introduce more strict assignment constraint. Also, time-involving facts increase the difficulty of the problem. We propose a novel resource assignment to minimize the average delay of requesting nodes, as well as the amount of failure of assignment for critical nodes in the system, denoted by *HAEP*. The framework is composed of three components, i.e., service center, mobile nodes, and resource providers. To avoid the disadvantage of local competition, resource providers will submit their occupancy status, while mobile nodes call the service center to declare their requirements so that the schedule could be done in the service center in a global view manner. The solution is built on the Hungarian algorithm, with the prediction and timescale, applied on a multidimension resource and requesters. The simulation shows that our work is very efficient compared with the three other usual methods.

Acknowledgment

This work is supported by the Scientific Research Foundation for the Returned Overseas Chinese Scholars, State Education Ministry of China, and Chinese Scholarship Council (201208330096). This work is also supported in part by NSF grants CNS 149860, CNS 1461932, CNS 1460971, CNS 1439672, CNS 1301774, ECCS 1231461, ECCS 1128209, and CNS 1138963.

References

[1] Li, Y. (2011). The application of improved early pre-consultation assessment in emergency treatment. *Guide China Med.* 9(8), 126–127.
[2] ChinaMobile. (2012). You have to be lucky to find a hospital that has room. ONLINE, http://labs.chinamobile.com/news/iot/86608, 2012.
[3] Times of India. (2014). Hospitals lack enoughbeds, admits government. ONLINE, http://timesofindia.indiatimes.com/city/patna/Hospitals-lack-enoughbeds-admits-government/articleshow/37798085.cms, 2014.
[4] Yao, S., zeng, Q., Peng, M., Ren, S., Chen, G., and Wang, J. (2014). Stop violence against medical workers in china. *J. Thoracic Dis.* 6(6), 141–145.

[5] Firstpost. (2012). Footballer in India dies due to lack of hospital beds. ONLINE, http://www.firstpost.com/sports/footballer-in-india-dies-due-to-lack-of-hospital-beds-372272.html, 2012.

[6] Hexun. (2012). Nearly 1000 calls perday for "120" emergency call. ONLINE, http://news.hexun.com/2012-12-14/149027241.html

[7] Krug, S., Siracusa, M., Schellenberg, S., Begerow, P., Seitz, J., Finke, T., and Schroeder, J. (2014). Movement patterns for mobile networks in disaster scenarios. In *WoWMoM*. IEEE, 2014, pp. 1–6.

[8] Chen, C.-Y., Chen, P.-Y., and Chen, W.-T. (2013). A novel emergency vehicle dispatching system. In *VTC Spring*. IEEE, pp. 1–5.

[9] Bondy, J., and Murty, U. (1976). *Graph theory with applications*, 1st ed. Elsevier Science Publishing Co. Inc.

[10] Xray. Assignment problem and Hungarian algorithm. Topcoder, https://www.topcoder.com/community/data-science/data-science-tutorials/assignment-problem-and-hungarian-algorithm/

[11] Jin, W. (2010). Analysis of in-patient difference between downtown and suburban in shanghai. *China Medical Herald*, 7(6), 150–151.

[12] Federgruen, A., and Groenevelt, H. (1986). The greedy procedure for resource allocation problems: necessary and sufficient conditions for optimality. *Operat. Res.,* 34(6), 909–918.

[13] Xu, M., Wong, T., Wong, S., Chin, K., Tsui, K., and Hsia, R. (2013). Delays in service for non-emergent patients due to arrival of emergent patients in the emergency department: a case study in Hong Kong. *J. Emerg. Med.,* 45(2), 271–280.

[14] Beraldi, P., and Bruni, M. E. (2009). A probabilistic model applied to emergency service vehicle location. *Eur. J. Operat. Res.,* 196, no. 1, pp. 323–331.

[15] Alsalloum, O. I., and Rand, G. K. (2006). Extensions to emergency vehicle location models. *Computers & OR*, 33, 2725–2743.

[16] Luo, J., Wang, J., and Yu, H. (2011). A dynamic vehicle routing problem for medical supplies in large-scale emergencies. In *2011 6th IEEE joint international information technology and artificial intelligence conference (ITAIC)*. IEEE, pp. 271–275.

[17] Senart, A., Bouroche, M., and Cahill, V. (2008). Modelling an emergency vehicle early-warning system using real-time feedback. *Int. J. Intell. Informat. Database Syst. (IJIIDS),* Special Issue on Information Processing in Intelligent Vehicles and Road Applications, 2(2), 222–239.

[18] Noori, H. (2013). Modeling the impact of vanet-enabled traffic lights control on the response time of emergency vehicles in realistic large-scale urban area. In *ICC. IEEE Computer Society*, pp. 526–531.

[19] Cardoen, B., Demeulemeester, E., and Beliën, J. (2010). Operating room planning and scheduling: A literature review. *Eur. J. Operat. Res.*, 201(3), 921–932.

[20] Lowery, J. C. (1996). Design of hospital admissions scheduling system using simulation. In *Winter Simulation Conference*. ACM, pp. 1199–1204.

[21] Reinus, W. R., Enyan, A., Flanagan, P., Pim, B., Sallee, D. S., and Segrist, J. (2000). A proposed scheduling model to improve use of computed tomography facilities. *J. Med. Syst.* 24(2), 61–76.

[22] Min, D., and Yih, Y. (2010). An elective surgery scheduling problem considering patient priority. *Computers & OR*, 37(6), 1091–1099.

[23] Lu, S., Ju, K., and Chon, K. H. (2001). A new algorithm for linear and nonlinear ARMA model parameter estimation using affine geometry [and application to blood flow/pressure data]. *IEEE Trans. Biomed. Eng.* 48(10), 1116–1124.

[24] Huang, Y. (2006). Using Markov model to predict the incidence of tuberculosis in taiwan for the next decade. Master's thesis, National Yang-Ming University.

[25] Kurs, A., Karalis, A., Moffatt, R., Joannopoulos, J., Fisher, P., and Solijacic, M. (2007). Wireless power transfer via strongly coupled magnetic resonances. *Science*, 317 (5834), 83–86.

[26] Xie, L., Shi, Y., Hou, Y., Lou, W., Sherali, H., and Midkiff, S. (2014). Multi-node wireless energy charging in sensor networks. *IEEE/ACM Transactions on Networks*.

[27] Xu, W., Liang, W., Hu, S., Lin, X., and Peng, J. (2015). Charging your smartphones on public commuters via wireless energy transfer. In *IPCCC. IEEE Computer Society*, pp. 1–8.

[28] Xu, W., Liang, W., Lin, X., Mao, G., and Ren, X. (2014). Towards perpetual sensor networks via deploying multiple mobile wireless chargers. In *ICPP*, pp. 80–89.

[29] Liang, W., Xu, W., Ren, X., Jia, X., and Lin, X. (2014). Maintaining sensor networks perpetually via wireless recharging mobile vehicles. In *LCN*. IEEE, pp. 270–278.

PART V

Multimedia Networks

14

User Experience Awareness Network Optimization for Video Streaming Based Applications

Hengky Susanto, ByungGuk Kim and Benyuan Liu

Computer Science Department, University of Massachusetts Lowell, Lowell, MA 01852, USA

Abstract

The growing popularity of video-based applications has put an enormous strain on the network and causes the network to be more prone to congestion. The study of congestion control and bandwidth allocation problems is often formulated into Network Utility Maximization (NUM) framework, and the existing solutions for NUM generally focus on single-layered applications. However, today's quality of video is often divided into several layers, where each layer provides a different level of quality enhancement. In this chapter, we study how multi-layered video-based applications impact network performance and pricing through NUM formulation, particularly traffic from video streaming. In our investigation, we design and implement a new multi-layered user utility model that leverages on studies of human visual perception. Then, using this new utility model to examine network activities, we demonstrate that solving NUM with multi-layered utility is intractable, and that rate allocation and network pricing may oscillate due to user behavior specific to multi-layered applications. To address this, we propose a new approach for admission control to ensure quality of service and experience. Our simulation results show that the proposed admission control mechanism can effectively eliminate the oscillation behavior and mitigate link bottleneck.[1]

---

[1]The conference version of this chapter [41] was accepted by the 24th IEEE International Conference on Computer Communications and Networks (IEEE ICCCN-2015).

Keywords: Congestion control, Real time, Networks multimedia.

14.1 Introduction

The explosive growth and popularity of video streaming-based applications has put immense pressure on network requirements and performance. The study of managing network congestion is commonly formulated into Network Utility Maximization (NUM) framework [1, 2]. The objective is to allocate bandwidth that maximizes total user utility, subject to network capacity constraints. Existing approaches usually model user satisfaction according to the network traffic elasticity of single-layered applications, as discussed in these literatures [1–6]. However, such model is insufficient to capture the characteristics of applications with multiple layers of quality, such as video streaming. This is because today's video streaming is divided into several layers, with lower layers containing low-resolution information and higher layers containing the fine information. The premise is that each layer is delivered separately, such that the vital layers may thus be transferred with guaranteed quality of service (QoS), while other layers that enhance the quality of the video could be sent as best effort. These characteristics allow users to be adaptive with their demand for quality of experience (QoE) and QoS, which is not reflected in single-layered applications. Therefore, to capture the characteristic of multi-layered video-based application, in our investigation we implement and design user utility function for multi-layered applications that incorporates studies from computer graphics [7–9]. Using this multi-layered user utility function, we study how multi-layered applications may dynamically impact the network traffic and pricing under the NUM framework, particularly for video streaming-based applications.

To better address the particularity of user utility function for multimedia applications, we design our utility function to reflect the characteristics of multi-layered encoding schemes, which is often used in video-based applications. That is, the level of user utility is just measured not only by the ability to meet the minimum required QoS, but also by the varying degrees of qualities associated with each encoding layer. The proposed user utility is guided by studies of human visual perceptions in the fields of computer graphics to ensure the accuracy of modeling the user's experience aspect. In essence, we derive three important insights: the unique adaptive nature of multimedia applications, users are willing to tolerate some level of disruption for the sake of better image quality, and human ability to detect improvement in image quality is not infinite; i.e., it reaches a certain point where human eyes are no longer able to detect further improvement in image quality. These insights contribute to our *staircase*-shaped user utility function that follows the law of diminishing returns. This utility function also illustrates that users may have different levels of QoE, which means users can be adaptive with their demand to achieve the desired QoE. Hence, we propose a model to encapsulate this user's demand adaptability to achieve the desired QoE. Furthermore, the model also considers the impact of user's willingness to pay (budget) for the desired QoE. We then use these models to investigate network activities by incorporating the multi-layered utility function into NUM framework. Our results show that the algorithm used to resolve NUM may not converge when users are actively seeking to meet their desired QoE, resulting in oscillation in bandwidth allocation and network price. Furthermore, the oscillation also causes frequent quality adaptation at the video application level and creates a visual *flickering effect*, which degrades QoE and users find the effect annoying [10]. Moreover, our results also show that the oscillation ripples to different parts of the network and causes bandwidth allocation to other users to oscillate too. This makes solving NUM problem with multi-layered user utility to become intractable; therefore, it may not have an optimal solution. To resolve this problem, we propose greedy-based network admission control to ensure that acceptable QoE and QoS are achieved. We also provide mathematical proofs to show that the proposed admission control scheme achieves convergence and prevents oscillation. Moreover, we also show that our solution also has characteristics of Nash Equilibrium when users have minimum to gain by changing their own bandwidth demand. Following a discussion of our proposed solution, we will address how the proposed solution can be implemented in practice and how it will benefit network traffic engineering.

This chapter is organized as follows. We begin by discussing previous related work. Following this, we present our major contributions: the proposed multi-layered user utility model and a discussion on how the new model impacts network activities in Sections 14.3 and 14.4, respectively. We introduce admission control in Section 14.5. The simulation results and discussion are presented in Sections 14.6 and 14.7, respectively, followed by concluding remarks.

14.2 Related Work

A number of proposals address NUM problems for multi-layered applications by incorporating rate video distortion minimization into user utility function. In [11], the authors' objective is to allocate bandwidth that minimizes video rate distortion, where lower rate of distortion increases higher user satisfaction. Likewise, in [12], peak signal-to-noise ratio (PNSR) is incorporated into user utility function in their NUM formulation. In [13], the authors develop network coding-based utility maximization model by integrating a taxation-based incentive mechanism to address how many layers each user should receive and how to deliver them. Fundamentally, these proposals follow the convention of single-layer utility function. That is, the layer used in the solution is predetermined and not adaptive, even though it is designed to support multi-layered applications.

Others propose to adopt a staircase shape or stepwise utility function. In [11], a staircase shape rate distortion-based utility function is incorporated into NUM frameworks in multicast network environment. Different weight factors are assigned to utility function, corresponding to the quality of the layer. The authors of [14] suggest an approach that connects different points in the staircase to produce a concave shape, which in turn provides some approximate rate change. However, it is still difficult to fine-tune the variables to achieve a tight approximation. Similarly, in [15], staircase shape utility function is approximated using the log function. The idea is that every level in the staircase is represented by a concave-shaped utility function and the authors assume the optimal solution falls within one of the steps of the staircase. In both [14, 15], the encoding layer is determined after bandwidth is allocated and user utility is measured according to the allocated bandwidth. In contrast, in our model, user utility is measured according to the quality of experience user has with the image quality, rather than plainly based on the amount of bandwidth allocated. In addition, our model considers a critical characteristic of multi-layered applications—the adaptive nature of such applications which adjusts its quality according to the traffic load in order to deliver the best QoE possible to users.

14.3 Multi-Layered User Utility Function

14.3.1 Foundations

Multi-layered user utility function is modeled according to layered encoding scheme to reflect how multimedia traffic is managed at the network and at the application level. This means video information is divided into several encoding layers in order to minimize the amount of bandwidth used, with lower layers containing low-resolution information and higher layers finer quality ones [16]. This strategy allows the video provider to provide a range in video qualities. This strategy is also known as hierarchical coding. Network forwards only the number of layers that the physical link can support and drops layers selectively at the bottleneck link. Thus, the hierarchical video encoding provides the foundation of staircase-like user utility function as illustrated in Figure 14.1.

Then, we incorporate knowledge of how human eyes perceive and evaluate image qualities in our utility model because QoE is significantly affected by our visual ability. A study in [17] shows as video is encoded with more layers, discerning differences in quality between images at similar levels of quality becomes progressively difficult as the quality improves. This is because human visual capacity is less sensitive to high-frequency image details and more sensitive to lower frequency [8, 9, 16]. Thus, we assert that human visual perception actually follows the law of diminishing returns; i.e., the benefits of offering higher quality diminishes as quality improves. Our user utility function incorporates these factors in the QoE, demonstrated by the progressively decreasing heights of the steps in the staircase-shaped function. In addition, supporting studies show that degradation in video is even less noticeable when there is a high degree of motion in the imagery [18], such as in action movies or sports videos, and especially when images are combined with good audio quality [17].

14.3.2 User Utility Function

By incorporating layered encoding scheme and limitations of human's visual perception, our user utility function for multi-layered applications is modeled as follows. Since the staircase in the user utility model obeys the law of diminishing returns, the height of the steps progressively flattens toward

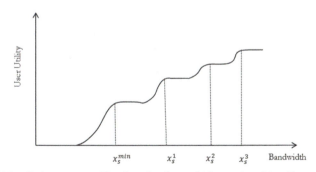

Figure 14.1 Staircase user utility function for multi-layered multimedia applications.

the maximum quality, similar to an upward moving escalator. Let Y be a set of video encoding layers, $Y = \{1, 2, 3, \ldots, |Y|\}$, and $y \in Y$ when $y = 1$, y is the lowest encoding layer, and when $y = |Y|$, y is the highest encoding layer with the maximum quality. Let x_s be the amount of bandwidth allocated to user s and x_s^y denotes the amount of bandwidth needed by user s to support the application quality at layer y_i, such that $x_s \geq x_s^y$, which means network allocates at least x_s^y to support the quality at layer y. Furthermore, let $U_{bw}(x_s^y)$ denote the user utility function of user s for the amount of bandwidth x_s at layer y, the relationship between user utility of a given layer can be illustrated as

$$\lim_{y \to \infty} U_{bw}(x_s^y) - U_{bw}(x_s^{y-1}) = 0.$$

Subsequently, by incorporating multi-layers of encoding into the user utility function, function $U_{bw}(x_s)$ is defined as follows:

$$U_{bw}(x_s^y) = \sum_{i=0}^{|Y|}(U_s^y(x_s^y)\,\psi_U(x_s^y)), \tag{14.1}$$

where $U_s^y(x_s^y)$ denotes user utility function given the required amount of bandwidth x_s^y to support the quality at layer y and $\psi_U(x_s^y)$ denotes the decay factor function at each layer and a good candidate function is $\psi_U(x_s^y) = e^{-\omega_y\,x_s^y}$, where ω_y is a positive variable used for normalization. Moreover, to ensure the minimum bandwidth required for the lowest encoding layer is met, condition $x_s \geq x_s^1$ must be met. Thus, $U_s^y(x_s) = \frac{1}{(1 + e^{-a x_s^y})}$. Next, we introduce $U_{cost}(x_s)$, the user utility function that measures user satisfaction over the cost of the service relative to user's willingness to spend m_s. Thus, $U_{cost}(x_s) = 1 - \frac{x_s \lambda_s}{m_s}$. Therefore, user utility function for multi-layered applications has the following properties:

1. $U_{bw}(x_s^y)$, $U_{cost}(x_s) \geq 0, \forall x_s,\ 0 \leq x_s^y \leq x_s \leq C_l$, and $U_{bw}^y(0) = 0$, $U_{cost}(0) = 1, \forall x_s,\ s$, where C_l is the capacity of link l.
2. $U_{bw}(x_s^y)$ is twice differentiable and follows the law of diminishing returns.
3. $U_{bw}(x_s^y)$ is *staircase-shape* like $\forall x_s, 0 \leq x_s^y \leq x_s, y = \max(Y|x_s^y \leq x_s)$, and $U_{bw}(x_s^y) \leq U_{bw}(x_s)$.
4. $U_{cost}(x_s)$ is linear.
5. $\frac{\delta U_{bw}(x_s^y)}{\delta x_s^y} < \infty$, for all $0 \leq x_s^y \leq x_s \leq c$.
6. $\lim_{x_s \to 0} \frac{\delta U_{bw}(x_s^y)}{\delta x_s^y} < \infty, \forall s,\ s \in S$ and for $x_s^y \leq x_s$.

7. For $\forall y, s, \quad y < y + 1, \quad x_s^y < x_s^{y+1}$, for $s \in S$ and $y \in Y$.

Then, the user utility function for multi-layered application is defined as follows.

$$U_s(x_s, x_s^y) = U_{bw}(x_s) + U_{cost}(x_s), \tag{14.2}$$

such that each user maximizes his/her own utility function by solving

$$\text{maximize} \quad U_s(x_s, x_s^y) \tag{14.3a}$$

$$\text{over} \quad x_s \geq x_s^y \geq 0. \tag{14.3b}$$

How user solves the problem (14.3) above will be discussed in the following section.

14.3.3 System Setup

Consider a network with a set of links L and a set of link capacities C over the links. Given a utility function $U_s(x_s, x_s^y)$ of user s with the allocated bandwidth of x_s, the NUM formulation becomes

$$\max \sum_{s \in S} U_s(x_s, x_s^y), \tag{14.4a}$$

$$\text{s.t.} \sum_{s \in l} x_s \leq C_l, \ \forall l \in L \tag{14.4b}$$

$$layer \ y < layer \ y + 1, \quad \forall y \in Y, \tag{14.4c}$$

$$\text{over} \ 0 \leq x_s^{min} \leq x_s, \quad \forall s \in S$$

where S denotes a set of users, x_s^{min} denotes the minimum bandwidth requirement of user s, and $s \in l$ means user who is transmitting data through link l. The constraint (14.4.c) is necessary to ensure that the higher layer depends on lower layer. In other words, a higher layer is only considered when lower layers are delivered.

Typically, the common solution to NUM is *subgradient-based method* [3], and the dual problem D to the primal problem of (14.4) is constructed as follows. $L(x, \lambda) = \sum_{s \in S} U_s(x_s, x_s^y) - \sum_{s \in S} \lambda_s x_s + \sum_{l \in L} \lambda_l C_l$, where $L(x, \lambda)$ is the Lagrangian form and λ is known as a set of Lagrangian multipliers λ_l, which is often interpreted as the link cost and $\lambda_s = \sum_{l \in r_s} \lambda_l$. The dual problem D is then defined as

$$\min D(\lambda), \quad s.t \ \lambda \geq \bar{0},$$

where the dual function $D(\lambda) = \max_{\overline{0} \leq x \leq x^{max}} L(x, \lambda)$. User decides the transmission rate $x_s(\lambda_s)$ at price λ_s by solving

$$x_s(\lambda_s) = \arg\max_{0 \leq x \leq x^{max}} (U_s(x_s, x_s^y)), \qquad (14.5)$$

where $x_s(\lambda_s)$ denotes bandwidth allocation at price λ_s. A subgradient projection method is used in [3]. Thus, the network on each link l updates λ_l on that link:

$$\lambda_l^{t+1} = \left[\lambda_l^t - \sigma^t \left(C_l - \sum_{s \in l} x_s \right) \right]^+, \qquad (14.6)$$

where $C_l - \sum_{s \in l} x_s$ is a subgradient of problem D for $\lambda_l^t \geq \lambda_{min} \geq 0$. Here, λ_{min} denotes the minimum price determined by the network. The time t denotes the iteration index, for $0 \leq t$, and $\sigma^{(t)}$ denotes the step size to control the trade-off between a convergence guarantee and the convergence speed, such that $\sigma_l^{(t)} \to 0$, as $t \to \infty$ and $\sum_{t=1}^{\infty} \sigma_l^{(t)} = \infty$. Price paid by users is $\lambda_s = \sum_{l \in s} \lambda_l$, where $l \in s$ means the link that is used by s to transmit data.

14.4 Adaptive User Demand

The authors of [19] claim that regardless of the algorithm, deciding the proper quality in dynamic network is essentially difficult because of a lack of information and transparency between OSI layers. Often, user experience at the application layer relies on users to decide and adapt to reach their desired quality. For that reason, we investigate how multi-layered user utility function mirrors user QoE and decision for bandwidth demand, and how these two factors impact network activities.

14.4.1 Adaptive Demand

The staircase like function allows user to be adaptive with their requirement to achieve the desired QoE, as long as the minimum demand is met. However, network does not distinguish between user utility of single- or multi-layered applications. The implication is that the bandwidth allocation which is sufficient for users of single-layer applications may not be sufficient forusers of multi-layered applications. This is because users using multi-layered applications may continue to demand more bandwidths to improve their experience. This is noted in HCI studies in [10] where viewers prefer jerking (or less smooth) video with better image quality over smooth but poor visual quality where objects in the video are not recognizable. Low tolerance for poor image quality may motivate users to demand for more bandwidths for better image quality. In other words, due to the dynamic nature of user demand, providing the minimum bandwidth requirement does not automatically maximize the aggregated user satisfaction level because there are varying levels of user satisfaction in a multi-layered environment. Conversely, a user may stop demanding for additional bandwidth as they consider cost of service, especially when a user is already experiencing a sufficiently high quality of service, when further improvement in image quality cannot be appreciated because our eyes are not able to detect the quality difference.

14.4.2 User's Desire for Better Quality

In order to investigate the impact of the adaptive nature of multi-layered utility on network activities, we first provide a model that encapsulates user's motivation to stay put with or scale up from the current quality. The rationale behind *user's desire for better quality* function $B(x_s, \ x_s^y)$ is described as follows. Intuitively, users are generally assumed to desire the best possible "value" for the money they pay. By this, it means a user may prefer to lower his/her requirement to achieve better perceived value. On the other hand, a user may demand more bandwidth for higher quality when

$$\beta_s \frac{x_s^{y+k}\lambda_s}{m_s} \leq B(x_s, \ x_s^y), \tag{14.7}$$

where β_s is a positive constant variable that indicates user's desire to save or spend money. Lower β_s means *higher* willingness to spend more money for additional bandwidth to achieve better quality at $y + k$. Otherwise, the user may stay put with the current quality at layer y. Additionally, positive variable k denotes the number of layers to be increased, m_s denotes user's budget or willingness to pay for the service, and λ_s is the network price. $(x_s^{y+k}\lambda_s)$ can be interpreted as the cost a user incurs for quality at layer $y + k$. Hence, the ratio $\frac{x_s^{y+k}\lambda_s}{m_s}$ shows that user's desire to for higher layer for better quality can be expressed through increasing m_s. Next, $B(x_s)$ is defined as follows.

$$B(x_s, \ x_s^y) = \frac{U_s^{y+k}(x_s^{y+k}) - U_s^y(x_s^y)}{x_s^{y+k}(\lambda_s + \lambda_s^{inc})}, \tag{14.8}$$

where λ_s^{inc} denotes the *range of price a user is willing to increase* for the desired quality. Equation (14.8) can be interpreted as user satisfaction gained over the cost users incurs from obtaining from additional k layers. By rearranging Equation (14.7), the lower bound for the additional price range λ_s^{inc} is given as

$$\lambda_s^{inc} \geq \left(\frac{m_s(U_s^{y+k}(x_s^{y+k}) - U_s^y(x_s^y))}{\beta_s \, \lambda_s (x_s^{y+k})^2} \right) - \lambda_s. \qquad (14.9)$$

As observed in (14.9), the relationship between β_s and λ_s^{inc} is that lower β_s (higher desire to spend) implies a greater range in the price a user is willing to pay. In other words, Equation (14.9) provides the lower bound for the additional cost a user must spend to achieve quality at layer $y + k$ at that given moment. Subsequently, given x_s^y, user may demand additional bandwidth to achieve higher quality if the new demand bandwidth x_s^{y+k} satisfies (14.7). Thus,

$$x_s = \begin{cases} x_s^{y+k}, & \text{If condition (14.7) is satisfied} \\ x_s, & \text{Otherwise} \end{cases}. \qquad (14.10)$$

That implies user may attempt to transmit data at x_s^{y+k} instead of x_s, for $x_s < x_s^{y+k}$, when the condition allows.

14.4.3 The Impact of Adaptive User Demand

According to the condition in (14.10), when (14.8) is not satisfied, then user must transmit data at x_s and stop demanding additional bandwidth x_s^{y+k}.

Proposition 14.4.1 When $x_s \geq x_s^y$, then $\frac{U_s(x_s, x_s^y)}{U_s^y(x_s^y)} = 1$, as $\lambda_s \to \infty$.

Proof. First, observe that

$$B(x_s, \, x_s^y) = \lim_{\lambda_s \to \infty} \frac{U_s^{y+k}(x_s^{y+k}) - U_s^y(x_s^y)}{x_s^{y+k}(\lambda_s + \lambda_s^{inc})} = 0$$

and $\lim\limits_{\lambda_s \to \infty} \beta_s \frac{x_s^{y+k} \, \lambda_s}{m_s} = \infty$, for $\beta_s > 0$. Since

$$\lim_{\lambda_s \to \infty} \sup \, \beta_s \frac{x_s^{y+k} \, \lambda_s}{m_s} = \infty,$$

notice that $B(x_s) < \infty$ for any λ_s. Hence, as $\lambda_s \to \infty$, $\beta_s \frac{x_s^{y+k} \lambda_s}{m_s} > B(x_s)$. Consequently, when the condition in (14.8) is no longer satisfied, user s

stops demanding additional bandwidth beyond x_s. As a result, user s settles with layer $y = \max(Y \mid x_s^y \leq x_s)$, such that $\lim_{\lambda_s \to \infty} \sup x_s^y = x_s$. Hence, $\frac{U_s(x_s, x_s^y)}{U_s^y(x_s^y)} = 1$, as $\lambda_s \to \infty$.

Proposition 14.4.1 shows that users eventually stop demanding more bandwidth when additional quality is not worth the additional cost. Next, we investigate whether the algorithm with multi-layered user utility function also converges. First, we prove the following statement.

Lemma 14.4.1 *Suppose that λ_l^* is an optimal solution for the dual problem $D(\lambda)$, where $\lambda_l^* > 0$ for link $\forall l$, $l \in L$, such that there exists a subgradient of $D(\lambda)$, $g(\lambda_l^*)$, at λ_l^*, where $g(\lambda_l^*) = 0$.*

Proof. We have λ_l^* as the minimizer of $D(\lambda)$, as $\forall l$, $l \in L$, there exists $g(\lambda_l^*)$ that satisfies

$$g(\lambda^*)^T \mid \lambda - \lambda^* \mid \geq 0, \quad \text{for } \forall \lambda, \ \lambda \geq \overline{0}. \tag{14.11}$$

If we take $\lambda = \lambda'$, where $\mid \lambda_l' - \lambda_l^* \mid = \in_l, \in_l > 0$. By (14.11), we have $g_l(\lambda_l^*) \in_l \geq 0$. Hence, when λ_l^* is the optimal solution, $g_l(\lambda_l^*) = 0$.

We have shown that there exists subgradient $g_l(\lambda_l^*) = 0$, $\forall l, l \in L$, when λ^* is an optimal solution for the dual problem $D(\lambda)$. Next, we investigate whether the algorithm converges with multi-layer user utility function.

Proposition 14.4.2 Suppose that λ_l^* is an *optimal solution* for the dual problem $D(\lambda)$, where $\lambda_l^* > 0$ for link $\forall l$, $l \in L$. If $D(\lambda)$ is differentiable at λ^*, then $x(\lambda^t)$ converges $x(\lambda^*)$ as λ^t converges to λ^*, for $t \to \infty$. Otherwise, $x(\lambda^t)$ and λ^t may not converge.

Proof. Certainly, when dual problem $D(\lambda)$ is differentiable at λ^*, $D(\lambda)$ has an unique subgradient at λ^*. Thus, $x(\lambda^*)$ is also unique. This means $x(\lambda)$ continues at λ^*, which implies that $x(\lambda^t)$ converges to $x(\lambda^*)$ and λ^t converges to λ^*, for $t \to \infty$. By Lemma 14.4.1, this includes when subgradient $g_l(\lambda_l^*) = 0$.

However, when $D(\lambda)$ is *not* differentiable at λ^*, the subgradient at λ^* is not unique. Thus, there exists a user with $x_s^{y_i+k}$, such that $x_s < x_s^{y+k}$. Given price at λ^t, by the condition in (14.10), $x_s(\lambda^t)$ is discontinued when condition (14.8) is *satisfied*. This implies the subgradient of $D(\lambda^t)$ at λ^t is not unique. Thus, by Equation (14.10), $x_s(\lambda^t)$ may not converge. Furthermore, since λ_l^{t+1} is a reflection of $\sum_{s \in S(l)} x_s(\lambda^t)$ in (14.6), λ_l may not converge either.

Lemma 14.4.2 *When $D(\lambda)$ is not differentiable at λ^*, there exists a link l that satisfies this following condition:*

$$\sum_{s\in l} x_s(\lambda_l^t) < C_l \text{ and Equation (14.7) is satisfied.} \qquad (14.12a)$$

$$\sum_{s\in l} x_s(\lambda_l^t) \geq C_l \text{ and Equation (14.7) is not satisfied.} \qquad (14.12b)$$

Proof. When $D(\lambda)$ is not differentiable at λ^t, Proposition 14.4.2 shows that there exists user s with λ_s^{inc} that satisfies condition (14.8). Thus, according to the condition in (14.10), $x_s(\lambda^t)$ discontinues at λ^t; then network ends up in condition (14.12a). Furthermore, user s may transmit data at rate x_s^{y+k} at time t, for $x_s < x_s^{y+k}$. Since $\sum_{s\in l} x_s(\lambda_l^t) \geq C_l$, user must transmit at $x_s(\lambda^t)$. Thus, $x_s(\lambda^t)$ also discontinues at λ^t and then network ends up in (14.12b).

Proposition 14.4.2 and Lemma 14.4.2 imply that the algorithm may not converge and the rate allocation may oscillate between the two cases in (14.10) as a result from users attempting to obtain additional bandwidth for better QoS. The oscillation is also an indication; there is a gap between the primal problem (14.4) and its dual problem D. The gap is driven by users' responses to the new prices advertized. That is, in one situation, users may feel the price is too high and decide not to demand additional bandwidth. In a different situation, the same users may demand additional bandwidth to achieve higher QoS when the price is acceptable to them. This makes solving the optimization problem with multi-layered user utility becomes intractable. Thus, there may be no optimal solution for the primal problem.

For this reason, we further investigate how this phenomenon may affect the network. Here, we divide users of multi-layered application into two categories: *Passive users* and *Active users*.

Definition 1. *Passive Users:* Users who accept the amount of bandwidth x_s allocated by the network and adjust the quality according to x_s, and achieve x_s^y by solving $y = \max(Y \mid x_s^y \leq x_s)$.

Definition 2. *Active Users:* Users who continue to try demanding additional bandwidth above the amount of bandwidth allocated to them.

These active users may cause oscillation as they change their transmission rate, which in turn affects the network pricing. They will stop demanding more bandwidth when they feel the additional quality is not worth the additional cost, or when the maximum quality is obtained. The question is therefore how the behavior of active users affects network activities.

14.4.4 The Ripple Effects of Active Users on Network

The following discussion addresses how the behavior of active users may affect the bandwidth allocation to passive users.

Lemma 14.4.3 *Suppose* $\sum_{s \in S(l)} x_s(\lambda_l^t)$ *on link* l *oscillates, for* $t \to \infty$, *then* λ_l^t *also oscillates.*

Proof. Assume that $x_s(\lambda_l^t)$ oscillates as $t \to \infty$ and let $z_l^t = \sum_{s \in S(l)} x_s(\lambda_l^t)$ on link l. Since λ_l^{t+1} is updated by Equation (14.6), for $\lambda_l^t \geq \lambda_l^{min} \geq 0$ and $z_l^{(t)} > 0$, when z_l^t increases, λ_l^{t+1} also increases; and when z_l^t decreases, λ_l^{t+1} also decreases.

Obviously, since Equation (14.6) is designed to respond to the traffic load in the network, the network price λ_l^t oscillates when the traffic load oscillates. In fact, Equation (14.6) is a feedback loop and λ_l^t continues to evolve as long as $\lambda_l^t \geq 0$. Hence, Lemma 14.4.3 shows that active users can cause oscillation in pricing. Subsequently, we explore the effect of this pricing oscillation on bandwidth allocation for passive users.

Proposition 14.4.3 Bandwidth allocation for passive users is affected by the changes in network price caused by active users.

Proof. Let set $\hat{S}'(l)$ denote a set of *active* users and $\hat{S}(l)$ be a set of *passive* users sharing link l, where $\hat{S}(l) + \hat{S}'(l) = S(l)$, $\hat{s} \in \hat{S}(l)$, and $\hat{s}' \in \hat{S}'(l)$, for $l \in L$. Assume at time $t - 1$, $x_{\hat{s}}(\lambda_l^{t-1})$ and $x_{\hat{s}'}(\lambda_l^{t-1})$ have converged at λ_l^{t-1}, for $\forall \hat{s}, \hat{s}'$. Suppose at t, users in $\hat{S}'(l)$ demand more bandwidth and transmit data at $x_{\hat{s}'}^{y+k}$, where $x_{\hat{s}'}^{y+k} > x_{\hat{s}'}(\lambda_l^{t-1}) \geq x_{\hat{s}'}^y$, for $y = \max(Y \mid x_{\hat{s}'}^y \leq x_{\hat{s}'}(\lambda_l^{t-1}))$, but for users in $\hat{S}'(l)$, $x_{\hat{s}}(\lambda_l^t) = x_{\hat{s}}(\lambda_l^{t-1})$. Now, we have $\sum_{\hat{s} \in \hat{S}(l)} x_s(\lambda_l^t) + \sum_{\hat{s}' \in \hat{S}'(l)} x_{\hat{s}'}^{y_i+k} > C_l$. Next, at $t+1$, λ_l^{t+1} is updated by solving (14.6). Then, user $\hat{s} \in \hat{S}(l)$ computes $x_{\hat{s}}(\lambda_l^{t+1})$ by solving (14.5). By rearranging (14.2), we have

$$\lambda_{\hat{s}}^{t+1} = \frac{m_{\hat{s}}(U_{Bw}(x_{\hat{s}}) - U_{\hat{s}}(x_{\hat{s}}, x_{\hat{s}}^y) - 1)}{x_{\hat{s}}(\lambda_l)}.$$

$\lambda_{\hat{s}}^{t+1}$ is the network price that must be paid by user $\hat{s}$ from $U_{cost(\hat{s})}(x_{\hat{s}})$ in (14.2). Hence, the relationship between $\lambda_{\hat{s}}$ and $x_{\hat{s}}(\lambda_l)$ can be illustrated as follows.

$$\lim_{x_{\hat{s}} \to \infty} \lambda_{\hat{s}}(x_{\hat{s}}) = \lim_{x_{\hat{s}} \to \infty} \frac{m_{\hat{s}}(U_{Bw}(x_{\hat{s}}) - U_{\hat{s}}(x_{\hat{s}}, x_{\hat{s}}^y) - 1)}{x_{\hat{s}}} = 0.$$

However, we have

$$\lim_{x_{\hat{s}} \to 0} \lambda_{\hat{s}}(x_{\hat{s}}) = \lim_{x_{\hat{s}} \to 0} \frac{m_{\hat{s}} \left(U_{Bw}(x_{\hat{s}}) - U_{\hat{s}}(x_{\hat{s}}, x_{\hat{s}}^y) - 1 \right)}{x_{\hat{s}}} = \infty.$$

Thus, when $\lambda_{\hat{s}(l)}^{t+1} < \lambda_{\hat{s}(l)}^{t}$, then $x_{\hat{s}}(\lambda_{\hat{s}(l)}^{t+1}) \geq x_{\hat{s}}(\lambda_{\hat{s}(l)}^{t})$. However, when $\lambda_{\hat{s}(l)}^{t+1} \geq \lambda_{\hat{s}(l)}^{t}$, then $x_{\hat{s}}(\lambda_{\hat{s}(l)}^{t+1}) < x_{\hat{s}}(\lambda_{\hat{s}(l)}^{t})$.

Proportion 14.4.3 implies that during excessive network congestion, oscillatory behavior exhibited by active users also impacts bandwidth allocation for passive users, which is consistent with the assumption that a heavier congestion leads to higher network price. As a result, passive users may end up with less bandwidth at a higher price. The *worst-case scenario* is when the oscillation of network price λ_l causes the bandwidth allocation for *passive user* $\hat{s}$ to oscillate between two cases: $x_{\hat{s}}(\lambda_l^t) \leq x_{\hat{s}}^{min}$ and $x_{\hat{s}}(\lambda_l^t) > x_{\hat{s}}^{min}$, where $x_{\hat{s}}^{min}$ is the minimum required bandwidth for minimum QoS. For these reason, the worst-case scenario also applies to single-layer user utility function.

Corollary 14.4.1 *Bandwidth allocation for users with* single-*layer utility function is also affected by the changes in network price caused by users with* multi-*layers utility function.*

Therefore, the actions of active users may have a negative impact on the bandwidth allocation for passive users, such that passive users may not receive sufficient bandwidth even to meet the minimum requirement. Additionally, the oscillation also causes the quality of video to oscillate creating the visual flickering effect at the user level that most people find annoying [10] and degrade user QoE. Therefore, we propose an admission control to assure QoE of users with multi-layered applications.

14.5 Admission Control

14.5.1 Admission Control Designed

In order to design an effective admission control (*adm ctrl*), network must decide the selection criteria to accept or reject users' requests for admission. In multi-layered user utility environment, user demand is adaptive and the long-term consequence of poor QoS is potential loss of future revenue. We assume that admission control is invoked at the occurrence of excessive network congestion and each candidate for admission is evaluated with function $\vartheta_s(x_s(\lambda_s))$ defined as follows.

$$\vartheta_s(x_s(\lambda_s)) = x_s^y \left(\frac{\widehat{\lambda}_s}{\lambda_s} \right)^{\delta_\lambda} + \delta_u \frac{U_s^y(x_s^y)}{\lambda_s \, x_s^y}, \qquad (14.13)$$

where δ_λ and δ_u are non-negative parameters that function as a weight: The increase in δ_λ implies that network puts more emphasis in revenue. Similarly, network places more weight in user utility when δ_u is increased.

Let λ_s be the network price decided by the network such that $\lambda_{min} \leq \lambda_s$ and $\widehat{\lambda}_s$ denote the *price user is willing to pay* for the desired service quality. Here, a static minimum price λ_{min} is determined by the network (for example, ISP). That is $\widehat{\lambda}_s = \lambda_s^{inc} + \lambda_s$, for $\lambda_s^{inc} \geq 0$. Observe that $lim_{\delta_\lambda \to \infty} \left(\frac{\widehat{\lambda}_s}{\lambda_s} \right)^{\delta_\lambda} = \infty$ when $\frac{\widehat{\lambda}_s}{\lambda_s} > 1$. However, $lim_{\delta_\lambda \to \infty} \left(\frac{\widehat{\lambda}_s}{\lambda_s} \right)^{\delta_\lambda} = 0$, for $\frac{\widehat{\lambda}_s}{\lambda_s} < 1$, as $\delta_\lambda \to \infty$. This means users with $\widehat{\lambda}_s > \lambda_s$ receive higher preference for admission when network places more emphasis in revenue. Additionally, since user utility function with multi-layers of quality follows the law of diminishing returns, network may consider $\frac{U_s^y(x_s^y)}{\lambda_s x_s^y}$ from Equation (14.13), which can be interpreted as user satisfaction over the cost for desired quality at layer y. Now we can formulate the admission control problem as the following optimization problem:

$$\max \sum_{s \in S} \vartheta_s(x_s^y) \, z_s, \qquad (14.14)$$

$$\text{s.t.} \sum_{s \in S(l)} x_s^y \leq C_l, \quad \text{for } \forall l, \, l \in L,$$

$$z_s \in \{0, 1\}, \qquad \text{for } \forall s, s \in S,$$

$$\text{over } x_s^y \geq 0, \qquad \text{for } \forall s, \, s \in S \text{ and } y \in Y,$$

where $z_s = 1$ if user s is selected, otherwise zero. The difficulty of solving this problem lies in the search for all possible combinations of $\vartheta_s(x_s^y)$, for $s \in S$. For this reason, problem (14.14) is reduced to the *0–1 Knapsack problem* [20], where each user must either be admitted or rejected. The network cannot admit a fraction of the amount of user's traffic flow or admit users above the available capacity. To ensure real-time performance and quick completion of the admission process, we propose a three-step heuristic greedy-based algorithm to solve (14.14).

Step one: Network determines the price λ_l of each link l. If $\lambda_l < \lambda_l^{min}$, then network sets the price $\lambda_l = \lambda_l^{min}$. This assures $\lambda_l^{min} \leq \lambda_l$. Next, network sends λ_s to user s, where $\lambda_s = \sum_{l \in s} \lambda_l$.

Step two: users submit a tuple of $\langle x_s^{y_i}, \widehat{\lambda}_s \rangle$, where $\widehat{\lambda}_s$ is the price user is willing to pay. Users respond to network after evaluating

$$x_s^y = \underset{0 \leq x_s^y \leq x_s^{|Y|}}{\arg \max} \left\{ \frac{U_s^y\left(x_s^y\right)}{x_s^y \lambda_s} \right\}.$$

Step three: Once network receives the necessary information, tuple $\langle x_s^y, \widehat{\lambda}_s \rangle$, from the entire users, network computes $\vartheta_s(x_s(\lambda_s)), \forall s \in S$, and invoke "*User Selection*" algorithm.

Let ϑ_{SET} denote a set of ϑ_s that is associated with user and $\hat{S}$ denote a set of users admitted into the network. In lines 1 and 2 of *user selection* algorithm, given ϑ_s^{max}, x_s^y is retrieved. In line 3, the algorithm verifies whether the link has sufficient capacity to provide at least x_s^y and that x_s^y has not been included from the previous run, and then, execute lines 4, 5, and 6. Next, ϑ_s^{max} is removed from the set in line 6. We assume that the network begins to provide service as soon as the user is admitted into network. The performance of this algorithm is determined by the number of links $|L|$, the number of users $|S|$, and the number of links in the path of each admitted user s that needs to be updated. Thus, the total running time is at most $O(|L|^2.|S|)$.

14.5.2 Convergence

In this section, we show Algorithm 14.1 achieves convergence. Let $U_s^*(x_s, x_s^y)$ be maximum user utility of user s, given the price λ_s that he/she is willing to pay and bandwidth allocation x_s allocated to him/her, such that $x_s^y \leq x_s$ and $\lambda_s \leq \widehat{\lambda}_s$. Moreover, every user $s \in \hat{S}$ that is selected through admission control of Algorithm 14.1 is guaranteed that he/she can achieve $U_s^*(x_s, x_s^y)$. In order to demonstrate that the algorithm converges, we first show that every link in the network is *feasible* [1, 2], which also means that the total demand for bandwidth in every link in the network does not exceed its capacity. Let $\hat{S}(l)$ be a set of selected users by Algorithm 14.1 whose traffic traverse through link l.

Theorem 14.4.1 $\forall l$, $l \in L$, Algorithm 14.4.1 *guarantees* $\sum_{s \in \hat{S}(l)} x_s \leq C_l$.

Proof. While considering a new user s' in set $\hat{S}$ for inclusion, line 3 in the *Algorithm 14.1* verifies whether additional bandwidth demand $x_{s'}$ of s' will cause the total aggregated demand to exceed any link capacity on path $r_{s'}$, for $\forall l \in r_{ss}$. When any of the link $l \in r_{ss}$ exceeds the link capacity

Algorithm 14.1 User selection
1. $\quad \vartheta_s^{\max} = \max\{\vartheta_{SET(l)}\}$
2. $\quad x_s^y = \text{get\_bandwidth}(\vartheta_s^{\max})$
3. $\quad$ If $(x_s^y + \sum_{\hat{s} \in l} \; x_{\hat{s}} \leq C_l,$ for $\forall l, \; l \in path \; r_s)$ and $(s \notin l)$ then
4. $\quad\quad$ Reserve l for user s, for $\forall l, \; l \in r_s$
5. $\quad\quad C_l = C_l - x_s^y$ for $\forall l, \; l \in r_s$
6. $\quad\quad \hat{S} = \hat{S} + s$
7. $\quad \vartheta_{SET} = \vartheta_{SET} - \{\vartheta_s^{\max}\}$
8. $\quad$ Repeat from line 1 until $|\vartheta_{SET}| = 0$ //until ϑ_{SET} is empty

C_l, user s' will not be included in $\hat{S}$, which is described in the following expression.

$$s' \in \hat{S} \mid \forall l \in r_{s'}, \quad x_{s'(l)} + \sum_{s \in \hat{S}(l)} x_s \leq C_l.$$

Otherwise, $s' \notin \hat{S}$. This completes the proof.

Theorem 14.5.1 demonstrates that admission control guarantees that every link is feasible. Next, we demonstrate that Algorithm 14.1 also achieves convergence.

Theorem 14.5.2 Algorithm 14.1 *achieves convergence.*

Proof. By Theorem 14.5.1, we have $\forall l \in L, \sum_{s \in \hat{S}(l)} x_s \leq C_l$. That is every link in the network is feasible. As a result, price λ_l^{t+1} determined by Equation (14.6) satisfies $\forall l \in L, \; \lambda_l^{t+1} \leq \lambda_l^t$. Moreover, since $\lambda_{min} \leq \lambda_s$, the price λ_l^t converges at $\lambda_{min}, t \to \infty$. As a result the price converges, which also implies bandwidth allocated to selected users also converges. Therefore, the Algorithm 14.1 converges.

Theorem 14.5.2 implies *Algorithm 14.1* converges when every link in the network is feasible. This is possible because Theorem 14.5.1 shows that *Algorithm 14.1* can guarantee that links in the network are feasible. So far, we have demonstrated how the algorithm achieves convergence. In the next discussion, we will endeavor to provide an insight why convergence is possible through Nash Equilibrium [21]. To address this question, we investigate the likelihood of selected users changing their bandwidth demand after they are selected by *Algorithm 14.1*. In the following discussion, we assume that price decided by network and the price ceiling that a user is willing to pay do not change. Additionally, we also assume that each user does not know the maximum price that other users are willing to pay.

Theorem 14.5.3 Algorithm 14.1 *achieves Nash Equilibrium.*

Proof. First, we demonstrate that user has minimal to gain by changing their demand for bandwidth. By definition described in Equation (14.1), users satisfaction $U_s(x_s, x_s^y)$ has diminishing return characteristic because

(i) $\lim\limits_{y \to \infty} U_{bw}(x_s^y) - U_{bw}(x_s^{y-1}) = 0,$ as described in Section 14.2.

(ii) $\lim\limits_{\lambda_s \to 0} U_{bw}(x_s) = 1.$

Therefore, when user s utility reaches $U_s(x_s, x_s^y) = U_s^*(x_s, x_s^y)$, demanding more bandwidth may not significantly improve his/her satisfaction of enjoying video streaming application. Following this argument, lowering the network price therefore may not encourage users to demand more bandwidth. Thus, no users benefit from demanding more bandwidth when $U_s(x_s, x_s^y) = U_s^*(x_s, x_s^y)$. Additionally, it is unlikely for users to reduce their bandwidth demand to achieve lower utility, such that $U_s(x_s, x_s^y) < U_s^*(x_s, x_s^y)$. For these reasons, there is a set of unique user utilities

$$U_1^*(x_1, x_1^y), U_2^*(x_2, x_2^y), \dots, U_{|\hat{S}|}^* \left(x_{|\hat{S}|}, x_{|\hat{S}|}^y \right),$$

that are associated with users that are selected by *Algorithm 14.1.* Hence, $U_s^*(x_s, x_s^y)$, $\forall s \in \hat{S}$, are the unique user utilities in Nash Equilibrium. This completes the proof.

Theorem 14.5.3 implies that bandwidth convergence is possible because users selected by *Algorithm 14.1* have characteristics of Nash Equilibrium when users have minimum to gain by changing their own bandwidth demand [21].

14.6 Simulation and Discussion

In this section, we present a demonstration of multi-layered utility function with admission control using a network shared by eight users, shown in Figure 14.2. The initial setup is listed in Table 14.1 and Table 14.2. Through this simulation written in C++, we demonstrate how user 3's switching between two layers may impact other users in the network and the pricing. We also show how implementing admission control improves network activities.

User 3 initially requests data transmission at 2, but demands bandwidth increase from 2 to 5 after iteration 70 when the network price is below his/her

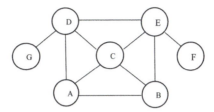

Figure 14.2 Network topology.

Table 14.1 Simulation setup

User	0	1	2	3	4	5	6	7
m_s	40	50	50	50	70	10	100	70
initial x_s^{min}	2	2	2	2	2	1	8	2
New x_s^{min}	–	–	–	5	–	–	–	–

Table 14.2 User route or path setup

U_0: ABCD	U_2: ABC	U_4: DEF	U_6: DG
U_1: ABCDE	U_3: AB	U_5: CDG	U_7: EF

threshold of 6 unit currency, from solving Equation (14.8). User 3 would reverse his demand back to 2 when the network price exceeds the threshold. Next, network eventually performs an admission control at iteration 300 by solving problem (14.14) with Algorithm 14.1.

We begin with our analysis of the most congested link, which is link AB that is shared by four users. These users are user 0, 1, 2, and 3. Prior to admission control, the network price (illustrated in Figure 14.3a) fluctuates whenever user 3 oscillates between two layers (Figure 14.3b). As a result, bandwidth allocation for users 0, 1, and 2 also oscillates, as illustrated in Figure 14.5a. At iteration 300, network implements admission control with *user selection* algorithm, resulting in the dismissal of user 0, and this in turn leads to the convergence of network price and bandwidth allocation for users 1, 2, and 3. In addition, without user 0, the network has additional bandwidth to meet the higher demand of user 3.

Next, we examine the ripple effects from the oscillation caused by user 3. The rate allocation in link BC shows a pattern similar to the allocation in link AB because the same flows (users 0, 1, and 2) traversing through link BC also traverse in link AB. Figure 14.6b shows that user 5 in link CD is *not* affected by the oscillation of users 0 and 1, which are reactions to the oscillation initiated by user 3. It is because the aggregated flow of the three users 0, 1, and 5, including the spike in rate allocation caused by the oscillation is still below

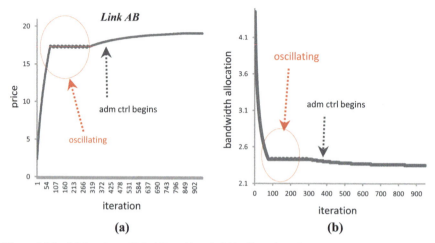

Figure 14.3   Link price at link AB and bandwidth allocation of users 0,1, and 2. "*adm ctrl*" denotes admission control.

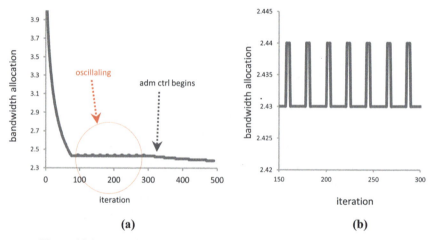

Figure 14.4   Zoom into (b) between iteration 0–500 and iteration 150–300.

the capacity constraint. Thus, the unused capacity in CD functions as a buffer for user 5, such that it allows users 0 and 1 to oscillate without affecting user 5. As demonstrated in Figure 14.6a, the rate allocation of users 5 and 6 in link DG converges and is not affected by the oscillation in CD. However, the oscillation from link AB is affecting user 4, such that his/her bandwidth allocation is also oscillating. This is because user 4 is sharing link DE with an oscillating flow that belongs to user 1, as illustrated in Figures 14.7a and 14.7b.

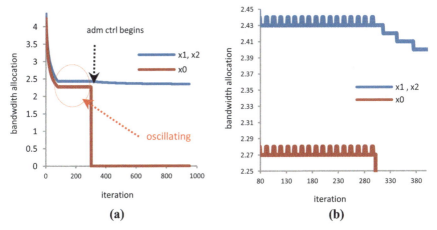

Figure 14.5   Impact of rate allocation to user 0, 1, and 2 in link AB. (b) magnifies the oscillating area in (a).

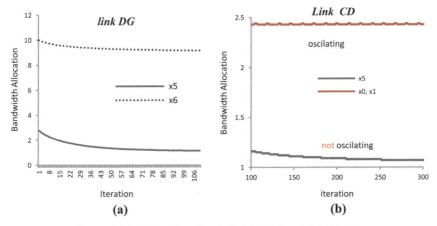

Figure 14.6   Rate allocation in link DG (a) and link CD (b).

User 3's oscillation has impacted link DE because the aggregated flow exceeds the maximum capacity. Thus, any spike in the transmission rate causes congestion in DE. That is, the aggregated flow in DE exceeds the capacity limit and it forces the network to hike the price at link DE. User 1 stops oscillating after the admission control is invoked at link AB, and this causes user 4 to also stabilize. Lastly, since user 7 is not sharing link EF with anyone, user 7 is not affected at all.

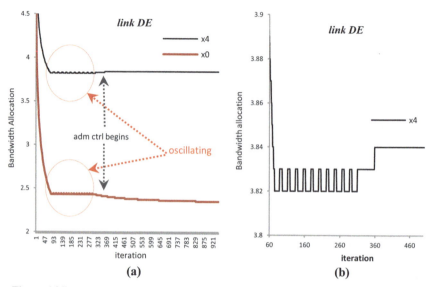

Figure 14.7   Rate allocation on link DE (a) and the oscilation area is maginified in (b).

The simulation shows that active users striving for better QoE may cause many ripple effects, causing rate allocation assigned to other users to oscillate. Furthermore, the ripple effects of bandwidth oscillation may spread from one specific link to other parts of the network. The oscillation and its ripple effects confirm that solving optimization involving adaptive QoE makes the optimization problem intractable. This is because there is a circular event of users continuously adjusting their transmission rate according to the fall and rise of the price. At the same time, price fluctuation follows and depends on the rise and fall in bandwidth demand. Additionally, the simulation also shows that higher throughput may result in higher risk of ripple effects, which lead to higher network instability. Such the ripple effects may cause more users experiencing the visual flickering effect and user QoE degradation, that is, price and bandwidth allocation oscillation at different part of the network. Admission control by the network is a viable approach to attend to and stop oscillation. This is because reducing the population in the network provides sufficient bandwidth for the admitted users to increase their demand until the desired QoE is achieved. The lessons learned in overcoming the oscillation are: Firstly, network may pick the higher value in the price oscillation as the network price; hence, users may have to settle with the QoE they can afford. Secondly, network must assure that it has sufficient bandwidth for admitted users to be able to achieve the desired QoE.

14.7 In Practice

In this chapter, we explicate the nature of the problem concerning optimizing multi-layered utility, specifically on how user desire for better quality may impact the network traffic and proposes a design of multi-layered utility model with the following benefits. Our multi-layered utility function may provide an upper bound for bandwidth allocation and a lower bound of how much video traffic can be reduced without significantly impacting user QoE. Secondly, our model provides useful guidance for the design of optimization algorithm for video streaming-based applications such that user QoE is maximized. Thirdly, the proposed model also provides the first step to understand the relationship between user QoE and network traffic management.

The following paragraphs describe how our model can be implemented in various scenarios. Applying admission control and selecting millions of incoming traffic flows in Internet may not be practical and effective because network topology in the Internet is too large to be managed by a single provider. However, we can scale down the implementation for smaller network topologies, such as private networks where video traffic is managed internally. Some examples of such networks are CDN providers that provide live video streaming [22], Google's Youtube datacenter [23], and local Internet providers that directly provide service to users. In such smaller private networks, computation of the selection of appropriate video traffic flows can be done quickly, and they also provide a more controlled environment compared to the Internet. Our proposed admission control is particularly useful for such types of environments. Therefore, although we cannot control the traffic in the Internet, implementation of our model improves traffic management and guarantees user QoE in internal networks. We also provide a general discussion on the implementation of our solution in secondary data market [24, 25]. Finally, we conclude this section with a discussion on how providers should decide per link cost.

Content Distribution Network (CDN). Figure 14.8 illustrates the high-level structure of CDN's network video delivery system [22]. A CDN's internal network consists of three logical parts: video sources, reflectors, and edge clusters. (i) *Video sources*: servers that retrieve videos from the originating sites (for example, YouTube.com and Hulu.com) into the CDN's system, (ii) *reflector*: servers that forward the video content internally; they act as intermediaries between video sources and the edge clusters, where each reflector can receive one or more streams from the video sources and can send those streams to one or more edge clusters. Note that a reflector is capable

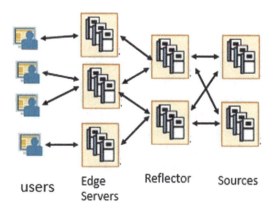

users   Edge   Reflector   Sources
         Servers

Figure 14.8   CDN's network structure, which consists of video sources, reflectors, and edge clusters.

of making multiple copies of each video stream received, and each copy can be sent to a different edge cluster. This feature enables the rapid replication of a stream to a large number of edge clusters to serve a highly popular video. (iii) *Edge clusters* directly serve end users. When a client's request for a particular channel arrives at an edge cluster, the edge cluster forwards it to a reflector, which in turn forwards it to a source; the content is returned via the reverse path. When multiple requests for the same content arrive at the same reflector server, only one request is forwarded upward. The end result is a distribution tree for each video from sources to end users. Video data stream is delivered between logical structures through CDN's internal network.

To meet users' minimum QoE requirements, the proposed admission control can be implemented in CDN's internal network, such that bandwidth oscillation can be avoided internally. Network between edge servers and reflectors is particularly important in assuring user's QoE requirement because the edge servers directly serve the end users and any oscillation in this part of network will be experienced by users. Thus, although traffic behavior in the Internet is unpredictable, guaranteeing good performance in CDN's internal network may improve the overall performance of video stream delivery to end users and in meeting users' requirements for QoE.

Datacenter Network. Today's datacenter architecture is typically a three-to-four layer multi-rooted tree topology [26, 27]. The leaves are the servers and the upper layers are populated by commodity switches, as illustrated in Figure 14.9. Such design allows multi-paths between two end points (severs) and at the same time provides load balancing and fault tolerance.

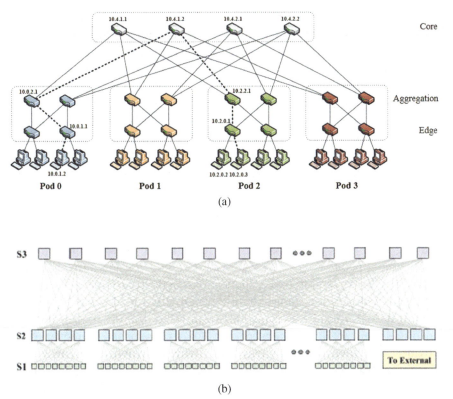

Figure 14.9 Examples of datacenter topology. (a) Fat-tree topology [26]. (b) Google's Clos topology [27].

Two examples of a typical datacenter network topologies are fat-tree [26] and Clos-based topology; the latter is adopted by *content providers* like Google [27] (including YouTube datacenter [28]) and Facebook [23]. In a datacenter that serves video applications, video data is streamed from one of the servers where the video is stored, through an internal datacenter network, to a gateway server before being finally delivered to the end users. One of challenges in managing an internal traffic in a datacenter is that bottleneck may occur when there are multiple video streams competing for the same resources [29]. Moreover, as we have shown previously, link bottleneck may lead to bandwidth oscillation, which may degrade user QoE. Our admission control can be considered to be implemented in a datacenter to mitigate bandwidth oscillation. This also assures that datacenter network provides supports to meet users QoE requirements.

Eyeball Internet service provider is an ISP that provides Internet access to individual users and is responsible for the last-mile connectivity [30]. As illustrated in Figure 14.10, video data is streamed from content providers (e.g., YouTube) to end users through transit ISP, which is responsible for the backbone network, and then finally through eyeball ISP before reaching end users. Since eyeball ISPs are directly serving end users, last-mile connection quality has a direct impact on user QoE. Thus, to avoid oscillation, our multi-layer utility model provides an estimation of the number of users that can be supported while meeting user QoE requirements. Additionally, our proposed admission control can be utilized to select which users eyeball ISP can support.

Secondary Data Market [31] is a concept of local Wi-fi providers (e.g., coffee shops, airports, and hotels) providing Internet access to end users by reselling bandwidth subscribed from ISP. As video streaming applications become more popular, these Wi-fi providers should consider providing supports for video streaming-based applications to their customers. Therefore, similar to previous case studies, multi-layered utility model may be utilized to estimate the amount of bandwidth required to provide proper QoS for video streaming application to meet user QoE requirements. Moreover, Wi-fi providers may leverage the proposed admission control to ensure that there is sufficient bandwidth to meet user QoE requirements. Similar approach can be applied to secondary market for mobile data [24, 25, 32] environment illustrated in Figure 14.11. This is a framework which allows an individual mobile Internet subscriber to resell their unused bandwidth to other users. Here, the subscriber may utilize our model to determine whether the subscriber has sufficient amount of bandwidth to provide acceptable QoE to their buyers that are buying bandwidth for video streaming. Moreover, the subscriber

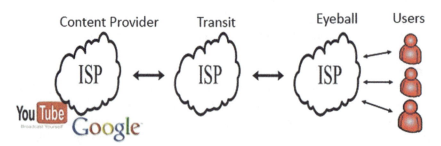

Figure 14.10   Video data is streamed from content provider (e.g., YouTube), to transit ISP, to eyeball ISP, and to end users.

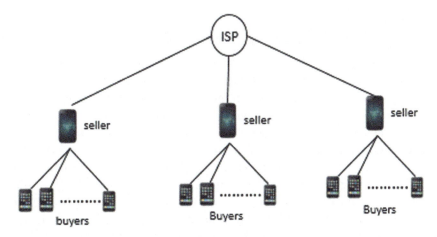

Figure 14.11 Secondary data market for mobile data network. Users may resale the unused bandwidth from their subscribed data plan to other users.

may also employ our admission control scheme to select a group of buyers that he/she can support, such that his/her customers' QoE requirements are met.

Discussion on link cost. Determining link cost or how much providers should charge per link generally depends on two factors: *operational cost* and *market pricing*. Operational cost is typically determined by the cost required to manage the infrastructure and perform administration tasks [33]. Examples of such costs include the cost of employing network administrators and the cost of electricity. As for market price, it is influenced by the dynamic and complex relationship between supply and user demand for bandwidth [24, 30]. The complexity of deciding a market price increases when there are dependencies between multiple parties involved in providing a service to users. For example, in secondary market, mobile Internet subscribers (users) are also the providers. This may create two layers of supply and demand which may lead to market price saturation [24]. Another example, amount of data an Eyeball ISP can deliver to its users are often determined by bandwidth availability at the transit ISP's network, as illustrated in Figure 14.10. Therefore, more bandwidth in Eyeball ISP's network does not translate to more supply when the network that belongs to transit ISP is experiencing heavy congestion. These concerns are, however, outside of the scope of this chapter. For more detailed discussions on operational cost and market pricing, please refer to [24, 30, 33].

14.8 Conclusion

General solutions for NUM problems in single-layered environment are not sufficient for traffic problems in multi-layered multimedia applications where user utility is adaptive. Our multi-layered user utility function incorporates insights from the fields of computer graphic and HCI, resulting in a user-utility function that considers human's natural visual ability to perceive changes in image quality, influencing their demand for desired QoE. This translates to a user utility demand that follows the law of diminishing returns. We demonstrate that the adaptive demand of some users causes oscillation that may ripple through the network, leading to lower aggregate QoE. Our study shows that optimization problem with users who constantly pursue better QoE makes the optimization problem intractable. This desire for better quality also causes the visual flickering effect at the video application level, which degrades user QoE. Thus, we propose a greedy-based solution for admission control, such that the balance between revenue and user satisfaction can be achieved. This allows the rate allocation algorithm to converge. However, even when the network seeks to maximize its revenue during the process of admission control, the algorithm may not yield the maximum revenue because of the nature of greedy algorithm. Furthermore, the efficiency of multi-layer utility function must be investigated, which will be addressed as part of our future work.

Acknowledgement

This research was supported in part by the NSF under Awards CNS-0953620 and CNS-1527303.

References

[1] Kelly, F. P. (1997). *Charging and rate control for elastic traffic. European Transaction on Telecommunication.*
[2] Kelly, F. P., Maullo, A., and Tan, D. (1998). Rate control in communication networks: shadow prices, proportional fairness and stability. *J. Operat. Res. Soc.*
[3] Lee, W., Mazumdar, R., and Shroff, N. B. (2005). Non-convex optimization and rate control for multi-class services in the Internet. *IEEE/ACM Trans. Netw.*, 13(4), 827–840.

[4] Susanto, H., and Kim, B. G. (2013). Congestion control with QoS and delay utility function. In *IEEE ICCCN*, Sep.

[5] Susanto, H., and Kim, B. G. (2014). Recovery of information lost by the least squares estimation in real-time network environment. In *IEEE NCA*.

[6] Susanto, H. (*2014*). From self-regulate to admission control in real-time traffic environment. *IEEE AINA*.

[7] Tham, C.-K., Jiang, Y., and Gan, Y.-S. (2003). Layered Coding for a Scalable Video Delivery System. In *IEEE/EURASIP*, Nantes.

[8] Fluckiger, F. (1995). *Understanding networked multimedia.*, Prentice Hall.

[9] Shirley, P. (2002). *Fundamentals of computer graphics*. A K Peters.

[10] McCarthy, J., Sasse, M., and Miras, D. (2004). Sharp or smooth?: comparing the effects of quantization vs. frame rate for streamed video. In *Conference on Human Factors in Computing Systems*.

[11] Zhu, X., Schierl, T., Wiegand, T., and Girod, B. (2011). Distributed media-aware rate allocation for video multicast over Wireless Networks. *IEEE Trans. Circuits Systems for Video Technology*, 21(9).

[12] An, C., and Nguyen, T. (2007). Analysis of utility functions for video. In *IEEE Intl. Conf. Image Process.*

[13] Hu, H., Guo, Y., and Liu, Y. (2011). Peer-to-peer streaming of layered video: efficiency, Fairness and incentive. *IEEE Transact. Circuits Syst. Video Technol.* 21(8).

[14] Kar, K., and Tassiulas, L. (2006). Layered multicast rate control based on Lagrangian relaxation and dynamic programming. *IEEE J. Commun.*, 24(8).

[15] Telebi, M., Khonsari, A., and Hajiesmaili, M. (2009). "Optimization bandwidth sharing for multimedia transmission supporting scalable video coding", *IEEE LCN*.

[16] Rao, K., Bojkovic, Z., and Milovanovic, D. (2002). *Multimedia Communication Systems*, 1st ed., Prentice Hall.

[17] Watson, A., and Sasse, M. (1997). Multimedia conferencing via multicast: determining the quality of service required by the end user. In *AVSPN*.

[18] Ward, M., Grinstein, G., and Keim, D. (2010). *Interactive data visualization*, A K Peters.

[19] Huan, T., Handigol, N., Heller, B., McKeown, N., and Johari, R. (2012). Network-congestion-aware video streaming: A rest-and-download approach, in *IMC*.

[20] Kleinberg, J., and Tardos, É. (2005). *Algorithm Design*, 1st ed., Addison-Wesley.

[21] Nash, J. (1950). "Equilibrium points in n-person games," PNAS.

[22] Nygren, E., Sitaraman, R., and Sun, J. (2010). The Akamai Network: A Platform for High-Performance Internet Applications. *ACM SIGOPS Operating System Review,* 44.

[23] Roy, A., Zeng, H., Bagga, J., Porter, G., and Snoeren, A. (2015). Inside the social network's (Datacenter) network. *ACM SIGCOMM.*

[24] Susanto, H., Kaushik, B., Liu, B., and Kim, B. G. (2014). Pricing and revenue sharing mechanism for secondary re-distribution of data service for mobile devices. In *IEEE IPCCC.*

[25] Zheng, L., et al. (2015). Secondary markets for mobile data: feasibility and benefits of traded data plans, in *IEEE INFOCOM.*

[26] Al-Fares, M., Loukissas, A., and Vahdat, A. (2008). A scalable, commodity data center network architecture. *ACM SIGCOMM.*

[27] Singh, A., et al. (2015). Jupiter rising: A decade of clos topologies and centralized control in google's datacenter network. *ACM SIGCOMM.*

[28] Adhikari, V., Jain, S., Chen, Y., and Zhang, Z. (2011). How do you 'Tube'? *ACM SIGMETRIX.*

[29] Ghobadi, M., Cheng, Y., Jain, A., and Mathis, M. (2012). Trickle: rate limiting youtube video streaming. *USENIX ATC.*

[30] T. B. Ma, R., Liu, J., and Misra, V. (2015). Evolution of the internet economic ecosystem. *ACM Trans. Network,* 23(1).

[31] Susanto, H., Liu, B., Kim, B. G., Zhang, H., and Fu, X. (2015). Pricing and revenue sharing in secondary market of mobile internet access. In *IEEE IPCCC.*

[32] Zhang, H., Liu, B., Susanto, H., Xue, G., and Sun, T. (2016). Incentive mechanism for proximity-based mobile crowd service systems. *IEEE INFOCOM.*

[33] Couch, A., Wu, N., and Susanto, H. (2005). Toward a cost model for system administration, in *USENIX Large Installation System Administration,* San Diago.

[34] Krishnan, S., and Sitaraman, R. (2012). Video stream quality impacts viewer behavior: inferring causality using quasi-experimental designs, in *Proceedings of the ACM Internet Measurement Conf.,* Boston.

[35] Wagner, J., and Frossard, P. (2009). Layer thickness in congestion-controlled scalable video. In *Multimedia Comput. Network.*

[36] Telebi, M., Khonsari, A., Hajiesmaili, M., and Jafarpour, S. (2009). A sub-optimal network utility maximization approach for scalable multimedia applications. *GLOBECOM.*

[37] Wang, X., and Schulzrinne, H. (2001). Pricing network resources for adaptive applications in a differentiated services network, in *INFOCOM*.

[38] Akhshabi, S., Begen, A., and Dovrolis, C. (2011). An experimental evaluation of rate-adaptation algorithms in adaptive streaming over HTTP, in *ACM MMSys Conference*.

[39] Ramadan, W., Dedu, E., and Bourgeois, J. (2011). Avoiding quality oscillations during adaptive streaming of Video. *J Digital Info. Wireless Commun.* 1(1).

[40] Akhshabi, S., Anantakrishnan, L., Constantine, D., and Ali, B. (2013). Server-based traffic shaping for stabilizing oscillating adaptive streaming players, in *NOSSDAV*.

[41] Susanto, H., Kim, B. G., and Liu, B. (2015). User experience driven multi-layered video based application. *IEEE ICCCN*.

15

METhoD: A Framework for the Emulation of a Delay-Tolerant Network Scenario for Media Content Distribution in Under-Served Regions

Adriano Galati[2], Sandra Siby[1], Theodoros Bourchas[2], Maria Olivares[2], Stefan Mangold[3] and Thomas R. Gross[1]

[1]Department of Computer Science, ETH Zurich, Switzerland
[2]Disney Research, Zurich, Switzerland
[3]Lovefield Wireless, Liebefeld, Switzerland

Abstract

Wireless communication is a cost-effective method of providing access to information for users in developing economies. We are interested in delay-tolerant networking (DTN) for distributing multimedia content to micro-entrepreneurs in under-served rural areas of South Africa. Buses equipped with WLAN-enabled devices (infostations), providing DTN connectivity, will be used to ferry data between urban and rural areas. Before the actual deployment, rigorous experiments must be performed to study the performance of the proposed network design. Such an activity is time-consuming and requires complex preparation. Simulators, while convenient, do not provide the most realistic results. Given the complexity of DTN testing, we try to find a middle-ground approach by building a mobility emulator test-bed, which we use to emulate the network scenario. In this chapter, we outline the design and implementation details of the network emulator. We then describe the experiments that were performed to study the impact of different network dynamics on content delivery and their results. The results from the experiments are used to improve our current network design.

Keywords: Delay-tolerant networks, media content distribution, mobility model, simulation, data analytics, socio-economic development.

15.1 Introduction

The global ICT statistics of the year 2014 by the International Telecommunications Union indicate higher penetration rates for mobile cellular and broadband subscriptions in developing countries as compared to fixed, wired, subscriptions [1]. Access to information is essential for economic growth and development in these areas. The lower costs associated with wireless networks make them an attractive means of providing access to information for developing economies.

However, there are challenges that arise when deploying connectivity solutions in these areas; the technologies are usually built keeping the developed world in mind. They are based on the assumption that there will be reliable end-to-end paths, steady connections, and power sources, which is not always the case in rural areas of developing countries [2].

Delay-tolerant networking (DTN) is an architecture aimed at providing communication in situations where end-to-end connectivity is not possible. To ease the development process of DTN applications, the DTN Research Group (DTNRG) [3] has defined an experimental network protocol for challenged networks known as bundle. A bundle is a protocol data unit of the DTN bundle protocol. Bundles are transported in a store–carry-forward manner by the nodes in the network. The layer that implements the bundle protocol operations acts as an overlay network over existing transport layers. The bundle protocol describes all the entities and operations of the bundle layer such as the description of the bundle, routing rules, processing, and security issues. The protocol specification is described in detail in RFC5050 [4]. DTNs may be a low-cost alternative to traditional wireless networks. The MOSAIC 2B project [5] aims to unleash business opportunities for micro-entrepreneurs in rural areas of South Africa by providing them with entertainment and educational media content. DTN is used as a method of delivery for this content. MOSAIC 2B uses DTN to take content from urban areas and distribute the content to recipients in the rural areas. However, before the actual deployment of the project, the entire network design must be validated and the software and hardware components must be tested to ensure that they work as expected. Simulators, while convenient, do not provide the most realistic results. Given the complexity of DTN testing, we try to find a middle-ground approach by building METhoD, a Mobility Emulator Testbed for DTNs, which we use to emulate the MOSAIC 2B scenario. In this chapter, which extends our previous reports [6, 7], we present the METhoD framework for the emulation of delay-tolerant networks and show network performance in the MOSAIC 2B settings before the actual implementation in South Africa.

An overview of the technical concepts of the MOSAIC 2B project is given in Section 15.2. Delay-tolerant networking and the DTN-enabled infostation are introduced in Section 15.3 and Section 15.4, respectively. The cinema-in-a-backpack kit is presented in Section 15.5. The METhoD is described in Section 15.6 and evaluated in Section 15.7. Section 15.8 describes an emulation of the South African DTN scenario, including evaluation results: System parameters are defined based on the test-bed results. Related work is compared in Section 15.9. Finally, Section 15.10 concludes the chapter.

15.2 MOSAIC 2B Overview

MOSAIC 2B is a research project aimed to provide business opportunities for micro-entrepreneurs living in rural South Africa by delivering multimedia content to them in a low-cost manner. Since cellular data access is usually unavailable or expensive in rural areas, content delivery will be performed using DTN. Figure 15.1 gives an overview of the project. Multimedia content will be delivered with the help of DTN-enabled mobile infostations. Infostations are battery-powered wireless local area network (WLAN)-enabled devices mounted in buses and bus depots (see Figures 15.2 and 15.3). We refer to the infostations placed in the bus depots as fixed infostations and the infostations placed in the buses as mobile infostations. Such infostations

Figure 15.1 MOSAIC 2B scenario; a DTN network between the bus station in the city of Pretoria and rural bus station in Kwaggafontein.

Figure 15.2   Mobile infostation (*left*) mounted in a bus (*right*).

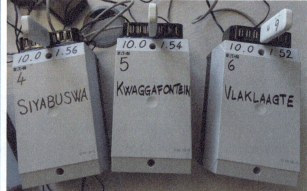

Figure 15.3   Fixed infostations ready for deployment.

act as peers that broadcast the content. The multimedia contents are archived at the fixed infostation located at the main bus depot in the city of Pretoria, which is within a 3G/LTE covered area. Every day, such a fixed infostation fetches a list of contents requested by the micro-entrepreneurs from a server located in the Internet cloud (micro-entrepreneurs are provided with a catalog of the available contents that they can order by sending an SMS to the server), finds such requests in the local archive, and injects them in the DTN network.

An advantage of this approach is that the Internet costs are highly reduced as only a few kilobytes will be downloaded every day, using the cellular network.

From the fixed station in Pretoria, data are sent to the entire DTN network once infostations are in radio range. The mobile infostations serve as intermediate relays (data mules), which carry the content between the server and the final destination, which in our case, is a fixed infostation installed in a bus depot located in a rural area. We have identified three bus depots of the PUTCO bus transportation, Siyabuswa, Vlaklaagte, and Kwaggafontein, about 135 km northeast of Pretoria, which serve several rural communities (see Figure 15.4).

Buses usually travel between the two locations in predetermined paths, for three hours in the morning and three hours in the evening. The buses travelling between rural and urban bus depots will act as carriers of data. This can be achieved by placing low-cost WLAN-enabled devices in the buses and bus stops at both locations (see Figure 15.5).

When mobile infostations come in contact with the fixed infostations, i.e., when buses arrive at the bus stop, data are forwarded to the mobile infostations via DTN (see Figure 15.5). The mobile infostations transport then the content between the infostation in Pretoria and another fixed infostation at the rural bus depots. Micro-entrepreneurs living in rural areas near Kwaggafontein, Vlaklaagte, and Siyabuswa are equipped with the cinema-in-a-backpack kit that allows them to obtain content from the final fixed infostation and screen it.

The infostations are wireless routers (see fixed and mobile infostations in Figures 15.2 and 15.3, respectively) equipped with a USB hub, battery supply, external memory, GPS receiver, and 3G dongle (the DTN-enabled infostation is presented in more details in Section 15.4).

The cinema-in-a-backpack kit (see Figure 15.6 (left)) carried by the entrepreneurs consists of the following components: a tablet, a projector, speakers, extra battery, and a backpack (see Section 15.5). The tablet allows the entrepreneur to request contents and download them from the fixed infostation at the rural bus depots.

Audio watermarking [8, 9] is used to detect copyright infringements. For this purpose, we have designed and implemented a software application that is able to capture location-based audio watermarks embedded in the soundtrack of the movies (see Figure 15.6 (right)). Such watermarks may contain, for example, GPS coordinates where the projection takes place. In our case, when the watermarks are detected, the mobile application displays the GPS coordinates where the micro-entrepreneurs are supposed to screen the media content and the MOSAIC 2B logo.

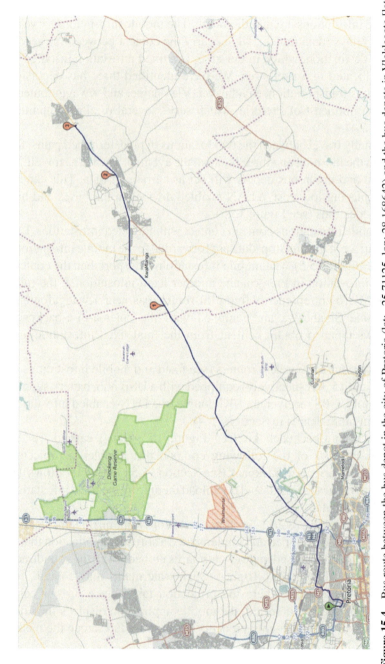

Figure 15.4   Bus route between the bus depot in the city of Pretoria (lat: −25.71125, lon: 28.168642) and the bus depots in Vlaklaagte [lat: −25.3796, lon: 28.846872], Kwagafontein [lat: −25.321132, lon: 28.925758], and Siyabuswa [lat: −25.125010, lon: 29.062170] (screenshot taken from http://www.openstreetmap.org).

Figure 15.5   Buses downloading content from the bus station in the city of Pretoria.

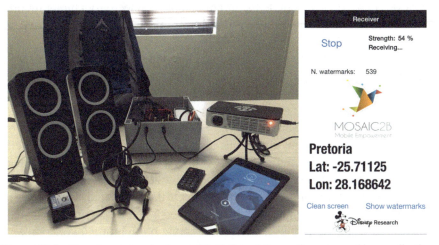

Figure 15.6   The cinema-in-a-backpack kit (*left*) and the audio watermarking application (*right*).

15.3 Delay-Tolerant Networking

Delay-tolerant networking [3] is an approach to communication systems that seeks to address technical issues in heterogeneous networks, such as lack of continuous network connectivity mainly due to mobility and limited power and of wireless communication devices. The acronym "DTN" has been often used to identify either delay- or disruption-, or disconnection-tolerant network, sometimes referring to one or the other without distinction. The IPNSIG (Inter-PlaNetary Internet Special Interest Group) was the pioneer in facing issues concerning delay experienced in transferring data between different planets of our solar system. The end systems must have a free line of sight to be able to communicate since radio waves cannot pass through large solid objects such as planets and moons. In such an environment, network protocols and algorithms, unlike the ones for terrestrial communications, must support delay. In this scenario, interruptions are somewhat predictable compared to unexpected disturbances that might occur in terrestrial networks, where disconnections can be caused by several factors, such as human mobility or obstacles, or natural disasters, such as earthquakes, seaquakes, floodings, and terrorist attacks.

DTNs [10] were conceived for networks in which patterns of connectivity are known or predictable, such as space communication systems (LEO satellite) [11–13], sparse mobile ad hoc networks [14], infostation-based systems [15], and carrier-based data collection in sensor networks [16]. However, they can also handle the unpredictable connectivity among mobile devices (e.g., PDAs) [17] and try to address most of the issues raised in networks lacking continuous connectivity.

Intermittent connectivity, long or variable delay, asymmetric data rates, high error rate, high mobility, unknown mobility patterns, energy, and storage exhaustion comprise just a few of the potential issues that make end-to-end communication unstable and unlikely in such networks. DTNs overcome such issues by adopting a so-called store–carry-forward paradigm. In these types of networks, any synchronous communication paradigm does not perform well. Basic synchronous systems rely on a connected path between sender and receiver, and they negotiate communication parameters (such as clocks) at the data link layer before communication begins. On the other hand, asynchronous systems may simply transmit with no negotiation with the receivers. This may be required when the parties are not in the same portion of network. In fact, networks may be partitioned because nodes may not be in range with one another due to their physical distance and/or their mobility.

15.4 DTN-Enabled Infostation

Infostations are wireless weatherproof boxes, each of which contains a WiFi router equipped with external memory storage, battery supply, GPS receiver, 3G dongle, and a mini-UPS (see Figure 15.7). We have selected the TP-Link

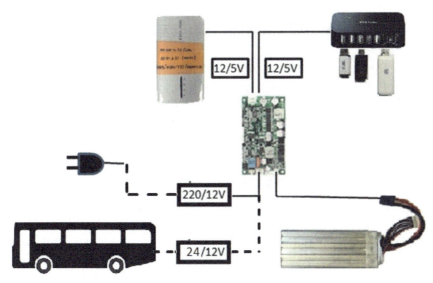

Figure 15.7 Diagram of the mobile infostation.

TLMR3040 ver. 2.0 WiFi router for the infostations. Since the memory of the router is not sufficient for our purposes, we connect a USB hub to the router and use it to accommodate external memory storage. In addition, the GPS receiver and the 3G dongle are also attached to the hub.

The GPS receiver is used to track mobility of the infostations. The 3G dongle allows the fixed infostation in Pretoria to fetch the list of requests of the micro-entrepreneurs from the server on a daily basis. Besides, it allows each infostation to send useful information, system and network performance metrics, and mobility traces (GPS coordinates) whenever cellular network is available. This provides some monitoring of the system and enables us to detect failures. All of the infostations have sufficient memory to store all the data sent by the source. The mobile infostations have been designed to be powered up by three different sources: the vehicle battery, the supplementary internal lithium ion polymer battery, and the power grid. Such power sources are connected to a mini-UPS (uninterruptible power supply), which provides instantaneous protection from input power interruptions by supplying energy stored in the supplementary battery. The mini-UPS can handle a maximum and minimum input operating voltage of 30 V and 6 V, respectively, and an input current limit of 10 A. The actual upper bound load current of the system is 4 A. We added two 12 V/5 V converters to connect the router and the USB hub to the mini-UPS. An AC/DC converter (INPUT: 100–240 VAC, OUTPUT: 12 V 5 A) is also provided to connect the mobile infostations to the power grid.

The infostations have been configured with an OpenWrt release [18], an embedded operating system based on the Linux kernel, and the IBR-DTN, a C++ implementation of the Bundle Protocol (rfc5050) [4] designed for embedded systems [19, 20]. IBR-DTN provides different routing schemes and supports the TCP and UDP convergence layers.

15.5 Cinema-in-a-Backpack Kit

Mobile cinema entertainment, possibly combined with educational content, will be the use case. Micro-entrepreneurs will be provided with a low-complex, small cinema-in-a-backpack system, which consists of the following components: a tablet, a pico-projector, speakers, and an extra battery (see the cinema-in-a-backpack kit in Figure 15.6 (left)). They can download the multimedia content from the DTN and screen it in remote villages by means of the tablet. The projector can be connected to the tablet via HDMI/VGA. A battery is provided for use when power supply is not available. The components of the cinema-in-a-backpack are shown in Figure 15.8. An AC/DC converter and a supplementary lithium-ion polymer battery (22.2 V 6200 mAh) are connected to a mini-UPS and packed in a weatherproof box. The actual upper bound load current of the system is 4.5 A. Such a box powers up both the pico-project and the speakers, whose power cables have been modified to be plugged in. Because of the common power source, a ground loop isolator has been used to filter the noise and to improve sound quality.

15.6 METhoD Framework

There are several methods of testing DTN setups before deployment. Real-life test-beds provide the most accurate results. However, creating and running an actual test-bed can be a time-consuming and costly affair, requiring constant supervision. In addition to this, test-beds involving mobility usually require large areas for producing node disconnections. A popular alternative is simulation. Most DTN applications are tested through simulators. Simulation can be easily and conveniently performed in the laboratory. However, most simulators make use of simple models and methods and do not capture various

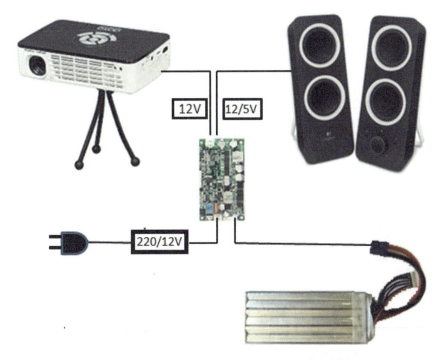

Figure 15.8 Diagram of the cinema-in-a-backpack kit.

aspects of real-life testing. A viable middle ground is an emulation approach. Rather than simulating the mobile nodes, we can use the actual wireless devices and replicate their movement in a test-bed. This would be an improvement over simulation since we use the same hardware and software components that we intend to use in the real-world deployment. The design of the METhoD architecture is driven by the following goals: mobility emulation, DTN layer communication, and node cluster isolation. The test-bed detects when mobile nodes are within the range of each other. When nodes are in contact, they can connect and exchange data at DTN level. Besides, it supports separate clusters of connected nodes. This allows groups of nodes communicating with each other in an intra-cluster fashion at the same time. Keeping these requirements in mind, a centralized controller is designed to coordinate the location and connectivity of all the nodes in the test-bed (see Figure 15.9). Every node has two connection interfaces: one interface is used to communicate with the controller, while the other one is used for communication with the other nodes in the test-bed. GPS traces, representing node mobility information, are fed

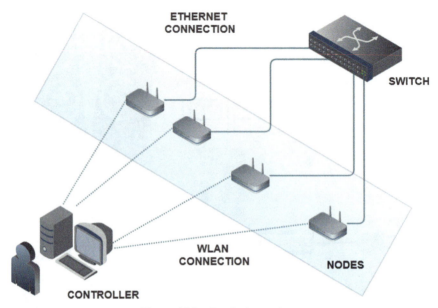

Figure 15.9   Test-bed overview.

as input into the test-bed. It can take GPS traces from existing datasets as input, provided that they are processed into the proper format that is accepted by the test-bed. However, for the emulation of the MOSAIC 2B scenario, GPS traces are generated to represent vehicles moving between Pretoria and Kwaggafontein. Such traces are used by the controller to calculate the time at which two nodes come in contact, so as to simulate their mobility. Connections and disconnections can be set up by changing the firewall rules of the devices on the fly. In this experiment, the bus route between the depot in Pretoria and Kwaggafontein has been considered as all of the bus routes between urban and rural bus depots show similar mobility pattern.

The mobile nodes are driven by the controller through the WLAN interface, while the data transfers among them take place via Ethernet cables connected to a 48-port switch. Figure 15.9 shows the overview of the METhoD. The aforementioned configuration of interfaces is done to study the effects of node connectivity without taking into account the nature of the wireless channel. However, METhoD works even if the interfaces are switched. The controller is connected to the same ad hoc network as the mobile nodes.

As shown in Figure 15.10, METhoD consists of four main components: the mobility trace generator, the mobility trace processor, the switching module,

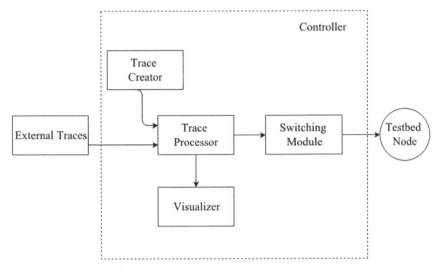

Figure 15.10 METhoD emulator framework.

and the visualizer. The trace generator is used to generate mobility traces. The trace processor takes the traces as input and produces connectivity traces. Such information is used by the switching module to run the actual test. The visualizer is a software tool to display the movement of the nodes. The functionality and implementation of the components are described in the following sections.

15.6.1 Trace Generator

METhoD provides a module to generate mobility traces of nodes. Mobility is simulated using GPS traces obtained by drawing paths on an OSM map (Open Street Map). GPS traces were obtained by drawing paths on an Open Street Map using the track creation software gpsies[1]. It generates a file containing the GPS coordinates of points on the drawn path. A drawback of using gpsies is that points can be placed on a map at a minimum distance which is too large for our purpose (20 meters when points are automatically generated). In order to generate traces of points that are very close to one another and to reduce the time taken to manually generate traces, the trace generator can take a base path and a distance and generate GPS traces for points placed on that

---

[1]Gpsies.com is a Web site that lets you find and download tracks for walking, running, hiking, cycling, driving, etc.

path at the specified distance. In addition, mobility traces can be generated for vehicles (nodes) of different speeds and start times. Such settings provide the flexibility to emulate different mobility scenarios.

15.6.2 Mobility Trace Processor

This module reads individual mobility traces and output connection information of each node. It achieves this by first extracting common timestamps between two nodes. Two nodes are never in contact if they have no matching timestamps. Then, it checks during which timestamps the nodes are within a certain radio range of each other. It collates this information into time intervals of connections and disconnections. Connectivity traces are also generated in JSON format, which contains node ID, location, and connection information for the visualizer.

15.6.3 Switching Module

The switching module performs several steps before the actual execution starts. First, it sets firewall rules on all the nodes. Then, it assigns the connectivity traces from the trace processor to all of them. Finally, the switching module reads the connectivity traces and spawns a new thread to handle it. The thread checks the connection status, and if the status specifies that two nodes are to be connected, it creates a firewall accept rule to open a connection. If a disconnection is specified, it removes the existing accept rule for the relevant node. Once all of the nodes are ready, the emulation can start.

15.6.4 Visualizer

To better visualize the movement of nodes, a simple visualizer is provided. The JSON file containing node ID, location, and connection information, generated by the trace processor, is submitted to the visualizer. The visualizer plots the nodes on an OSM map. Figure 15.11 shows two screenshots while reproducing the MOSAIC 2B scenario.

15.7 Validation

Before using the test-bed to perform experiments for MOSAIC 2B, we have validated METhoD with the UMass dataset [21]. This dataset has been selected because, as for MOSAIC 2B, it also involves mobility of buses. The dataset consists of connectivity traces of buses travelling in the UMass

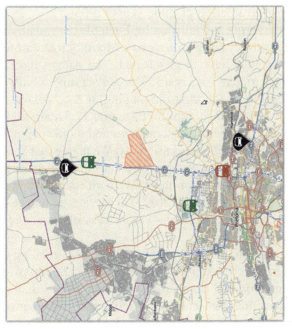

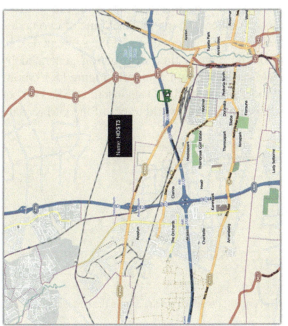

Figure 15.11 Snapshots of the MOSAIC 2B scenario in the visualizer of the METhoD emulator.

Amherst campus over a period of three years (2005–2007). The data were collected during the spring and fall months. Each file in the dataset represents connectivity of the buses during a single day. Mobility traces of fifteen buses travelling in the campus have been selected. The dataset has connectivity information over a period of 17.5 hours. The contact times of all of the nodes have been extracted by monitoring the pinging of each device with all the others. Although the experiment runs for the entire 17.5 hours and everything worked as expected, we noticed that sometimes nodes did not ping each other. In addition, some devices reboot occasionally. We notice that such nodes are the ones that have the highest number of connections, that is, they cannot handle many processes running on them. We confirm this by rerunning the entire experiment without launching the DTN services to reduce the number of processes, and then the experiment was successfully completed multiple times without any issues.

These results indicate that the number of processes running on a device needs to be taken into account during the deployment in South Africa. Besides, the METhoD emulator has been cross-evaluated by using four different metrics: intercontact time, contact duration, number of nodes with n-sightings, and number of pairs with n-contacts. Figures 15.12 and 15.13 show the contact duration and the intercontact time [22–25] based on ping activity for fifteen nodes. The contact duration between two nodes is the time during which they are in contact with each other. The intercontact time of two nodes is the time between two consecutive contact durations of the same nodes. Figure 15.12 shows the CCDF of the contact time distribution generated by METhoD against the actual UMass distribution. Similarly, Figure 15.13 shows the CCDF of the intercontact times of the METhoD against the UMass. In both plots, the distributions overlap with the real-world traces of the UMass dataset.

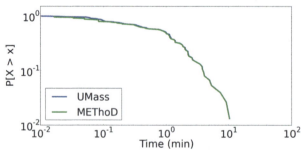

Figure 15.12   CCDF of contact durations obtained from the UMass dataset and the METhoD emulator.

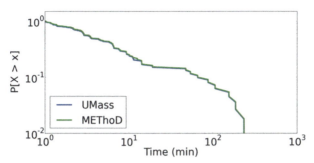

Figure 15.13 CCDF of intercontact durations obtained from the UMass dataset and the METhoD emulator.

The plot in Figure 15.14 presents the number of nodes with respect to how many times they were sighted, while the plot in Figure 15.15 presents the number of pairs with respect to how many times they got in contact with each other, distinguishing between the UMass real-world traces and the METhoD-generated traces. Both of the two plots show similar distributions. In Figure 15.14, all of the devices but one express the same number of sightings, and in Figure 15.15, all of the pairs but one present the same number of contacts.

Such results show that the METhoD distributions are very good approximations of the UMass distributions, indicating that the connection setup occurs almost immediately, with negligible latency. The latency overhead, namely the

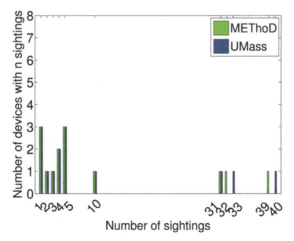

Figure 15.14 UMass dataset and the METhoD emulator.

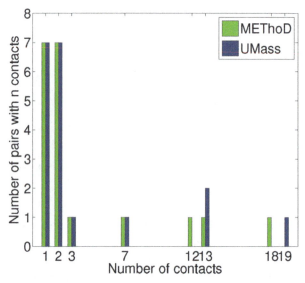

Figure 15.15   UMass dataset and the METhoD emulator.

excess time taken by METhoD to change the firewall rules with respect to the real traces, is 0.5%.

15.8 MOSAIC 2B Emulation

Experiments to study the performance of the MOSAIC 2B network were reported previously [6, 7]. However, those experiments had shortcomings; outdoor experiments were time-consuming and required constant supervision while the indoor experiments involved stationary nodes and were not true representations of the MOSAIC 2B scenario. To perform an in-depth analysis of the system, we built METhoD, a test-bed that allows us to replicate the target scenario. We use the METhoD to reproduce an environment similar to the one in MOSAIC 2B and run experiments (see Figure 15.16). In this way, we analyze the network performance in such settings and eventually identify and fix issues before the real deployment.

15.8.1 Experimental Setting

Since datasets of mobility traces of buses between Pretoria and Kwaggafontein are not available, we create our own traces using the test-bed trace creator module provided by the METhoD framework. As an initial MOSAIC 2B

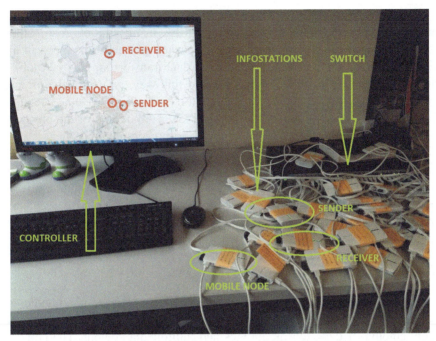

Figure 15.16 Testing real devices in the test-bed while tracking them on the visualizer map.

scenario, we assume that buses wait for 15 minutes at the Pretoria stop, travel towards the Kwaggafontein stop, and wait there for 15 minutes. Buses have a constant speed of 60 Km/h. The route between Pretoria and Kwaggafontein is the one followed by the buses during the actual South African experiments. Mobility traces for five buses carrying the mobile infostations between two fixed infostations at the end points are generated. The time between two consecutive buses leaving an end point was one hour. The connection range between two infostations is derived using the Haversine formula [26–28], which results to be about thirty meters. Such a distance guarantees a stable connection in real-world scenarios, where obstacles and the nature of the surrounding environment affect the communication link. The infostations are configured with OpenWrt, bleeding edge version (Barrier Breaker), a Linux-based operating system for embedded devices. They are also configured with IBR-DTN [19, 20], a C++ implementation of the bundle protocol targeted at embedded systems running OpenWrt. Since we want to observe data transfer at the DTN layer, we set up METhoD to turn on the IBR-DTN daemon on all the nodes. Initially, all of the nodes are set up with epidemic routing [29].

To make the scenario more realistic, the link speed of the Ethernet interface has been changed from 100 Mb/s to 10 Mb/s using ethtool. Also, other tools that can control parameters such as packet loss and link speed in an Ethernet connection can be used. Here, all data transfers among the mobile nodes take place via Ethernet interface. In this way, we study the effects of connectivity without taking the uncertain nature of wireless into account. The test-bed does not emulate the physical layer. Emulation of the physical layer would involve a lot of computation and processing since factors like signal propagation, fading, and interference, have to be modeled and tailored toward specific scenarios. Moreover, current methods indicate that even after modeling these characteristics, performance of the system in the emulation can be quite different from real-world performance. Trying to produce a realistic model for physical layer emulation is an interesting direction to take for the improvement of the test-bed.

15.8.2 Emulation with a Single Movie

Initially, the source in Pretoria sends 1 GB data as a single bundle to Kwaggafontein bus station. The MOSAIC 2B scenario considers transfer of large amount of data, that is, movies and multimedia content. The content is therefore forwarded to the mobile infostations mounted on the buses and carried to Kwaggafontein where it is forwarded to the final recipient. However, the contact time between the sender and the mobile nodes is not enough to transfer 1-GB bundles. Hence, reactive fragmentation occurs and the mobile nodes get part of the data. Notice that, only 1-GB data file is considered for transmission, and all of the mobile nodes receive about 800 MB of it.

Figure 15.17 shows the transmission activity at the receiver during connection with each mobile node. Based on our previous results [6], which show proactive fragmentation expressing higher delivery times, the data file is provided as a whole bundle.

The first mobile node transfers the entire fragment to the receiver. Meanwhile, the second mobile node obtains a fragment and reaches the receiver. At this point, the data bundle stored at the receiver is forwarded back to the mobile node. This happens because both of the nodes store fragments of the original bundles identified by different signatures. The receiver assumes that the mobile node does not have its fragment, received by the first carrier, and forwards it to the second one. Such a two-way data exchange behavior is seen in all the following meetings between the receiver and the mobile nodes. Notice that, in the last two trips, both nodes try to send data to each other,

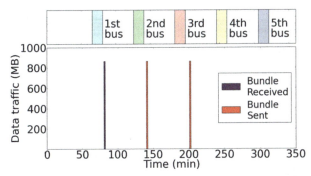

Figure 15.17 Data traffic at the receiver. 1-GB movie transfer with no proactive fragmentation and receiver with Epidemic routing.

and ultimately, only a few hundred kilobytes are sent while in contact, which cannot be clearly seen in the plot. From this initial experiment, we observe that epidemic routing cannot be set on the receiver. One option would be to set up such a fixed node with default routing that will allow transfer of data bundles only to the intended recipient.

Figure 15.18 shows the bundle activity at the receiver, which is set up with default routing, during connection with each mobile node. It shows that 15 minutes is not enough for the mobile nodes to obtain the entire 1-GB bundle from the sender. Hence, reactive fragmentation occurs, and the fixed infostation in Kwaggafontein gets several fragments of the movie. Each mobile node transfers roughly the same initial part of the entire bundle. If the contact time is not sufficient to transfer the entire bundle, the DTN node may fragment a bundle cooperatively when a bundle is only partially transferred.

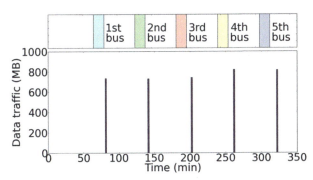

Figure 15.18 Data traffic at the receiver. 1-GB movie transfer with no proactive fragmentation and receiver with default routing.

A reactive fragmentation process occurs after an attempted transmission has taken place. A DTN node may also divide application data into multiple smaller blocks and transmit each block as an independent bundle. This approach is used primarily when contact volumes are known in advance. In the MOSAIC 2B scenario, this is exactly the case, where buses follow a predefined timetable.

In this first experiment, having no proactive fragmentation raises a new issue. Since the contact time is not sufficient to transfer the entire bundle, the receiver gets multiple overlapping large fragments. Eventually, it will receive the remaining parts when the same buses will be back in Kwaggafontein.

Despite longer delays, proactive fragmentation can help decrease over-lapping of data. Since mobile nodes carry fragments of the initial 1-GB data bundle, the receiver cannot distinguish between them so as to identify overlapping parts. DTN nodes must receive the entire bundles to be able to identify them. Enabling proactive fragmentation and splitting up the data file into smaller bundles, such that the forwarding time of a bundle is shorter than the contact duration between sender and receiver, will reduce retransmission of data. In this case, disconnections will not occur while transferring the first bundle; namely at least the first bundle will not be fragmented. Thus, future nodes carrying the same bundles stored by the receiver will not be forwarded again. For that, we enable proactive fragmentation and split the 1-GB movie to be sent to Kwaggafontein into smaller bundles of 100 MB each. In our previous work [6], we investigate the influence that bundles with different payload size and the number of bundles have on the performance of the network. Figure 15.19 shows the bundle activity at receiver when proactive fragmentation is enabled. Unlike the distribution in Figure 15.18, from the

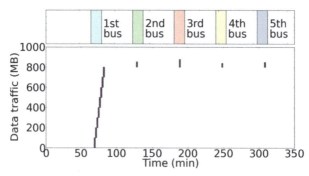

Figure 15.19   Data traffic at the receiver. 1-GB movie transfer with 100 MB bundle size.

second contact onward, bundles are not sent from the beginning. The proactive fragmentation enables the mobile node to send bundles not yet at the receiver. All of the nodes after the first one transmit only a small amount of data. This is because all of the nodes are at the sender for the same amount of time and receive the bundles in the same order. All the mobile nodes take approximately eight bundles. This means that when the nodes after the first one come to the receiver, they transmit only a fragment of a bundle. This is because each fragment of bundle is created from a reactive fragmentation, and it is not recognized as one of those bundles generated by the sender. Therefore, each mobile node produces a fragmented bundle if the contact time is not sufficient to transfer all the content. Thus, the efficiency we gain from using proactive fragmentation does not really help us since new bundles are never forwarded.

To ensure that the sender at the Pretoria bus station sends different sets of bundles to different mobile nodes, we make use of a scheduler to decide which bundles are to be sent at every contact. Such a scheduler provides two simple scheduling methods, random and FIFO. Figure 15.20 shows the effect of a random distribution of bundles by the sender, at the receiver side. It has a better performance with respect to the results in Figures 15.18 and 15.19; all of the bundles get delivered to the five contacts. However, since the process of picking bundles is random, there is a chance that the same bundles could be picked during a contact, leading to a redundancy. In the FIFO distribution, the bundles are sorted and sequentially forwarded to the carriers. The process is repeated when it reaches the end of the list of bundles. Figure 15.21 shows the effect of a FIFO distribution of bundles by the sender, at the receiver side. All the bundles are delivered within three contacts.

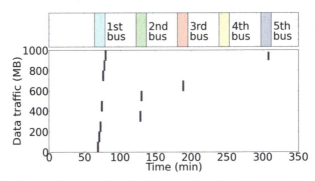

Figure 15.20 Data traffic at receiver. 1-GB movie transfer with 100 MB bundle size and random distribution of bundles by the sender.

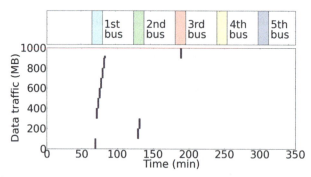

Figure 15.21   Data traffic at receiver. 1-GB movie transfer with 100 MB bundle size and FIFO distribution of bundles by the sender.

In this case, we can better control which bundles are sent at which contact. Out of these first experiments, to get some quantitative results on which method performs better, we experiment with multiple movies in the following section.

15.8.3 Emulation with Multiple Movies

Since the MOSAIC 2B scenario involves sending multiple movies to recipients, we investigate the behavior of the system when more than one movie is sent. For that, three 1-GB movies are to be sent to Kwaggafontein. The devices are set up with proactive fragmentation of 100 MB size. After the bundles were created and stored on the sender, the test was kicked off. We first run the experiment without any scheduling at the sender side. The bundle activity at the receiver is shown in Figure 15.22 (left). After the first contact with the mobile infostations, only a fragment of the same bundle of the same movie is sent to the receiver at each contact. This means that, irrespective

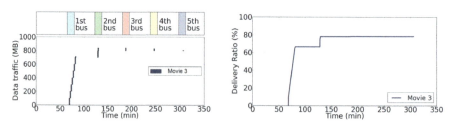

Figure 15.22   (*Left*) Data traffic at receiver. Three 1-GB movies, 100-MB proactive fragmentation, default routing at receiver. No bundle scheduling. Bundles of only one movie are sent to the receiver by the mobile node. (*Right*) Delivery ratio.

of how many mobile nodes are there, they all receive the same bundles. Hence, unless the mobile node returns to the sender a second time, new bundles will not be received by the receiver.

The same experiment is also repeated with random and FIFO bundle distributions. Only routing schemes and scheduling methods provided by IBR-DTN are allowed. However, based on underlying settings, scheduling methods can be implemented at application level and easily plugged in METhoD. As shown from the results of the random distribution of bundles in Figure 15.23 (left) with respect to Figure 15.22 (left), a simple random distribution strategy can increase the efficiency. The FIFO distribution is the most efficient, as shown in Figure 15.24 (left). It leads to the delivery of the largest amount of data. This is because there is greater control over which bundles are chosen by the sender to transmit to the mobile node at each contact.

The delivery ratios of the three experiments are also presented next to each plot in Figures 15.22, 15.23, and 15.24, respectively. None of the three movies are delivered in one journey when there is no scheduling in place because all nodes get the same bundles. The random distribution has a much better

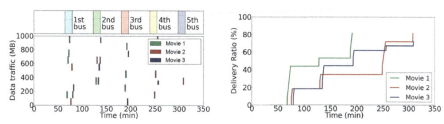

Figure 15.23 (*Left*) Data traffic at receiver. Three 1-GB movies, 100 MB proactive fragmentation, default routing at receiver. Random scheduling. Bundles of different movies are now sent to the receiver by the mobile node. (*Right*) Delivery ratio.

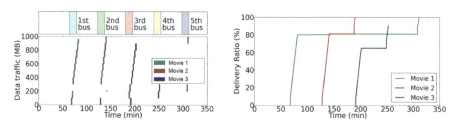

Figure 15.24 (*Left*) Data traffic at receiver. Three 1-GB movies, 100 MB proactive fragmentation, default routing at receiver. FIFO scheduling. Bundles of different movies are now sent to the receiver by the mobile node. (*Right*) Delivery ratio.

performance and manages to deliver approximately 80% of all the movies. Finally, the FIFO distribution has the best performance; it almost succeeds to deliver all of the three movies. More precisely, it delivered two movies and 90% of the third one.

15.9 Related Work

Previous work on the MOSAIC 2B project mainly consists of initial experiments to determine the performance of the system [6, 7]. They describe the effect of varying bundle sizes on delivery time under different network topologies. These works help assess the system, but the experiments were time-consuming. The test-bed we propose aims to facilitate the experimentation process in the laboratory and builds up on the results produced from [6, 7].

Even though majority of the tests in DTN-related work have been performed on real test-beds or using simulators, efforts have been taken to create realistic emulators for mobile scenarios. RAMON [30] is an emulator that tries to imitate realistic wireless characteristics in high-speed systems. A moving node is emulated using an access point, a computer, an attenuator, and an antenna. Speed is emulated by changing the signal strength according to a path loss equation. RAMON also uses a propagation model tailored toward indoor and microcellular environments. While RAMON is useful in the emulation of the physical layer, it does not scale well. New models have to be created for every scenario and attenuators have to be placed strategically for the most accurate readings.

An on/off-based emulation approach is followed in [31]. The authors claim that a tight interaction between the application and the MAC/PHY layers is missing in existing emulation methods. Their idea is to have a grid of nodes running emulation clients. An emulation server would trigger operations. Mobility is emulated by migrating a running application from one node to the other based on a mobility pattern. In other words, on obtaining a trigger from the emulation server, the emulation client on the node would take a snapshot of the state of the running application on the node, stop it, and migrate it to the next node on the path, where the application would be restarted. This method satisfies the interaction criterion, but migration of the application introduces latency, and the emulation method does not work well in high-speed scenarios, where fast migrations would be required. Besides, the size of the snapshots impacts on the performance of the system as it creates further delay. Finally, channel quality fluctuations cannot be emulated by this method.

A large-scale spatial switching emulation method is introduced in [32] to tackle large-scale topologies. They make use of the m-ORBIT grid test-bed and a split-stack architecture to emulate mobility. The application and network layers of a node run on a single machine, the virtual node. The physical and link layers of the grid nodes are used to emulate mobility. Communication between the virtual and grid nodes occurred via tunneling over gigabit Ethernet. The virtual nodes keep changing tunnels to different grid nodes based on the mobility patterns. Communication between two grid nodes representing two different mobile nodes is done over the wireless interface. Additive white Gaussian noise is introduced into the test-bed to disconnect nodes despite their close proximity. This approach is able to emulate mobility on a larger scale and has a much simpler noise model. However, it could not handle high speed and the tunneling introduces some amount of latency. [30, 31] employs complex designs to emulate mobility in the physical layer. [32], while using a simpler noise model, makes use of tunneling. These decisions resulted in scalability issues. Our design circumvents these issues by forgoing the physical layer emulation and utilizing a less complex emulation method.

With regard to content dissemination, works like [33–35] propose solutions to adapt the BitTorrent protocol to ad hoc networks. [36] develops a content dissemination protocol for vehicular ad hoc networks called SPAWN. This is based on a strategy known as rarest-closest first, where a content piece's distribution is determined by a neighboring node's proximity and possession of that piece. [37] has a similar approach for opportunistic networks, where pieces were disseminated based on their prevalence in the network. The METhoD framework can be used to experiment with strategies based on this prior work and comparisons can be made with our current distribution methods.

DakNet [38] was one of the first projects to propose the use of existing transport infrastructure to bring connectivity to developing regions in India and Cambodia at a lower cost. It was a wireless ad hoc network where buses equipped with mobile access points (MAP) would transmit data to WiFi-enabled kiosks in villages when they were in contact. A MAP consisted of a PC, an antenna, an amplifier, and a power supply. The main uses of DakNet were to transport land records between urban and rural areas and to provide asynchronous connectivity to the Internet. The MOSAIC 2B project is in a similar vein as DakNet. However, we make use of delay-tolerant networking for transmission of larger amounts of data. Not only this, we use different equipments (Section 2.4) that we hope will bring down the costs further.

KioskNet [39] attempted to improve the service provided by rural kiosks in a low-cost manner. These kiosks are a means to provide Internet connectivity at a low cost, but technical problems lead to unreliable service. In KioskNet, buses acted as ferries that carried data to and from kiosks to a gateway that had reliable Internet connection. Both kiosks and ferries were equipped with low-cost single-board computers. Delay-tolerant networking was used—the DTN2[2] implementation was modified for the scenario, and an overlay called OCMP (Opportunistic Communication Management Protocol) that runs on top of DTN was plugged in. KioskNet also addressed the issue of cost by allowing recycled PCs the ability to run their controller software. KioskNet was deployed in Indian and Ghana. The work on KioskNet later led to the development of VLink—a delay-tolerant method to transmit data via USB keys. Our approach is quite similar to KioskNet since we use buses as ferries and a DTN method of communication. Where we differ is in the hardware and DTN software we use and our aim of delivering large amounts of media content to the end users.

15.10 Conclusion

Our DTN for media content distribution using bus transportation in rural South Africa unleashes business opportunities for micro-entrepreneurs in such regions. Before the actual deployment, the entire system must be evaluated during the design process. We have developed METhoD, a mobility emulator test-bed for DTNs. METhoD is able to emulate the mobility of the devices and their connections on the DTN layer in real time, using port firewalls to mimic the device proximity and hence wireless connectivity taken from synthetic location traces. We analyze network performance and identify possible issues. The experiments we conducted provide interesting insights into the behavior of the devices and the whole network. We observe that setting up proactive fragmentation at the sender side leads to more efficient use of the contact duration between two nodes. Besides, since bundles are always sent in the same order from the sender, an increase in the number of mobile nodes has no positive effect on the amount of data delivered. A bundle distribution strategy is necessary for higher delivery ratios. A FIFO distribution of bundles is a better option than a random strategy. The experiments also show that the devices have a limitation when it comes to large number of processes and we must ensure that unnecessary processes are not running in the background.

---

[2]Delay Tolerant Networking Research Group. DTN Reference Implementation, August 2006. http://dtn.sourceforge.net/DTN2/doc/manual/

Acknowledgement

We thank Johannes Morgenroth and the team of Lars Wolf at Techn. Univ. Braunschweig, Germany, for provisioning software and support of their IBR-DTN. We also thank the Disney Lab Artists, Maurizio Nitti, and Alessia Marra, for providing artworks.

References

[1] The World in 2014, ICT Facts and Figures. *ITU Technical report, Geneva, Switzerland*, 2014.

[2] Enabling Network Connectivity in Developing Nations. *NYUAD Research Report*, 2012.

[3] Delay-Tolerant Networking Research Group (DTNRG). https://irtf.org/dtnrg. Retrieved 12/2015.

[4] Scott, K., and Burleigh. S. (2007). Bundle Protocol Specification. RFC 5050 (Experimental), November.

[5] MOSAIC 2B: Mobile Empowerment. http://mobile-empowerment.org. Retrieved 3/2015.

[6] Galati, A., Bourchas, T., Siby, S., and Mangold. S. (2014). System Architecture for Delay Tolerant Media Distribution for Rural South Africa. In the *Proceedings of the 9th ACM International Workshop on Wireless Network Testbeds, Experimental Evaluation & Characterization (WiNTECH 2014)*, pp. 65–72.

[7] Galati, A., Bourchas, T., Siby, S., Frey, S., Olivares, M., and Mangold. S. (2014). Mobile-Enabled Delay Tolerant Networking in Rural Developing Regions. In the *Proceedings of the IEEE Global Humanitarian Technology Conference (GHTC 2014)*, pp. 699–705.

[8] Frigg, R., Corbellini, G., Mangold, S., and Gross, T. R. (2014). Acoustic data transmission to collaborating smartphones: An experimental study. *11th Annual IEEE Conference on Wireless On-demand Network Systems and Services (WONS)*, pp. 17–24.

[9] Frigg, R., Mangold, S., and Gross. T. R. (2013). Multi-channel acoustic data transmission to ad-hoc mobile phone arrays. *ACM SIGGRAPH 2013 Mobile*, p. 1.

[10] Fall, K. (2003). A delay-tolerant network architecture for challenged internets. *Proceedings of the 2003 Conference on Applications, Technologies, Architectures, and Protocols for Computer Communications*, pp. 27–34.

[11] Interplanetary Internet Special Interest Group (IPNSIG). http://www.ipn sig.org. Retrieved 12/2015.

[12] Burleigh, S., Cerf, V., Durst, B., Hooke, A., Rumeau, R., Scott, K., Travisand, E., and Weiss, H. (2001). The interplanetary internet: The next frontier in mobility. *IPN Special Interest Group.*

[13] Durst, R. C., Feighery, P. D., and Scott, K. L. (2003). Why not use the Standard Internet Suite for the Interplanetary Internet? *IPN Special Interest Group.*

[14] Li, Q., and Rus, D., (2000). Sending messages to mobile users in disconnected ad-hoc wireless networks. In *Proceedings of the 6th Annual International Conference on Mobile Computing and Networking*, MobiCom '00, pp. 44–55, New York, NY, USA. ACM.

[15] Frenkiel, R. H., Badrinath, B., Borras, J., and Yates, R. (2000). The infostations challenge: balancing cost and ubiquity in delivering wireless data. *IEEE Personal Communications*, 7(2), 66–71.

[16] Shah, R. C., Roy, S., Jain, S., and Brunette. W. (2003). Data Mules: Modeling a Three-Tier Architecture for Sparse Sensor Networks. In *Proceedings of the First IEEE Sensor Network Protocols and Applications*, pp. 33–41.

[17] Chen, X., and Murphy. A. L. (2001). Enabling disconnected transitive communication in mobile ad-hoc networks. In *Workshop on Principles of Mobile Computing*, pp. 30–41.

[18] OpenWrt, Wireless Freedom. https://www.openwrt.org. Retrieved 12/2015.

[19] Pottner, W., Schildt, S., Morgenroth, J., and Wolf. L. (2011). IBR-DTN: A light-weight, modular and highly portable Bundle Protocol implementation. *Electronic Communications of the EASST*, pp. 1–11.

[20] A modular and lightweight implementation of the bundle protocol. http://trac.ibr.cs.tu-bs.de. Retrieved 2/2015.

[21] Burgess, J., Gallagher, B., Jensen, D., and Levine, B. N. (2006). Maxprop: Routing for vehicle-based disruption-tolerant networks. In *INFOCOM 2006. 25th IEEE International Conference on Computer Communications*. Proceedings, pp. 1–11, April.

[22] Chaintreau, A., Hui, P., Crowcroft, J., Diot, C., Gass, R., and Scott. J. (2007). Impact of human mobility on opportunistic forwarding algorithms. *IEEE Trans. Mob. Comput.*, 6:606–620.

[23] Chaintreau, A., Hui, P., Crowcroft, J., Diot, C., Gass, R., and Scott. J. (2005). Pocket switched networks: Real-world mobility and its

consequences for opportunistic forwarding. *Technical Report UCAM-CL-TR-617, University of Cambridge*, February.

[24] Hui, P., Chaintreau, A., Scott, J., Gass, R., Crowcroft, J., and Diot, C. (2005). Pocket switched networks and human mobility in conference environments. In *Proceedings of the ACM SIGCOMM*.

[25] Leguay, J., Lindgren, A., Scott, J., Friedman, T., and Crowcroft, J. (2006). Opportunistic content distribution in an urban setting. *Proceedings of the 2006 SIGCOMM Workshop on Challenged networks, CHANTS '06*, pp. 205–212.

[26] Vanajakshi, L., Subramanian, S. C., and Sivanandan. R. (2009). Travel time prediction under heterogeneous traffic conditions using global positioning system data from buses. *IET Intelligent Transport Systems*, 3 (1), 1–9.

[27] Montavont, J., and Noel, T. (2006). Ieee 802.11 handovers assisted by GPS information. In *Wireless and Mobile Computing, Networking and Communications, 2006. (WiMob'2006). IEEE International Conference on*, pages 166–172. IEEE.

[28] Nguyen, T., and Szymanski, B. K. (2012). Using location-based social networks to validate human mobility and relationships models. In *Advances in Social Networks Analysis and Mining (ASONAM), 2012 IEEE/ACM International Conference on*, pages 1215–1221. IEEE.

[29] Vahdat, A., Becker, D. et al. (2000). Epidemic routing for partially connected ad hoc networks. Technical report, Technical Report CS-200006, Duke University.

[30] Hernandez, E., and Helal, A. S. (2002). RAMON: Rapid-Mobility Network Emulator. In *Proceedings of the 27th Annual IEEE Conference on Local Computer Networks, LCN '02*, pages 809–820, Washington, DC, USA, IEEE Computer Society.

[31] Yoon, H., Kim, J., Ott, M., and Rakotoarivelo, T. (2009). Mobility emulator for dtn and manet applications. In *Proceedings of the 4th ACM International Workshop on Experimental Evaluation and Characterization*, WINTECH '09, pp. 51–58, New York, NY, USA. ACM.

[32] Ramachandran, K., Kaul, S., Mathur, S., Gruteser, M., and Seskar, I. (2005). Towards large-scale mobile network emulation through spatial switching on a wireless grid. In *Proceedings of the 2005 ACM SIGCOMM Workshop on Experimental Approaches to Wireless Network Design and Analysis, E-WIND '05*, pp. 46–51, New York, NY, USA. ACM.

[33] Rajagopalan, S., and Chien-Chung, S. (2006). A cross-layer decentralized bittorrent for mobile ad hoc networks. In *Mobile and Ubiquitous Systems: Networking & Services, 2006 Third Annual International Conference on*, pp. 1–10. IEEE.

[34] Gaddam, N., and Potluri. A. (2009). Study of bittorrent for file sharing in ad hoc networks. In *Wireless Communication and Sensor Networks (WCSN), 2009 Fifth IEEE Conference on*, pp. 1–6. IEEE.

[35] Sbai, M. K., Salhi, E., and Barakat, C. (2010). P2p content sharing in spontaneous multi-hop wireless networks. In *Communication Systems and Networks (COMSNETS), 2010 Second International Conference on*, pp. 1–10. IEEE.

[36] Nandan, A., Das, S., Pau, G., Gerla, M., and Sanadidi, MY (2005). Co-operative downloading in vehicular ad-hoc wireless networks. In *Wireless On-demand Network Systems and Services, 2005. WONS 2005. Second Annual Conference on*, pp. 32–41. IEEE.

[37] Belblidia, N,. Dias de Amorim, M., Costa, L. H. M. K., Leguay, J., and Conan, V. (2011). Pacs: Chopping and shuffling large contents for faster opportunistic dissemination. In *Wireless On-Demand Network Systems and Services (WONS), 2011 Eighth International Conference on*, pp. 9–16, Jan.

[38] Pentland, A., Fletcher, R., and Hasson. A. (2004). Daknet: Rethinking connectivity in developing nations. *IEEE Computer Society*, 37:78–83.

[39] Guo, S., Falaki, M. H., Oliver, E. A., Ur Rahman, S., Seth, A., Zaharia, M. A., and Keshav, S. (2007). Very low-cost internet access using kiosknet. *ACM Computer Communication Review*, 37, 95–100.

PART VI

Network Optimization

16

On the Routing of Kademlia-type Systems

Stefanie Roos[1], Hani Salah[2] and Thorsten Strufe[1]

[1]TU Dresden, Dresden, Germany
[2]TU Darmstadt, Darmstadt, Germany

Abstract

The family of Kademlia-type systems represents the most efficient and most widely deployed class of Internet-scale distributed systems. Its success has caused plenty of large-scale measurements and simulation studies, and several improvements have been introduced. Kademlia's use of parallel and non-deterministic lookups, however, so far has prevented any concise formal analysis. We introduce a comprehensive formal model of the routing of the entire family of systems that is validated against both simulations and real-world measurements. In particular, we extend our previous work by including the effect of churn into the model. Our evaluation additionally shows that several of the recent improvements to the protocol in fact are counter-productive and identify preferable designs with regard to routing overhead and robustness to failures.

Keywords: Routing, kademlia, performance enhancements.

16.1 Introduction

Distributed Hash Tables (DHTs) received considerable attention during the last decade. On an abstract level, DHTs allow the mapping of objects to nodes in a completely decentralized and highly dynamic network on the basis of IDs, such that both the number of nodes contacted during object retrieval and the connections maintained by each node increase logarithmically with the network size. As a consequence, DHTs are candidate solutions for large-scale distributed data storage as well as for decentralized resilient communication systems.

In practice, only variants of the Kademlia DHT [1] have been deployed successfully, attracting millions of users in the file-sharing applications BitTorrent and eMule [2, 3]. Even in networks of such an enormous size, the discovered routes are generally in the order of 3–4 hops [4–6]. Additionally, Kademlia's redundant routing tables combined with an iterative parallel lookup scheme make it particularly suitable for dynamic environments due to its inherent redundancy.

Despite the considerable attention Kademlia received from both research and industry, the impact of the design parameters on the routing performance is poorly understood. Measurements only offer insights on deployed systems, whereas simulations do not scale beyond several ten thousands of nodes.

In order to assess different design choices, a concise model for the complete hop count distribution is needed, which covers all the existing Kademlia implementations as well as allowing for a huge variety of modifications. The model is required to give a close bound on the hop count distribution based on the routing table structure and the routing algorithm, and it has to consider the impact of churn and routing table incompleteness, while still being computationally efficient.

We model routing in Kademlia as a Markov chain with a multidimensional state space (Section 16.3). Our derivation provides extremely tight upper and lower bounds on the hop count distribution and covers a wide range of overlay topologies. Analyzing the topologies of deployed systems, we find that they do not always outperform the original protocol. Furthermore, the analysis of the parameter space enables us to derive guidelines for design decisions (Section 16.7).

The computation of the hop count is efficient, requiring $\mathcal{O}(npolylog(n))$ basic operations and $\mathcal{O}(polylog(n))$ storage space. Given such a moderate increase, networks of up to $1B$ nodes can easily be analyzed.

The model is verified in two ways: First, the initial model for a static environment is verified by simulations (Section 16.5). Secondly, the extended model, allowing for churn and routing table incompleteness, is compared to measurements made by Stutzbach and Rejaie for the KAD network [4], resulting in an error rate of 2.7% for the average hop count, in contrast to 5.5% provided by their analytic model (Section 16.6). The chapter extends upon [7] by including a more detailed theoretical analysis and considering dynamic systems in addition to the simplified static scenario.

16.2 Kademlia-type Systems

In this section, we give a short overview of the concepts Kademlia is based on, before presenting various studies on modeling and analyzing P2P routing, with a focus on Kademlia.

16.2.1 Introducing Kademlia

Kademlia [1] is a structured peer-to-peer (P2P) system. Nodes and objects are assigned IDs from the same b-bit ID space and the distance between two IDs is defined as the XOR of their values. Kademlia implements key-based routing and storage of key-value (ID-object) pairs. The nodes at the closest distance to an object's ID are responsible for storing it.

Each node v maintains a routing table to store the IDs and addresses of other nodes, without keeping persistent network connections to them. In Kademlia, the neighbors, also called *contacts*, are stored in a tree-like routing table structure. The level in the tree a contact w of v is stored at reflects the common prefix length of v and w. At most k contacts are stored at each level, making up the so-called k-*bucket*. Information stored in the routing table may be outdated, or *stale*, when the respective nodes have left the system.

Kademlia implements greedy routing: To route a message from node v to a target ID t (for the storage or retrieval of objects), v sends parallel lookup requests to the α-known contacts that are closest to t. Every queried contact that is online replies with the set of β nodes that are locally known as being closest to t, thus extending v's set of candidate contacts. This process is iterated until the lookup does not produce any contacts closer to t than previously have been discovered, or a time-out is caught. The original Kademlia publication suggests to use $k = 20$ and $\alpha = 3$.

Kademlia proved highly efficient and reliable, and thus has frequently been modified, generating a broad family of Kademlia-type systems. Each adaptation mainly adjusts the given parameters or the routing table structure. The current mainline implementation of BitTorrent (MDHT), for example, integrates a Kademlia-type DHT for node discovery. uTorrent, the most popular client implementing MDHT, is implemented using 8-buckets, $\alpha = 4$, and $\beta = 1$ [6]. To reflect the fractions of the ID space that are covered at different levels, and hence to increase the distance reduction at each hop, variable bucket sizes k_i are introduced in iMDHT [6]. They are chosen to be 128, 64, 32, and 16 for the buckets at levels $i \in (0..3)$, respectively, and 8 for all lower levels. The variation used in the highly popular eDonkey file-sharing network, KAD, adds multiple buckets per level, grouping contacts according to the first l bits after the first diverging bit. This way the *bit gain*, i.e., the difference between the common prefix length of the current hop and the next hop to t, is at least l. Choosing k to be 10, the implementation contains buckets for all 4-bit prefixes at level 0 (containing contacts that share no common prefix with v), and one bucket for each of the sub-prefixes 111, 110, 101, 1001, and 1000 at all remaining levels. Thus, the guaranteed distance reduction is 3 bits for 75% of the targets IDs and 4 bits for the remaining 25%. By default, KAD implements $\alpha = 3$ and $\beta = 2$.

Figure 16.1 illustrates the routing table structures of the aforementioned three systems.

16.2.2 Analyzing P2P Routing

Motivated by the success and popularity of Kademlia-type systems, a large number of studies over the past few years [5, 6, 8–11] focused on the routing performance of these systems. For instance, Crosby and Wallach [11] measured the lookup latency in MDHT and Azureus (the DHT that is used by the Vuze BitTorrent client). They observed a high latency, which they attributed to the high ratio of stale contacts in the routing tables. Complementing these measurement studies, there has been significant work in improving the lookup performance, for instance, by (i) adapting the lookup parameters at runtime according to the number of expected lookup response messages [10], (ii) coupling the lookup with the content retrieval process [5], (iii) modified caching of content [12, 13], (iv) integrating geographical proximity in the neighbor selection [14, 15], or (v) using recursive lookups [16]. These improvements, however, are mainly evaluated based on simulations and do not yield insight into the impact of isolated design adaptations.

There is a vast related work on obtaining asymptotic bounds on the complexity of routing algorithms [17–20] when arranging nodes in d-dimensional lattices. The methods applied for deriving asymptotic bounds in such abstract settings have also been applied for exemplary P2P systems [1, 21–23]. However, the results are usually restricted to showing that the worst-case complexity is $\mathcal{O}(n)$ for a network of order n. They thus fail to consider the impact of the parameter on the performance and consequently do not facilitate choosing suitable parameters. Few studies deriving exact formulas (e.g., [4, 24, 25]) commonly only consider the average hop count and use simplified assumptions such as a bijective mapping from identifiers to nodes. In addition, they are of limited accuracy when compared to measurements or

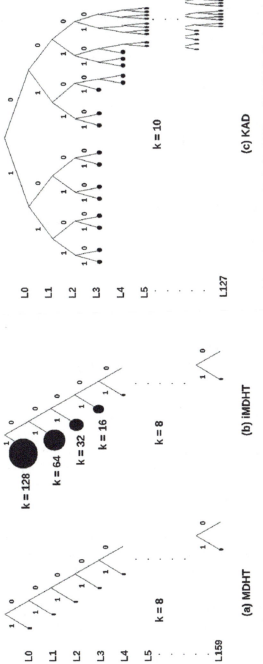

Figure 16.1 Routing table structures of three Kademlia-type systems (adapted from [8]): (a) MDHT, (b) iMDHT, and (c) KAD. Each routing table encodes the distances in the identifier space: the top part stores the farthest contacts, while the bottom part stores the closest ones. Each leaf represents a bucket of contacts.

simulations [4]. In particular, [24, 25] describe P2P routing using a Markov chain approach similar to our model but are restricted to systems without parallelism. Similarly, the only previously derived formula for the KAD implementation [4] is restricted to the non-parallel case. While it allows parametrization of the bucket size k and can (to some extent) be extended to include different routing table structures, the impact of the parameters α and β remains unclear. In addition, as the formula is only concerned with the average hop count, it also fails to provide further insight into the hop count distribution. It hence does not allow for the choice of sensible time-out duration and termination criteria for more sophisticated, possibly time-critical, applications.

In this chapter, we model Kademlia-like systems as a stochastic process, similar to [24, 25] but with considerable less restrictions. In particular, our model (i) includes parallelism, (ii) allows more diverse routing table structures than [4], and (iii) considers the hop count distribution rather than only the average hop count.

16.3 Model

The *hop count* refers to the number of edges on the shortest path that has been traversed during the lookup process. Each routing step (i.e., *hop*) is a *transition* from a set of queried contacts to either another set of queried contacts or routing termination.

In our model, a *state* is defined by the common prefix lengths of the currently known closest contacts with the target. That is, the state space of the Markov chain consists of α-dimensional integer vectors. The *initial distribution* I corresponds to the common prefix lengths of the closest α contacts in the requester's routing table. A hop in the routing corresponds to a *transition* from one α-dimensional vector of common prefix lengths to a either a second vector of common prefix lengths or routing termination.

With the *initial distribution* I and the *transition matrix* T, the common prefix length distribution of the nodes queried in the i-th step is $H_i = T^{i-1}I$. As a consequence, the *cumulative hop count distribution* can be obtained from H_i as the fraction of queries that have reached the terminal state. Due to the Markov property, our model fails to cover the improbable, but technically possible, event that nodes other than those returned from the most recent query is chosen to be contacted because the most recent query has not provided α distinct closer nodes. We overcome this insufficiency by computing T^{up} and T^{low}, which provide an upper and lower bounds, respectively, on the fraction of terminated queries. In the following, we first derive the distribution of closest entries in a routing table in Section 16.3.3, which allows us to derive I in Section 16.3.4 and T in Section 16.3.5.

We summarize the different parameters governing Kademlia-type systems and hence our model in Table 16.1.

Note that in this section, we focus on a stable system without failures. In Section 16.6, we extend the model to deal with network dynamics and failures.

16.3.1 Assumptions

We model a query for an ID of an existing node. Our basic model relies on the following assumptions, which allow a very general, application-independent view. We first state the assumptions, before elaborating on their motivation and impact on the validity of the model.

1. There are no stale contacts in the routing tables.
2. Nodes do not fail nor do they drop messages.
3. Buckets are maximally full; i.e., if a bucket contains $k_1 < k$ values, there are exactly those k_1 nodes in the region the bucket is responsible for.
4. Node IDs are uniformly and independently distributed over the whole identifier space.
5. Routing table entries are chosen independently.
6. If the distance between a node and the target ID is 0, then the node's routing table contains the target.
7. The lookup uses strict parallelism; i.e., a node awaits all answers to its queries before sending additional ones.

Table 16.1   Important notation

b	Bit-length of IDs
α	Degree of parallelism
β	Number of returned contacts
k_i	Bucket size on level i
L_{ij}	Fraction of buckets with 2^{i-j} IDs on level i
S_α	State space of distance sets
I	Initial distribution
T	Transition matrix
T^{up}/T^{low}	Transition matrix upper/lower bound
X_l	l-th state of Markov chain
D	Distance distribution
γ	Denotes either α or β
C_γ	Distribution of closest γ neighbors
$\mathbb{B}(n,p)$	Binomial distribution with parameters n, p
$F_{d,l}$	Distribution of random ID with distance $d - l$

Assumptions 1, 2, and 3 can be summarized as the assumption of a steady-state system, without churn or failures. However, we extend the model in Section 16.6 to allow for churn and bucket incompleteness. Note that for applications such as critical infrastructures and large data centers, churn is basically nonexistent and the failure rate can assumed to be negligible. Assumption 4 is given in general, since the ID is usually chosen randomly or as a hash of some identifying value, e.g., the IP address. Assumption 5 holds in as far as that nodes discover contacts initially by searching for their own ID which should result with contacts close to their own ID independently of the starting point. However, nodes encountered and potentially added during routing tend to have a higher than average in-degree. By this, the chance that a node in one routing table is present in another is slightly higher than the chance that a random node is contained in a routing table. Still, the probability should be negligible for large networks, as indicated from the agreement of our model with real-world measurements. Assumption 6 considers the case that multiple nodes share the same ID, which can happen by Assumption 4 (independent choice), but is highly unlikely in practice and hence only a theoretical construct to simplify the derivations. Assumption 7 is consistent with various implementations, whereas others allow interleaving queries as well as more than α concurrently outstanding answers. Steiner et al. present an analysis of how the latency can be enhanced by immediately reacting to a query [26]. However, it is not possible to always select the closest of all returned contacts, resulting in a higher hop count and number of contacted nodes. Consequently, strict parallelism is optimal with regard to our metric of interest, the hop count.

16.3.2 Model Overview

In the following, the idea is formalized, defining the parameters governing the routing and the states of the Markov chain.

The common prefix lengths of the closest nodes to the target ID t is used to characterize the routing process. Because routing is commonly modeled as a monotonously decreasing process that converges to a distance of 0, we define the distance of two nodes w and v to be

$$
\begin{aligned}
dist(w, v) &= b - commonprefixlength(id(w), id(v)) \\
&= \lfloor \log_2 XOR(id(w), id(v)) \rfloor + 1,
\end{aligned}
\tag{16.1}
$$

where b is the ID space size and $id(v)$ denotes the b-bit ID of node v. We here use distance to refer to $dist$ rather than the XOR distance, unless stated otherwise.

The state of a query is either $\emptyset$, denoting a terminated query, or the distance of the currently queried α nodes to the target t. Formally, the state space is given by

$$S_\alpha = \{\emptyset\} \cup S'_\alpha \quad \text{with} \tag{16.2}$$
$$S'_\alpha = \{x = (x_1, \ldots, x_\alpha) \in \mathbb{Z}^\alpha_{b+1} : x_j \leq x_{j+1}, j = 1 \ldots \alpha - 1\}.$$

Aiming to reduce the number of states and consequently the storage and computation cost, we assume the vector of distances to be sorted in ascending order.

It remains to define the parameters influencing the hop count. We characterize a Kademlia-type system by the ID space size b, the routing parameter α and β, as well as the routing table parameters k and routing table structure L, which determines the number of buckets per level as well as how the ID space is mapped to those buckets.

Definition 16.3.1 A $\mathcal{K}(b, \alpha, \beta, k, L)$-system is a Kademlia-type system with the following properties:

- A b-bit identifier space is used for addressing.
- α parallel iterative queries are sent for each lookup.
- Each queried node answers with at most β contacts closer to the target than itself.
- The d-th entry k_d of the vector $k \in \mathbb{N}_0^{b+1}$ gives the bucket size for nodes with distance d to the routing table owner (i.e., the bucket size at level $b - d$).
- The i-th row of the matrix L gives the distribution of the guaranteed bit gain at distance i to the routing table owner; i.e., the entry $L_{ij} = \frac{x}{2^i}$ is defined by the number x of IDs with distance i that are sorted in buckets covering a region of 2^{i-j} IDs each.

Furthermore, the network order n influences the hop count distribution. Note that in most Kademlia-type systems, such as MDHT and KAD, k is constant. Similarly, the matrix L is commonly sparse. For instance, in MDHT only one bucket is used for each common prefix length, so $L_{i1} = 1$ for $i = 0 \ldots b$ and $L_{ij} = 0$ in all other cases. KAD is more complicated: $L_{b4}^{KAD} = 1$, $L_{i3}^{KAD} = 0.75$, and $L_{i4}^{KAD} = 0.25$ for $i < b$ determine the routing table structure in the KAD system. This is due to resolving at least 4 more bits on the top level, and splitting into buckets with prefixes 111, 110, 101 (75% of IDs), as well as 1001 and 1000 (25% of IDs) for all lower levels.

In the following, we derive the distribution of closest contacts in a node's routing table, which is essential for computing both the initial distribution and the transition matrix.

16.3.3 Distribution of Closest Contacts

We are interested in the distribution of the closest $\gamma \in \{\alpha, \beta\}$ contacts to a target t in a routing table of a node v. Let the random variable X_0 with values in $\mathbb{Z}_{b+1}$ be the distance of v to t. The random variable X_1 with values in S_γ gives the state characterizing the closest neighbors. In the following, we derive the probability $P(X_1 = s|X_0 = d)$.

By Assumption 6, the case $d = 0$ is trivially given by

$$P(X_1 = s|X_0 = 0) = \begin{cases} 1, & s = \emptyset \\ 0, & s \neq \emptyset \end{cases}. \tag{16.3}$$

So, from now we consider $d > 0$. The success probability is determined by the distribution for the guaranteed bit gain L_d defined by the d-th row of the matrix L and the additional bit gain dependent on the bucket size k_d.

$$
\begin{aligned}
P(X_1 = s|X_0 = d) \\
= \sum_{l=0}^{b} P(X_1 = s|X_0 = d, L_d = l)P(L_d = l) \\
= \sum_{l=0}^{b} P(X_1 = s|X_0 = d, L_d = l)L_{dl}.
\end{aligned}
\tag{16.4}
$$

It remains to obtain $P(X_1 = s|X_0 = d, L_d = l)$.

We start by determining the probability to reach the state $\emptyset$. Recall that r's routing table has k_d-buckets of nodes that differ in the $b-d$-th bit. Let x be the number of candidate nodes to be in the bucket, i.e., the number of nodes in the respective part of the ID space. If the bucket contains less than k_d contacts, by Assumption 3, then t is one of them with probability $q_m = 1$. Otherwise, with m candidates, t is contained in the bucket with probability $q_m = \frac{k_d}{m+1}$, the likelihood that t is one of the k_d nodes selected among $m+1$ nodes. If a node has distance at most $d-l$ to t, there are 2^{d-l} IDs it can potentially have, making up a fraction $\frac{2^{d-l}}{2^b} = 2^{d-l-b}$ of all IDs. The number of nodes X within a fraction 2^{d-l-b} of the ID space is binomially distributed, $X \sim B(n-2, 2^{d-l-b})$[1], by

---

[1] $n - 2$ nodes because t and v are excluded.

Assumption 4. So the probability that t is contained in the routing table is computed as

$$P(X_1 = \emptyset | X_0 = d, B_d = l) = \sum_{m=0}^{n-2} P(X = m) q_m$$

$$= \sum_{m=0}^{k_d} \binom{n-2}{m} \left(2^{d-l-b}\right)^m \left(1 - 2^{d-l-b}\right)^{n-2-m} \tag{16.5}$$

$$+ \sum_{m=k_d+1}^{n-2} \binom{n-2}{m} \left(2^{d-l-b}\right)^m \left(1 - 2^{d-l-b}\right)^{n-2-m} \frac{k_d}{m+1}.$$

If, on the other hand, t is not contained in the routing table, we need to derive $P(X_1 = (\delta_1, \ldots, \delta_\gamma) | X_0 = d, L_d = l)$ for all $(\delta_1, \ldots, \delta_\gamma) \in S_\gamma$. The distribution of distances within one bucket is needed. The probability that a contact has a certain distance corresponds to the fraction of IDs with this distance. Consequently, the cumulative distribution function $F_{d,l}$ of the distance of one randomly chosen contact in a bucket of contacts with distance at most d–l is given by

$$F_{d,l}(x) = \min \left\{ 1, \frac{2^{\lfloor x \rfloor}}{2^{d-l}} \right\} \tag{16.6}$$

for $x \geq 0$. Knowing the distance distribution of a random contact, one can derive the distance of the γ closest contacts. First, we rewrite the vector $(\delta_1, \ldots, \delta_\gamma)$, grouping identical values. This later allows us to treat the number of nodes with the same distance as a binomially distributed random variable. More specifically, the transformation M is applied to X_1 in order to obtain tuples $M_1, \ldots, M_{\gamma'} \in \mathbb{Z}^2$, so that the first component of $M_i = (y_i, c_i)$ is the i-th smallest value in $(\delta_1, \ldots, \delta_\gamma)$ and c_i is the number of occurrences of y_i in $(\delta_1, \ldots, \delta_\gamma)$. For reasons of presentation, we set $M_0 = (y_0, c_0) = (-1, 0)$. As a result, an equivalent expression for the probability distribution of X_1 is the following:

$$P(X_1 = (\delta_1, \ldots, \delta_\gamma) | X_0 = d, L_d = l)$$
$$= P(M(X_1) = ((y_1, c_1), \ldots, (y_{\gamma'}, c_{\gamma'})) | X_0 = d, L_d = l)$$
$$= (1 - P(X_1 = \emptyset | X_0 = d, L_d = l)) \tag{16.7}$$
$$\cdot \prod_{i=1}^{\gamma'} P(M_i = (y_i, c_i) | X_0 = d, L_d = l, X_1 \neq \emptyset,$$
$$M_0 = (y_0, c_0), \ldots, M_{i-1} = (y_{i-1}, c_{i-1})).$$

It remains to compute each factor in Equation (16.7). We first treat the case $i < \gamma'$, for which we have to determine the probability that (1) all $C_{i-1} = k_d - \sum_{j=1}^{i-1} c_j$ bucket entries with distance at least $y_{i-1} + 1$ are at distance at least y_i to the target, and (2) there are *exactly* c_i such entries. More precisely, event (2) conditioned on event (1) corresponds to the event that a binomially distributed random variable with C_{i-1} trials and success probability $p_i = \frac{F_{d,l}(y_i) - F_{d,l}(y_i - 1)}{1 - F_{d,l}(y_i - 1)}$ has exactly c_i successes. Note that the number of trials C_i and the denominator $1 - F_{d,l}(y_i - 1)$ result from conditioning on $M_0, \ldots, M_{i-1}$ and event 1), respectively. Using the above terminology, we get:

$$
\begin{aligned}
P\big(M_i &= (y_i, c_i) \mid X_0 = d, L_d = l, X_1 \neq \emptyset, \\
& M_0 = (y_0, c_0), \ldots, M_{i-1} = (y_{i-1}, c_{i-1})\big) \\
&= P\big(M_i(1) \geq y_i \mid X_0 = d, L_d = l, X_1 \neq \emptyset, \\
& M_0 = (y_0, c_0), \ldots, M_{i-1} = (y_{i-1}, c_{i-1})\big) \\
& \cdot P\big(M_i = (y_i, c_i) \mid X_0 = d, L_d = l, X_1 \neq \emptyset, \\
& M_0 = (y_0, c_0), \ldots, M_{i-1} = (y_{i-1}, c_{i-1}), M_i(1) \geq y_i\big) \\
&= \left(\frac{1 - F_{d,l}(y_i - 1)}{1 - F_{d,l}(y_{i-1})}\right)^{C_{i-1}} \binom{C_{i-1}}{c_i} p_i^{c_i} (1 - p_i)^{C_i}.
\end{aligned}
\tag{16.8}
$$

For the γ'-th distinct value, the probability that there are *at least* $c_{\gamma'}$ equal values rather than exactly $c_{\gamma'}$ values is derived. There might be other contacts with the same distance in the bucket, which are not part of the chosen α contacts. So, we have:

$$
\begin{aligned}
P\big(M_{\gamma'} &= (y_{\gamma'}, c_{\gamma'}) \mid X_0 = d, L_d = l, X_1 \neq \emptyset, \\
& M_1 = (y_1, c_1), \ldots, M_{i-1} = (y_{\gamma'-1}, c_{\gamma'-1})\big) \\
&= \left(\frac{1 - F_{d,l}(y_{\gamma'})}{1 - F_{d,l}(y_{\gamma'-1})}\right)^{C_{\gamma'-1}} \\
& \quad \left(1 - \sum_{j=0}^{c_{\gamma'}-1} \binom{C_{\gamma'-1}}{j} p_{\gamma'}^{j} (1 - p_{\gamma'})^{C_{\gamma'-1-j}}\right).
\end{aligned}
\tag{16.9}
$$

By Equations (16.8), (16.9) and the fact that

$$
1 - \frac{F_{d,l}(y_j) - F_{d,l}(y_j - 1)}{1 - F_{d,l}(y_j - 1)} = \frac{1 - F_{d,l}(y_j)}{1 - F_{d,l}(y_j - 1)},
$$

Equation (16.7) can be simplified to

$$P\big(X_1 = (\delta_1, ..., \delta_\gamma)|X_0 = d, L_d = l\big)$$

$$= \left(\sum_{i=1}^{\gamma'-1} \binom{C_{i-1}}{C_i} (F_{d,l}(y_i) - F_{d,l}(y_i - 1))^{c_i}\right) \qquad (16.10)$$

$$\cdot \left((1 - F_{d,l}(y_{\gamma'}))^{C_{\gamma'}-1} - \sum_{j=0}^{c_{\gamma'}-1} \binom{C_{\gamma'}-1}{j}\right.$$

$$\left.\cdot (F_{d,l}(y_{\gamma'}) - F_{d,l}(y_{\gamma'} - 1))^j (1 - F_{d,l}(y_{\gamma'}))^{C_{\gamma'}-1-j}\right).$$

We can now determine the missing term $P(X_1 = s|X_0 = d, B = l)$ in Equation (16.4), which completes the derivation of the closest contacts distribution.

16.3.4 Derivation of *I*

The derivation of the initial distribution I requires the closest contact distribution as derived above and the distribution of X_0. For any state $s \in S$ with initial probability $I(s)$, we have the following:

$$I(s) = \sum_{d=0}^{b} P(X_1 = s|X_0 = d)P(X_0 = d). \qquad (16.11)$$

The probability that a random requesting node r has distance d to t corresponds to the fraction of IDs with this distance; hence,

$$P(X_0 = d) = \begin{cases} \frac{1}{2^b}, & d = 0 \\ \frac{2^{d-1}}{2^b}, & d > 0 \end{cases}. \qquad (16.12)$$

16.3.5 Derivation of *T*

The derivation of T is more complex, but it is based on similar concepts as earlier steps. Let A_0 be the random variable for the current state, and A_1 be the next state. The transition probability $P(A_1 = s|A_0 = s_0)$ is derived for all $s_0, s \in S$. The probability of the transition from s_0 to s is given by first considering all possible sets of $\alpha\beta$ returned contacts for state s_0. For each set of returned contacts, the probability distribution over the set of distinct contacts needs to be derived.

We start by considering case $A_0 = \emptyset$, for which

$$P(A_1 = s | A_0 = \emptyset) = \begin{cases} 1, & s = \emptyset \\ 0, & s \neq \emptyset \end{cases}$$

holds. The remaining entries of T are of the form $P(A_1 = s | A_0 = (d_1, \ldots, d_\alpha))$, where the next state s is either $\emptyset$ or a vector consisting of the distances of the α closest nodes queried in the next step. The probability $P(Z^j = s^j | A_0(j) = d_j)$ for the state $Z^j = s_j = (s_j^1, \ldots, s_j^\beta)$ of the closest β nodes in the routing table of the j-th queried node v_j is given by Equation (16.4). By Assumption 5, routing tables are chosen independently, so that

$$P\left(Z^1 = s^1, \ldots, Z^\alpha = s^\alpha | A_0 = (d_1, \ldots, d_\alpha)\right)$$
$$= \prod_{j=1}^{\alpha} P\left(Z^j = s^j | A_0(j) = d_j\right). \tag{16.13}$$

For each of the α considered contacts, the probability of termination is obtained from Equation (16.5), using the bucket size k_{d_j} and the d_j-th row of the matrix L. The probability to terminate in the next step is then given as the complement of the event that none of the parallel lookups terminates, i.e.,

$$P\left(A_1 = \emptyset | A_0 = (d_1, \ldots, d_\alpha)\right)$$
$$= 1 - \prod_{j=1}^{\alpha} \left(1 - P\left(Z_j = \emptyset | A_0(j) = d_j\right)\right). \tag{16.14}$$

If routing does not terminate, then it remains to obtain the closest α contacts from a set of $\alpha\beta$ returned contacts. Let $\Gamma = (s_1^1, \ldots, s_\beta^1, \ldots, s_\beta^\alpha)$ be the distances of the returned contacts. Due to the Markov property, we can only determine upper and lower bounds on the distance of replacement contacts from earlier steps or the requester's routing table. All known but not contacted nodes have distance at least d_α, so that for an upper bound on the success probability, we minimize the distance of a replacement contact by $K^{up} = d_\alpha$. In contrast, for a lower bound on the success probability, $K^{low} = b$ is chosen, corresponding to a replacement node with maximal distance to t. In the following, definitions and formulas specific to the upper bound are identified by the superscript up, whereas the superscript low characterizes the lower bound. We use $*$ to mean either low or up.

Denote by

$$U^*(\Gamma) = \{u = (u_1^1, \ldots, u_\beta^\alpha) : u_i^j \in \{s_i^j, K^*\}\}$$

all possible sets of distances of distinct contacts given the distances $(s_1^1, \ldots, s_\beta^\alpha)$. So, in general, we obtain the transition probabilities as

$$P\big(A_1 = (\delta_1, \ldots, \delta_\alpha) | (A_0 = (d_1, \ldots, d_\alpha)\big)$$
$$= \sum_\Gamma \sum_{u \in U_\delta(\Gamma)} P^*(u|\Gamma) \qquad (16.15)$$
$$\prod_{j=1}^\alpha P\big((Z^j(1), \ldots, Z^j(\beta)) = (s_1^j, \ldots, s_\beta^j) | A_0(j) = d_j\big)$$

with $U_\delta(\Gamma) = \{u \in U^*(\Gamma) : top_\alpha(u) = (\delta_1, \ldots, \delta_\alpha)\}$. In the following, we derive the probability $P^*(u|\Gamma)$ for each $u \in U^*(\Gamma)$. The underlying idea is to first find a maximal set Y^* of definitive distinct contacts and then iteratively determine for each remaining element the probability to be distinct from all elements in Y^* as well as contacts queried earlier during routing. The probability $P^*(u|\Gamma)$ for $u \in U(\Gamma)$ in Equation (16.15) is inductively computed, conditioning on Y^*. More precisely, we transform

$$P^*(u = (u_1^1, \ldots, u_\beta^\alpha)|\Gamma)$$
$$= P^*(u_1^1|Y^*, Z)P^*(u_2^1|Y^*, \Gamma, u_1^1) \qquad (16.16)$$
$$\cdots P^*(u_\beta^1|Y^*, \Gamma, u_1^1, \ldots, u_{\beta-1}^1)P^*(u_1^2|Y^*, \Gamma, u_1^1, \ldots, u_\beta^1)$$
$$\cdots P^*(u_\beta^\alpha|Y^*, \Gamma, u_1^1, \ldots, u_{\beta-1}^\alpha)$$

and determine each factor. For a distance a and a queried contact v_j, let y_a^j be the set of pairs (j, i), so that $s_j^i = a$. All entries in one set y_a^j are distinct, and so $y_a^{max} = argmax\{|y_a^j| : j = 1 \ldots \alpha\}$ contains the maximal number of contacts with distance a that are guaranteed to be unique. So all contacts in $Y^{low} = \bigcup_{a=0}^{d_\alpha-1} y_a^{max}$ are unique and have not been contacted before because d_1 is the minimal distance of all nodes contacted up to this point. In contrast, for the upper bound, earlier steps are not considered for computing the probability of a contact to be distinct, i.e., $Y^{up} = \bigcup_{a=0}^b y_a^{max}$.

The probability that the i-th node returned by v_j and having distance s_i^j to t is identical to an previously contacted node is given as the ratio of contacted nodes at distance s_i^j and all nodes at distance s_i^j. Consequently, we first compute the number of nodes $count(s_i^j, Y^*, \Gamma, u_1^1, \ldots, u_{i-1}^j)$ at distance s_i^j that have been contacted and may be identical if only nodes from the current set of returned contacts are considered, i.e., for the upper bound or if $s_i^j < d_1$, let

$$count\left(s_i^j, Y^*, \Gamma, u_1^1, \ldots, u_{i-1}^j\right) = |y_{s_i^j}^{max}| + |\{(\gamma, \mu) : s_i^j$$

$$= s_\mu^\gamma = u_\mu^\gamma, \gamma < j\}| - |\{\mu : s_i^j = s_\mu^j, u_\mu^j = r, \mu < i\}| \tag{16.17}$$

be the number of returned contacts that are potentially identical to the i-th returned contact of j-th queried node v_j. These consists of all returned contacts that have been decided to be unique up to this point. The subtraction follows from the fact that v_j's returned contacts are distinct. So if a different contact returned by v_j is identical to some contact w, we know that the i-th contact is not identical to w. On the other hand, if we are considering the lower bound and $s_i^j \geq d_1$, we set

$$count(s_i^j, Y^*, \Gamma, u_1^1, \ldots, u_{i-1}^j) = \alpha b, \tag{16.18}$$

since each parallel lookup is guaranteed to terminate after maximally b steps. The non-contacted number of nodes X_i^j at distance s_j^i is $B(n - \alpha\beta, 2^{s_i^j - 1 - b})$ distributed. Using the above terminology,

$$P^*(u_i^j = s_i^j | Y^*, \Gamma, u_1^1, \ldots, u_{\beta-1}^{j-1}, u_1^j, \ldots u_{i-1}^j)$$

$$= \sum_{m=0}^{n-\alpha\beta} P(X_i^j = m) \frac{m}{m + count(s_i^j)} \tag{16.19}$$

$$= \sum_{m=0}^{n-\alpha\beta} \binom{n - \alpha\beta}{m} \left(2^{s_i^j - 1 - b}\right)^m \left(1 - 2^{s_i^j - 1 - b}\right)^{n - \alpha\beta - m}$$

$$\frac{m}{m + count(s_i^j, Y^*, \Gamma, u_1^1, \ldots, u_{i-1}^j)}$$

for $(j, i) \notin Y^*$ and by construction

$$P^*(u_i^j = s_i^j | Y^*, \Gamma, u_1^1, \ldots, u_{\beta-1}^{j-1}, u_1^j, \ldots u_{i-1}^j) = 1 \tag{16.20}$$

if $(j, i) \in Y^*$. Inserting Equations (16.19) and (16.20) in Equation (16.16), the remaining term $P(u|\Gamma)$ in Equation (16.15) is determined. Equation (16.15) now gives the transition probabilities for general queries with the goal of finding a lower bound or upper bound on the hop count distribution. This completes our derivation of T^{low} and T^{up}.

16.3.6 Summary

We have modeled the hop count distribution in a Kademlia-type system as a Markov chain with an α-dimensional state space corresponding to the α contacted nodes in each step. We derived an initial distribution I on the closest contacts in the requester's routing table and transition matrices T^{low} and T^{up} for upper and lower bounds on the hop count distribution. The fraction of queries that need at most i steps is consequently bounded by computing the distributions H_i^{up} and H_i^{low} and choosing the entry corresponding to $\emptyset$.

Note that there are various possibilities to map the transition probabilities in Equation (16.15) to entries in the matrices. Any bijective mapping from S to $\mathbb{Z}_{|S|}$ (i.e., the row/column index) is suitable. On the basis of such mapping, we analyze the storage and computation complexity in the next section.

16.4 Model Complexity

In the first part of this section, we determine the space and computation complexity of deriving the hop count distribution. Finding that the complexity is at least $\mathcal{O}(b^\alpha)$, an evaluation of the accuracy of smaller ID spaces than the common 128 or 160 bits is considered.

16.4.1 Space Complexity

We assume that the whole matrix T needs to be stored, without any memory enhancements.

Lemma 16.4.1 *The storage complexity for computing the hop count distribution of a $\mathcal{K}(b, \alpha, \beta, k, L)$-system is $\mathcal{O}\left(\frac{1}{(\alpha!)^2} b^{2\alpha}\right)$.*

Proof. The storage complexity is dominated by the matrix $T \in \mathbb{R}^{|S|^2}$. Consequently, $|S|$ needs to be determined.

$$|S| = |\{\emptyset\} \cup \{s \in \mathbb{Z}_{b+1}^\alpha : s_j \leq s_{j+1}, j = 1 \ldots \alpha - 1\}|$$

$$= 1 + \sum_{i_\alpha=0}^{b} \sum_{i_{\alpha-1}=0}^{i_\alpha} \cdots \sum_{i_1=0}^{i_2} 1$$

$$= \mathcal{O}\left(\int_0^b \int_0^{x_\alpha} \cdots \int_0^{x_2} 1 dx_1 dx_2 \ldots dx_\alpha\right)$$

$$= \mathcal{O}\left(\frac{1}{\alpha!} b^\alpha\right)$$

The size of the matrix T is S^2 and by this the space complexity is $\mathcal{O}\left(\frac{1}{(\alpha!)^2}b^{2\alpha}\right)$ as claimed.

16.4.2 Computation Complexity

We bound the computation complexity in terms of both the number of nodes n and the number of bits b.

Lemma 16.4.2 *The computation complexity is linear with regard to the network order n, and polynomial with regard to the bit number b. More precisely, the complexity is of order $\mathcal{O}(nb^{\alpha(\beta+2)})$.*

Proof. We need to analyze the computation costs for the initial distribution I, the transition matrix T, and the matrix multiplication for obtaining P_i.

Note that in the case of both I and T, the computation of $\gamma \in \{\alpha, \beta\}$ closest neighbor distribution is essential. The success probability given in Equation (16.5) can be determined in $\mathcal{O}(n)$ if binomial coefficients are computed iteratively. Note that this has to be done for $d = 1, ..., b$, resulting in a cost of

$$H_\emptyset = \mathcal{O}(n \cdot b). \tag{16.21}$$

These probabilities can be precomputed and stored, as can the values of the cumulative distributions $F_{d,l}$ and the binomial coefficients used in Equation (16.10). Using iterative computations of powers and binomial coefficients, the cost of these computations is

$$H_P = \mathcal{O}(b^3 + \max\{\alpha, \beta\}^2). \tag{16.22}$$

The term b^3 for the CDF computations follows because there are $\mathcal{O}(b^2)$ functions (one for each $d, l \in \mathbb{Z}_{b+1}$), each taking up to b distinct values. Assuming precomputation, one evaluation of Equation (16.10) has computation cost $\mathcal{O}(\gamma\kappa)$ for $\kappa = \max\{k_d : d = 0...b\}$. To see this, consider that the sum from 1 to $\gamma' - 1$ has at most $\gamma - 1$ summands that are products of terms $F_{d,l}(y_i) - F_{d,l}(y_i - 1)$ and binomial coefficients. The remaining factor for the last term has at most γ summands, each consisting of at most $k_d + 1$ factors. Note that for each pair d, l, there are $d - l$ distances a node in the respective bucket can have. Therefore, the number of evaluations for a given distance d is, similarly to Lemma 4.1, bounded by

$$\sum_{l=1}^{d} \sum_{\delta_\alpha=0}^{d-l} \sum_{\delta_{\alpha-1}=0}^{\delta_\alpha} \cdots \sum_{\delta_1=0}^{i_2} 1$$

$$= \mathcal{O}\left(\int_0^d \int_0^{y-z} \int_0^{x_\alpha} \cdots \int_0^{x_2} 1 dx_1 dx_2 \ldots dx_\alpha dz \right) \qquad (16.23)$$

$$= \mathcal{O}\left(\int_0^d \frac{1}{\gamma!}(d-z)^\gamma dz \right) = \mathcal{O}\left(\frac{1}{(\gamma+1)!} d^{\gamma+1} \right).$$

For the initial distribution, the computation cost is hence given by

$$H_I = \mathcal{O}\left(\frac{1}{(\alpha+2)!} b^{\alpha+2} \alpha \kappa \right), \qquad (16.24)$$

summarizing over $d = 1 \ldots b$. The additional computation cost of the requester's distance distribution X_0 in Equation (16.12) is $\mathcal{O}(b)^2$, which is clearly dominated by the computation cost of closest neighbor distribution.

In contrast, for computing the transition matrix, one first has to consider all possible $\alpha\beta$ returned values for each state $A_0 = (d_1, \ldots, d_\alpha)$. The number of returned sets is determined based on Equation (16.23) with $\gamma = \beta$.

$$\mathcal{O}\left(\sum_{d_\alpha=0}^{b} \frac{1}{(\beta+1)!} d_\alpha^{\beta+1} \cdots \sum_{d_1=0}^{d_2} \frac{1}{(\beta+1)!} i_1^{\beta+1} \right)$$

$$= \mathcal{O}\left(\frac{1}{((\beta+1)!)^\alpha} \int_0^b x_\alpha^{\beta+1} \cdots \int_0^{x_2} x_1^{\beta+1} dx_1 \ldots dx_\alpha \right)$$

$$= \mathcal{O}\left(\frac{1}{((\beta+1)!)^\alpha} \int_0^b x_\alpha^{\beta+1} \cdots \int_0^{x_3} \frac{1}{\beta+2} x_2^{2\beta+3} dx_2 .. dx_\alpha \right)$$

$$= \mathcal{O}\left(\frac{1}{((\beta+1)!)^\alpha} b^{\alpha\beta+2\cdot\alpha} \prod_{j=1}^{\alpha} \frac{1}{j\beta+2\cdot j} \right)$$

$$= \mathcal{O}\left(\frac{1}{((\beta+1)!)^\alpha} b^{\alpha(\beta+2)} \prod_{j=1}^{\alpha} \frac{1}{(\beta+2)j} \right).$$

For each set of returned contact, all possible distinct sets have to be evaluated. This results in a factor $\mathcal{O}\left(2^{\alpha\beta}\right)$. The probability of each contact being distinct is determined requiring at most $\mathcal{O}(n)$ operations by Equation (16.19).

---

[2]As usual, we assume that powers of two are calculated iteratively.

However, it is possible to precompute the probabilities for all distances d and $(\alpha-1)\beta+1$ values of count (Equations (16.17) and (16.18)). So an additional cost of $\mathcal{O}(n\alpha\beta 2^{\alpha\beta})$ per set of returned contacts is needed. Furthermore, for each of these combinations the function T_α is applied, which is a factor of $\mathcal{O}(\alpha^2\beta)$. The total complexity of transition matrix computation is hence

$$H_T = \mathcal{O}\left(\frac{1}{((\beta+1)!)^\alpha}b^{\alpha(\beta+2)}\prod_{j=1}^{\alpha}\frac{1}{(\beta+2)j}\beta\kappa n\alpha\beta 2^{\alpha\beta}\right). \qquad (16.25)$$

It remains to determine the overhead of matrix multiplication. The target is definitely reached after at most $b+1$ steps, since the distance of the first lookup decreases by at least one in each step. Each matrix multiplication takes $|S|^2$ operations. By the proof of Lemma 4.1, the complexity of the matrix operations is hence

$$H_M = \mathcal{O}\left(|S|^2 b\right) = \mathcal{O}\left(\frac{1}{(\alpha!)^2}b^{2\alpha+1}\right). \qquad (16.26)$$

Summarizing overall computation costs, the total complexity is

$$H_\emptyset + H_P + H_I + H_T + H_M$$
$$= \mathcal{O}\left(nb + b^3 + \max\{\alpha,\beta\}^2 + \alpha\kappa\frac{1}{(\alpha+2)!}b^{\alpha+2}\right.$$
$$+\frac{1}{((\beta+1)!)^\alpha}\prod_{j=1}^{\alpha}\frac{1}{(\beta+2)j}\kappa\alpha\beta^2 2^{\alpha\beta}nb^{\alpha(\beta+2)}$$
$$\left.+\frac{1}{(\alpha!)^2}b^{2\alpha+1}\right)$$

basic operations by Equations (16.21), (16.22), (16.24)–(16.26). Treating κ, α, and β as constant factors gives the claimed complexity.

16.4.3 Reducing the ID Space Size

From Lemmas 4.1 and 4.2, we can see that both the storage and the computation complexity are polynomial in the bit-size b for a relative high-degree polynomial. In contrast, the dependence on the network order n is only linear for the computation complexity, whereas the storage complexity is independent of n. Though the dependence on α and β is exponential, both can be assumed to be small. For instance, if $\alpha = 3$, the number of entries in the matrix T can be precisely computed as

$$|\{F\} \cup \{(s_1, s_2, s_3) \in \mathbb{Z}_{b+1}^3 : s_1 \leq s_2 \leq s_3\}|^2$$

$$= \left(1 + \sum_{i_3=0}^{b} \sum_{i_2=0}^{i_3} \sum_{i_1=0}^{i_2} 1\right)^2$$

$$= \left(1 + \sum_{i_3=1}^{b+1} \frac{i_3(i_3 + 1)}{2}\right)^2$$

$$= \left(1 + 0.5 \cdot \sum_{i_3=1}^{b+1} (i_3^2 + i_3)\right)^2$$

$$= \left(1 + 0.5 \cdot \left(\frac{(b+1)(b+2)(2b+3)}{6} + \frac{(b+1)(b+2)}{2}\right)\right)^2$$

$$= \left(1 + \frac{(b+1)(b+2)(2b+6)}{12}\right)^2 .$$

In practice, $b = 128$ and $b = 160$ are typically used, corresponding to the length of MD-5 and SHA hashes. For these sizes, the matrix T has $134,062,161,025$ and $502,058,690,721$ entries, respectively. Assuming 32 bit float numbers, this amounts to roughly 500 GB and $1,870$ GB, respectively.

Consequently, the computations are too expensive to present an alternative to extensive simulations. However, the dependence on the actual ID space size can be expected to be small, at least if the number of IDs is decisively higher than the number of nodes. The following Lemma provides an upper bound on the influence of b.

Lemma 16.4.3 *Consider two Kademlia-type systems* $K = \mathcal{K}(b, \alpha, \beta, k, L)$ *and* $\tilde{K} = \mathcal{K}(\tilde{b}, \alpha, \beta, \tilde{k}, \tilde{L})$, *such that*

- $\tilde{b} < b$.
- $k[0..\tilde{b}] = \tilde{k}$; *i.e., the vector* $\tilde{k}$ *contains exactly the* $\tilde{b} + 1$ *first entries of* k.
- $L[0..\tilde{b}][0..\tilde{b}] = \tilde{L}$; *i.e., the matrix* $\tilde{L}$ *is the square matrix containing the first* $\tilde{b} + 1$ *first entries of* L.

The fraction of terminated queries after i hops in K and $\tilde{K}$ differs by at most

$$|P^*(i) - \tilde{P}^*(i)| \leq 1 - \sum_{j=0}^{\kappa} \binom{n}{j} p^j (1-p)^{n-j} \qquad (16.27)$$

with $p = 2^{-\tilde{b}} - 2^{-b}$, $\kappa = \min\{k_d : d = 0 \ldots b\}$, and $* \in \{up, low\}$.

Proof. Note that the two systems entail different results if there are at least κ nodes that share a common prefix of length at least $\tilde{b}$ with the target t, but not with a common prefix length of b. Recall that the query is considered successful in the next step after reaching a node with common prefix length $\tilde{b}$ by Assumption 6 (see Section 16.3.1) in $\tilde{K}$, but not necessarily in K. The probability that two nodes share a common prefix of length $\tilde{b}$ to $b - 1$ is $p = 2^{-\tilde{b}} - 2^{-b}$. The number of nodes with this property is hence $B(n, p)$-distributed. The claim follows directly.

Based on Equation (16.27), we can consider the trade-off between accuracy and computation speed in terms of the network order n.

Theorem 16.4.4 *When the error is supposed to be bounded by δ for some $C > 0$, $\tilde{b} = \log_2(n\frac{2}{\ln 1/\delta})$ achieves the required accuracy. Consequently, the storage complexity is $\mathcal{O}\left(\log^{2\alpha} n\right)$ and the computation complexity is $\mathcal{O}\left(npolylog(n)\right)$ for a constant error δ.*

Proof. For $\kappa = 0$ in Equation (16.27), we can determine an upper bound on the minimal value for $\tilde{b}$ to achieve an error of less than δ. Set $C = \frac{2}{\ln 1/\delta}$ in the following. From

$$\delta \leq 1 - (1-p)^n < 1 - (1 - 2^{-\tilde{b}})^n,$$

it follows that

$$\tilde{b} \geq \log_2 \frac{1}{1 - \delta^{1/n}}.$$

It remains to show that for n large enough,

$$\frac{1}{1 - \delta^{1/n}} < Cn$$

Rewriting results in $\delta < \left(1 - \frac{1}{Cn}\right)^n$. Because $\left(1 - \frac{1}{Cn}\right)^n$ converges to $e^{-1/C}$, there exists n, such that

$$\left(1 - \frac{1}{Cn}\right)^n > e^{-2/C} = e^{-\ln 1/\delta} = \delta$$

Therefore, for n large enough $\tilde{b} \geq \log_2\left(n\frac{2}{\ln 1/\delta}\right)$ ensures that $|P^*(i) - \tilde{P}^*(i)| \leq \delta$.

The storage complexity is $\mathcal{O}\left(\log^{2\alpha} n\right)$ by Lemma 4.1 with $\tilde{b} = \mathcal{O}\left(\log n\right)$, the computation complexity of $\mathcal{O}\left(npolylog(n)\right)$ follows from Lemma 4.2.

Note that by using Equation (16.27) rather than the approximation in Theorem 4.4, the bound on $\tilde{b}$ can be further reduced. However, we have shown in Theorem 4.4 that the number of bits needed for a certain accuracy grows at most logarithmically in the network size.

In this section, we have seen that the storage complexity of our analytical solution is polylog in the number of nodes, whereas simulations require at least a linear overhead. The computation complexity is slightly higher than linear, but simulations are bound to require a similar cost for establishing the routing tables, not even considering the actual routing performance. Our results in Sections 16.5 and 16.7 show that our model can easily compute hop count distributions for large network sizes.

16.5 Verification and Scalability

In this section, we compare the model with static simulations as well as with real-world measurements.

16.5.1 Model Verification

We considered three routing table structures: MDHT, iMDHT, and KAD, described in Section 16.2. As for the routing parameters, we focus on the two settings which are used by widely used Kademlia implementations: ($\alpha = 3$, $\beta = 2$) and ($\alpha = 4$, $\beta = 1$).

The error rate is chosen as $\delta = 0.001$, which can be expected to be far below the confidence intervals length of simulation results. By Equation (16.27), the numbers of bits needed for the desired accuracy are 14 ($100K$, KAD), 15 ($100K$, MDHT and iMDHT), and 21 ($10M$). The first value is lower for KAD than MDHT and iMDHT, because KAD has a bucket size of 10 rather than 8.

For validation, we use the simulation framework GTNA [27], a simulation framework for network analysis. GTNA offers a variety of metrics for analyzing graphs, as well as an easily extensible routing algorithm interface. We extended the framework by adding the MDHT, iMDHT, and KAD systems, as well as the Kademlia routing algorithm.[3] We chose GTNA rather than an event-based simulator for various reasons. Most importantly, our initial model does not consider churn, failure, and varying latencies between peers. Including such behavior in the simulation environment will inevitably lead to derivations of the analytic results and the observed hop counts. However, it is not possible to easily distinguish between these derivations and actual faults in the model. Using a network simulator without enabling real network conditions is an overhead with regard to storage space, computation time as well as implementation complexity. GTNA can easily scale to $1M$ nodes, which is hard to achieve by an event-based simulator, e.g., OverSim [28].

We generate overlay topologies as follows: Each node v is given a 128-bit identifier, and its routing table is constructed by first randomizing the list of nodes. Nodes in the list are considered iteratively and added to v's routing table if there is an empty slot in the corresponding bucket. In this way, maximally full buckets are realized, which cannot be guaranteed by the real-world protocol. The routing algorithm progresses step-wise, always querying α nodes and processing all answers, before contacting the next set of nodes. We generated 20 topologies uniformly at random, and routed to five distinct, randomly selected, target nodes from each node's routing table.

Tables 16.2 and 16.3 show the resulting cumulative hop count distributions for a $100K$-node network with two different parameter settings ($\alpha = 3$, $\beta = 2$ and $\alpha = 4$, $\beta = 1$, respectively). Both the upper and lower analytic bounds are shown for the three systems and the two sets of routing parameters. Furthermore, the 95% confidence interval of the simulations is given. Upper and lower bounds are extremely close (at most an absolute difference of 0.2%), and both are within the confidence interval of the simulations. The negligible difference between the upper and lower bounds can be expected, seeing that the success probability for the first two steps is identical and most routes terminate within three hops. The parameters $\alpha = 3$, $\beta = 2$ (Table 16.2) and $\alpha = 4$, $\beta = 1$ (Table 16.3) are hard to distinguish, but the later achieves a slightly higher success rate for each hop. We discuss the impact of the routing algorithms as well as the routing table structure in Section 16.7. All in all, the results show a strong agreement between the model and the simulations, indicating that both derivation and implementation are indeed correct.

---

[3]The code is available at: https://github.com/stef-roos/GTNA/tree/grouting

Table 16.2 Model vs. Simulation: Cumulative hop count distribution of MDHT, iMDHT, and KAD (100K nodes, $\alpha = 3$, $\beta = 2$)

| | MDHT | | | iMDHT | | | KAD | | |
| | | Model | | | Model | | | Model | |
Hops	Simulation	Lower Bound	Upper Bound	Simulation	Lower Bound	Upper Bound	Simulation	Lower Bound	Upper Bound
0	0	0	0	0	0	0	0	0	0
1	0.001157	0.001157	0.001157	0.003218	0.003237	0.003237	0.0061686	0.006177	0.006177
2	0.043913	0.043973	0.043973	0.159117	0.158934	0.158934	0.4946125	0.490293	0.490293
3	0.450753	0.449902	0.449903	0.879459	0.877503	0.877504	0.9999939	0.999989	0.999989
4	0.962199	0.961339	0.961338	0.999866	0.999838	0.999838	1	0.999999	0.999999
5	0.999951	0.999947	0.999946	1	0.999999	0.999999	1	0.999999	0.999999
6	1	0.999999	0.999999	1	0.999999	0.999999	1	0.999999	0.999999

Table 16.3 Model vs. Simulation: Cumulative hop count distribution of MDHT, iMDHT, and KAD (100K nodes, $\alpha = 4$, $\beta = 1$)

| | MDHT | | | iMDHT | | | KAD | | |
| | | Model | | | Model | | | Model | |
Hops	Simulation	Lower Bound	Upper Bound	Simulation	Lower Bound	Upper Bound	Simulation	Lower Bound	Upper Bound
0	0	0	0	0	0	0	0	0	0
1	0.001141	0.001157	0.001157	0.003218	0.003237	0.003237	0.006188	0.006177	0.006177
2	0.045975	0.045936	0.045936	0.167163	0.167168	0.167168	0.516323	0.512319	0.512319
3	0.459710	0.457843	0.457867	0.896495	0.893489	0.894811	0.999997	0.999996	0.999996
4	0.966182	0.964492	0.964534	0.999934	0.999916	0.999931	1	0.999999	0.999999
5	0.999975	0.999967	0.999968	1	0.999999	0.999999	1	0.999999	0.999999
6	1	0.999999	0.999999	1	0.999999	0.999999	1	0.999999	0.999999

16.5.2 Scalability

In this section, the hop count distribution is analytically computed for networks of $10M$ nodes. Besides showing that the model easily scales, it is essential to see that the divergence between upper and lower bounds is acceptable for giving meaningful performance bounds.

As shown in Tables 16.4 and 16.5, upper and lower bounds remain extremely close, differing at most by 0.5% in the case of MDHT using $\alpha = 4$ and $\beta = 1$.

Table 16.4   Cumulative hop count distribution of MDHT, iMDHT, and KAD ($10M$ nodes, $\alpha = 3, \beta = 2$)

	MDHT		iMDHT		KAD	
	Model		Model		Model	
Hops	Lower Bound	Upper Bound	Lower Bound	Upper Bound	Lower Bound	Upper Bound
0	0	0	0	0	0	0
1	1.689E-5	1.689E-5	3.768E-5	3.768E-5	9.500E-5	9.500E-5
2	0.001110	0.001110	0.003456	0.003456	0.022689	0.022689
3	0.030272	0.030272	0.100601	0.100601	0.656786	0.656786
4	0.299162	0.299163	0.658978	0.658979	0.999958	0.999958
5	0.846644	0.846645	0.989410	0.989410	0.999999	0.999999
6	0.996839	0.996839	0.999991	0.999991	0.999999	0.999999
7	0.999997	0.999997	0.999999	0.999999	0.999999	0.999999
8	0.999999	0.999999	0.999999	0.999999	0.999999	0.999999

Table 16.5   Cumulative hop count distribution of MDHT, iMDHT, and KAD ($10M$ nodes, $\alpha = 4, \beta = 1$)

	MDHT		iMDHT		KAD	
	Model		Model		Model	
Hops	Lower Bound	Upper Bound	Lower Bound	Upper Bound	Lower Bound	Upper Bound
0	0	0	0	0	0	0
1	1.689E-5	1.689E-5	3.768E-5	3.768E-5	9.501E-5	9.501E-5
2	0.001149	0.001149	0.003576	0.003576	0.023688	0.023688
3	0.029770	0.029770	0.099622	0.099653	0.670569	0.670569
4	0.281589	0.281603	0.648982	0.649768	0.999983	0.999983
5	0.826464	0.826517	0.989528	0.989840	0.999999	0.999999
6	0.996271	0.996281	0.999994	0.999995	0.999999	0.999999
7	0.999997	0.999997	0.999999	0.999999	0.999999	0.999999
8	0.999999	0.999999	0.999999	0.999999	0.999999	0.999999

In general, we see that lookups terminate fastly in the KAD system, and faster in iMDHT than MDHT, for all considered network sizes and routing parameters. Note that the difference between the two routing algorithms is more noticeable than for $100K$ nodes. Furthermore, $\alpha = 3$, $\beta = 2$ achieves a higher success rate for MDHT from the third hop onward.

16.5.3 Real-World Measurements

After validating the model in a controlled environment, we compare our results with real-world measurements. Because real-world KAD routing tables have been shown to contain a lot of stale entries as well as missing entries [4, 5], it is expected that the results of our model differ from the measured ones. It remains to quantify the difference between the static model and the dynamic real-world system. In practice, hop counts have been measured in KAD of $1M$ nodes [4]. The measured average hop count of 3.08 is reasonably higher than our prediction of 2.81. Note that we compare the results of our model to the result of [4] rather than [5, 26] or [6], because the first considers locating files with a high number of replicates rather than source-destination lookups and the latter only gives latencies. The network has been found to have about 10% of stale contacts as well as about 15% of missing entries, which are bound to slow down the routing process. As a consequence, obtaining reasonable bounds on deployed networks requires enhancing our model to deal with churn and routing tables incompleteness. In the next section, we give an initial approach for dealing with failures and routing table incompleteness, which closes the gap between the model and the reality.

16.6 Extending the Model

In this section, we exemplary show how to modify the model to account for stale entries and bucket incompleteness.

The model treats non-responding contacts as follows: With a probability of $1 - p$, a queried node is online and returns new contacts following the distribution described in Section 16.3; otherwise, there are no returned values for this contact. If less than α distinct contacts are returned altogether, the remaining distance values are chosen as the highest distance d_α of the currently queried nodes (for an upper bound on the hop count) and the overall maximal distance of b bits (lower bound), analogously to Section 16.3. Formally, the state Off is added to characterize an unresponsive node. Assume that X_j is the distance of the j-th closest contact v_j, and L_d is the guaranteed bit gain as in Section 16.3. Then, the probability distribution of contacts Y_j returned by v_j is given by

$$P(Y_j = s | X_j = d, L_d = l) =$$

$$\begin{cases} p, & s = \textit{Off} \\ (1-p)P(Y_j = s | X_j = d, L_d = l, s \neq \textit{Off}), & s \neq \textit{Off} \end{cases}. \quad (16.28)$$

$P(Y_j = s | X_j = d, B = l, s \neq \textit{Off})$ is determined in Equations (16.3) and (16.5). One more change has to be made with regard to the model in Section 16.3. The lower bound on the success rate relies on the fact that the distance to the target decreases in every step. This cannot be guaranteed because of fallbacks due to failures. For this reason, a hops-to-live counter, htl, is added, which is the maximal number of routing steps until the query is aborted. As a result, at most $count = htl \cdot \alpha$ nodes can be contacted during routing, rather than the bounds provided by Equations (16.17) and (16.18). Note that this change only influences the lower bound on the success probability. For the upper bound, Equation (16.17) still applies because all earlier steps are disregarded.

Our modification regarding bucket incompleteness is simple: We reduce the bucket size to a distance-dependent factor $c[d]$ of its actual value. An accurate model is to use level-dependent distributions on the actual number of nodes per bucket; however, it is unlikely to obtain representative data while designing the systems, so that aggregates offer probably a similar accuracy and reduce the complexity in terms of both comprehensibility and actual computation cost.

The remainder of this section deals with the comparison of the extended model to measurements. Without considering bucket incompleteness, the average hop count is bounded by 2.90 and 2.91 (with $htl = 7$, which achieves 99.7% of successfully terminated queries) for a stale entry rate $p = 0.1$ in a KAD network with $1M$ nodes. This is still considerably lower than the bound of 3.08 observed in [4]. However, their measurements reveal that there are in average about 1.5 missing entries per bucket. Indeed, the average hop count is increased to 3.0 for both upper and lower bounds if the bucket size is reduced to $c[d] = 0.9$ for $d = b - 9...b$ and $c[d] = 0.8$ for $d < b - 9$ of its actual value (in rough agreement with the per-level averages in [4]). The remaining difference can be explained by using averages for the missing entries rather than the actual distribution. Despite the expected discrepancy between theory and measurements, our model more than halves the error rate from 5.5%, provided in [4] when integrating all of the above in their analytic model, to merely 2.67%. In conclusion, our model can be extended to include failure rates and bucket incompleteness, as long as rough estimates of these parameters are known.

16.7 Lessons Learned

In this section, we analyze the influence of the routing parameters α and β as well as the routing table structure on the average hop count and the resilience to stale entries. For this purpose, we denote the routing algorithm with parameters α and β by $R_{\alpha,\beta}$. Furthermore, MDHT and KAD refer to the routing table structures in the corresponding systems, independent of the routing algorithm. Networks of order $n = 2^i \cdot 1000$ are evaluated for $i = 0...20$; i.e., our model scales easily up to $1B$ nodes. As in Section 16.5, the error rate is chosen as $\delta = 0.001$. Only upper bounds on the hop count are presented in favor of readability. However, results for the lower bounds are very similar and entail the same conclusions.

For all considered parameters, the hop count has been shown to increase at most logarithmically with $n = 2^i$ [1]. However, if the number of contacts per level is limited by k, the expected out-degree increases by k whenever the number of nodes doubles, so that the average shortest path length is of order $d = \mathcal{O}\left(\frac{i}{\log i + \log k}\right)$ (solving $(ik)^d = 2^i$) rather than $i = \log n$. Due to the extremely short routes observed in large networks, the hop count can be expected to follow a similar dependence. Indeed, the sub-logarithmic routing complexity is visualized as a slight curvature in the log plot (see Table 16.6), which is more noticeable if the number of contacts per level is higher.

We start our evaluation of real-world systems by analyzing if the change from $R_{3,2}$ to $R_{4,1}$ in MDHT actually decreases the average hop count. In addition, we also consider the influence on the KAD routing table structure. In order to evaluate the impact of churn, the fraction f of stale contacts is

Table 16.6 Impact of different routing parameters on the average hop count for $1K$, $1M$, $1B$ nodes

	MDHT				KAD			
	$\alpha=3,\beta=2$		$\alpha=4,\beta=1$		$\alpha=3,\beta=2$		$\alpha=4,\beta=1$	
n	f = 0	f = 0.2	f = 0	f = 0.2	f = 0	f = 0.2	f = 0	f = 0.2
1K	2.259971	2.352179	2.236963	2.326310	1.7140267	1.738595	1.714027	1.718999
1M	4.190581	4.516146	4.199727	4.661209	2.9042929	3.069776	2.891574	3.129235
1B	6.123873	6.702001	6.242113	7.318988	4.1790780	4.535189	4.187086	5.043731

varied between 0.0 and 0.2. It can be expected that for smaller networks, the use of a higher value for α actually increases the success rate because more routing tables are considered in each hop. However, when the network size increases, the high fallback in the case of duplicates or failures for $\beta = 1$ is bound to decrease the performance, so that there is a threshold above which $R_{3,2}$ achieves shorter routes. The advantage of $R_{3,2}$ is bound to be more obvious when the fraction of stale entries is high because of the increased use of fallback contacts.

Indeed, in the absence of churn, about half a million nodes are needed for $R_{3,2}$ to have a lower hop count for MDHT, but for KAD the threshold is only reached at half a billion nodes. As expected, $R_{3,2}$ deals with churn more effectively due to the high number of returned contacts to choose from. Assuming a stale entry rate of 20%, the average hop count in MDHT is increased by up to 8% for $R_{3,2}$, but more than 12% for $R_{4,1}$ assuming 8 million participants. The relative performance degradation increases with the network size due to the the longer routes, so that for $1B$ nodes, a divergence of up to 21% occurs.

Interestingly, for common real-world networks of $1M$ to $10M$ nodes, the change from $R_{3,2}$ to $R_{4,1}$ would have decreased the hop count in the KAD system for a very low stale entry rate, but not in MDHT, which actually introduced this change. Note that the number of contacted nodes per hop is not increased by a high value for β, but grows linearly with α. Overall, an increased number of returned contacts β is preferable to an increased degree of parallelism α in dynamic networks with a non-negligible stale entry rate.

The second part of this section deals with the impact of multiple buckets per level. For the evaluation, we first compare KAD to a Kademlia with the same limited number of contacts per level, i.e., $k_b = 80$ contacts on the first level and $k_i = 50$ for $i < b$ on all lower levels, denoted Kademlia80:50. Furthermore, the 5 buckets on the same level are responsible for different sized fractions of the ID space in KAD. In order to evaluate the influence of such a skewed split, a KAD (called KAD4) with buckets 111, 110, 101, and 100 for all lower levels, together with the respective version of Kademlia (Kademlia80:40) is analyzed. Furthermore, churn and routing table incompleteness are also included in the evaluation, using the parameters from Section 16.6: The stale entry rate is chosen as 0.1, and the bucket size is reduced to 0.9 of its actual value for the first 10 levels and 0.8 for the remaining levels.

Expectations on the hop count can be derived from the expected bit gain per level: KAD offers a slightly lower bit gain on all levels but the first (See [4], Equation (16.6)). In contrast, KAD4 offers a slightly higher expected bit

Table 16.7 Impact of different routing table structures on the average hop count for $1K$, $1M$, $1B$ nodes

n	KAD	Kademlia 80:50	KAD4	Kademlia 80:40	KAD Fail	Kademlia 80:50 Fail	KAD4 Fail	Kademlia 80:40 Fail
$1K$	1.7140267	1.707412	1.739189	1.737457	1.739338	1.731071	1.763333	1.758744
$1M$	2.9042929	2.901251	2.956258	2.964655	3.021491	2.998661	3.099768	3.087984
$1B$	4.1790780	4.171878	4.310453	4.324178	4.444238	4.407622	4.587251	4.569379

gain than Kademlia80:40 on all levels. So, we can expect KAD to have a worse performance in terms of the average hop count than Kademlia80:50. In contrast, KAD4 is bound to outperform Kademlia80:40 for networks of a sufficiently large size, at least in the absence of churn. Note that the average degree is lower in KAD/KAD4, because the one bucket in the respective Kademlia version is full if all KAD buckets are full, but not vice versa. Consequently, resilience to node failures should be higher in Kademlia.

The results not only agree with the above expectations, but also show that the influence of multiple buckets is small, as shown in Table 16.7. In the absence of churn, the advantage of Kademlia80:50 is barely noticeable, being always in the order of 0.006 to 0.007 hops. On the other hand, when considering churn and bucket incompleteness, Kademlia80:50 has an advantage of up to 0.04 hops due to the higher fraction of queries that terminate in the first few hops. The difference between KAD4 and Kademlia80:40 is even less, at most 0.002 hops. In the absence of churn, KAD4 indeed achieves a slightly reduced hop count, whereas Kademlia80:40 has a similarly small advantage when considering churn. Note that multiple buckets per level reduce the average routing table size by about 7 (KAD) and 2 (KAD4). Given that routing tables usually contain hundreds to thousands of contacts, such a small constant storage advantage is negligible.

All in all, our results indicate that multiple buckets reduce the routing table size slightly, but only have a positive effect on the average hop count if the churn rate is low and the ID space is split equally. Indeed, the observed advantage of an equal split between multiple buckets can also be achieved by a modified replacement algorithm in one bucket, which prefers contacts that enhance the diversity of the IDs in the bucket.

16.8 Conclusion

We introduced a scalable accurate computation of the hop count distribution in Kademlia-type systems—the only widely deployed structured P2P systems. Both simulations and measurements validated our model. Furthermore, we demonstrated the utility of our model by analyzing common design decisions in Kademlia-type systems, showing that returning $\beta > 1$ contacts per query is essential for achieving shorter routes both in static and in dynamic environments. In addition, we found that having multiple buckets per level does not necessarily increase the performance but degrades the resilience, and suggests a modified replacement strategy to combine the higher resilience of a single bucket per level with the increased bit gain of multiple buckets.

Whereas the model covers all common routing table structures, alterations in the routing process, such as interleaving queries and recursive routing as well as a vulnerability analysis of the systems in face of attacks, remain future work.

References

[1] Maymounkov, P., and Mazières, D. (2002). Kademlia: a peer-to-peer information system based on the XOR metric. In *IPTPS*.

[2] Junemann, K., et al. (2011). Towards a Basic DHT Service: Analyzing Network Characteristics of a Widely Deployed DHT. In *GridPeer*.

[3] Salah, H., and Strufe, T. (2013). Capturing connectivity graphs of a large-scale P2P overlay network. In *HotPOST*.

[4] Stutzbach, D., and Rejaie, R. (2006). Improving lookup performance over a widely-deployed DHT. In *INFOCOM*.

[5] Steiner, M., et al. (2010) Evaluating and improving the content access in KAD. *Peer-to-peer networking and applications*.

[6] Jimenez, R., et al. (2011). Sub-Second Lookups on a Large-Scale Kademlia-Based Overlay. In *P2P*.

[7] Roos, S., Salah, H., and Strufe, T. (2015). Determining the hop count in kademlia-type systems. In *ICCCN*.

[8] Wang, P., et al. (2008). Attacking the kad network. In *SecureComm*.

[9] Steiner, M. et al. (2007). A Global View of KAD. In *IMC*.

[10] Falkner, J., et al. (2007). Profiling a Million User DHT. In *IMC*.

[11] Crosby, S., and Wallach, D. (2007). An analysis of bittorrent's two kademlia-based DHTs. Technical report, Rice University.

[12] Einziger, G., Friedman, R., and Kibbar, E. (2013). Kaleidoscope: Adding Colors to Kademlia. In *IEEE P2P*.

[13] Wang, C. et al. (2006). DiCAS: An efficient distributed caching mechanism for P2P systems. *TPDS*.

[14] Kaune, S. et al. (2008). Embracing the peer next door: Proximity in kademlia. In *IEEE P2P*.

[15] Castroand, M. et al. (2002). Exploiting network proximity in distributed hash tables. In *FuDiCo*.

[16] Heep, B. (2010). R/Kademlia: Recursive and topology-aware overlay routing. In *IEEE ATNAC*.

[17] Kleinberg, J. (2000). The small-world phenomenon: An algorithmic perspective. In *STOC*.

[18] Singh Manku, G. et al. (2004). Know thy neighbor's neighbor: The power of lookahead in randomized P2P networks. In *STOC*.

[19] Lebhar, E., and Schabanel, N. (2004). Almost optimal decentralized routing in long-range contact networks. In *ICALP*.

[20] Fraigniaud, P., and Giakkoupis, G. (2009). The effect of power-law degrees on the navigability of small worlds. In *PODC*.

[21] Stoica, I. et al. (2001). Chord: A scalable peer-to-peer lookup service for internet applications. In *SIGCOMM*.

[22] Rowstron, A., and Druschel, P. (2001). Pastry: Scalable, Decentralized Object Location, and Routing for Large-scale Peer-to-Peer Systems. In *Middleware*.

[23] Malkhi, D., Naor, M., and Ratajczak, D. (2002). Viceroy: A scalable and dynamic emulation of the butterfly. In *PODC*.

[24] Spognardi, A., and Di Pietro, R. (2006). A formal framework for the performance analysis of P2P networks protocols. In *IPDPS*.

[25] Rai, I., et al. (2007). Performance Modelling of Peer-to-Peer Routing. In *IPDPS*.

[26] Steiner, M. et al. (2008). Faster Content Access in KAD. In *P2P*.

[27] Schiller, B., et al. (2010). GTNA: A framework for the graph-theoretic network analysis. In *Proceedings of Springsim*.

[28] Baumgart, I. et al. (2009). OverSim: A Scalable and Flexible Overlay Framework for Simulation and Real Network Applications. In *P2P*.

17

Access Efficient Bloom Filters with TinySet

Gil Einziger and Roy Friedman

Computer Science Department, Technion, Haifa 32000, Israel

Abstract

Bloom filters are a very popular and efficient data structure for approximate set membership queries. However, Bloom filters have several key limitations as they require 44% more space than the lower bound, their operations access multiple memory words, and they do not support removals.

This work presents TinySet, an alternative Bloom filter construction that is more space efficient than Bloom filters for false-positive rates smaller than 2.8% and accesses only a single memory word and partially supports removals. TinySet is mathematically analyzed and extensively tested and is shown to be fast and more space efficient than a variety of Bloom filter variants. TinySet also has low sensitivity to configuration parameters and is therefore more flexible than a Bloom filter.

Keywords: Bloom filter, compact hash table, network services approximate set membership.

17.1 Introduction

Approximate set membership data structures offer a memory- and computation-efficient set representation. They obtain this efficiency by trading exact answers to membership queries with approximate results. In particular, negative answers by these data structures are always correct, while positive answers are correct with a probability of $1 - \varepsilon$, where ε is a performance parameter; the smaller ε is, the larger the data structure is. These are attractive when the available space is limited.

The most popular example of such structures is Bloom filter [1]. Bloom filters (and variants) are extensively used in network caching [2, 3], routing and prefix matching [4, 5], security [6–8], and many more [9].

Bloom filters are also implemented in some very successful widely deployed systems. For example, Mellanox's IB Switch System [10] uses Bloom filters in order to provide monitoring capability. Google's Big Table [11] and Apache Cassandra [12] employ them in order to avoid performing disk lookups for non-existing data. In these systems, Bloom filters are stored in main memory, and their content approximates that of a significantly larger disk. If the data are stored on disk, we are guaranteed of a positive reply from the Bloom filter and therefore never miss the disc content. False positives cause redundant accesses to the disk. Yet, as the false-positive ratio is relatively low, these are acceptable. The main benefit comes from a negative answer, as these answers are always correct. That is, on a negative result, these systems avoid a disk access entirely.

Variants of Bloom filters are also suggested in order to reduce communication in MapReduce [13] and Google Chrome [14].

BitCoin [15], a very successful peer-to-peer currency, uses Bloom filters in order to expedite transaction verification [16]. In addition, Bloom filters are extensively used in distributed cache architectures. Famous examples include Summery Cache [2] and Squid [17]. They can also be used in order to manage a single cache [18] as well as to expedite the lookup latency of Kademlia [19]—the distributed hash table that is used by the bitorrent protocol.

Bloom filters are also used in distributed routing; for example, OceanStore [20], a distributed storage system, uses a cluster of Bloom filters called *Attenuated Bloom Filters* in order to route requests to their destination. Further, routing using Bloom filters is suggested in the context of publish/subscribe [21–23] where Bloom filters determine whether matching subscribers may exist in a certain direction.

Despite their enormous popularity and success, Bloom filters have several key limitations. First, they require $\approx 44\%$ more space than the theoretical lower bound [24]. Second, each access requires calculating a number of hash functions proportional to ε, which is time-consuming. Third, each hash function calculation is followed by a memory access to a random location, which is cache unfriendly. Finally, Bloom filters do not support remove operations at all. These drawbacks motivate the search for alternatives.

17.1.1 Our Contribution

In this chapter, we introduce TinySet, a novel data structure for approximate membership queries. TinySet combines flavors from both compact hash tables [25, 26] and blocked Bloom filters [27, 28] in order to provide a combination of good properties, all in a single data structure.

In particular, every TinySet operation only effects a single- and fixed size memory block, dynamically changes its configuration, and achieves more flexibility than a Bloom filter. Unlike Bloom filters, TinySet also supports removals, although these degrade the performance over time. Moreover, TinySet is also faster than Bloom filters for read operations and requires a smaller memory space for interesting false-positive rates. In particular, the most space-efficient TinySet configuration discussed in this chapter requires less space than Bloom filters when the required probability of false positives is lower than 2.8%. We believe that since Bloom filters are extensively used by so many systems, TinySet can have a significant impact on the performance of applications in many domains.

A high-level intuition about TinySet is given in Figure 17.1. In this example, 3 independent blocks are drawn. Each block possibly uses a slightly different encoding method. An arriving item is hashed and inserted to one of these blocks. Prior to the insertion, the block reconfigures itself in order to represent the stored items as accurately as possible. In this case, the first block is under-loaded, as it contains a lower-than-average number of items. It can thus achieve a lower false-positive rate. The second block is a typical block that contains an average number of items and achieves an average false-positive rate. Finally, the third one is overloaded and contains more items than initially intended. It therefore yields a higher false-positive rate. The crux is that unlike previous suggestions, TinySet efficiently utilizes a different configuration for each individual block. The overall false-positive rate is an average of many blocks with varying accuracy, and extreme loads are rare.

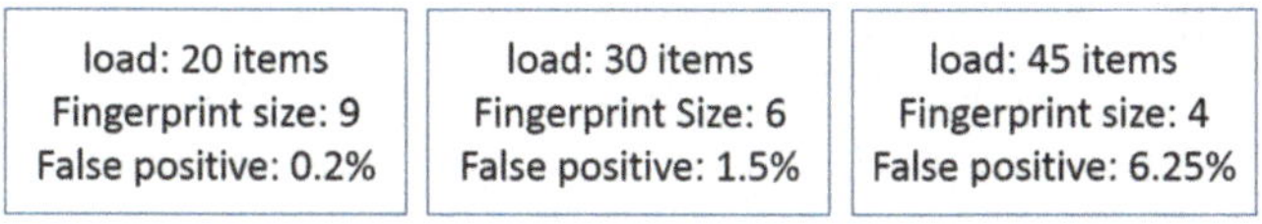

load: 20 items	load: 30 items	load: 45 items
Fingerprint size: 9	Fingerprint Size: 6	Fingerprint size: 4
False positive: 0.2%	False positive: 1.5%	False positive: 6.25%

Figure 17.1 A high-level overview of TinySet. The structure is partitioned into many fixed size blocks. Each block is dynamically configured according to the actual load placed on it.

Our last contribution is studying the integration of TinySet with TinyTable [29], a recent hash table-based counting Bloom filter construction. This includes the algorithmic aspects of performing such integration and exploring its benefits as well as limitations compared to each of TinySet and TinyTable in isolation.

Paper Roadmap

An overview of background and related work is described in Section 17.2. TinySet is presented in Section 17.3 followed by an analysis in Section 17.4. A comprehensive performance study of TinySet including a comparison of Bloom filters, d-left hashing, rank-indexed hashing, and TinyTable appears in Section 17.5. We conclude with a discussion in Section 17.6.

17.2 Background and Related Work

17.2.1 Bloom Filter Variants

Over the years, many data structures were suggested in order to improve different aspects of Bloom filters. For example, *compressed Bloom filters* [30] use compression techniques in order to achieve optimal space efficiency at the expense of calculation speed.

Blocked Bloom filter (BlockedBF) [27, 28] partitions a Bloom filter into many fixed size blocks, each containing an independent Bloom filter. An arriving item is first hashed to a block and is then inserted to the Bloom filter of that block.

Since blocked Bloom filters only access a single block, they typically achieve higher throughput and consume less power [31]. Unfortunately, the unequal load placed on each block makes this suggestion less space efficient than bloom filters.

Alternatively, *Balanced Bloom filters (BalancedBF)* [31] improve space efficiency with load balancing techniques. That is, a small fraction of the items are not inserted to the blocked Bloom filter and are separately maintained in an *overflow list* that is implemented with TCAM memory. However, even when employing these techniques, BalancedBF is still less space efficient than a Bloom filter. Moreover, this technique can similarly augment the accuracy of TinySet that also benefits from a more balanced load.

Counting Bloom filters (CBF) [20] enhance Bloom filters in order to support removals. Alas, these are significantly less space efficient when compared to standard Bloom filters even in sophisticated, state-of-the-art implementations [32–35]. Despite their space inefficiency, however, removal functionality is essential to many problems, and counting Bloom filters are therefore extensively used. An interesting kind of a counting Bloom filter is the *Inverted Bloom filter* [36] that can also associate values with the stored items.

On a more theoretical note, a space-optimal counting Bloom filter is suggested in [37]. This approach is based on compact hash tables and sophisticated encoding. Although it is asymptotically optimal, it is not very attractive in practice due to its high overheads. The authors do suggest a practical variant that does not support removals or access memory efficiently.

17.2.2 Hash Table-Based Bloom Filters

In principle, a hash table can be used as an alternative for Bloom filters and their variants. In particular, it is well known that when the set of items is static and known in advance, we can calculate a *perfect hash function*. Since this function hashes all the items in the set without collisions, a simple array can be used to store fingerprints of the items in the set [9].

More formally, a perfect hash function $P : S \rightarrow [n]$ hashes each element in S to a unique location in an array of size n, where each entry of the array stores the fingerprint of the item that hashes to that location. In order to check whether $x \in S$, $P(x)$ is calculated and the fingerprint of x is compared to the fingerprint stored at $P(x)$. Since any element x s.t. $x \notin S$ hashes to a certain location in the array, the false-positive probability in this case is the same as the probability that fingerprints of two different items are identical. That is, if the fingerprint size is $\lceil \log(\frac{1}{\varepsilon}) \rceil$ bits, the false-positive probability is ε. Yet, for many practical applications, perfect hashing is impractical, motivating the search for a different solution.

The first fingerprint hash table that was suggested is called *d-left hashing* [25]. The idea behind d-left hashing is to use a balanced allocation approach. That is, the hash table is partitioned into multiple equally-sized subtables, where new elements are placed in the least-loaded subtable. Balanced allocation allows d-left hash tables to be dimensioned statically so that overflows are unlikely and the average load per block is close to the maximum load.

Similarly, cuckoo hashing was also employed in creating efficient Bloom filters [38], with the idea of calculating a perfect hash function using the power of two choices. The main drawback of cuckoo hashing is that an insertion can cause a large number of memory accesses in order to terminate.

Rank-indexed hashing [26] has an alternative approach. Instead of balancing the load between subtables, rank-indexed hashing allocates block extensions to overflowing blocks. Statistical multiplexing is used in order to bound the number of required extensions, and the optimal configuration is discovered with an exhaustive search.

TinyTable is another compact fingerprint hash table [29]. TinyTable differs from rank-indexed hashing by the way it handles overflows. In TinyTable, an overflowing bucket is extended on expense of its neighboring bucket. This overflow handling mechanism allows TinyTable to be more space efficient than rank-indexed hashing. When comparing TinyTable with TinySet, the former supports counting functionality and unlimited deletions with guaranteed false positives regardless of the load. Yet, TinyTable's update operations take constant time and memory only with high probability. Also, TinyTable is slightly less space efficient than TinySet, so when no counting functionality is needed and when the number of deletions is relatively small, TinySet is more attractive. In Section 17.3.7 below, we show an integration of TinySet and TinyTable that enables TinyTable to offer improved accuracy for under-utilized buckets. Conversely, for a fixed average false-positive ratio, this translates to better memory efficiency.

17.3 TinySet: Dynamic Fingerprint Resizing

17.3.1 Motivation and Overview

Our goal is a very space-efficient Bloom filter variant that is relatively simple to understand, implement, and configure. We would like our data structure to use a single hash function, access a fixed size memory, and degrade performance gracefully as the load increases (like a regular Bloom filter).

We use a similar structure to a blocked Bloom filter. That is, the data structure is partitioned into many fixed size independent blocks. Unlike blocked Bloom filters, however, each of these blocks is not a Bloom filter, but a compact representation of a chain-based hash table [39]. In a normal chain-based hash table, collisions are handled by chaining all items whose hashed values collide. That is, the hash function can be viewed as mapping an item to a given chain, or in other words, returning the index of the chain in which the item is supposed to be located. An item is inserted by adding it to the corresponding chain, and it is looked up by scanning the chain pointed to by the hash function.

However, pointers are too expensive in our context. Therefore, we suggest a simple and yet efficient (pointer free) encoding instead, as elaborated below. Further, since the load placed on each specific block fluctuates and cannot be anticipated in advance, we dynamically reconfigure the hash table in order to provide attractive false-positive rates. That is, each block has a different configuration according to the load placed on it, as illustrated in Figure 17.1. One of the novelties of this chapter is a method to maintain this additional configuration without explicit counters.

17.3.2 Basic Block Structure

We use a single hash function $H \rightarrow B \times L \times R$, where B is the block number, L is the index of the chain within that block, and R is the *remainder* (or *fingerprint*) to be stored in that block. Unlike traditional hash tables, this value contains pseudo-random bits and not the actual key of the item. A block is a continuous, fixed size memory, and we would like to use it as a compact chain-based hash table. We therefore suggest an efficient coding technique.

The first set of bits in a block contains a fixed size index (I), as illustrated in Figure 17.2. This index has a single bit per chain in the hash table. If the chain is empty, this bit is unset and vice versa. The rest of the bits in the block are treated as an array (A). This array stores fingerprints extended by an additional indexing bit called the *last bit*, which indicates whether this fingerprint is the last in its chain. Empty chains consume no space in the array. Non-empty chains are stored ordered by their chain number.

Our block supports three operations: add, remove, and query. The add operation updates the state of the block to include an additional item. Similarly, the remove operation removes an instance of a specific item, while the query operation indicates whether or not an item is contained in the block. As with Bloom filters, a negative answer is always correct, but a positive one only indicates a fingerprint match and has a false-positive probability.

Figure 17.3 describes a flowchart of these operations. As can be observed, the initial phases are the same for every operation. Specifically, in order to perform an operation on a certain item (T), we first apply the hash function to T. This generates the block number, the chain index, and the fingerprint $(B \times L \times R)$.

The second step is to access the specific block using the block number. In our design, all the blocks are of fixed size and are continuously aligned in memory. We can therefore simply calculate the block offset and access the block. We then check whether or not the chain we seek is empty. To do so, we access the appropriate bit in the index.

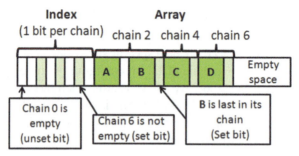

(a) A block with 8 chains and 4 items. Three of the chains are non-empty (2, 4, and 6) and one of the chains has more than one item (chain 2).

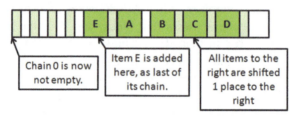

(b) An example of an addition to the previous block; in this case, item E is inserted into the first chain (0). As a result, chain 0 is now not empty (bit 0 is set in index). Since chain 0 is the first non-empty chain, E is inserted into the first location in the array, shifting other items to the right.

Figure 17.2   Basic block indexing technique: The first 8 bits are used for index, set bits are marked with light green, and unset bits are white. The rest of the bits are allocated to an array that stores remainders with an additional bit that indicates whether that remainder is last in chain. Remainders are colored dark green and are marked with letters for easy reference. Let us emphasize that there are *no pointers* in this encoding!

The index is also of fixed size and is placed in the first bits of the block. The bit of the i'th chain is always at offset i. For query and remove operations, encountering an unset bit at this stage allows us to finish the operation as we already know that the item in question is not present. Specifically, a query operation returns false and a remove returns an indication that the desired item was not found.

If we did not terminate early, we first calculate the *Logical Chain Offset (LCO)*. Recall that in the array, non-empty chains are stored ordered by chain index. The logical chain offset tells us how many non-empty chains are stored before the requested chain. For example, assume that we wish to calculate the LCO of chain 5. Out of chains 0 to 4, only chains 1 and 4 are non-empty. Hence, the logical chain offset of 5 is 2 as there are two non-empty chains

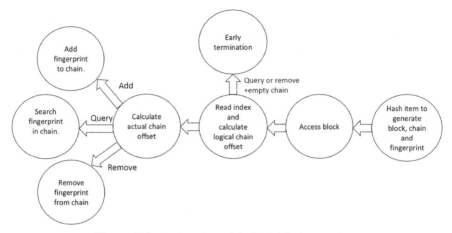

Figure 17.3 A Flowchart of the basic block operations.

smaller than 5. The offset calculation can be implemented very efficiently with a *rank* operation on the index. Specifically, a $rank(I, c)$ operation returns the number of set bits before the c'th bit. This operation can be implemented efficiently by combining a bit count and a bitwise and instructions (both are very efficient). In our example, we calculate $rank(I, 5) = 2$.

Since non-empty chains can have more than a single item, our next step is to find the *Actual Chain Offset (ACO)*. ACO is the offset in the array where the chain is stored. To do so, we scan the array and count last in their chain items until we reach the logical offset. At that offset, the desired chain starts (or should start in the case of add to an empty chain).

An initial observation is that the actual offset is always larger or equal to the logical one ($ACO \geq LCO$), as non-empty chains by definition have at least a single item (and may have more). Moreover, although the scan operation may seem inefficient, we are interested in blocks that are relatively small and store only a moderate amount of items to begin with. For these parameters, this operation is also cache friendly, so we expect good performance (and indeed obtain it as reported in Section 17.5.1).

Once the ACO is discovered, we can access the required chain. From this point, each operation is different. In particular, a query scans the chain comparing the item's fingerprint to the ones stored in the chain. Add and remove operations are slightly more complicated and require shifting items in order to keep the chains ordered. They are inherently slower, but the overheads are dominated by the block size.

In order to add an item, we first calculate the ACO as detailed above. We then shift all the fingerprints from that offset until the end of the block has a single place to the right, and insert the new fingerprint at the ACO offset. Finally, if the appropriate bit in I is unset, we set it and mark the new item as last of its chain.

The remove operation is very similar. We first calculate the ACO and then shift all items from the ACO until the end of the block has a single place to the left. Finally, if the removed item was marked as last, we either mark the previous item as last in its chain or mark the entire chain as empty. These operations can be implemented in a simple manner. In particular, we first examine the previous last bit. If that bit is set, then the removed item was first in its chain. If it is also marked as last, we update the index to indicate that the chain is now empty. Similarly, if the previous last bit is unset, the previous item belongs to the same chain as the removed item. If the removed item is marked as last, we can simply mark the previous item as last. No update to the index is necessary here since the chain is still not empty. Finally, if the removed item is not last in its chain, we can simply remove it.

A detailed explanation of the memory layout and indexing technique of a single block is found in Figure 17.2. Figure 17.2a describes a block with an 8-bit index. Indeed, the first 8 bits are dedicated to the index. On reading the index, we can understand that this block has 3 non-empty chains (2, 4, and 6). In the array, the chains are always stored sorted by chain number, and therefore, the first chain is number 2. For the first item (A), the last bit is unset, and therefore, A is not the last of its chain. All other items are last in their chains.

Figure 17.2b illustrates the case where a new item (E) is inserted to chain 0. In this case, since chain 0 becomes the first non-empty chain in the block, it is stored at the first location. We shift all the items one place to the right and store E in the first location. Finally, we set E as last of its chain since chain 0 was empty prior to the addition. The remove operation is exactly the opposite. To remove E, we have to shift all items to the right (effectively erasing E) and also clear the index bit of chain 0, basically returning the state to that of Figure 17.2a.

17.3.3 Variable Fingerprint Size

Although the block structure we described is individually very space efficient, it cannot be used efficiently. In particular, a block should be able to accommodate many items since the load fluctuates with some blocks expected to contain a large number of items. Unfortunately, configuring all the blocks to contain many items results in a lot of wasted space since the majority of blocks are expected to be only average loaded.

Our approach is to start with a large fingerprint size that has very high accuracy. We then gradually reduce the fingerprint size as the block becomes more crowded. In order to do so, we only need a counter that counts the number of fingerprints in every block. When accessing a block, if there are X fingerprints stored in that block and we know $BlockBitSize$, the number of bits in the block, we can calculate their maximal possible size: $size = \lfloor \frac{BlockBitSize}{X} \rfloor$.

When adding a new item, we need to check whether adding the new item should reduce the size of the fingerprints. Formally, when $\lfloor \frac{BlockBitSize}{X} \rfloor \neq \lfloor \frac{BlockBitSize}{X+1} \rfloor$, a block resize operation is called. It reduces the size of all stored fingerprints in the array in order to make room for an additional fingerprint.

An example of the block resize algorithm appears in Figure 17.4. In this example, each block has 8 chains and the array is allocated an additional 24 bits. Since the block contains the same items as in the previous example, we only show the memory alignment of items (the logical structure is the same as previously). In Figure 17.4a, the block contains only 4 items, and therefore, each item is allocated 6 bits: 5 bits for fingerprints and 1 bit to indicate if it is last in chain. Figure 17.4 describes the state of the table after an additional item (E) is inserted. Since the number of items is now 5, each item can only be 4 bits long. This means that fingerprints can only be 3 bits long.

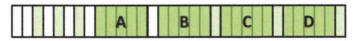

(a) Before adding E, 4 items in block. We have 24 bits, so we allocated 5 bits per fingerprint +1 additional bit to indicate whether a fingerprint is last in chain.

(b) After adding E, 5 items in block, we now allocate 3 bits per fingerprint + 1 additional bit. Note that not all the space is claimed as we cannot increase the size of all fingerprints.

Figure 17.4 Dynamic fingerprint size according to load.

17.3.4 Two Fingerprint Sizes in One Block

Taking a closer look at Figure 17.4b reveals additional unused bits. We can therefore potentially achieve a slightly better accuracy for the same size. The reason for the unused space is that we use the same size for every item in the block. This is when there are not enough unused bits in order to increase the size of all items by 1.

We propose to allow two item sizes per block by calculating how many of the fingerprints can be extended by 1 bit. Specifically, we calculate a second value $mod = X \bmod BlockBitSize$. In workloads that only add items, fingerprints are only shifted to the right and we can always store the first mod fingerprints along with an additional bit.

In order to access the i-th item, the calculation changes in the following way: If $i > mod$, we add mod bits, as the first mod fingerprints are 1 bit larger, whereas for $i \le mod$, we add i bits. Similarly, the size of the extracted fingerprint is $size + 1$ for $i < mod$ and $size$ otherwise.

Figure 17.5a illustrates the memory alignment of the block before and after the addition of E. Before adding E, we used $size = 5$, $mod = 0$. This means that all the fingerprints are of size 4. However, after we add E, $size = 3$ and $mod = 4$. That is, we are able to store the first 4 fingerprints with $size = 4$. Further, notice that all the bits in the array are now utilized. As long as no removals are present, this is now always the case for our blocks. This optimization makes the resize operations more frequent, as the fingerprints are now resized after any addition. However, these operations are usually less complex since in most cases, only a few fingerprints actually change their size.

(a) Before adding E, we have 24 bits in the array and 4 items. Therefore, all items are of size 5 + last bit.

(b) After adding E, we have to split our 24 bits between 5 items. We therefore use $size = 3$, and $mod = 4$. The first 4 fingerprints are now longer than before.

Figure 17.5   Adding items with size and mod.

17.3.5 Removing Items

Unfortunately, since we only store fingerprints of items, we cannot extend them after a removal. However, we can still support removals. Upon a removal, we update the index and shift fingerprints as usual, yet we do not reduce the number of stored items. This way, the size and mod of the existing items are calculated correctly. In order to perform an addition, we first check whether there are removed items. If there are such items, we perform an addition without resizing the fingerprints and incrementing the number of items per block. If there are no removed items, we perform the regular add operation, which downsizes the fingerprints and increases the number of items.

We now suggest a simple method of checking whether there are removed items in the block. It relies on the observation that if there are no removed items, the bitwise array is full, and the last item in the array is always last in its chain. Therefore, the last bit of the array is always set when there are no removals, meaning that testing this bit is a quick indicator of the state of the block. After a removal, since we shift items to the left, the last location in the array is zeroed and the following add operation can reclaim this space.

Also, note that when removals are involved, the number of items per block counter-monitors the maximal amount of stored items in this block during the operation of the data structure. For brevity, when we discuss the number of stored items in a block, we use the term *actual capacity* to describe how many useful items are stored in a block and the term *logical capacity* to describe how many items the block is configured for. Under constant removals and additions, the actual capacity remains the same, but the logical one slowly increases and TinySet becomes less space efficient.

There is an additional delicate point to consider when removals are present regarding the two sized fingerprints per block optimization. For example, in Figure 17.6, item B is removed and as a result item D is shifted from a non-mod-extended location to an extended one. In order to read D correctly, we need to add a 0 bit to D. However, when querying for D, the mod bit may be 1. Therefore, padding D with 0 may result in a false negative that we wish to prevent. Thus, when aligning the items, we only treat the extended bit of a stored item if it is 1. Since naturally only 50% of these bits are 1, supporting remove operations makes the two item sizes per block optimization $\approx 50\%$ less efficient. In conclusion, a TinySet that supports removals requires slightly more space for the same accuracy.

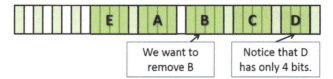

(a) Before removing B, in this example, size = 3 and mod = 4, notice therefore that D is only 3 bits + last bit.

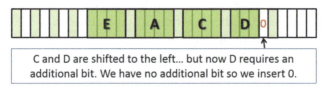

(b) Removing B caused us to shift left C and D; however, this time D should be mod extended. We cannot make D longer, so we extend it with 0 instead.

Figure 17.6   Removing an item with size and mod.

17.3.6 Implicit Size Counters

While conceptually simple, maintaining an explicit item counter is a bit wasteful in situations where the block size is small. We therefore suggest a method to completely eliminate size counters. The idea is to always align the "isLast" bits at the beginning of the array so that we can read them knowing the item size. This idea is illustrated in Figure 17.7. Notice that the indexing method is the same, but the location of the bits is changed.

We can now calculate the number of stored items simply by counting last bits until we reach the last non-empty chain. This may seem a bit wasteful at first, but we can also calculate the ACO while doing so. For example, in Figure 17.7, there are 4 non-empty chains. We therefore count bits in the

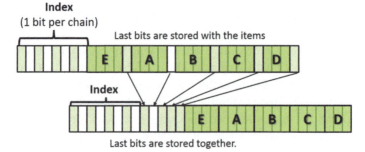

Figure 17.7   Implicit size counters memory alignment.

array until we reach the 4th set bit. This happens after 5 bits, and therefore, there are 5 items in the array. Moreover, while calculating the size, we can also calculate the ACO. In particular, if we wish to access chain 5, we first calculate the logical chain offset. In Figure 17.7, this offset is 2. That is, after seeing 2 set last bits, we can write down the offset as the ACO. In our example, the ACO of chain 5 is 3. Hence, in the same operation, we both calculate the number of items in the block and the ACO.

17.3.7 Integration with TinyTable

As mentioned before, TinyTable has a different mechanism for overflowing buckets. Instead of reducing the fingerprint size like in TinySet, in TinyTable, the overflowing bucket is resized at the expense of a neighboring bucket.

Since a TinyTable bucket can overflow into the neighbor and doing so shifts the neighbor bucket, their memory offsets are not constant as in TinySet. Instead, there is a need to keep track of the current memory offset of each bucket. To do so, TinyTable uses a counter called *Anchor Distance (AD)* that is attached to each bucket. The AD counter-measures how many times the bucket was shifted. Thus, using the AD counter and the original bucket location enables accessing the bucket.

Figure 17.8 illustrates an example of this. In Figure 17.8a, Bucket 1 was received 14 different items due to variable load. Since initially each bucket was only allocated space for 10 fingerprints, Bucket 1 overflowed into Bucket 2. In this example, Bucket 2 only received 6 items and therefore did not overflow when Bucket 1 took space for 4 more items. In order to know where to find Bucket 2 in the memory, we read its AD counter and discovered that it is 4.

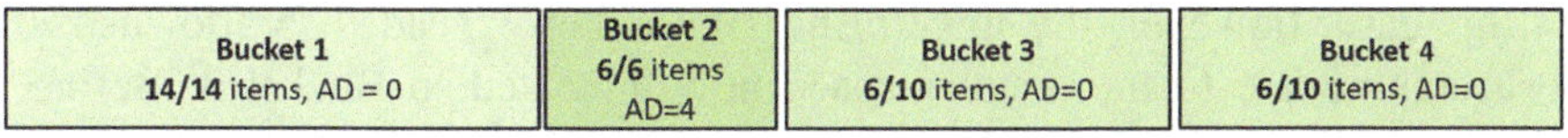

Bucket 1 14/14 items, AD = 0	Bucket 2 6/6 items AD=4	Bucket 3 6/10 items, AD=0	Bucket 4 6/10 items, AD=0

a) Bucket 1 overflowed and expanded into Bucket 2, which did not overflow. Buckets 3 and 4 did not overflow.

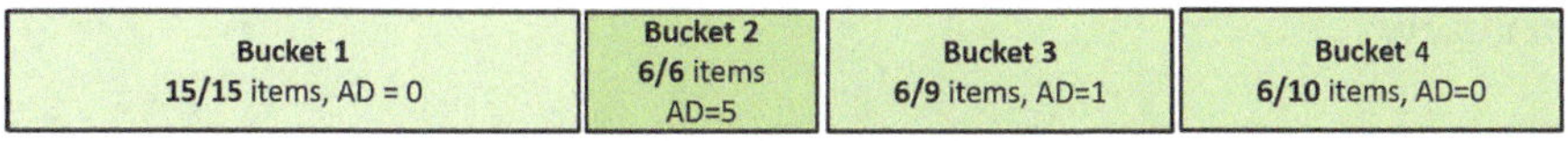

Bucket 1 15/15 items, AD = 0	Bucket 2 6/6 items AD=5	Bucket 3 6/9 items, AD=1	Bucket 4 6/10 items, AD=0

(b) Following an addition to Bucket 1, Bucket 1 overflows and expands into Bucket 2, which also overflows and expands into Bucket 3 that does not overflow.

Figure 17.8 An example of the overflow mechanism in TinyTable. Buckets are resized to fit the actual load.

This means that we add 4 to the offset of Bucket 2, in this case, after the original offset (10 fingerprints) + the AD counter (4 fingerprints).

Figure 17.8b describes what happens when an additional item is added to Bucket 1. In this case, Bucket 1 overflows again into Bucket 2 and doing so it shifts Bucket 2 one place to the right and increment Bucket 2's AD counter. However, since the size of Bucket 2 was reduced to 5 and it has 6 fingerprints to store, Bucket 2 overflows into Bucket 3. Luckily, Bucket 3 does not overflow, and the update process is finished.

The complexity of update operations in TinyTable depends on the amount of free space in the table. More free space makes the update operation less complex on average. The parameter $\alpha = \frac{allocatedSpace}{FreeSpace}$ is a performance parameter that can be tuned to the required operation speed. A large α makes the table faster but less space efficient, and the authors identify $\alpha = 1.1$ and $\alpha = 1.2$ as good tradeoff configurations.

Integrating the two approaches can generate an even more space-efficient solution. Empty spaces in TinyTable can be used to store longer fingerprints that will gradually decrease their size until the bucket is full. From that moment on, we use the minimal fingerprint size and the bucket extends into other buckets according to TinyTable. For example, in Figure 17.8a, both Bucket 3 and Bucket 4 have room for 10 fingerprints but only store 6. Therefore, the free space in these buckets can be used to store longer fingerprints, which reduces the average false positives of the entire table. Once a bucket looses space, if the loosing bucket is not full, TinySet will downsize the fingerprints to fit the newly available space.

For example, in Figure 17.8b, Bucket 1 further expands and now contains the minimal fingerprint size and so is Bucket 2 (as it is full). Thus, Bucket 1 expands into Bucket 2 and Bucket 2 expands into Bucket 3. In this case, however, the fingerprints of Bucket 3 need to be downsized to accommodate to the smaller space now allocated to Bucket 3. Before, they were 6 fingerprints with room for 10, but now Bucket 3 stores 6 fingerprints with space for only 9. Hence, TinyTable can become a more efficient Bloom filter when incorporating TinySet's fingerprint resizing approach.

17.3.8 Final Overview

We conclude the presentation with a high-level overview of TinySet operation. Figure 17.9 presents the ultimate work flow. The basic operation and concepts are similar to that of the basic block. The main differences are in the gray

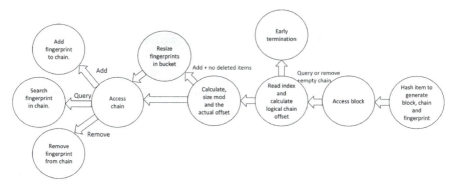

Figure 17.9 Flowchart of TinySet.

stages. In particular, before accessing the chain, we now calculate the size and mod in addition to the actual chain offset.

Since add operations now downsize fingerprints, we perform a resize operation before executing an add operation. If there are removed items in the block, we simply reclaim a previously removed item and do not perform the resize.

17.4 Analysis

17.4.1 Memory Overheads

For a chain-based fingerprint hash table with an average chain length of λ and a (fixed) fingerprint size of $\log(\varepsilon)$ bits, the false-positive ratio is $\lambda \cdot \varepsilon$ [26]. Therefore, if we pick $\lambda = 1$, then each chain is on average 1 item long and the expected false-positive rate is ε. This false-positive rate is optimal up to the indexing overheads of the block [26].

In TinySet, we also require a single bit per item and an additional bit per chain. When $\lambda = 1$, our indexing cost is exactly 2 bits per item. A single TinySet block is therefore optimal up to an additive factor of 2 bits per item. Similarly, if the load of all blocks is perfectly balanced, then all individual blocks are configured the same and TinySet is only 2 bits per item from being optimal.

Each TinySet block contains a fixed size index (typically 64 or 32 bits) and a fixed size array. Additionally, TinySet blocks may contain an item counter that has to be sized so that overflows are unlikely. This counter can be eliminated with the implicit counter-optimization. The per item bit cost of TinySet is simply $\frac{BlockSize + CounterSize}{itemsPerBlock}$.

17.4.2 Variable-Sized Fingerprints

In order to calculate the accuracy of TinySet, we average the accuracy for each block. We start by calculating the false-positive rate of a single block with r fingerprints and L chains. Consider $\lambda_r = \frac{r}{L}$ the local λ of a block with r fingerprints. We first analyze the more simple case where all fingerprints in a single block are of the same size.

For each possible fingerprint size $S_1, \ldots, S_k$, we calculate the maximal number of fingerprints that can be stored with this size, $C_1, \ldots, C_k$, subject to that block bit size. That is, if there are $BlockSize$ bits allocated for the array, $C_i = \lfloor \frac{BlockSize}{S_i+1} \rfloor$.

Denote p_r the probability that there are exactly r fingerprints in a block. Assuming we already know p_r, we can calculate the contribution of each size to the overall error. This impact is: $Error(size = S_i) = \sum_{r=C_{i-1}+1}^{r=C_i} p_r \cdot \frac{\lambda_r}{2^{S_i}}$.

In order to calculate p_r, we consider the classic balls and bins experiment, where n balls are thrown randomly into m bins. The probability that a certain bin contains exactly r balls is:

$$
\begin{aligned}
p_r &= \binom{m}{k} \left(\frac{1}{n}\right)^r \left(1 - \frac{1}{n}\right)^{n-r} \\
&= \frac{1}{r!} \frac{m\,(m-1)\ldots(m-r+1)}{n^r} \left(1 - \frac{1}{n}\right)^{m-r} \\
&\approx \frac{e^{-m/n}(m/n)^r}{r!}
\end{aligned}
$$

We therefore conclude that the false-positive rate is

$$
FP = \sum Error(size = S_i).
$$

17.4.3 Variable-Sized Fingerprint with Mod

Next, we analyze the case of two item sizes in the same block. Denote m_r the probability of a fingerprint to be modulo-expanded when r fingerprints are stored in the block ($m_r = \frac{mod_r}{r}$). We sum up the contribution to the error of each possible size in a similar way as before:

$$
\begin{aligned}
&Error(size = S_i) \\
&= \sum_{r=C_{i-1}+1}^{r=C_i} p_r \cdot (1 - m_r) \cdot \frac{\lambda_r}{2^{S_i}} + \sum_{r=C_{i-2}+1}^{r=C_i-1} p_r \cdot (m_r) \cdot \frac{\lambda_r}{2^{S_i}}.
\end{aligned}
$$

The false-positive rate remains

$$FP = \sum Error(size = S_i).$$

17.4.4 Overflows

TinySet has no significant overflow problem, since we can size the items counter reasonably. Yet, overflow is unavoidable in very extreme cases, where so many fingerprints are inserted to a block that there is no more room to allocate even a single bit per item. We denote the maximal number of fingerprints a block can contain by Z_{max}.

Denote X_i, $0 \leq i \leq B$, the number of items inserted to the i'th block, and denote by n the average number of fingerprints a block stores. Denote O the overflow indicator variable: $O = 1$ if there is an overflow, and $O = 0$ otherwise.

$$\Pr[O = 1] = \Pr[(\max X_i) > Z_{max}]$$

$$\leq \sum_{i=1}^{B} \Pr[X_i > Z_{\max}] = B \times Binotail\left(n, \frac{1}{B}, W_3\right)$$

where $Binotail(N, P, K)$ denotes the tail probability $\Pr[Y > k]$ and Y has the distribution $Binomial(N, P)$. To get a sense for this, in the above example, the probability of a single-block overflow is smaller than 10^{-200}. Therefore, even in extremely big TinySets with billions of fingerprints, this probability is negligible. When size counters are used, we need to intelligently pick their size so that the overflow probability is arbitrarily small and overheads are low. To do so, we perform the same calculation. In particular, if we pick the items counter to be 7 or 8 bits, the overflow probability of a single block with average load of 64 items is less than 10^{-13} and 10^{-23}, respectively. The overhead for 8-bit counters is only 0.125 bits per item in the 64 items per block case. Even in this case, it is a significant improvement over previous works that required a different configuration for each overflow probability.

17.5 Results

We start by comparing our work to rank-indexed hashing, Bloom filters, and d-left hashing. TinySet requires the following parameters for configuration: the block size, the number of chains per block (L), and the desired average

Table 17.1 Time to add/query 1 million items (seconds)

	Bloom Filter	TinySet_32I	TinySet_64I
1%	0.65/0.69	0.28/0.05	0.34/0.08
0.1%	0.96/0.98	0.28/0.05	0.34/0.08
0.01%	1.33/1.29	0.28/0.05	0.35/0.08

block load(λ) that determines how many blocks to create. Our measurements are calculated with the implicit size counters optimization (Section 17.3.6). If this optimization is not invoked, counter-overheads should also be accounted for. Bloom filters and rank-indexed hashing both require knowing the expected number of elements, while rank-indexed hashing requires additional parameters. Whenever we present rank-indexed hashing or Bloom filters, we configure them optimally for the corresponding data point. When we present d-left hashing, we configure it using the configuration suggested by its authors (8 fingerprints, 4 subtables, expected load of 6).

17.5.1 Operation Speed

In this section, we study the performance of TinySet compared to an open-source Bloom filter implementation[1]. Our experiment goes as follows: We measure the time it takes to add 1 million items to each of the constructions. We then perform a single query for each contained item, repeat this 10 times, and average the results. We compare the Bloom filter to TinySet 32I and TinySet 64I; the former has 32 items on average per block and the latter 64. The experiment was run on an Intel i7 working at 3.2 GHz. The computer also has 32 GB RAM, so all the data structures easily fit in main memory.

Our results are listed in Table 17.1. As can be observed, both TinySet configurations are faster than a Bloom filter for the false-positive range. We also note that the difference is more dramatic for low false-positive rates. As anticipated, since the complexity of TinySet operations depends on the block size, TinySet 32I is significantly faster than TinySet 64I.

Figure 17.10 sheds additional insight about the dynamics of the performance in the case of 0.1% false-positive rate. In these measurements, we averaged 10 runs of each algorithm for a 1-million items benchmark. As can be observed in Figure 17.10a, the add operation is initially orders of magnitude faster than the Bloom filter as the TinySets are empty and the block management overheads are low. As TinySet becomes more crowded,

---

[1]The project can be found at https://code.google.com/p/java-bloomfilter/

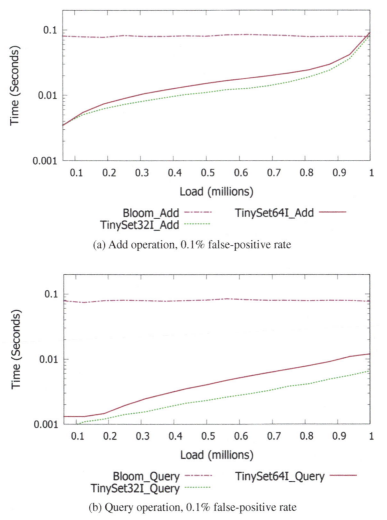

(a) Add operation, 0.1% false-positive rate

(b) Query operation, 0.1% false-positive rate

Figure 17.10 Time to perform 62 k operations as a function of the load: at 1 million items, TinySet 64I has 64 items on average per block and similarly TinySet 32I has on average 32 items per block.

the add operation gets slower, until the blocks are nearly full and both configurations are slightly slower than the Bloom filter. For the query operation, however, the increase in run speed is more moderate, and both configurations are significantly faster than the Bloom filter. We note that although the decoding of our blocks requires linear complexity, their average size is small and in practice they are decoded quickly.

Table 17.2   Required storage (in bits) per element for the same false positive rate (without removals)

False Positive	TinySet 64I (L = 64, λ = 1) Rank	Bloom Filter	Lower Bound (Information Theory)	Comparison vs Rank	vs Bloom Filter	vs Lower Bound
1%	9.1	10.6 9.6	6.4	−14%	−5%	+38%
0.1%	12.8	14.4 14.4	10	−11%	−11%	+28%
0.01%	16.6	18.2 19.1	13.3	−9%	−13%	+25%

17.5.2 Space/Accuracy Tradeoff

Table 17.2 compares TinySet (64 items per block, $\lambda = 1$) to both Bloom filters and rank-indexed hashing. The Bloom filter is optimally configured, and rank-indexed hashing is configured according to the configuration suggested by its authors. For perspective, we also added to the table the lower bound for this problem [24]. As can be observed, this configuration of TinySet is more compact than both Bloom filters and rank-indexed hashing.

In Table 17.3, we compare TinySet against a variable increment counting Bloom filter (VI-CBF) [35], d-left, and rank-indexed hashing that supports removals. We note that this is not an entirely fair comparison, as removals gradually make TinySet less space efficient. To complete the picture, Section 17.5.4 quantifies what happens to TinySet's efficiency when removing items. As can be observed, TinySet is initially 18–27% more spaced efficient than the best alternative for the range. Yet, as this advantage degrades over time, we conclude that it may be attractive as a CBF only for workloads with limited number of removals.

In Figure 17.11, we configured TinySet to contain an average of 40 items per block ($\lambda = 0.625$). Fixing this parameter, we analyzed the impact of the number of bits allocated per item on the accuracy using both our analysis and a simulation exercising the real code. The experiment was repeated 100 times and averages are reported. As can be seen, the analysis is accurate and TinySet is superior to both rank-indexed hashing and Bloom filters. For Bloom filters, the break-even point is ∼7.9 bits per item; from this point on, TinySet offers a better tradeoff than Bloom filters. To quantify this, we note that for 19 bits per element, this TinySet configuration is over 4 times more accurate than the best alternative.

17.5.3 Flexibility

Unlike other hash table constructions, TinySet is very flexible, and as the load increases, TinySet behaves in a similar way to a Bloom filter. Further, similar to a Bloom filter, TinySet can easily accommodate more items than anticipated

Table 17.3 Required storage (in bits) per element for the same false positive rate (with removals)

False Positive	TinySet 64I (L = 64, λ = 1)	Rank CBF	d-left CBF	VI-CBF	Comparison vs Rank CBF	vs d-left CBF	vs VI-CBF
1%	9.5	13	17.6	25	−27%	−46%	−62%
0.1%	13.2	16.8	22.3	37.8	−21%	−41%	−65%
0.01%	16.9	20.6	26.4	50	−18%	−36%	−66%

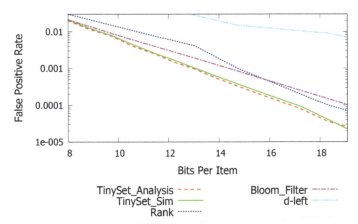

Figure 17.11 Space/Accuracy tradeoff for different algorithms.

and its accuracy degrades gracefully. In such cases, other hash table solutions typically overflow.

Figure 17.12 describes an experiment where we configure both Bloom filter and rank-indexed hashing optimally for a specific amount of items (38 K items). Rank-indexed hashing is configured with 1-K blocks, and similarly TinySet is configured with 1-K blocks. Overall, all constructions are allocated exactly the same amount of space. We insert items to the data structures and evaluate their accuracy as the load increases. We continue doing so even after the anticipated load is achieved.

As can be observed, TinySet offers better space accuracy tradeoff than both rank-indexed hashing and the Bloom filter. While at the anticipated load, all constructions offer a very similar space/accuracy tradeoff and rank-indexed hashing is less accurate throughout the experiment and cannot continue without overflowing as the load increases. TinySet remains more accurate than the Bloom filter until the end of the experiment, with a false-positive rate of 3.3%. If we configure a Bloom filter optimally to the eventual load, we should use two fewer hash functions. In this case, the Bloom filter can be configured to be slightly more accurate than TinySet.

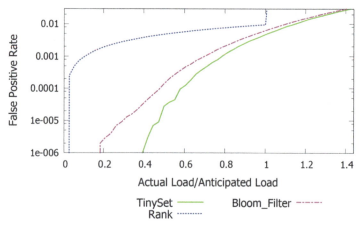

Figure 17.12   Performance under increasing load.

17.5.4 Removals

As we stated above, frequent removals degrade the space efficiency, as we cannot increase the size of fingerprints after removal. However, TinySet can reasonably support a moderate number of removals and still provide competitive accuracy. In the following experiments, we test the total amount of items stored in TinySet compared to the logical amount that remains the same. That is, once TinySet is full, at each step we add an item and remove an item. We tested two removal patterns, a sliding window where the oldest entry is removed and a random removal pattern.

Figure 17.13 illustrates the results of these measurements. As can be observed, the behavior of TinySet under these two workloads is almost identical. At the beginning, removed items are infrequent, and therefore, we usually add new items. Indeed, we see a sharp increase in the number of stored items. In particular, after 50% of the items are replaced, TinySet contains ≈11% removed items. However, over time, removed items become more frequent, and after 100% removals, 16% of the items stored are removed items. Taken to the extreme, after 1000% of the items are replaced (the number of remove operations is 10 times the number of stored items), only ≈35% of the items are removed items. Since TinySet is a very space efficient to begin with, it can remain competitive for a while and may also be attractive as a counting Bloom filter for some applications.

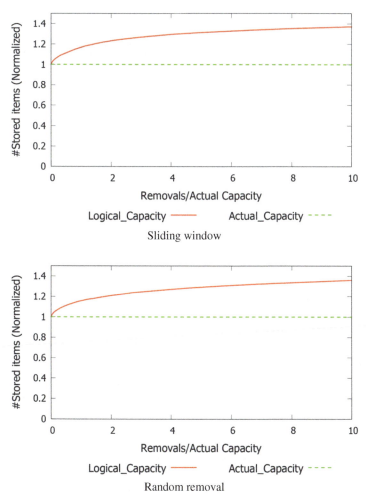

Figure 17.13 Effect of removals on TinySet memory space utilization.

17.5.5 Integration with TinyTable

In this section, we evaluate TinySet 64I, which has on average 64 items per bucket. We compare it to TinyTable with $\alpha = 1.1$ and $\alpha = 1.2$ as suggested in [29] as well as with the integrated scheme of both (as explained in Section 3.7). In all these suggestions, there are on average 64 items per bucket and $\lambda = 1$.

Our results are listed in Table 17.4. As expected, integrating TinySet into TinyTable improves space efficiency, and the unified design is more space efficient than either of them independently. For the tested range, false

Table 17.4   Memory consumption of TinySet when integrated with TinyTable

False Positive	1%	0.1%	0.01%
TinyTable alpha = 1.2	10.3	14.2	18.2
+TinySet	9.0	12.6	16.1
TinyTable alpha = 1.1	9.4	13.2	16.8
+TinySet	9.0	12.5	15.8
TinyTable 1.025	9.1	12.6	16
+TinySet	8.9	12.3	15.7
TinySet 64I	9.1	12.8	16.6
Perfect Load Balance	8.4	12	15.3

positives can be reduced by as much as 75%. To evaluate the quality of the integrated scheme, we also present in the line titled "Perfect Load Balance" the performance of TinySet when coupled with a perfect load balancer. As can be observed, the unified scheme comes very close to what can be achieved with a perfect load balancer. Thus, for the tested range, the potential for further space efficiency is limited to at most 0.5 bits per item (unless a better indexing technique is found).

Unfortunately, TinyTable uses a (potentially) unbounded addition and removal process, which only offers a high probability bound on its completion time. In that sense, the combined scheme is not superior to TinySet, which always access a single fixed size bucket. However, for applications in which the complexity limitations of TinyTable are acceptable, it does make a very space-efficient solution.

17.6 Conclusions and Discussion

In this work, we have introduced TinySet, an alternative Bloom filter construction that combines several appealing properties, namely access efficiency, speed, space efficiency, and partial support for removals. To the best of our knowledge, it is the only mechanism that provides all of these at once. Interestingly, TinySet is both faster and more space efficient than plain Bloom filters, especially for query operations. We also demonstrated that TinySet is more space efficient than previously suggested techniques.

TinySet's access efficiency comes from the fact that its operations only access a single fixed size block, which can be configured to match a single cache line. We achieve this by employing a novel indexing technique that dynamically downsizes the stored fingerprints as the load increases. Since the load fluctuates between the blocks, each one has a slightly different local configuration. When no removals are present, each block is 100% utilized in order to provide the best possible accuracy.

While removals gradually degrade TinySet's space efficiency (since we have no way of making fingerprints longer once the load decreases), we have showed that in practice, TinySet remains relatively space efficient for a large number of remove operations under different removal patterns.

We showed that TinySet is more flexible than a Bloom filter, as it can start with very long fingerprints and gradually downsize them as the load increases. Bloom filters, on the other hand, cannot dynamically change the number of hash functions they use. Notice that TinySet's flexibility also prevents it from overflowing, unlike other hash table-based constructions.

Finally, we showed that TinySet can be integrated with TinyTable to generate an even more space-efficient solution than either can offer. Yet, that solution only provides probabilistic termination guarantees.

References

[1] Bloom. B. H. Space/time trade-offs in hash coding with allowable errors. *Commun. ACM 1970*.

[2] Fan, L., Cao, P., Almeida, J., and Broder, A. Z. (2000). Summary cache: A scalable wide-area web cache sharing protocol. *IEEE/ACM Trans. on Netw.*

[3] Rottenstreich, O., and Keslassy, I. (2012). The bloom paradox: When not to use a bloom filter? In *IEEE INFOCOM 2012*.

[4] Dharmapurikar, S., Krishnamurthy, P., and Taylor, D. E. (2003). Longest prefix matching using bloom filters. *ACM SIGCOMM 2003*.

[5] Song, H., Hao, F., Kodialam, M., and Lakshman, T. V. (2009). IPV6 lookups using distributed and load balanced bloom filters for 100 GBPS core router line cards. In *IEEE INFOCOM 2009*.

[6] Artan, N. S., Sinkar, K., Patel, J., and Chao, H. J. (*2007*). Aggregated bloom filters for intrusion detection and prevention hardware. In *IEEE GLOBECOM 2007*.

[7] Bonomi, F., Mitzenmacher, M., Panigrahy, R., Singh, S., and Varghese, G. (2006). Beyond bloom filters: From approximate membership checks to approximate state machines. In *ACM SIGCOMM 2006*.

[8] Geravand, S., and Ahmadi, M. (2013). Bloom filter applications in network security: A state-of-the-art survey. *Computer Networks 2013*.

[9] Broder, A., and Mitzenmacher, M. (2002). Network applications of bloom filters: A survey. In *Internet Mathematics*, 2002.

[10] Mellanox ib qdr 324p switch system – overview.

[11] Chang, F., Dean, J., Ghemawat, S., Hsieh, W. C., Wallach, D. A., Burrows, M., Chandra, T., Fikes, A., and Gruber. R. E. (*2008*). Bigtable: A distributed storage system for structured data. *ACM Trans. Comput. Syst. 2008.*

[12] Lakshman, A., and Malik, P. (2010). Cassandra: A decentralized structured storage system. *ACM SIGOPS 2010.*

[13] Lee, T., Kim, K., and Kim, H.-J. (2012). Join processing using bloom filter in mapreduce. In *ACM RACS 2012.*

[14] http://blog.alexyakunin.com/2010/03/nice-bloom-filter-application.html

[15] Nakamoto, S. (2008). Bitcoin: A peer-to-peer electronic cash system. *Consulted*, 2008.

[16] https://bitcoinfoundation.org/blog/?p=16

[17] Squid. http://www.squid-cache.org/

[18] Einziger, G., and Friedman, R. (2014). TinyLFU: A highly efficient cache admission policy. In *Euromicro PDP*, 2014.

[19] Einziger, G., Friedman, R., and Kantor, Y. (2014). Shades: Expediting kademlia's lookup process. In *International Conference on Parallel Processing (Euro-Par)*. 2014.

[20] Kubiatowicz, J., Bindel, D., Chen, Y., Czerwinski, S., Eaton, P., Geels, D., Gummadi, R., Rhea, S., Weatherspoon, H., Weimer, W., et al. (2000). Oceanstore: An architecture for global-scale persistent storage. *ACM Sigplan Notices 2000.*

[21] Einziger, G., and Friedman. R. (2014). Postman: An elastic highly resilient publish/subscribe framework for self sustained service independent P2P networks. In *Springer SSS 2014.*

[22] Jerzak, Z., and Fetzer, C. (2008). Bloom filter based routing for content-based publish/subscribe. In *Proc. of the 2nd International Conference on Distributed Event-based Systems*, DEBS, pp. 71–81. ACM, 2008.

[23] Jokela, P., Zahemszky, A., Esteve Rothenberg, C., Arianfar, S., and Nikander, P. (2009). Lipsin: Line speed publish/subscribe internetworking. *ACM SIGCOMM 2009.*

[24] Carter, L., Floyd, R., Gill, J., Markowsky, G., and Wegman, M. (1978). Exact and approximate membership testers. In *ACM STOC 1978.*

[25] Bonomi, F., Mitzenmacher, M., Panigrahy, R., Singh, S., and Varghese, G. (2006). An improved construction for counting bloom filters. In *European Symposium on Algorithms 2006.*

[26] Hua, N., Zhao, H. C., Lin, B., and Xu, J. (2008). Rank-indexed hashing: A compact construction of bloom filters and variants. In *IEEE ICNP 2008.*

[27] Putze, F., Sanders, P., and Singler, J. (2007). Cache-, hash- and space-efficient bloom filters. In *WEA*, pp. 108–121, 2007.

[28] Qiao, Y., Li, T., and Chen, S. (2011). One memory access bloom filters and their generalization. In *IEEE INFOCOM 2011*.

[29] Einziger, G., and Friedman, R. (2015). Counting with TinyTable: Every bit counts! Technical Report CS-2015-04, Computer Science Department, Technion.

[30] Mitzenmacher, M. (2001). Compressed bloom filters. In *ACM PODC 2001*.

[31] Kanizo, Y., Hay, D., and Keslassy, I. (2013). Access-efficient balanced bloom filters. *Comput. Commun.*, 36(4), 373–385.

[32] Ficara, D., Di Pietro, A., Giordano, S., Procissi, G., and Vitucci, F. (2010). Enhancing counting bloom filters through huffman-coded multilayer structures. *IEEE/ACM Trans. Network*, 18(6), 1977–1987.

[33] Huang, K., Zhang, J., Zhang, D., Xie, G., Salamatian, K., Liu, A., and Li, W. (2013). A multi-partitioning approach to building fast and accurate counting bloom filters. In *IEEE IPDPS 2013*.

[34] Li, L., Wang, B., and Lan, J. (2010). A variable length counting bloom filter. In *ICCET 2010*.

[35] Rottenstreich, O., Kanizo, Y., and Keslassy, I. (2012). The variable-increment counting bloom filter. In *IEEE INFOCOM 2012*.

[36] Goodrich, M. T., and Mitzenmacher, M. (2011). Invertible bloom lookup tables. *CoRR*, abs/1101.2245.

[37] Pagh, A., Pagh, R., and Rao, S. S. (2005). An optimal bloom filter replacement. In *ACM-SIAM SODA*.

[38] Lumetta, S., and Mitzenmacher, M. (2007). Using the power of two choices to improve bloom filters. *Internet Mathematics*, 4(1), 17–33.

[39] Corman, T. H., Leiserson, C. E., and Rivest, R. L. *Introduction to Algorithms*. MIT Press.

18

Maximum Correntropy-Based Distributed Estimation of Adaptive Networks

**Amir Rastegarnia[1], Azam Khalili[1], Wael M. Bazzi[2]
and Saeid Sanei[3]**

[1]Department of Electrical Engineering, Malayer University,
Malayer 65719-95863, Iran
[2]Electrical Engineering Department, American University in Dubai,
P. O. Box 28282, Dubai, UAE
[3]Department of Computer Science, University of Surrey,
Surrey GU2 7XH, UK

Abstract

Adaptive multi-agent networks have received much attention due to their ability in decentralized real-time data processing and optimization. In this study, we consider solving the distributed estimation problem using an adaptive network strategy where the objective is to collectively estimate a parameter from streaming noisy measurements. The conventional adaptive networks exhibit good performance in the presence of Gaussian noise but their performance decreases in the presence of impulsive noise. The aim of present work was to propose new robust adaptive networks, based on both incremental and diffusion strategies that alleviate the effect of impulsive noise. To this end, we move beyond mean square error (MSE) criterion and recast the estimation problem in terms of the maximum correntropy criterion (MCC). We use stochastic gradient ascent and some useful approximations to derive adaptive networks that are appropriate for distributed implementations. The resultant algorithms have the computational simplicity of the popular least mean squares (LMS) algorithm, along with the robustness that is obtained by using higher order moments. The performance of the algorithms are compared with those of some popular approaches relying on MSE criterion.

Keywords: Adaptive networks, Diffusion, Distributed estimation, Impulsive noise, Incremental, Maximum correntropy criterion.

18.1 Introduction

Recently, due to the growth of network-based applications, there has been great interest in the development of efficient in-network processing algorithms. Some relevant applications include wireless sensor networks (WSNs) [1, 2], Internet-of-things (IoT) [3], peer-to-peer (P2P) networks [4], robot swarms [5], cognitive radio [6], modeling brain cortical connectivity [7], and modeling of self-organization in biological networks [8–10]. Several strategies for decentralized information processing problems have been proposed. The most notable among them are the ones constructed upon the notions of consensus and adaptive networks. In the consensus strategies, the objective is to derive useful mechanisms to enforce agreement among cooperating nodes [11–17]. Original implementations of the consensus strategy relied on the use of two time scales: one time scale for the collection of measurements across the nodes and another time scale to iterate sufficiently enough over the collected data to attain agreement [11]. The main problem of such mentioned methods is that the framework does not allow the network to undertake a continuous learning. This issue motivated the development of adaptive networks. An adaptive network is a collection of interconnected nodes collaborate with each other to perform real-time information processing and optimization [18]. They are known as excellent solutions for decentralized information processing problems.

So far several strategies for distributed processing over adaptive networks have been introduced in the literature. The first proposed schemes were based on incremental strategies and least mean square (LMS) learning rule [19, 20]. The incremental strategies require a Hamilton cycle through the nodes so that the information is transformed from one node to a neighboring node around the network and back to the initial node (see Figure 18.1). Several variations have been developed afterward, including incremental recursive least squares (RLS) [21], incremental affine projection algorithm [22, 23], quantized incremental LMS algorithm [24], incremental LMS with norm constraint [25], incremental LMS with variable step-size [26], and incremental augmented complex LMS algorithm [27].

For networks with a few nodes or where the deployment of the nodes is controlled, incremental (cyclic) strategies are very relevant [28]. However, these techniques require the determination of a cyclic path that runs across all nodes, which is generally an NP-hard problem [29]. Besides, incremental

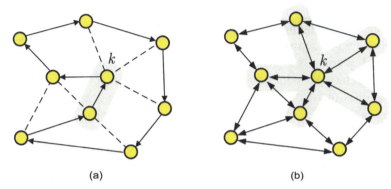

Figure 18.1 Different adaptive network strategies: (a) incremental and (b) diffusion.

solutions are sensitive to node and link failures [30]. On the other hand, in diffusion strategy the nodes communicate with all of their neighbors as dictated by the network topology (see Figure 18.1). Diffusion strategies are attractive since they are scalable, robust, and enable continuous adaptation and learning. Different diffusion adaptive networks are also available in the literature. Some examples include diffusion LMS strategies [31, 32], diffusion RLS [33], diffusion Kalman filters and smoothers [34], sparsity-aware diffusion LMS [35], variable tap-length diffusion algorithms [36], partial-update diffusion network [37], and mobile diffusion network [38, 39]. In places where the number of objectives is more than one, a number of multi-task strategies have also been introduced [40, 41]. Other research works focus on the performance analysis of diffusion strategies [42–45].

It is worth mentioning that besides the incremental and diffusion strategies, single-scale consensus implementations have also introduced in several recent works [14–16]. However, as it is shown in [46], diffusion networks have faster converge rates and also reach lower mean-squared deviation than consensus networks. For this reason, in this work we focus on development of robust algorithms based on the adaptive network strategies.

Most of the mentioned adaptive networks are based on the second order statistics where the final objective is to minimize a cost function with the mean square error (MSE) criterion. The optimality of MSE-based algorithms relies heavily on the Gaussian noise assumption which has finite second order statistics. In some circumstances, the noise may not follow Gaussian distribution and a better approximation for the actual noise distribution requires higher order statistics. Unfortunately, the MSE-based adaptive networks perform poorly under such situations, especially when the data are corrupted by impulsive noise that arises from nature or man-made sources [47–49]. For example, in WSNs outliers in sensed data may be caused due to compromised or malfunctioning sensor nodes.

To address this issue, recently some methods based on information theoretic metrics have been introduced as cost functions and optimsed using some learning rules in adaptive filters. For example, in [50] a diffusion adaptive network optimization based on the minimum error entropy (MEE) has been developed. The main problem of MEE-based adaptive networks is their high computational complexity as evaluation of the error entropy involves a double sum over the samples. On the other hand, the adaptive filters that rely on maximum correntropy criterion are able to exploit higher order moments of the data with a complexity as low as that of the LMS algorithm [51–53].

Thus, in this study we focus on the problem of distributed estimation with adaptive networks in the presence of impulsive noise which is modeled as a two-component Gaussian mixture and is not a Gaussian variate anymore. We develop efficient algorithms, with both of incremental and diffusion strategies, that are robust to measurements that are corrupted by impulsive noise (outliers). In the proposed algorithms, we move beyond MSE and use the maximum correntropy criterion (MCC) as the cost function. For both strategies, we first formulate the estimation problem as a suitable optimization problem. Then, we use stochastic gradient ascent and useful approximations to make the algorithms appropriate for distributed implementation. The resultant algorithms have the computational simplicity of the popular LMS algorithm, along with the robustness that is obtained by using higher order moments [51, 52].

Notation: Throughout the Chapter, we adopt small boldface letters for vectors and bold capital letters for matrices. We also use $(\cdot)^{\mathrm{T}}$ to denote the transpose of a vector or a matrix, $\mathrm{Tr}\,[\cdot]$ for the trace of matrix, and $\mathbb{E}\,[\cdot]$ for the statistical expectation.

18.2 Background

18.2.1 Cooperative Strategies

Let us denote a connected[1] *ad hoc* network with N nodes as a set $\mathcal{N} = \{1, 2, \ldots, N\}$. The network is deployed to estimate a vector $\mathbf{w}^o \in \mathbb{R}^{L \times 1}$ from

---

[1]In a connected network, there exists (possibly) a multi-hop communication path connecting two nodes (sensors).

streaming measurements. At every time instant $i = 0, 1, 2, \ldots$, every node k observes time sequences $\{d_{k,i}, \mathbf{u}_{k,i}\}$, where $d_{k,i}$ represents the reference signal, and $\mathbf{u}_{k,i}$ denotes a $1 \times M$ regression vector with covariance matrix $\mathbf{R}_{\mathbf{u},k} = \mathbb{E}[\mathbf{u}_{k,i}^T \mathbf{u}_{k,i}] > 0$. The data at node k are assumed to be related via the linear model:

$$d_{k,i} = \mathbf{u}_{k,i}\mathbf{w}^o + n_{k,i} \tag{18.1}$$

where $n_{k,i}$ is a zero mean, i.i.d. observation noise that is independent of other signals and has variance $\sigma_{n,k}^2$. The linear regression model in (18.1) appears in many practical applications such as spectrum sensing, target tracking, and source localization [18].

Using the MSE criterion, the unknown parameter of interest $\mathbf{w}^o$ can be obtained by solving the following optimization problem:

$$J_{\mathrm{MSE}}(\mathbf{w}) \triangleq \sum_{k=1}^{N} \mathbb{E}\left[|d_{k,i} - \mathbf{u}_{k,i}\mathbf{w}|^2\right] \tag{18.2}$$

Remark 1 *In correntropy-based adaptive filtering, the general objective is to maximize a cost function which is defined as the correntropy between the desired signal and filter output.*

As we mentioned earlier, different cooperative strategies can be used to solve the optimization problem like (18.2). For example, the MSE-based incremental strategy is described by the following update equation

$$\mathbf{w}_{k,i} = \mathbf{w}_{k-1,i} + \mu_k \mathbf{u}_{k,i}^T(d_{k,i} - \mathbf{u}_{k,i}\mathbf{w}_{k-1,i}) \tag{18.3}$$

where $\mu_k > 0$ is step-size parameter. The incremental LMS algorithm works as follows: At every iteration i, each node utilizes the local data $\{d_{k,i}, \mathbf{u}_{k,i}\}$ and local estimate $\mathbf{w}_{k-1,i}$ received from the previous node in the cycle, to update its own local estimate $\mathbf{w}_{k,i}$. Since cooperation is established by a Hamilton cycle, we have $\mathbf{w}_i = \mathbf{w}_{N,i}$.

The other widely used strategies are consensus strategies in which each node performs a local estimation and fuses its estimate with those of its neighbors so that all nodes converge to the same estimate as the number of iterations increases. The consensus strategy for the optimization of (18.2) can be described in the following form [49]:

$$\mathbf{w}_{k,i} = \mathbf{w}_{k,i-1} - \mu_{k,i} \sum_{l \in \mathcal{N}_k'} b_{k,l}(\mathbf{w}_{k,i-1} - \mathbf{w}_{l,i-1})$$
$$+ \alpha_{k,i}\mathbf{u}_{k,i}^T(d_{k,i} - \mathbf{u}_{k,i}\mathbf{w}_{k,i-1}) \tag{18.4}$$

where $\{b_{k,l}\} > 0$ are used to give different weights to different neighbors, and $\mathcal{N}_k'$ denotes the neighbor nodes of node k excluding itself, i.e., $\mathcal{N}_k' = \mathcal{N}_k/\{k\}$. It should be noted that in most consensus implementations, the step-sizes $\alpha_{k,i}$ are required to satisfy

$$\sum_{i=0}^{\infty} \alpha_{k,i} = \infty, \quad \sum_{i=0}^{\infty} \alpha_{k,i}^2 < \infty \qquad (18.5)$$

which means that the step-size sequence should to vanish as $i \to \infty$. The other solution for (18.2) is diffusion strategies which were derived in [31–33]. The diffusion LMS obtains the solution through a two-step iteration consisting of *adaption* and *combination* steps. According to the order of these two steps, diffusion LMS is classified into the Combine-then-Adapt (CTA) algorithm and the Adapt-then-Combine (ATC) algorithm. The ATC form of the diffusion strategy is described by the following update equations

$$\psi_{k,i} = \mathbf{w}_{k,i-1} + \mu_{k,i}\mathbf{u}_{k,i}^{\mathrm{T}}(d_{k,i} - \mathbf{u}_{k,i}\mathbf{w}_{k,i-1})$$
$$\mathbf{w}_{k,i} = \sum_{l\in\mathcal{N}} a_{k,l}\psi_{l,i} \qquad (18.6)$$

The above strategy consists of two steps. The first step of (18.6) involves local adaptation, where the node uses its own data $\{d_{k,i}, \mathbf{u}_{k,i}\}$ to update its weight estimate from $\mathbf{w}_{k,i-1}$ to an intermediate value $\psi_{k,i}$. The second step of (18.6) is a consultation (combination) step where the intermediate estimates $\{\psi_{l,i}\}$ from the neighborhood of node are combined through weights $\{a_{k,l}\}$ to obtain the updated weight estimate $\mathbf{w}_{k,i}{}^2$. The CTA algorithm can be described by the following update equations [33]:

$$\psi_{k,i-1} = \sum_{l\in\mathcal{N}} a_{k,l}\mathbf{w}_{l,i-1}$$
$$\mathbf{w}_{k,i} = \psi_{k,i-1} + \mu_{k,i}\mathbf{u}_{k,i}^{\mathrm{T}}(d_{k,i} - \mathbf{u}_{k,i}\psi_{k,i-1}) \qquad (18.7)$$

It is shown in [33] that in general, the ATC outperforms CTA strategy as it combines the updated estimates from all neighbors, while CTA combines the *existing* estimates $\{\mathbf{w}_{l,i-1}\}$ with a *local* adaptation step.

---

[2]$\{a_{l,k}\}$ satisfies the following properties: If two nodes l and k are not linked, then their corresponding entry $\{a_{k,l}\}$ is zero. The weights on the links arriving at node k add up to one.

18.2.2 Correntropy

Correntropy is a localized similarity measure between two scalar random variables X and Y defined by

$$C_\sigma(X, Y) \triangleq \mathbb{E}[\kappa_\sigma(X - Y)] = \iint_{x,y} \kappa_\sigma(x - y) f_{X,Y}(x, y) dx dy \quad (18.8)$$

where $\kappa_\sigma(\cdot)$ is a shift-invariant Mercer kernel, with the kernel width $\sigma > 0$ and $f_{X,Y}(x, y)$ denotes the joint probability distribution function of (X_1, X_2). The most widely used kernel in correntropy is the complex Gaussian kernel which is given by

$$\kappa_\sigma(\varepsilon) = \frac{1}{\sqrt{2\pi}\sigma} \exp(\frac{-|\varepsilon|^2}{2\sigma^2}) \quad (18.9)$$

As it is shown in [51] correntropy is positive, symmetric $(C_\sigma(X, Y) = C_\sigma(Y, X))$, and bounded $(0 < C_\sigma(X, Y) < 1/\sqrt{2\pi}\sigma)$. Moreover, it involves all the even moments of the random variable $D = X - Y$, i.e.,

$$C_\sigma(X, Y) = \frac{1}{\sqrt{2\pi}\sigma} \sum_{j=0}^{\infty} \frac{-1^j}{2^j j!} \mathbb{E}\left[\frac{(X - Y)^{2j}}{\sigma^{2j}}\right] \quad (18.10)$$

As shown in (18.10), correntropy can be interpreted as a generalized correlation function between two random variables X and Y. Comparing correntropy with MSE, we can conclude that as similarity metrics they are assessing similarity in different ways: Correntropy is local, whereas MSE is global meaning that all the samples in the joint space contribute appreciably to the value of the similarity metric, while the locality of correntropy means that the value is primarily dictated by the kernel function along the $x = y$ line [51, 52]. Thus, we can use the localization provided by the kernel width to reduce the effects of outliers in the measured data. Note that the metrics such as MSE that rely only on the second order moment can easily get biased in such conditions [29].

In practice, the joint pdf $f_{X,Y}(x, y)$ is unknown and only finite number of samples $\{x_t, y_t\}$, $t = 1, 2, \ldots, L$ from X and Y are available. Thus, a sample estimator for correntropy can be defined as

$$\hat{C}_\sigma(X, Y) = \frac{1}{L} \sum_{t=1}^{L} \kappa_\sigma(x_t - y_t)^2 \quad (18.11)$$

18.2.3 Impulsive Noise Model

Here the impulsive noise is modeled as a two-component Gaussian mixture. Due to its use in modelling complex distributions, the Gaussian mixture model (GMM) has wide applications in WSNs, such as source localization, target tracking, and target classification [24, 27, 28]. To generate the impulsive noise at node k, we can assume that the noise term in (18.1) is given by [24]

$$n_{k,i} = g_{k,i}^{(1)} + b_{k,i} g_{k,i}^{(2)} \tag{18.12}$$

where $g_{k,i}^{(1)}$ and $g_{k,i}^{(2)}$ are independent, zero mean Gaussian noise with variances $\sigma_{g,1}^2$ and $\sigma_{g,2}^2$, respectively, and $b_{k,i}$ is a switch sequence of ones and zeros which is modeled as an independent and identically distributed Bernoulli random process with occurrence probabilities $\mathrm{prob}(b_{k,i} = 1) = p$. Note that the variance of $g_{k,i}^{(2)}$ is chosen to be very much larger than that of $g_{k,i}^{(1)}$ so that when $b_{k,i} = 1$, a large amplitude impulse is generated.

18.3 Derivation of Adaptive Networks under MCC

In this section, we derive MCC-based adaptive networks. We begin with incremental cooperation and then consider the diffusion cooperation.

18.3.1 Incremental MCC Algorithm

Consider a network with N nodes with a cyclic path established across the network. Using the MCC, our objective is to develop a mechanism that would allow the nodes to cooperate with each other (according to an incremental strategy) in order to solve the following estimation problem:

$$\mathbf{w}^o = \arg \max_{\mathbf{w}} J_{\mathrm{mcc}}(\mathbf{w}) \tag{18.13}$$

given

$$J_{\mathrm{mcc}}(\mathbf{w}) \triangleq \sum_{k=1}^N \mathbb{E}\left[\exp\left(\frac{-(d_{k,i} - \mathbf{u}_{k,i}\mathbf{w})^2}{2\sigma^2}\right)\right] \tag{18.14}$$

Using (18.11) the *instantaneous* cost function in (18.14) becomes

$$\hat{J}_{\mathrm{mcc}}(\mathbf{w}, i) = \frac{1}{L\sigma\sqrt{2\pi}} \sum_{k=1}^N \sum_{j=i-L+1}^i \exp\left(\frac{-(d_{k,j} - \mathbf{u}_{k,j}\mathbf{w})^2}{2\sigma^2}\right) \tag{18.15}$$

Now, let us denote by $\mathbf{w}_i^g$ as *global* estimate for the optimal solution of (18.13). A gradient ascent approach to solve the optimization problem (18.13) is given by

$$\mathbf{w}_i^g = \mathbf{w}_{i-1}^g + \beta \nabla_{\mathbf{w}} (\hat{J}_{\text{mcc}}(\mathbf{w}_{i-1}^g)) \tag{18.16}$$

where $\beta > 0$ is a step-size parameter and $\nabla_{\mathbf{w}}$ denotes the gradient with respect to $\mathbf{w}$. Computing the gradient at $\mathbf{w}_{i-1}^g$ and replacing it in (18.16), we obtain

$$\mathbf{w}_i^g = \mathbf{w}_{i-1}^g + \mu \sum_{k=1}^{N} \sum_{j=i-L+1}^{i} \mathbf{u}_{k,j}^T \zeta_{k,j} \exp\left(\frac{-\zeta_{k,j}^2}{2\sigma^2}\right) \tag{18.17}$$

where $\mu = \frac{\beta}{L\sigma^3\sqrt{2\pi}}$ and $\zeta_{k,i} = d_{k,i} - \mathbf{u}_{k,i}\mathbf{w}_i^g$. Clearly, the solution given by (18.17) is not a distributed solution for (18.13) since this implementation requires the nodes to have access to the global information $\mathbf{w}_i^{g\,3}$. To have an adaptive distributed solution, we apply the following modifications to (18.17):

1. Use only the most recent data ($L = 1$).
2. Split the update Equation (18.17) into N separate steps, whereby each step adds one term to the summation. Doing so we obtain

$$\mathbf{w}_{1,i} \leftarrow \mathbf{w}_{i-1}^g \tag{18.18a}$$

$$\mathbf{w}_{k,i} = \mathbf{w}_{k-1,i} + \mu \mathbf{u}_{k,i}^T \zeta_{k,i} \exp\left(\frac{-\zeta_{k,i}^2}{2\sigma^2}\right) \tag{18.18b}$$

$$\mathbf{w}_i^g \leftarrow \mathbf{w}_{N,i} \tag{18.18c}$$

3. Replace $\zeta_{k,i}$ with the local error $e_{k,i}^{\text{inc}}$ defined as

$$e_{k,i}^{\text{inc}} \triangleq d_{k,i} - \mathbf{u}_{k,i}\mathbf{w}_{k-1,i} \tag{18.19}$$

to obtain

$$\mathbf{w}_{1,i} \leftarrow \mathbf{w}_{i-1} \tag{18.20a}$$

$$\mathbf{w}_{k,i} = \mathbf{w}_{k-1,i} + \mu \mathbf{u}_{k,i}^T e_{k,i}^{\text{inc}} \exp\left(\frac{-(e_{k,i}^{\text{inc}})^2}{2\sigma^2}\right) \tag{18.20b}$$

$$\mathbf{w}_i \leftarrow \mathbf{w}_{N,i} \tag{18.20c}$$

where $\mathbf{w}_{k,i}$ denotes the local estimate for the optimal solution of (18.13) available at node k and time instant i.

The given Equation in (18.20), which will be referred as MCC-based incremental algorithm can be easily implemented by a network.

---

[3] We call it global as its evaluation requires the entire network information (See (18.17)).

18.3.2 Diffusion MCC Algorithms

Consider a connected network and assume that every node $k \in \mathcal{N}_k$ has some initial estimate ψ_k. Then, using the MCC, we can formulate at node k the problem of estimating the weight vector w as follows:

$$
\max_{\mathbf{w}} \left(\mathbb{E}\left[\exp\left(-\frac{(e_{k,i}^{\mathrm{diff}})^2}{2\sigma^2} \right) \right] - \delta \sum_{l \in \mathcal{N}_k} c_{k,l} \| w - \psi_l \|^2 \right), \tag{18.21}
$$

where $\delta > 0$ is a regularization parameter which makes a balance between the actual cost and the constraint, and $c_{k,l}$ are some weighting coefficients that add up to one. Note that the local error $e_{k,i}^{\mathrm{diff}}$ is defined as

$$
e_{k,i}^{\mathrm{diff}} = d_{k,i} - \mathbf{u}_{k,i} \mathbf{w}_{k,i-1} \tag{18.22}
$$

Note that the first term in the cost function, i.e., $\mathbb{E}\left[\exp\left(-\frac{(e_{k,i}^{\mathrm{diff}})^2}{2\sigma^2} \right) \right]$ is the MCC, and the second term $-\delta \sum_{l \in \mathcal{N}_k} c_{k,l} \| w - \psi_l \|^2$ is used to penalize the distance between the solution $\mathbf{w}$ and the prior information represented by the available local estimates $\{\psi_l\}$. To obtain a recursion for the estimate of $\mathbf{w}$ at node k, we apply stochastic gradient ascent approach (at the prior iteration $\mathbf{w}_{k,i-1}$) to obtain

$$
\mathbf{w}_{k,i} = \mathbf{w}_{k,i-1} + \frac{\eta}{2\sigma^2} \mathbf{u}_{k,i}^{\mathrm{T}} \exp\left(-\frac{(e_{k,i}^{\mathrm{diff}})^2}{2\sigma^2} \right) e_{k,i}^{\mathrm{diff}}
$$
$$
+ \eta\delta \sum_{l \in \mathcal{N}_k} c_{k,l} (\mathbf{w}_{k,i-1} - \psi_l) \tag{18.23}
$$

Algorithm 18.1 MCC-based incremental algorithm

---

1. Initialization: $\mathbf{w}_{1,0} = \mathbf{0}$
2. **for** $i = 1, 2, \ldots$ **do**
3.   **for** $k = 1$ to N **do**
4.   receive $\mathbf{w}_{k-1,i}$ from previous node.
5.   compute $e_{k,i}^{\mathrm{inc}}$ as $e_{k,i}^{\mathrm{inc}} = d_{k,i} - \mathbf{u}_{k,i} \mathbf{w}_{k-1,i}$.
6.   update $\mathbf{w}_{k,i}$ using (18.20).
7.   **end for**
8. $\mathbf{w}_{1,i+1} = \mathbf{w}_{N,i}$.
9. **end for**

---

where $\eta > 0$ is the step-size parameter. We can accomplish the update (18.23) in two steps by generating an intermediate estimate $\phi_{k,i}$ as

$$\phi_{k,i} = \mathbf{w}_{k,i-1} + \frac{\eta}{2\sigma^2} \mathbf{u}_{k,i}^{\mathrm{T}} \exp\left(-\frac{(e_{k,i}^{\mathrm{diff}})^2}{2\sigma^2}\right) e_{k,i}^{\mathrm{diff}} \tag{18.24}$$

$$\mathbf{w}_{k,i} = \phi_{k,i} + \eta\delta \sum_{l \in \mathcal{N}_k} c_{k,l}(\mathbf{w}_{k,i-1} - \psi_l) \tag{18.25}$$

Now, we need to recast (18.24) and (18.25) into approximated forms which are amenable to distributed implementation. To this end, we consider the following approximations, as often used in solving diffusion adaptation problems:

- Replace ψ_l in (18.25) by the intermediate estimate $\phi_{l,i}$ that is available at node at time i.
- Replace $\mathbf{w}_{k,i-1}$ in (18.25) by the intermediate estimate $\phi_{k,i}$.

This leads to

$$\phi_{k,i} = \mathbf{w}_{k,i-1} + \frac{\eta}{2\sigma^2} \mathbf{u}_{k,i}^{\mathrm{T}} \exp\left(-\frac{(e_{k,i}^{\mathrm{diff}})^2}{2\sigma^2}\right) e_{k,i}^{\mathrm{diff}} \tag{18.26}$$

$$\mathbf{w}_{k,i} = \phi_{k,i} + \eta\delta \sum_{l \in \mathcal{N}_k} c_{k,l}(\phi_{k,i} - \phi_{l,i}) \tag{18.27}$$

If we define the new coefficients $c'_{k,l}$ as

$$c'_{k,k} = (1 - \eta\delta + \eta\delta c_{k,k}), \quad c'_{k,l} = \eta\delta c_{k,l} \tag{18.28}$$

then (18.26) and (18.27) can be rewritten as

$$\phi_{k,i} = \mathbf{w}_{k,i-1} + \frac{\eta}{2\sigma^2} \mathbf{u}_{k,i}^{\mathrm{T}} \exp\left(-\frac{(e_{k,i}^{\mathrm{diff}})^2}{2\sigma^2}\right) e_{k,i}^{\mathrm{diff}}$$

$$\mathbf{w}_{k,i} = \sum_{l \in \mathcal{N}_k} c'_{k,l}\phi_{l,i} \tag{18.29}$$

which is the MCC-based ATC algorithm.

If we simply reverse the order by which the incremental split was done in (18.23), then the CTA version (18.29) can be obtained in a similar manner as

$$\phi_{k,i-1} = \sum_{l \in \mathcal{N}_k} c'_{k,l}\mathbf{w}_{l,i-1} \tag{18.30}$$

$$\mathbf{w}_{k,i} = \phi_{k,i-1} + \frac{\eta}{2\sigma^2} \mathbf{u}_{k,i}^{\mathrm{T}} \exp\left(-\frac{z_{k,i}^2}{2\sigma^2}\right) z_{k,i} \tag{18.31}$$

Algorithm 18.2 MCC-based ATC algorithm
1.   Initialization: $\mathbf{w}_{\mathbf{k},\mathbf{0}} = \mathbf{0}$ for $k \in \mathcal{N}_k$
2.   **for** $i = 1, 2, \ldots$ **do**
3.   **for** $k = 1$ to N **do**
4.   compute $e_{k,i}^{\text{diff}}$ as $e_{k,i}^{\text{diff}} = d_{k,i} - \mathbf{u}_{k,i}\mathbf{w}_{k,i-1}$.
5.   compute $\phi_{k,i}$ using (18.29).
6.   update $\mathbf{w}_{\mathbf{k},\mathbf{i}}$ using (18.29).
7.   **end for**
8.   **end for**

where $z_{k,i} = d_{k,i} - \mathbf{u}_{k,i}\phi_{k,i-1}$.

In the next section, we attempt to show the performance of the proposed algorithm.

18.4 Simulation Results

In this section, some simulation results are presented to evaluate the performance of the proposed algorithms. To this end, we consider two different experiments as follows.

18.4.1 Experiment 1

We consider distributed networks with $N = 12$ nodes as shown in Figure 18.2. The unknown parameter $\mathbf{w}^o$ of length $L = 4$ is randomly generated so that $\|\mathbf{w}^o\|^2 = 1$. We assume that $\mathbf{u}_{k,i}$'s are white Gaussian regression data with covariance matrices $\mathbf{R}_{u,k}$, with eigenvalue spread[4] $\chi(\mathbf{R}_{u,k}) = 4$, where the data profile $\text{Tr}[\mathbf{R}_{u,k}]$ is shown in Figure 18.3. We adopt uniform step-size $\mu = 0.1$ for the MCC-based with kernel size $\sigma = 2$ in the proposed MCC-ATC algorithms. Note that as our aim is to compare the steady-state performance of the proposed MCC-based algorithms with MSE-based algorithms, the parameters are selected such that the algorithms have almost the same initial convergence rate[5]. We use metropolis weights [18] for the adaptation matrix.

---

[4]This spread is defined as the ratio between the largest and the smallest eigenvalues.

[5]If our aim is to **implement** the MCC-based algorithms for a specific application, we can select one the followings ways to adjust the free parameters; (1) select a suitable step size (which guarantees the stability) and adaptively adjust the kernel size, for example, using the following rule [54].

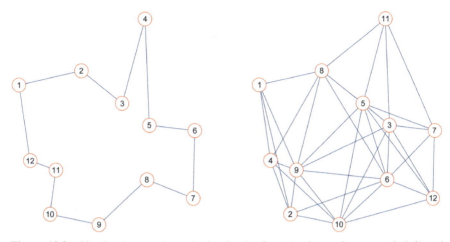

Figure 18.2 Simulated network topologies for the first experiment: incremental (*left*) and diffusion (*right*).

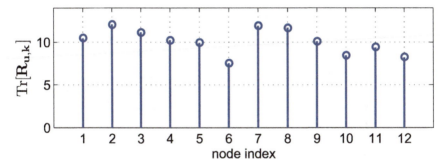

Figure 18.3 Data profile in $\mathrm{Tr}[\mathbf{R}_{u,k}]$.

The required parameters for the measurement noise are $\sigma_{g,1}^2 = 0.01$ and $\sigma_{g,2}^2 = 1000\,\sigma_{g,1}^2$. Each curve is obtained by averaging over 100 independent experiments. We examine the network performance by the global average mean square deviation (MSD) which is defined, respectively, as

$$\sigma_{k,i} = \sigma_{k,i-1} + \eta \left(\frac{(e_{k,i}^{\mathrm{diff}} - e_{k,i-1}^{\mathrm{diff}})^2}{\sigma_{k,i-1}^3} - \frac{1}{\sigma_{k,i-1}} \right)$$

where $\sigma_{k,i}$ is the kernel-size at node k and iteration i; (2) select a kernel size and adjust the step size value by scanning for the best result.

$$\text{MSD} = \frac{1}{N} \sum_{k=1}^{N} \mathbb{E} \left[\|\mathbf{w}^o - \mathbf{w}_{k,i}\|^2 \right] \qquad (18.32)$$

MSD is widely used as a metric to assess the deviation of the vector of diffusion adaptive filter parameters in almost all the corresponding literatures for cooperative networks. In places where the filter is used as a discriminator, such a metric is the best parameter to assess the classifier.

In the first simulation, we compare the traditional MSE-based incremental algorithm [20] with the proposed MSE-based incremental algorithm. The global average MSD for different algorithms is shown in Figure 18.4 for occurrence probability $p = 0.02$. As it is clear from Figure 18.4, the performance of MSE-based incremental algorithm is strongly affected by the impulsive noise, while the proposed algorithm is able to overcome this problem. Specially, the steady-state error is clearly improved by means of our proposed method. In the next simulation, we compare the performance of MSE-based ATC algorithm with that of the proposed MCC-based ATC algorithm for $p = 0.02$, where the results are shown in Figure 18.5.

We can see from Figure 18.5 that for diffusion strategy the proposed algorithm has better performance. Figure 18.6 compares the performance of

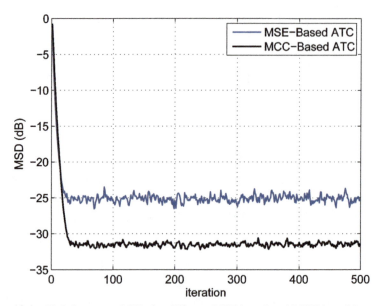

Figure 18.4 Global average MSD for different MSE-based and MCC-based incremental adaptive networks in the presence of impulsive noise.

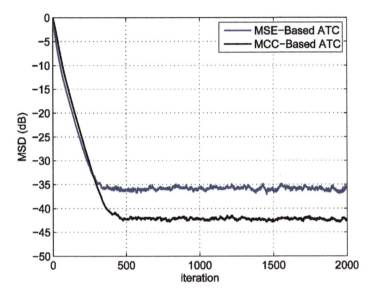

Figure 18.5 Global average MSD for different MSE-based and MCC-based diffusion adaptive networks in the presence of impulsive noise.

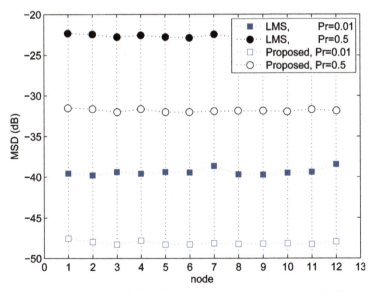

Figure 18.6 Performance of MSE-based ATC algorithm with the proposed MCC-based ATC algorithm for $\mu = 0.05$ and $p = 0.01$ for different kernel size values.

MSE-based ATC algorithm with the proposed MCC-based ATC algorithm for different kernel size values for $\mu = 0.05$ and $p = 0.01$. We see that as the kernel size σ increases, the performance of MCC-based ATC algorithm deceases, and for $\sigma \to \infty$, the proposed MCC-based algorithm will approach to the LMS algorithm.

18.4.2 Experiment 2

In this experiment, we compare the performance of different algorithms for estimating a scalar field. So, let us consider a geographical region $\mathcal{X} = [0\ 1] \times [0\ 1] \in \mathbb{R}^2$ and a scalar field f defined over this region as

$$f(x_1, x_2) = \frac{0.7}{\sqrt{1 + 20(x_1 - 0.15)^2 + 20(x_2 - 0.15)^2}}$$
$$+ \frac{1}{\sqrt{1 + 15(x_1 - 0.7)^2 + 20(x_2 - 0.7)^2}} \qquad (18.33)$$

Note that (18.33) has been used in [55] as an example of a scalar field, and we used it here to compare our proposed algorithms. We assume that nodes are deployed randomly at start and remain fixed in their locations. We consider a network with $N = 45$ nodes that is placed in the field where each node k knows its own current position $\mathbf{p}_k$. At time i a noisy version of the scalar field is captured by node k as $d_{k,i} = f(\mathbf{p}_k) + v_{k,i}$. We assume that there are 8 nodes in the network that generate outliers with $\sigma_{v,k}^2 = 1$, while for other nodes we have $\sigma_{v,k}^2 = 0.001$. The objective of this network is to find optimum weight $\mathbf{w}^o$ which gives the best approximation for f at the point $\mathbf{x} \in \mathcal{X}$ as follows:

$$\hat{f}(\mathbf{x}) \approx \mathbf{h}(\mathbf{x})\mathbf{w}^o \qquad (18.34)$$

where $\mathbf{h}(\mathbf{x}) = [h_1(\mathbf{x}), h_2(\mathbf{x}), \ldots h_M(\mathbf{x})]$ is a row vector which contains the (linearly independent) set of basis functions for $\mathcal{X}$. In this work, we consider $M = 36$ and Gaussian basis functions which are given as

$$h_l(\mathbf{x}) = \exp\left(-20\|\mathbf{x} - \mathbf{m}_l\|^2\right) \qquad (18.35)$$

where $\{\mathbf{m}_l\}$ are evenly distributed over $\mathcal{X}$. Note that the regression data for this setup are given as $\mathbf{u}_{k,i}\mathbf{h}(\mathbf{p}_{k,i})$. For this experiment, we used a kernel size $\sigma = 3$, with $\mu = 0.05$ and $\mu = 0.2$ for MSE-based and MCC-based algorithms, respectively. The MSD curve is generated by averaging over 100 independent runs. Figure 18.7 compares the performance of the MSE-based

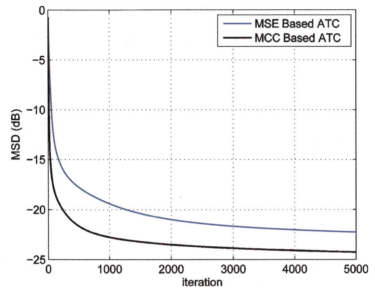

Figure 18.7 MSD performance for MSE-based and MCC-based diffusion adaptive networks for scalar field estimation.

ATC algorithm with that of the proposed MCC-based ATC algorithm, where we can observe that the proposed algorithm has better MSD performance. Note further that the MSD curves are smooth as we have assumed the nodes are fixed.

18.5 Conclusion

In this chapter, we proposed MCC-based adaptive networks for distributed estimation in an environment where the noise is impulsive. To alleviate the effect of impulsive noise, in the proposed algorithms we modified the cost function in a way to consider the higher order moments of the error. The resultant algorithms have the computational simplicity of the popular LMS algorithm. Numerical examples showed that the proposed algorithms outperform the existing MSS-based adaptive networks. Although we tested the proposed algorithms only in two different scenarios, it is obvious that we can apply them in a wide range of applications where the adaptive networks have already been used, such as frequency estimation in power systems, robust principal component analysis in WSNs, and source localization using mobile robots.

References

[1] Barbarossa, S., Sardellitti, S., and Di Lorenzo, P. (2013). Distributed Detection and Estimation in Wireless Sensor Networks. In Rama Chellappa and Sergios Theodoridis, eds., *Academic Press Library in Signal Processing*, Vol. 2, *Communications and Radar Signal Processing*, 329–408.

[2] Schizas, I. D., Mateos, G., and Giannakis, G. B. (2009). Distributed LMS for consensus-based in-network adaptive processing. *IEEE Trans. Signal Process.*, 57 (6), 2365–2382.

[3] Stankovic, J. A. (2014). Research Directions for the Internet of Things. *IEEE Internet Things J.*, 1(1), 3–9.

[4] Datta, S., Bhaduri, K., Giannella, C., Wolff, R., and Kargupta, H. (2006). Distributed data mining in peer-to-peer networks. *IEEE Internet Comput.*, 10(4), 18–26.

[5] McLurkin, J., and Yamins, D. (2005). Dynamic Task Assignment in Robot Swarms, in *Robotics: Sci. Syst.*, 8.

[6] Di Lorenzo, P., Barbarossa, S., and Sayed, A. H. (2012). Decentralized resource assignment in cognitive networks based on swarming mechanisms over random graphs. *IEEE Trans. Signal Process.*, 60(7), 3755–3769.

[7] Eftaxias, K., Sanei, S., and Sayed, A. H. (2013). Modeling brain cortical connectivity using diffusion adaptation. *Proc. IEEE ICASSP*, 959–962, Vancouver, Canada.

[8] Li, J. and Sayed, A. H. (2012). Modeling bee swarming behavior through diffusion adaptation with asymmetric information sharing. *EURASIP J. Adv. Signal Process.*, 2012:18, doi:10.1186/1687-6180-2012-18.

[9] Khalili, A., Rastegarnia, A., Islam, M. K., and Yang, Z. (2013). A bio-inspired cooperative algorithm for distributed source localization with mobile nodes. In *Annual International Conference of the IEEE Engineering in Medicine and Biology Society*, Osaka, Japan, Jan., 3515–3518.

[10] Monajemi, S., Ong, S. H., and Sanei, S. (2014). Advances in bacteria motility modelling via diffusion adaptation,. *Proceedings of European Signal Processing Conference, EUSIPCO 2014*, Portugal.

[11] Xiao, L., Boyd, S., and Lall, S. (2006). A space-time diffusion scheme peer-to-peer least-squares-estimation. In *Proc. Inf. Process. Sensor Netw. (IPSN)*, Nashville, TN, Apr., 168–176.

[12] Bertrand, A., and Moonen, M. (2011). Consensus-based distributed total least squares estimation in ad hoc wireless sensor networks. *IEEE Trans. Signal Process.*, 59(5), 2320–2330.

[13] Stankovic, S. S., Stankovic, M. S., and Stipanovic, D. M. (2011). Decentralized parameter estimation by consensus based stochastic approximation. *IEEE Trans. Autom. Control*, 56(3), 531–543.

[14] Mateos, G., Schizas, I. D., and Giannakis, G. B. (2009). Distributed recursive least-squares for consensus-based in-network adaptive estimation. *IEEE Trans. Signal Process.*, 57(11), 4583–4588.

[15] Schizas, I. D., Mateos, G., and Giannakis, G. B. (2009). Distributed LMS for consensus-based in-network adaptive processing. *IEEE Trans. Signal Process.*, 57(6), 2365–2382.

[16] Kar, S., and Moura, J. M. F. (2009). Distributed consensus algorithms in sensor networks with imperfect communication: link failures and channel noise. *IEEE Trans. Signal Process.*, 57(1), 355–369.

[17] Schizas, I. D., Ribeiro, A., and Giannakis, G. B. (2008). Consensus in ad hoc WSNs with noisy links – Part I: distributed estimation of deterministic signals. *IEEE Trans. Signal Process.*, 56(1), 350–364.

[18] Sayed, A. H. (2014). Adaptation, learning, and optimization over networks. *Foundations and Trends in Machine Learning*, 7(4–5), 311–801, NOW Publishers, Boston-Delft.

[19] Lopes, C. G., and Sayed, A. H. (2006). Distributed processing over adaptive networks. In *Proc. Adaptive Sensor Array Processing Workshop*, MIT Lincoln Lab., Lexington, MA, Jun.

[20] Lopes, C. G., and Sayed, A. H. (2007). Incremental adaptive strategies over distributed networks. *IEEE Trans. Signal Processing*, 55(8), 4064–4077.

[21] Sayed, A. H. and Lopes, C. G. (2007). Adaptive processing over distributed networks. *IEICE Trans. Fund. Electron., Commun. Comput. Sci.*, E90-A(8), 1504–1510.

[22] Li, L., and Chambers, J. A. (2008). A new incremental affine projection based adaptive algorithm for distributed networks. *Signal Process.*, 88(10), 2599–2603.

[23] Abadi, M. S. E., and Danaee, A.-R. (2014). Low computational complexity family of affine projection algorithms over adaptive distributed incremental networks. *AEU – Int. J. Elect. Commun.*, 68(2), 97–110.

[24] Rastegarnia, A., Tinati, M. A., and Khalili, A. (2010). Performance analysis of quantized incremental LMS algorithm for distributed adaptive estimation. *Signal Processing*, 90(8), 2621–2627.

[25] Liu, Y., and Tang, W. K. (2014). Enhanced incremental LMS with norm constraints for distributed in-network estimation. *Signal Process.*, 94(0), 373–385.

[26] Bin Saeed, M., Zerguine, A., and Zummo, S. (2010). Variable step-size least mean square algorithms over adaptive networks. In *Information Sciences Signal Processing and their Applications (ISSPA), 2010 10th International Conference on*, May, 381–384.

[27] Khalili, A., Rastegarnia, A., Bazzi, W., and Yang, Z. (2015). Derivation and analysis of incremental augmented complex least mean square algorithm. *Signal Process.*, IET, 9(4), 312–319.

[28] Cattivelli, F. and Sayed, A. H. (2011). Analysis of spatial and incremental LMS processing for distributed estimation. *IEEE Trans. on Signal Process.*, 59(4), 1465–1480.

[29] Ram, S., Nedic, A., and Veeravalli, V. (2007). Stochastic incremental gradient descent for estimation in sensor networks. In *Signals, Systems and Computers, 2007. ACSSC 2007. Conference Record of the Forty-First Asilomar Conference on*, 582–586.

[30] Khalili, A., Tinati, M. A., and Rastegarnia, A. (2011). Steady-state analysis of incremental LMS adaptive networks with noisy links. *IEEE Trans. Signal Process*, 59(5), 2416–2421.

[31] Lopes, C. G., and Sayed, A. H. (2008). Diffusion least-mean squares over adaptive networks: Formulation and performance analysis. *IEEE Trans. Signal Process.*, 56(7), 3122–3136.

[32] Cattivelli, F., and Sayed, A. (2010). Diffusion LMS strategies for distributed estimation. *Signal Process., IEEE Trans.*, 58(3), 1035–1048.

[33] Cattivelli, F. S., Lopes, C. G., and Sayed, A. H. (2008). Diffusion recursive least-squares for distributed estimation over adaptive networks. *IEEE Trans. Signal Process.*, 56(5), 1865–1877.

[34] Cattivelli, F. S., and Sayed, A. H. (2010). Diffusion strategies for distributed Kalman filtering and smoothing. *IEEE Trans. Autom. Control*, 55(9), 2069–2084.

[35] Di Lorenzo, P., and Barbarossa, S. (2014). Distributed least mean squares strategies for sparsity-aware estimation over Gaussian Markov random fields. In *Acoustics, Speech and Signal Processing (ICASSP), 2014 IEEE International Conference on*, May, pp. 5472–5476.

[36] Zhang, Y., Wang, C., Zhao, L., and Chambers, J. A. (2015). A spatial diffusion strategy for tap-length estimation over adaptive networks. *IEEE Trans. Signal Process.*, 63(17), 4487–4501.

[37] Arablouei, R., Werner, S., Huang, Y.-F., and Doðançay, K. (2014). Distributed least mean-square estimation with partial diffusion. *IEEE Trans. Signal Process.*, 62(2), 472–484.

[38] Tu, S. -Y., and Sayed, A. H. (2011). Mobile adaptive networks. *IEEE J. Sel. Topics Signal Process.*, 5(4), 649–664.

[39] Bazzi, W. M., Lotfzad Pak, A., Rastegarnia, A., Khalili, A., Yang, Z. (2015). Formulation and steady-state analysis of diffusion mobile adaptive networks with noisy links. *IET Signal Process.*, 9(9), 631–637.

[40] Chen, J., Richard, C., and Sayed, A. H. (2015). Diffusion LMS over multitask networks. *IEEE Trans. Signal Process.*, 63(11), 2733–2748.

[41] Monajemi, S., Sanei, S., Ong, S. H., and Sayed, A. H. (2015). Adaptive regularized diffusion adaptation over multitask networks. *Proc. of IEEE Workshop on Machine Learning for Signal Processing (MLSP)*, Boston, USA.

[42] Rastegarnia, A., Bazzi, W., Khalili, A., and Chambers, J. A. (2014). Diffusion adaptive networks with imperfect communications: link failure and channel noise. *IET Signal Process.*, 8(1), 59–66.

[43] Zhao, X., Tu, S.-Y., and Sayed, A. H. (2012). Diffusion adaptation over networks under imperfect information exchange and non-stationary data. *IEEE Trans. Signal Process.*, 60(7), 3460–3475.

[44] Khalili, A., Tinati, M. A., Rastegarnia, A., and Chambers, J. A. (2012). Steady-state analysis of diffusion LMS adaptive networks with noisy links. *IEEE Trans. Signal Process.*, 60(2), 974–979.

[45] Xie, S. and Li, H. (2013). Distributed LMS estimation over networks with quantised communications. *Int. J. Control*, 86(3), 478–492.

[46] Tu, S. Y., and Sayed, A. H. (2012). Diffusion strategies outperform consensus strategies for distributed estimation over adaptive networks. *Signal Process., IEEE Trans.*, 60(12), 6217–6234.

[47] Panigrahi, T., Panda, G., and Mulgrew, B. (2014). Error Saturation Non-linearities for Robust Incremental LMS over Wireless Sensor Networks. *ACM Transaction on Sensor Network*, 11(2), 1–20.

[48] Majhi, B., Panda, G., and Mulgrew, B. (2009). Robust identification using new Wilcoxon least mean square algorithm. *IEEE Electronics Letters*, 45(6), 334–335.

[49] Ma, W., Qu, H., Chen, B., Zhao, J., and Duan Robust, J. Diffusion maximum correntropy criterion algorithm for distributed network estimation. *arXiv:1508.01903*.

[50] Li, C., Shen, P., Liu, Y., and Zhang, Z. (2013). Diffusion information theoretic learning for distributed estimation over network. *IEEE Trans. Signal Process.*, 61(16), 4011–4024.

[51] Liu, W., Pokharel, P. P., and Principe, J. C. (2007). Correntropy: Properties, and applications in non-gaussian signal processing. *IEEE Trans. Signal Process.*, 55(11), 5286–5298.

[52] Singh, A., and Principe, J. (2009). Using correntropy as a cost function in linear adaptive filters. In *Neural Networks, 2009. IJCNN 2009. International Joint Conference on*, June, pp. 2950–2955.

[53] Zhao, S., Chen, B., and Principe, J. C. (2011). Kernel adaptive filtering with maximum correntropy criterion. In *Int. Joint Conf. Neural Networks (IJCNN)*, pp. 2012–2017.

[54] Singh, A. (2010). Cost Functions For Supervised Learning Based on a Robust Similarity Metric, M. S. Thesis, Department of Electrical and Computer Engineering, University of Florida, Gainesville. May.

[55] Bergamo, Y., and Lopes, C. (2012). Scalar field estimation using adaptive networks, in *Acoustics, Speech and Signal Processing (ICASSP), 2012 IEEE International Conference on*, March, pp. 3565–3568.

19

InfoMax: A Transport-Layer Paradigm for the Age of Data Overload

Jongdeog Lee[1], Akash Kapoor[1], Md Tanvir Al Amin[1], Zeyuan Zhang[1], Radhika Goyal[1], Tarek Abdelzaher[1], Zhehao Wang[2] and Ilya Moiseenko[3]

[1]Department of Computer Science, University of Illinois at Urbana-Champaign, Champaign, 61801, Illinois, USA
[2]REMAP, University of California at Los Angeles, Los Angeles, 90095, California, USA
[3]Department of Computer Science, University of California at Los Angeles, Los Angeles, 90095, California, USA

Abstract

Traditional transport-layer protocols, such as TCP and UDP, envisioned a communication architecture with receivers who need *all* (or at least *most*) of the data originating at the respective senders. In the emerging age of data overload, a more common application data-access pattern will increasingly be that of consumers who need a *representative sampling* of data at producers, leading to a paradigm shift in transport protocols. InfoMax implements this abstraction on top of information-centric networks (ICNs). In this chapter, we present the design of InfoMax, a new transport-layer protocol that allows applications to request a representative sampling of a named data set. InfoMax exploits the recently proposed information-centric network paradigm that gives hierarchical names to data items, as opposed to IP addresses. Assuming that named objects that share a longer branch in the name space are semantically more similar, InfoMax has the property of minimizing semantic redundancy among the data items selected for delivery, hence offering the best coverage of the requested data set with the fewest delivered items. This chapter discusses the design of InfoMax, its experimental evaluation, and example applications.

Keywords: Summarization, Representative sampling, Customization, Configurable granularity, Marginal utility, Information utility, Shortest-shared-postfix-first.

19.1 Introduction

This chapter[1] describes a new transport-layer communication abstraction, underlying data dissemination protocol that allows a consumer to request a *sampling* of a named data set at a *configurable granularity*. InfoMax is a new transport paradigm for the age of data overload that embraces lossy information transfer to substantially reduce transmitted data volume while meeting application information needs. The protocol is implemented on top of a named-data-networking (NDN) stack—an instance of information-centric networks.

The work is motivated by the proliferation of sensors (and "things") connected to the Internet and the advent of social networks that democratize information broadcast, propelling us into a world of data overload at an accelerating rate. Indeed, over 90% of the world data were produced within the last decade [1]. As the amount of digital data grows exponentially, a gap arises between the over-abundant supply and actual application demand. The growing disparity between the amounts of data generated and what suffices to meet application information needs suggests that an increasingly important function in the future will be one of data sampling (as a means of summarizing large data sets).[2] The chapter shows that *information-centric networks* facilitate the development of generic transport protocols that have the effect of sampling data in a consumer-controlled manner, achieving two desirable properties: First, the consumer gets to decide on the volume of data it receives (i.e., it decides on the degree of summarization of the requested data set). Second, for the given volume of data delivered, the protocol minimizes information loss (resulting from nondelivery of the remaining part of the data set). We expect that this protocol for delivery of a representative sampling of requested data will be of increasing importance to a growing range of data-centric applications.

Many applications can benefit from the above sampling paradigm. For instance, a small subset of speed measurements obtained from vehicles on a

---

[1]This work was supported in part by NSF grant CNS 13-45266 and CNS 10-40380.

[2]In this chapter, we use the word "sampling" or "sub-sampling" in the conventional general sense of selecting representative examples of objects from a larger set of similar objects.

given street would be sufficient to estimate average speed of traffic on that street. Similarly, a few pictures of the same collapsed building are all that is needed (in the aftermath of some disaster) to understand that it collapsed. InfoMax favors data that is less redundant by diversifying data sampling. Note that applications may have different *sampling needs*. For instance, while in the traffic example, the number of speed samples that are sufficient to estimate speed on a street is independent of the total data available from each street, in a tourism application one might want to select representative examples of good and bad reviews (e.g., of some hotel) proportionally to their frequency of occurrence. In this case, selecting reviews in proportion to the number of good and bad entries would give a more representative summary. Also, an application can have its own notion of data importance. For example, perhaps data from bicycles should be ignored in computing average traffic speed. The sampling protocol should be able to accommodate such differences, so as not to deliver less important or irrelevant data to consumers. Indeed, *customizability* of representative sampling is an important design factor in InfoMax to support the various sampling needs.

Our protocol is implemented on top of the *named-data-networking* (NDN) protocol stack [2]. NDN is a network-layer protocol that assigns hierarchical names to data objects instead of hosts. This architectural decision has two important implications from the perspective of InfoMax design. First, the network and transport layers are *aware of application-level object boundaries*. This is because objects have explicit names that the network is aware of. Second, since a hierarchical naming scheme is used, inferences can be made about *semantic overlap between named objects based on the distance between the corresponding names in the name tree*. These two insights allow us to develop an information-loss-minimizing protocol for *representative sampling* of application-level data objects.

Unlike the IP paradigm, which is *push-based* (where a sender decides what a receiver gets), InfoMax lies atop of ICN, which is *pull-based*. The consumer pulls data from the network by name. Hence, InfoMax offers a primitive for requesting a data set specified by a name prefix. In this chapter, we find it is useful to logically break up each data name into a prefix and a postfix. The prefix is a name of a data set, often associated with a producer, whereas the postfix is the additional part of the name that denotes an individual data item within the data set. A consumer supplies a name prefix to InfoMax, denoting the root of a content tree that it wishes to sample. The request signifies that the consumer wishes to receive a sampling of the named tree. The consumer does not need to know what actual objects are available in the tree and does not need to know the internal structure of tree name space below the requested prefix. It just needs the data under the tree at a configurable level of (extractive) summarization.[3] The consumer controls the amount of information transfer; it can pull less content or more content, until its information needs are satisfied. Hence, besides specifying the name prefix that decides the data set, the consumer controls how much information is actually pulled from the network.

The novelty of InfoMax lies in the order that objects belonging to the requested tree are transferred. InfoMax assumes that objects that share a longer postfix within the tree have more semantic overlap. Hence, the marginal utility of sending an object that shares a longer postfix with a previously delivered one is *lower* than the marginal utility of sending an object that shares a shorter postfix. This leads to a *shortest-shared-postfix-first* transmission order for objects in the requested tree. This order is shown to maximize the marginal utility of delivered objects and hence, the total accumulated utility for a given delivered data volume. Said differently, data delivery can be stopped at any time, short of receiving the whole tree, while ensuring the property that the information loss resulting from early termination is minimized overall possible object delivery orders (for the same total delivered content size). Since, in a pull-based architecture, it is the consumer that controls when delivery stops; the consumer gets to decide when they have had data at a fine-enough granularity. The idea is reminiscent of receiver-driven multicast [3], implemented in an application-independent fashion.

The shortest-shared-postfix-first is the default delivery order of InfoMax. It maximizes semantic diversity of delivered items. Selecting diverse items, however, should be done with care in order not to accidentally favor outliers and less important data as discussed in earlier examples. InfoMax allows producers to define three customizable dimensions of extractive summarization that determine the best delivery order for their data. These dimensions are object weight (*weight*), the utility accrual function (*aggregate*), and tree branch sampling policy (*branching order*). In a producer's name space, *weight* is a function that assigns importance to individual data item. The utility accrual function, *aggregate*, defines how utilities (i.e., weights) of objects are to be combined. *Branching order* decides the sampling policy of different branches based on their accrued total utility. For example, it may specify that all branches

---

[3]Extractive summarization refers to summarization of a data set that is obtained by choosing a subset of all its items.

whose accrued utility is above a threshold be sampled equally in a round-robin fashion. Alternatively, it may specify that branches be sampled proportionally to their subtree size or utility. The ability to customize object weights, utility accrual functions, and sampling policies offers applications flexible control over the manner, in which extractive summarization is performed on data.

The IETF Information-Centric Networking Research Group (ICNRG) enumerates several research challenges [4], brought about by the information-centric paradigm, including the *naming* of data objects; the challenge is addressed in this work. While naming can be left to the user, recent work also suggests ways to automate it, including the proposed *nameless objects* [5]. We offer a different approach that incorporates semantic information (on partial redundancy) in object names.

The rest of the chapter is organized as follows. The design and implementation of the new protocol are described in Section 19.2. Section 19.3 describes the experimental setup and explains the evaluation results on a nation-wide test bed. Section 19.4 describes two prospective application examples. We discuss related work in Section 19.5 and conclude this chapter with a summary and discussion of future work in Section 19.6.

19.2 Design and Implementation

In this section, we discuss the InfoMax abstraction, design decisions, and implementation on top of the named-data-networking stack.

19.2.1 The InfoMax Information Summarization Abstraction

The InfoMax protocol is designed to sample large sets of mostly homogeneous and partially redundant data. An example application class whose data fit this category is social networks. For instance, Twitter serves data (collected from a large population) as JSON objects; each object is a tweet together with its metadata. Similarly, Instagram and Flickr serve images; each image includes a picture file and EXIF metadata. These data sets are large and syntactically homogeneous, but amenable to hierarchical clustering based on content similarity. For example, recent work suggested a hierarchical clustering framework for tweets [6] that significantly speeds up indexing to retrieve tweets that match arbitrary keywords. The hierarchy arises because some tweets are more similar than others, as can be measured by various text similarity metrics. At the top of the hierarchy are tweet clusters that have no text overlap. Subsequent levels refine each parent cluster into sibling subsets with progressively more overlap until leaves are reached, representing individual tweets. If a name space reflected the above hierarchy, tweets whose names have a longer shared postfix would be more similar.

Cyber-physical systems offer another application class that gives rise to syntactically homogeneous data objects that semantically fall into hierarchical structures. For example, a campus building management system might collect data on temperature in different campus spaces. Such data may be organized hierarchically by building, floor, room, and temperature sensor [7]. Clearly, data from sensors in the same room would be more similar than data from different floors or buildings. If the data were named hierarchically as described above, items that share a longer name postfix would generally be more correlated (being from the same or close physical spaces). With the rise of visions, such as the Internet of Things, and the increased ubiquity of sensory instrumentation, the amount of generated sensor data is going to increase making cyber-physical data sets more common on networks.

The InfoMax protocol is designed as a general protocol for requesting a sampling of such data sets at a receiver-driven degree of granularity. Name space design is clearly a key consideration when using InfoMax, in that it influences summarization. The examples above illustrate that the name space in important category applications lends itself naturally to a hierarchical organization that is suitable for InfoMax. The main API calls provided by the InfoMax producer and consumer, along with their uses, are summarized in Tables 19.1 and 19.2, respectively.

Table 19.1　InfoMax producer APIs

InfoMaxProducer (Name prefix): Constructor for the InfoMax producer object, `producer`. It is called by an application to initialize the producer and specify a name prefix associated with this producer. The prefix specifies a content name space that the producer should make available on the network.

Produce (Name postfix, Data data): A method of the producer, `producer->produce` to add data objects to its tree. The call specifies the data object that needs to be added along with its position in the tree.

Delete (Name postfix): A method of the producer, `producer->delete` to delete the specified tree branch.

Table 19.2 InfoMax consumer APIs

InfoMaxConsumer (Name prefix): Similar to the InfoMaxProducer, this is a constructor for the InfoMax consumer object, `consumer`. A prefix is specified that the consumer is interested in. It creates the receiving end for all content that matches this prefix.

Consume (): This method, `consumer->consume`, waits to receive the object from the network that match the prefix requested by the consumer (or timeout). The method may be called multiple times to receive the set of matching objects. The InfoMax protocol continues to retrieve objects until the application stops calling this method. The application at the receiver can continue to retrieve objects until the desired level of detail is met.

19.2.2 Assumptions and Properties

The InfoMax protocol was designed to (i) minimize overhead, (ii) make the best use of the underlying NDN stack, and (iii) offer a general sampling function that is application-independent and *minimizes* information loss (compared to receiving the entire set), given the fraction of retrieved objects.

Definition 1. The name space assumption: The fundamental assumption that InfoMax makes regarding an application's name space is that data objects have hierarchical names that respect the following property: The longer the shared name postfix between two objects, measured in the number of shared name segments (i.e., tree levels), the lower the marginal utility of receiving the second object.

In accordance with the above assumption, objects are retrieved in a *shortest-shared-postfix-first* manner. Below, we offer a proof sketch of why this retrieval policy minimizes information (utility) loss when all objects are otherwise equal. In this case, the utility of individual objects does not depend on the object itself. The proof sketch will also assume that the marginal utility of retrieving a second object depends only on the length of the shared postfix between them. This leads to the following statement:

Theorem *Shortest-shared-postfix-first retrieval order minimizes consumer's information utility loss when information utility is defined as a function of the length of shared postfix.*

Proof Sketch. Let us define *information utility* of an object at the consumer as a measure of value that the consumer gets from receiving the object. Consider a name prefix, advertised by a producer, that names a tree, T, containing a total of N data objects. Each object $D_i \in T$ has a marginal information utility, U_i, when received. Given the *name space assumption* mentioned above, U_i is a function of the length of the shared postfix between D_i and previously received objects. For objects sharing the same postfix length, utility is further a function of expressed object weights (or weight aggregates for intermediate nodes). This function decides the order in which ties are broken (when many objects share the same shared postfix length). Let us denote the length of the largest such shared postfix between object D_i and previously delivered objects by S_i. Hence, $U_i = f(S_i)$, where f is a nonincreasing function. Shortest-shared-postfix-first implies that the sent object will minimize S_i over the remaining (unsent) object set, $R \subset T$. Hence, it will maximize $f(S_i)$. Since $f(S_i)$ is maximized for each sent object, it is possible to prove by induction over the number of received objects, K, that the least-shared-postfix-first algorithm, together with the weight-based tie-breaking rule, maximizes the sum of $f(S_i)$ over the K objects received in InfoMax order, among all possible sums of K objects. Hence, the utility sum over the remaining (unsent) object set, R, is minimum, which is defined as the *information utility loss* should transfer stop after K objects. Accordingly, InfoMax minimizes information loss for any K.

The above proof offers a rational for the basic InfoMax retrieval order. Clearly, in practice other factors need to be accounted for, such as the different application sampling needs described in the introduction, thereby violating optimality. Two observations should thus be made here. First, the above proof hinges on the assumptions that (i) U_i is *only* a function of S_i, denoted $U_i = f(S_i)$ in the proof, with the weight-based tie-breaking rule used as a second-order optimization criterion, and (ii) the function f is nonincreasing. Second, the optimality result does not hold for more general utility functions. For example, it does not hold when marginal utilities are a function of the *identities* of received objects. It is notoriously hard to figure out utility values in practice. Hence, it may be counter-productive to consider optimality results for more nuanced utility models.

19.2.3 The InfoMax Protocol

This subsection describes details of the InfoMax implementation that includes underlying producer–consumer APIs previously developed in NDN [8].

19.2.3.1 Producer and consumer APIs in NDN

The InfoMax protocol [9] is implemented on top of producer and consumer APIs of NDN [8]. It offers APIs to produce or consume a single data object. *Consumer→consume* issues an interest packet that requests data by name and *Producer→produce* issues a data packet that includes actual data matching the interest packet. A consumer can send an interest packet to retrieve a named object from the network. NDN will route that interest to the nearest node has a matching object. If it finds a match in the cache of a router, the router will forward the data packet. Otherwise, the interest packet will reach the original producer. The object will then be returned following the reverse route. Routers on the reverse path cache objects that they see.

Three retrieval functions are provided in the original NDN's producer–consumer framrwork [8]: *simple data retrieval (SDR), reliable data retrieval (RDR)*, and *unreliable data retrieval (UDR)*. *SDR* retrieves a single NDN data packet. Packet size is configurable by the producer, but the maximum size is limited to 8 KB. Both *RDR* and *UDR* allow retrieving larger data objects, breaking them into multiple data packets through data segmentation. Similar to the difference between TCP and UDP, *RDR* is used to reliably retrieve data, while *UDR* does not provide reliability. InfoMax retrieves (a sampling of) entire content trees, not single data objects. Hence, it complements the aforementioned data retrieval protocols, which can be chosen depending on how individual objects are to be transmitted. This is depicted in Figure 19.1, where the InfoMax Data Retrieval *(IDR)* protocol sits on top of the other three and can use any of them to retrieve the individual objects. By default, InfoMax exploits *RDR* since each representative sample should ideally be reliably delivered.

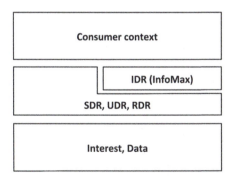

Figure 19.1 InfoMax implementation in NDN consumer context.

19.2.3.2 Enforcing the InfoMax order

InfoMax uses the shortest-shared-postfix-first retrieval order for objects in the requested tree. In this chapter, we assume that content has a *single producer*. However, this content can be cached at multiple places in the network. The first design question is: Who should maintain the InfoMax shortest-shared-postfix-first order? When requesting a content tree that matches a prefix, the InfoMax consumer does not know what content exists in the tree. Hence, the consumer cannot request objects in the tree in the correct order. It is therefore the producer's responsibility to compute a retrieval order for the tree.

Native NDN allows consumers to send interests in partially specified name prefixes (e.g., the root of the desired content tree) with an exclusion filter that specifies objects (matching the prefix) that the consumer already received. Unfortunately, this approach does not specify which object to retrieve next from the tree. Any matching object that is not excluded may be returned. Also, exclusion filters take space in a packet making this approach less scalable.

A simple design would have the producer keep track of how many objects were delivered so far to each consumer. When the next request arrives from a consumer, the next object matching that consumer's interest would be delivered. The design would require producers to keep per-client state, which undesirable. Furthermore, any design that requires recipients of consumers' NDN interests to know the *next* object to send in InfoMax order (e.g., based on previously received objects) would not work in the presence of network caching. Caches (unless they run InfoMax code as well) will not know the correct InfoMax order and hence may not serve the correct next object. To keep caches from having to implement InfoMax, consumers should always ask for the *full* object name, as opposed to a partially specified name prefix.

Together, the requirement that producers maintain the shortest-shared-postfix-first list for their content (since they know their content best) and the requirement that consumers ask for objects by their exact full name lead to a simple design, where:

- *At the producer*: The transport entity sorts objects in the tree (under the exported name prefix) in a shortest-shared-postfix-first delivery order. This sorted sequence (of object names) is recorded, broken into objects, called *lists*. Individual lists are numbered and made available by the transport entity to be retrieved sequentially by consumers.

- *At the consumer*: Given a tree prefix, say $root$, to retrieve objects from, the transport entity requests the aforementioned numbered lists using default names, such as $root/list/1$, $root/list/2$, etc. (In our implementation, the default list names contain special characters not used in naming data objects, in order not to clash with data object names.) Once a list is retrieved, the transport entity at the consumer parses it for data object names and requests the listed objects by exact name in the order they appear on the list.

The approach allows for both list objects and data objects to be cached. The cache need not run InfoMax code. It simply responds to interests in named list and data objects, either serving them or forwarding the interest further toward a producer.

The size of the lists turns out to be an important design parameter that significantly impacts communication performance. A small list size could lead to frequent exchange of lists in order to determine which data objects are to retrieve next, thus leading to more network traffic. A large list size can increase overhead if the consumer does not need many data objects in their tree summary. Since the network-layer packet in NDN is 8 KB long, we fixed our list size to 100 object names, which is roughly 8 KB in size. Hence, list objects take one full NDN packet. This is suitable for transfer of large numbers of data objects, which is the common case in the data-intensive applications that motivate InfoMax. Whenever the consumer needs more data objects, it can ask for the next list and will get names of the next 100 objects.

19.2.3.3 Handling dynamic updates

The approach mentioned above is suitable for retrieving primarily static content. If the tree at the producer is updated often, two problems occur. First, extra overhead is paid in recomputing the lists that maintain the sorted order of objects in the tree. Second, if the lists are updated while some consumers are retrieving content, retrieval of inconsistent list versions may result in an incorrect data object retrieval order.

To reduce the overhead of recomputing the lists (as well as the number of list versions generated and hence potential cache pollution with these versions), the protocol controls the frequency at which the lists are recomputed. For example, the transport entity at the producer may recompute lists when a new object is added or 10 minutes after the last recomputation, whichever is *longer*. This imposes a minimum list update period and hence bounds overhead.

To prevent inconsistent list versions from jeopardizing delivery order to a consumer, a limited number of past versions may be maintained at the producer's transport entity. The transport entity at the consumer notes the version number of the first list it receives, $root/list/1$ and, in the rest of the data transfer, requests lists of the same version number.

Having said the above, we should note that InfoMax envisions data that do not change often. For example, once sensor data objects are collected, they are stored for a long period of time without modification. Similarly, once users issue tweets or images, they live in the blogosphere without modification until obsolete. The most common case of updates is to add new objects or new branches to the tree. We recommend that in applications featuring frequent updates, the first branch under the consumer's prefix $root$ should refer to a time window. Hence, each window, a new tree is constructed from objects in that window. That tree then remains fixed for the rest of time, once a new window starts. A consumer can request specifically the window(s) that they need under $root$.

Note that name space design is in general an important application-level issue for applications that serve data over information-centric networks. Application designers must consider other design implications as well, such as security and ease of routing. Those implications may be orthogonal to the naming hint suggested above, as they tend to affect the composition of the prefix $root$ itself, even more so than the topology of the descending tree under it.

Figure 19.2 depicts an example flow of the InfoMax protocol. In step 1, an application offers a new name space by creating an InfoMax producer. In step 2, a consumer is created who would like to receive a sampling of this content. In step 3, the consumer's transport entity sends an NDN interest in the first list object (called a "diversity interest" in the figure). The interest is routed by NDN to whoever has this list. Unless the content is already cached somewhere, the NDN interest packet makes it to the producer. On receiving this packet at the producer, the NDN layer invokes a callback to the InfoMax transport entity. The entity sends back the first list with the names of data objects that the consumer should ask for (step 4). On receiving this list, the consumer requests the data objects (step 5). These requests are fulfilled by whoever has the content. In the figure, the content is fulfilled again by the producer (step 6). The consumer may request more content, repeating steps 3–6 until satisfied or until all content under the tree is collected.

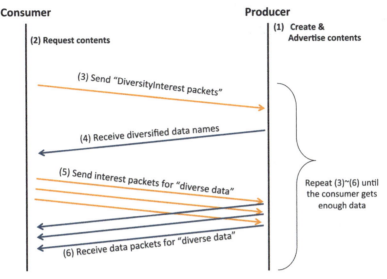

Figure 19.2 Communication flow of InfoMax.

19.2.4 An Approximate Transmission Ordering Algorithm

An efficient implementation of the shortest-shared-postfix-first delivery algorithm is needed to reduce InfoMax overhead. We developed a breadth-first traversal algorithm that operates in a round-robin fashion to diversify data objects sampled from a tree, as shown in Algorithm 19.1.

The algorithm starts with all branch counts initialized to zero and then traverses the tree from the root node taking, at each level of the tree, the least visited branch and incrementing its count. If there is a tie, it compares importance of candidates and selects one with the highest weight. In this subsection, we only focus on the default diversification order, where all nodes are assumed to have the same importance. Once a leaf node is reached, the object at the leaf is returned and its name appended to the output list. The traversal starts again from the root for each subsequent object. The algorithm terminates when all leaves have been visited.

The list generated in the above process approximates shortest-shared-postfix-first. This is because an object that has the shortest shared postfix with previously listed ones must descend from some branch, B, not yet traversed, whose siblings have been traversed already. The count of that branch would therefore be zero. When the algorithm reaches the parent node at branch, B, say node P, it will take branch B because it has the smallest count (specifically,

---

Algorithm 19.1   Top-down shortest-shared-postfix algorithm

---

simplePrioritizer(Node root)

L := empty list;

while *unvisited leaf nodes in tree* **do**

> L.append(getNextNode(root));

end

return L;

Node getNextNode(Node currentNode)

currentNode.count += 1;

if *currentNode == leafNode* **then**

> **return** currentNode;

end

for *child = currentNode.children* **do**

>> **if** *child has the smallest count among siblings with unvisited leaf nodes* **then**
>>
>>> getNextNode(child);
>>
>> **else if** *child has the biggest weight among siblings with unvisited leaf nodes* **then**
>>
>>> getNextNode(child);
>>
>> **else**
>>
>>> getNextNode(leftMostChild);
>>
>> **end**

end

---

its count is zero). Hence, the algorithm will correctly choose the branch leading to a shortest-shared-postfix object. The problem here is that the greedy nature of the algorithm might cause it to take a different branch earlier on, hence never reaching node P. This occurs when there is a tie in the counts somewhere closer to the root of the tree, causing the algorithm to descend into a different subtree (i.e., away from P).

For example, in Figure 19.3, the optimal transmission order that minimizes the length of the shared postfix would be $A/B/D/H$, $A/C/E$, $A/C/F$, $A/C/G$, $A/B/D/I$, and then $A/B/D/J$. However, Algorithm 19.1 generates a different order ($A/B/D/H$, $A/C/E$, then $A/B/D/I$, . . .),

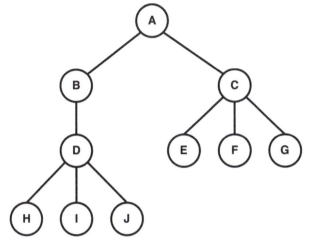

Figure 19.3 An example where Algorithm 19.1 does not generate the optimal results.

since it alternatively traverses the branches (i.e., data objects cannot be consecutively retrieved from the same branch). Prior work eliminates this problem and achieves optimality, but the algorithm becomes relatively complicated [10]. The lighter-weight algorithm is recommended to mitigate the computational bottleneck on the producer side. We investigate the efficacy of this algorithm in the evaluation section.

An alternative implementation of this algorithm changes the *top-down* traversal to a *bottom-up* one as described in [10]. The pseudocode for the latter algorithm is shown in Algorithm 19.2. It is fairly similar to merge sorting. Conceptually, it divides the given tree into n sublists that contain only one leaf node. Then, it repeatedly merges the sublists to produce new sublists sorted in the InfoMax order until there is only one list remaining. The default InfoMax order fairly selects branches from left to right when merging the sublists. Since the sublists are merged at each intermediate node, each subtree has its own InfoMax-ordered sublist. This algorithm generates all the lists sorted in the InfoMax order for all subtrees.

Both the top-down and bottom-up algorithms generate the same order. The top-down algorithm is faster, whereas the bottom-up algorithm generates lists for subtrees as well, as opposed to only one list for the root tree. If an application only needs to sample the whole tree (i.e., it does not need sampling of subtrees), then employing the top-down algorithm is better choice. Otherwise, the bottom-up implementation should be used.

Algorithm 19.2   Bottom-up shortest-shared-postfix algorithm

mergePrioritizer(Node N)

L := empty list;

if *N == leaf* **then**
 | L.append(N);
 | **return L**;
end

for *each children c of N* **do**
 | L[i] := mergePrioritizer(c);
end

L := merge(L[], N);

merge(List L[], Node N)

for *each children c of N* **do**
 | aggregate[i] := c.getWeight();
end

order := branchingOrder(aggregate[]);

List := empty list;

while *L[] are nonempty list* **do**
 | Pop front item from L[] in order;
end

return List;

19.2.5 Customizing InfoMax Order

Although the described algorithms successfully diversify data delivery, they do not consider importance of individual data objects. Applications may assign different weights to different data items. The question becomes one of accommodating both diversity and importance. The following parameters are defined to offer this flexibility:

- *weight*: defined for each object and denotes the utility of the individual data item.
- *aggregate*: a function that computes the aggregate utility of a node given the utility of each of its children.
- *branching order*: the sampling policy defined by the order in which branches are to be sampled, as a function of their respective utilities.

Weight is application defined via an API we offer for this purpose. Once weight of content (leaf nodes) is defined, applications can choose how to aggregate individual object weights to determine importance of their parent. Standard options are sum, maximum, minimum, or average. Applications can alternatively define their own aggregation functions, if needed. Once aggregate utility of each branch is computed, applications can choose a branching order. The default InfoMax sampling will select branches from left to right in a round-robin fashion. Other policies include skipping branches whose aggregate utility is below a threshold, or sampling branches proportionally to their utility or subtree size. Applications are further allowed to create their own policy for sampling the branches as a function of their utility.

Table 19.3 shows several variations of InfoMax with different combinations of above three customization dimensions. In all cases, the output is a sequence that gives rise to list objects. These are the lists that the transport entity at the consumer will request in order to determine the order in which individual objects are to be pulled from the producer. Below are example orders produced by the aforementioned customization:

- *Diversity only*: It is the default InfoMax order that samples branches equally in a round-robin fashion, regardless of weights.
- *proportional to aggregates*: This order selects more samples the bigger clusters, proportionally to their size.
- *high aggregates*: This samples branches equally, most important first.
- *aggregates above threshold*: This order samples only those branches that have higher weights than a threshold.

Each order above can meet different application sampling needs. For example, when sampling positive and negative reviews of a product, it may be appropriate to return a number of samples of each category that is proportional to the total size of the category. However, when sampling reported traffic speed on different streets, contributed by individual vehicles, it may be more appropriate to return the same number of samples from each street (regardless of the number of contributing vehicles). Finally, in applications where a natural

Table 19.3 Variations of InfoMax order

Retrieval Patterns	Weight	Aggregation	Branching Order
Diversity only	*	*	Least visited left first
Proportional to aggregates	1	Sum	Proportional to sum
High aggregates	From application	Max	Least visited Highest first
Aggregates above threshold	From application	Avg	Least visited above threshold

measure of importance exists (e.g., the number of retweets in Twitter), it may be desirable to skip items that are below a given importance values.

We demonstrate two applications in this chapter that benefit from different InfoMax orders. The first application is visual tourism, providing sample pictures for a city that users want to visit. In this case, users want to retrieve semantically less redundant images in order to quickly cover all landmarks of the city. *Above threshold* is recommended, in this case to filter out less informative images such as "selfies." If an image is regarded as selfie, the image will be not included in the sample list. For simplicity's sake, we consider the images with a front face as selfies. The second application is Twitter news that summarizes tweets on emerging events. *High aggregates* are selected for this particular application to balance diversity and importance of tweets.

19.3 Evaluation

In this section, we evaluate InfoMax performance. This includes evaluating the overhead, investigating the impact of network caching, and assessing the degree to which the algorithm approaches the optimal shortest-shared-postfix-first transmission order. The experiments were done on a nation-wide test bed implementing the NDN protocol. There are 17 institutions participating in the NDN test bed in the world. The test bed map can be found in the NDN site [11]. These nodes are created for research purposes. Hosts can be connected to NDN routers by configuring and executing an NDN forwarding Daemon [12]. We used the above test bed for following experiments.

19.3.1 Transmission Overhead

In order to measure end-to-end delay between a producer and a consumer, we created a producer in the Midwest (connecting to the UIUC NDN router) and a consumer in California (connecting the UCLA NDN router) as shown in Figure 19.4.

To evaluate the extra overhead of InfoMax, we then created an additional consumer (also at UCLA) that is "clairvoyant" in that it knows exactly the topology of the producer's data tree. This consumer therefore simply requests data objects one at a time by name using RDR, bypassing the InfoMax transport. This clairvoyant consumer is thus expected to show the best achievable performance in terms of latency and throughput.

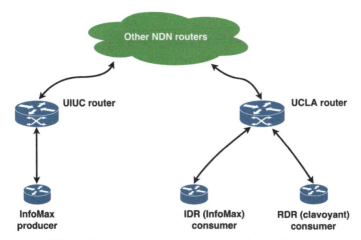

Figure 19.4 Topology for measuring end-to-end delay.

We perform the experiment varying the number of requested data objects. In Figure 19.5, the x-axis denotes the number of data objects requested and the y-axis denotes latency per object. Considering that InfoMax exchanges additional packets and parses the list packets for data object names, we expect the base (clairvoyant) client to outperform InfoMax. However, the added latency is not significant. Moreover, this extra latency could mitigated by successful NDN caching as will be shown in the following experiment.

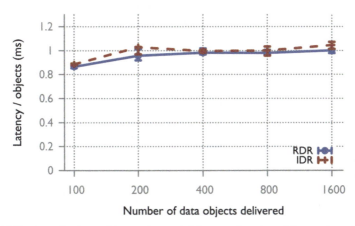

Figure 19.5 Amortized per-object end-to-end delay as a function of the number of objects.

19.3.2 Scaling Delivery

A key benefit of NDN is in-network caching. InfoMax is carefully designed to leverage the NDN router cache. To demonstrate, we measure transmission rate at the producer side while increasing the number of InfoMax consumers. If NDN caching works well, the load on the producer should not change. It should remain the same as with only one consumer. Figure 19.6 shows the network topology used in this experiment. As before, the InfoMax producer is in the Midwest (connected to the UIUC NDN router). Four consumers are logically connected to different NDN routers across the test bed (UCLA, University of Michigan, Colorado State University, and VERISIGN laboratory).

Each consumer is programmed to request 100 data objects, repeatedly. We compare what happens when cache is enabled and disabled, respectively. The resulting transmission rate of the producer is shown in Figure 19.7. In the figure, the x-axis and the y-axis represent the number of consumers and the total transmission rate at the producer, respectively. The solid line in Figure 19.7 shows the producer's transmission rate in the presence of caching. Note that, it remains almost same as the number of consumers increases. This is because of successful NDN caching. Next, we disabled the cache function by setting cache time-to-live to be very short (1 ms). As shown by the dotted line in Figure 19.7, the resulting transmission rate of the producer increases

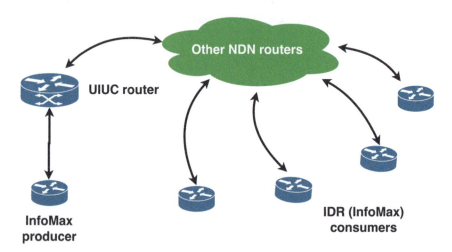

Figure 19.6   Topology for testing NDN caching with multiple consumers.

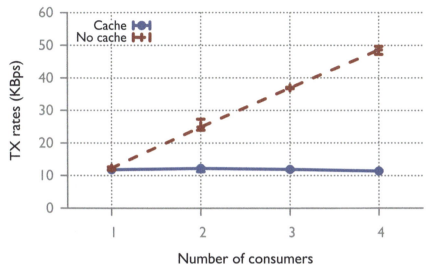

Figure 19.7 Transmission rate of a producer with multiple consumers.

almost linearly with the number of consumers in this case. The experiments show that InfoMax does not increase the producer's transmission bandwidth needs.

19.3.3 Shortest-Shared-Postfix-First Ordering

InfoMax provides an approximate implementation of the shortest-shared-postfix-first order that approaches optimal behavior with low overhead. In order to show it, we first devise an experiment to measure how close InfoMax comes to an optimal shortest-shared-postfix-first algorithm. The optimal algorithm was described in the author's prior work and is used as the comparison point [10]. For this experiment, a random tree generator is developed to create a diverse set of trees by controlling the total number of nodes, N, and the maximum number of children, C, under one parent. Different trees are created (with $N = 1600$ and $C = 20$), and the output of an optimal implementation of shortest-shared-postfix-first is compared to the approximate Infomax algorithm. Figure 19.8 shows the result of this comparison, where the x-axis denotes the number of objects transmitted, and the y-axis denotes the average number of shared segments (tree levels) between the current object name and previously sent data.

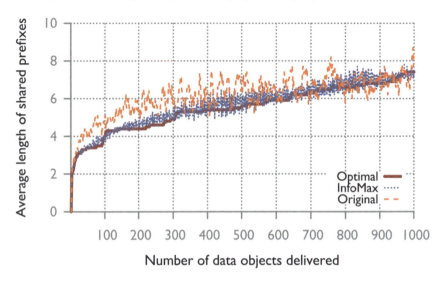

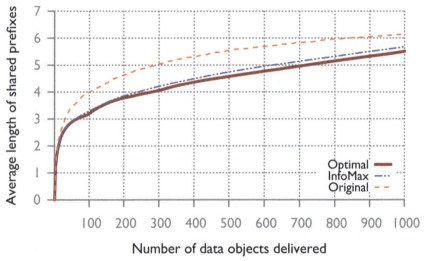

Figure 19.8 Average length of shared postfix (in number of shared name segments) as a function of objects transmitted. Raw data (*up*) and smoothed data (*down*).

As plotted in the graph, the number of shared segments in the name postfix increases monotonically as more objects are transmitted, indicating that diverse objects are transmitted first (i.e., those with a smaller postfix overlap) followed by progressively more redundant objects (with a higher postfix overlap). In application terms, the degree of redundancy increases as more objects are sent. Note also that the InfoMax curve is very close to the optimal, whereas the regular NDN curve (which, for the purposes of this experiment, corresponds to a leftmost child first traversal) is significantly higher meaning that more redundant objects are sent earlier, thereby reducing transmitted information value.

We perform the same experiments with two different sets of trees created by our generator. Both sets have the same number of nodes in a tree ($N = 1600$), but different maximum number of children ($C = 5, 100$). Trees having smaller C are deeper and those having bigger C are wider and shallower. Figure 19.9 shows that deeper trees gain more from our algorithm in that the separation between the InfoMax and NDN curves is larger. This is because leaf nodes in a deeper tree have longer names, so the impact of a shortest-shared-postfix-first strategy is more pronounced. In the limit, if the tree depth is equal to 1, there would be no difference among the algorithms.

As discussed in Section 19.2, we provide two types of algorithms for diversification: a top-down and a bottom-up. Figure 19.10 shows computation time of both algorithms for a different number of data objects. Computation time for the top-down algorithm linearly increases with the number of leaf nodes (N) in a tree as its complexity is $O(H \cdot N)$, where H is the height of a tree. Compared to this, the bottom-up algorithm grows slightly faster because its complexity is $O(N \log N)$ inherited from merge sorting. The bottom-up algorithm, however, generates ordered lists for all the subtrees at once, whereas the top-down algorithm generates a list for only one tree. In this sense, the bottom-up algorithm could be better because a tree with N leaf nodes can have at most $N - 1$ subtrees by induction. Hence, if an application needs samples for all subtrees, the bottom-up algorithm can actually outperform the top-down algorithm.

19.3.4 Customized Ordering

InfoMax also allows accommodating individual object weights. In Figure 19.11, the result of different orders is presented that trade off importance and diversification. The s-axis shows the number of data objects delivered, whereas the y-axis shows their cumulative weight in percentage ot total. As expected, the best result is *importance only* and the worst result is *diversity only* because one only considers importance and the other only considers diversity. Other orders fall somewhere in between. Although *high aggregates* and *aggregates above threshold* accumulate utility slower than the optimal result, they do much better result than *diversity only*. Hence, they offer diversification while largely respecting importance.

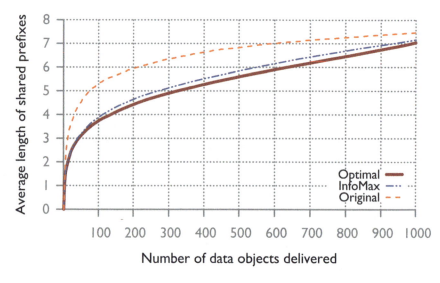

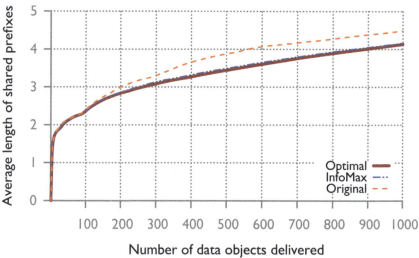

Figure 19.9   Average length of shared postfix (in the number of shared name segments) as a function of objects transmitted for deep trees (*up*) and wide trees (*down*).

19.4 Application Examples

InfoMax, as previously discussed, relieves the consumer's application from having to figure out how best to sample a producer's abundant data. It also relieves the producer from having to understand consumers' data summarization

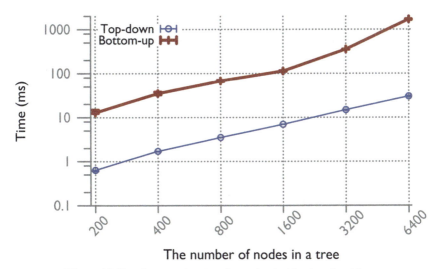

Figure 19.10 Computation time for each prioritization algorithm.

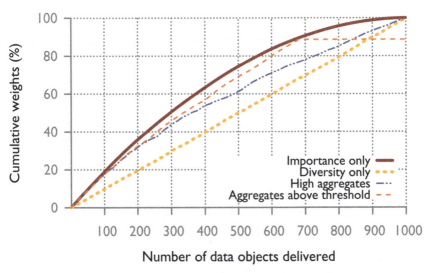

Figure 19.11 Result on importance for different orders.

needs. The producer is simply responsible for naming and organizing its data objects, such that object categories with less similarity branch out earlier in the name space. Everything else follows naturally, as illustrated in the examples below.

19.4.1 Visual Tourism

As a simple example, we created a tourism application. It allows consumers to browse pictures of points of interest at destinations of choice. The pictures are organized hierarchically first by destination (e.g., "Rome" and "Paris"), then by category of points of interest (e.g., "Ancient Landmarks," "Renaissance Landmarks," "Modern Landmarks," and "Religious Landmarks"), then by landmarks that belong to each category, and finally pictures of each landmark.

Suppose a consumer, a tourist queries for ideas for a possible itinerary in Rome. The client software issues an InfoMax request to download pictures under the "/Tourism/Rome" prefix. Figure 19.12 shows the first 6 images retrieved. We observe that the pictures are spread across various branches in the content tree, including ancient, middle age, and religious landmarks. They give a diverse view of Rome overall. The number of pictures in the overall data set exceeds 3000, with over 150 pictures of each single landmark. A random retrieval of these pictures would likely yield much redundancy. Instead, by following diverse top-level branches first, the shortest-shared-prefix-first algorithm automatically retrieved a diverse set of pictures. No application semantics were needed to achieve this diversification.

Next, the consumer wishes to take a closer look at Rome's ancient landmarks. The client software queries InfoMax for "/Tourism/Rome/Ancient."

Figure 19.12   Top-level query—overview of Rome.

The results are shown in Figure 19.13 Again, they come from different branches at this level of the tree, showing different ancient landmarks. Finally, the consumer wishes to see the Colosseum. The software queries InfoMax for "Tourism/Rome/Ancient/Colosseum." At this level, only Colosseum pictures are retrieved as shown in Figure 19.14.

Figure 19.13 Second-level query—scenes from ancient Rome.

Figure 19.14 Query on the Colosseum.

Two points are notable in this example. First, the retrieved pictures, in each case, constituted a sampling of a data set that contained *orders of magnitude more matching entries* to the consumer's specific query, most of which were partially redundant. Among these *matching* entries, an appropriately diverse small sampling was returned. Second, such diversification of retrieved information occurred with no use of application-specific information. It was simply an artifact of following the generic shortest-shared-postfix-first retrieval order by the InfoMax transport entity.

19.4.2 Twitter Search

Next, we describe a Twitter-based news-feed application utilizing InfoMax transport [13]. In this application, our producer crawls Twitter for tweets on events that it wants to report a Twitter-based news-feed on. These tweets have various degrees of similarity. Some are plain retweets without new information, some are retweets with added information, some provide information related to an event from possibly different perspectives, and some are entirely different. As a result, these tweets can be hierarchically clustered by similarity as described by a subset of the authors in prior work [6]. The resulting hierarchical structure is used to generate a content tree in the NDN name space. This content tree is served by the InfoMax producer, offering news about ongoing events at a configurable level of granularity to consumers.

Specifically, we collected tweets on recent ISIS attacks in the Iraqi Kurdish town of Kirkuk. Around 200 K tweets were collected (from January 31, 2015, to Febuary 11, 2015) matching the keywords "ISIS," "Kirkuk," and "Kurdish" or the geographic locations around Kirkuk, Iraq. To prioritize what may be important for the consumer, siblings sharing a common postfix were ordered left to right in a decreasing order of the number of their descendants. (This takes advantage of the generic InfoMax tie-breaking rule, described earlier, that takes the leftmost branch of those involved in a tie.

Consider a consumer who specifies that they want a 10-tweet summary of the aforementioned news feed. Following the InfoMax shortest-shared-postfix-first order, the producer serves content from ten different top-level branches of the content tree. The resulting output is shown in Table 19.4. (Only English tweets are selected for readability in this example.)

After reading the result, the consumer is interested in the details of a particular event. The client software allows it makes a follow-up query to retrieve more results on a given tweet (similar to retrieving results on the Colosseum in the tourism example). This is implemented by having the

Table 19.4   Top-level Kirkuk news feed

S1	RT @Mojahedineng: #PMOI #Iran Iran: Three Kurdish teachers arrested to prevent teachers protest gathering: http://t.co/DQnqfniFyQ
S2	RT @DerekStoffelCBC: Let remember journalist Kenji Goto as a brave, humane man, not an #ISIS victim. http://t.co/Ih4Teuz33e #CBC
S3	RT @MEAIndia: Beginning the day with good news. 11 Indian nurses from Kerala, evacuated from Kirkuk, to return home from Erbil on 7 February
S4	RT @peopleofearth: A Kurdish fighter walks with his child in the streets of Kobani, Syria after they recaptured it from ISIS militants.
S5	RT @Alladin_Al: Y didnt any news Chanel broadcast the muslims being burned in Burma? Or are they only focused on their beloved "ISIS"
S6	RT @grasswire: A Kurdish marksman stands atop a building as he looks at the destroyed Syrian town of #Kobane. Photo @Kilicbil #AFP
S7	RT @salar_dd: #Turkish special forces member: We have launched attacks on #Kurdish villages; killed babies then blamed #PKK on TV
S8	RT @itirerhart: @BBCr4today: Kurdish victory in Kobane: Islamic State militants have been driven out of the Syrian town.
S9	RT @_Kurda_: #Peshmerga Destroyed Huge Islamic Jihadist Armor in #Kirkuk by MIlan Rockets #TweitterKurds #Kobane #JordanianPilot
S10	RT @TheDarlingBeast: 1988, #Kurdish children lined up to be slaughtered under Iraqi regime. Look at that innocent smile, so unaware.

software ask InfoMax for the subtree of this tweet's parent. (Note that the software knows the name of the parent node because it already retrieved the tweet in question from the producer, including its full name in the producer's name space). Tables 19.5 and 19.6 show 10 tweets on topics $S8$ and $S2$ from Table 19.4, respectively. Observe that these subsequent queries return much more targetted results.

The second data set "OccupySandy" contains around 50 K tweets from November 04, 2012, to November 14, 2012, matching the keyword "Occupysandy" or the geographic locations around New York City. It contains tweets related to the Occupy Sandy relief effort to help victims of Hurricane Sandy. Similar to the first example, Table 19.7 provides a quick automatic summary of the OccupySandy data set. Tables 19.8 and 19.9 show the output of drill-down

Table 19.5 Kirkuk: More on Kobane attack

S8	RT @itirerhart: @BBCr4today: Kurdish victory in Kobane: Islamic State militants have been driven out of the Syrian town.
D1	VIDEO: Awaiting return to war-torn Kobane http://t.co/3YeFlCPvvv Days after Kurdish forces drove Islamic State militants from Kobane
D2	Kurdish forces engage in sporadic battles with Islamic State militants around the Syrian town of Kobane, as they fight to expel IS
D3	Kurdish forces have driven Islamic State (IS) militants from Kobane, activists say, ending a four-month battle. . . http://t.co/4RJVgha2jA
D4	SYRIAN Kurdish fighters have seized dozens of villages from Islamic State jihadists around the town of Kobane. http://t.co/7oLYiwlKC2
D5	Iraqi Kurdish forces retook an oil station Saturday that had been seized by Islamic state militants, but there. . . http://t.co/oM5UjkuYv8
D6	"Earlier gains were fueled by reports of Islamic State militants striking at Kurdish forces" talking oil prices http://t.co/GsCQU1pDol
D7	The Kurds Kurdish forces engaged in sporadic battles with Islamic State jihadists around the Syrian town of Kobane on Saturday
D8	Islamic State moves into south-west of Syrian Kurdish town – The Malaysian Insider (themalaysianinsider) http://t.co/3El3TuiO8m
D9	RT @indianews: After Kobani, Islamic State may target Kurdish town of Afrin next: Western and Arab powers that . . .
D10	[Xinhua News] IS militants flee surrounding of Syrian Kurdish city: activists: The militants of the Islamic St. . . http://t.co/ep5ebx1ghb

queries for tweets $S3$ and $S4$ from Table 19.7. Note that, the results of these queries present more detail on each of the respective topics. For example, the tweets in Table 19.8 articulate the schedule of free food distribution for victims, while the tweets in Table 19.9 articulate the use of an Amazon wedding registry to donate resources to the victims.

The point to emphasize here is that we were able to control the level of specificity of query results simply by following the shortest-shared-postfix-first transmission order (starting with different nodes in the tree). This order was efficiently approximated in the InfoMax transport layer that is completely unaware of application semantics. The only interface each application used in the above examples was to request content from subtrees, matching a specific

Table 19.6   Kirkuk: More on Kenji Goto

S2	RT @DerekStoffelCBC: Let remember journalist Kenji Goto as a brave, humane man, not an #ISIS victim. http://t.co/Ih4Teuz33e #CBC
D1	My condolences RT @DerekStoffelCBC Lets rmmbr Kenji Goto as brave man, not #ISIS victim. http://t.co/NIuoBLHEve #CBC http://t.co/qjJzoWWTG1
D2	ISIS exhibited the video by which Kenji Goto is killed. http://t.co/6KahTbaFYr http://t.co/oiutIpdCIi
D3	Kenji Goto's wife announced the declaration which requests to free him in ISIS. http://t.co/FY0Z1oKUbg http://t.co/LDmLjJfkMS
D4	Free Kenji Goto. #KenjiGoto #ISIS
D5	kenji goto death official execution by isis: http://t.co/DW5Gt72x5w @YouTube
D6	RT @Acosta: POTUS statement on ISIS killing of Kenji Goto http://t.co/IWAtUE7ypw
D7	Caroleina2 @NewsHour Kenji #Goto R.I.P. #ISIS are depraved bloodthirsty maniacs. Not part of humanity.
D8	Rest in peace Kenji Goto. ISIS does not represent Islam, period.
D9	RT @Yunghi: "Kenji Goto's reporting is voice of humanity in times of atrocity" https://t.co/hFKdisfRDt #kenjigoto #isis
D10	RT @rcampbelltokyo: More samples from Kenji Goto's blog! #IAMKENJI #ISIS #JapaneseHostage. https://t.co/CXsxCCl3xX

prefix in the producer's name space. Very importantly, not all content matching the consumer's query was retrieved but only a very small sampling in each case. The size of the sample was determined by the consumer. By following the InfoMax retrieval order, the samples were properly diversified to offer a broad coverage of each query using the small number of retrieved objects. We believe this diversified subsampling capability will be of increasing value to many emerging data-centric applications, where data over-abundance is the norm.

19.5 Related Work

With the growing demand on data oriented applications, many recent proposals advertized moving from host- to content-oriented networking. Information-centric networks, such as NDN, espouse named content rather than named hosts, as a central abstraction. The result retains the simplicity and scalability of IP but offers better security, delivery efficiency, and disruption tolerance.

Table 19.7   Top-level OccupySandy news feed

S1	RT @morningmoneyben: As a sometime critic of the Occupy movement, have to say they are out BIG TIME helping w/Sandy relief, huge credit to them #occupysandy
S2	RT @JimGaffigan: Before the Red Cross and FEMA came to help @OccupySandy was there. Thanks! http://t.co/nlyOS7qt
S3	RT @VeganLunchTruck: Serving FREE hot #Vegan food, fresh donuts Friday 12:00–6:00ish 192 beach 96th street rockaway beach @occupysandy #sandyrelief
S4	RT @TheAtlantic: How @OccupySandy is using Amazon's wedding registry to collect donations for storm victims. http://t.co/PjUwo8te
S5	RT @OccupyWallStNYC: "Capable of summoning an army with the posting of a tweet" @NYTimes http://t.co/kSskJe54 #OccupySandy
S6	Urgent: need A Lot of thermals+ponchos for #Rockaways for Weds storm. Deliver to 5406 4th Ave or 520 Clinton Ave in BK. @OccupySandy #Sandy
S7	RT @whoisMGMT: Hurricane Sandy devastated the coastal areas(cont. – http://t.co/aLtkP2fZ) @OccupySandy @wavesforwater @RockawayHelp #ro . . .
S8	RT @ofthespirit: you know things are changing when you get an official email from the city of new york telling u to volunteer through @Occupysandy
S9	RT @OneLoveOccupy: On the ground with #occupysandy – more effective than the Red Cross? http://t.co/kVTBaX10 via @slate
S10	RT @OccupySandy: Drug store offering free meds RT @Jamester85: @OccupySandy it's awesome everyone is doing their part to help out. . http . . .

Our chapter leverages NDN [14] to offer a transport-layer solution that reduces redundancy in a representative category of modern data sets. The problem of data redundancy was described initially by a subset of the authors in Photonet [15, 16]. A protocol was designed to prioritize images forwarding and replacement in a disruption-tolerant network depending on the degree of similarity (or dissimilarity) among them. Minerva [17] presented an

Table 19.8 OccupySandy: More on the schedule of free food distribution

S3	RT @VeganLunchTruck: Serving FREE hot #Vegan food, fresh donuts Friday 12:00–6:00ish 192 beach 96th street rockaway beach @occupysandy #sandyrelief
D1	Bless y'all! MT @VeganLunchTruck Wednesday serving FREE hot Vegan food, donuts @ 192 beach 96th St Rockaway Beach @occupysandy #sandyrelief
D2	@OccupySandy We'll be serving hot chili, water, and cookies on the parking lot on Beach 88 Rockaway Beach Blvd at 1PM! Pls RT!
D3	RT @xrisfg: Rockaway Home Base of @OccupySandy, Beach 113th Street, Rockaway Beach, Queens http://t.co/33WMMnPA
D4	@occupysandy do you have any sand bags for my grandma on beach 101 street in rockaway
D5	RT @InterOcc: Do NOT send hot food or anything to Far Rockaway after 5 pm. No deliveries whatsoever after this time #curfew #danger #sandyrelief
D6	@mediagirl333 @ABFalecbaldwin @OccupySandy International Psychic Medium http://t.co/148ifUW9 free reading now.
D7	@amazon How about free shipping, tech support for @OccupySandy http://t.co/v7sVUn1n
D8	@justinstoned @OccupySandy @holageorge @groamerica download the mp3 version of the song for free. .. pass it on. .
D9	RT @orenstark: @VeganLunchTruck @occupysandy saving human and non-human animals simultaneously!!!! #Vegan
D10	RT @lirael_abhorsen: NYPD were policing bbq serving in Coney while city workers wouldn't go into building I was canvassing w/o police gu . . .

information-centric programming paradigm and toolkit for social sensing that generalizes PhotoNet to arbitrary content types. Related ideas that exploit a hierarchical data organization were presented in recent work [18, 19].

Our chapter falls into the category of saving resource by reducing data transmission volume. Much work was done in this area over the decades, from compressing data [20–22] to selectively sending a subset [23, 24]. Our research is closer to the latter category. It is unique in exploiting a generic

Table 19.9   OccupySandy: More on the use of Amazon wedding registry for donation

S4	RT @TheAtlantic: How @OccupySandy is using Amazon's wedding registry to collect donations for storm victims http://t.co/PjUwo8te
D1	RT @annawiener: A rare moment of unbridled enthusiasm about Amazon: @occupysandy is using its wedding registry to collect donations (!)
D2	RT @GregChase: Creative: @occupysandy using wedding registry on Amazon to coordinate donations for #SandyRelief http://t.co/c8QrXKya
D3	Brilliant RT @OccupyWallStNYC RT @NYCSandyNeeds Genius. RT @TheAtlantic: How @OccupySandy is using Amazon's wedding . . . http://t.co/VG8TzmeQ
D4	@occupysandy are you using Amazon wedding registry to coordinate donation requests? Are deliveries coming? New registry how often?
D5	Great: RT @TheAtlantic How @OccupySandy uses Amazon's wedding registry to collect donations for storm victims http://t.co/5xL5sqqs
D6	RT @rachaelmaddux: Shop @OccupySandy "wedding registry," have supplies shipped straight to hurricane victims: http://t.co/QQe6ibBT
D7	@OccupySandy has set up a wedding registry on Amazon for anyone who wants to donate supplies. http://t.co/7VmVLDWZ @EcoWatch
D8	RT @gregpak: (h/t to @RNonesuch_OH for the scoop on the @occupysandy wedding registry: http://t.co/Dg6viHS7)
D9	RT @askdebra: If you don't know about the @occupysandy amazon gift registry, it's an innovative crowdfunding idea: http://t.co/11zLTyfi
D10	@OccupySandy I'm on the wedding registry team. @sandy registry temp down, but Amazon registry is on fire!!!

scheme that sits in a transport layer and does not require application-specific knowledge. An advantage of this approach is that different applications can use it, instead of relying on their own application-specific solutions.

InfoMax extends a previous transport-layer abstraction, called the *Information Funnel* [10], proposed earlier by a subset of the authors. While the latter was geared for data *collection* by a single consumer from multiple producers, InfoMax is geared for the complementary case of data *dissemination*, where multiple consumers pull data from the same producer, possibly at different levels of summarization. Together, the Information Funnel and the InfoMax dissemination protocol may constitute the two building blocks of many data-intensive services. The funnel will collect data into repositories from multiple sources, whereas InfoMax will serve it to multiple eventual consumers at different degrees of detail.

19.6 Conclusions and Future Work

In this chapter, we proposed a new transport-layer protocol to subsample big data sets. This protocol is built on top of the NDN and exploit its hierarchical name structure. By observing that data objects with a longer shared postfix have more semantic similarity, InfoMax produces (an approximation of) the shortest shared postfix order among transmitted data objects to minimize data redundancy. It allows multiple consumers to retrieve data at different configurable levels of detail from one producer while maximally leveraging NDN caching to minimize producer overhead. InfoMax has been deployed on a nation-wide test bed. Analysis of long-term experiences with this deployment is deferred to a subsequent publication.

References

[1] Elijah, E. (2014). Information output is growing exponentially: 3 strategies for laboratory data management. http://accelrys.com/, May 2014, [Online; posted 22-May-2014].

[2] Zhang, L., Afanasyev, A., Burke, J., Jacobson, V., kc claffy, Crowley, P., Papadopoulos, C., Wang, L., and Zhang, B. (2014). Named data networking. *SIGCOMM Computer Communication Review (CCR)*, July 2014.

[3] McCanne, S., Jacobson, V., and Vetterli, M. (1996). Receiver-driven layered multicast. *SIGCOMM Comput. Commun. Rev.*, 26(4), 117–130, Aug. 1996. [Online]. Available: http://doi.acm.org/10.1145/248157.248168

[4] Kutscher, E. D., Eum, S., Pentikousis, K., Psaras, I., Corujo, D., Saucez, D., Schmidt, T., Hamburg, H., and Waehlisch, M. (2016). ICN Research Challenges. https://tools.ietf.org/html/draft-irtf-icnrg-challenges-06, March.

[5] Tschudin, C., and Wood, C. (2016). File-Like ICN Collection (FLIC). https://www.ietf.org/id/draft-tschudin-icnrg-flic-00.pdf, April.

[6] Amin, M. T., Li, S., Rahman, M. R., Seetharamu, P., Wang, S., Abdelzaher, T., Gupta, I., Srivatsa, M., Ganti, R., Ahmed, R., and Le, H. (2015). SocialTrove: A self-summarizing storage service for social sensing. In *Proc. International Conference on Autonomic Computing (ICAC'15)*, July.

[7] Shang, W., Ding, Q., Marianantoni, A., Burke, J., and Zhang, L. (2014). Securing building management systems using named data networking. *Network, IEEE*, 28(3), 50–56, May.

[8] Moiseenko, I., Wang, L., and Zhang, L. (2015). Consumer/producer communication with application level framing in named data networking. In *Proc. 2nd International Conference on Information–Centric Networking (ACM ICN'15)*, September.

[9] "Consumer-Producer-API," https://github.com/named-data/Consumer-Producer-API

[10] Wang, S., Abdelzaher, T., Gajendran, S., Herga, A., Kulkarni, S., Li, S., Liu, H., Suresh, C., Sreenath, A., Wang, H., Dron, W., Leung, A., Govindan, R., and Hancock, J. (2014). The information funnel: Exploiting named data for information-maximizing data collection. In *Proc. International Conference on Distributed Computing in Sensor Systems*, May.

[11] Washington University in St. Louis, "NDN testbed," http://ndnmap.arl.wustl.edu

[12] NFD developers, "NFD tutorial," http://named-data.net/doc/NFD/current/INSTALL.html

[13] "Apollo Twitter Search iOS client built on InfoMax," https://github.com/infomaxndn/ApolloExample

[14] Jacobson, V., Smetters, D. K., Thornton, J. D., Plass, M. F., Briggs, N. H., and Braynard, R. L. (2009). Networking named content. *SIGCOMM*, December 2009.

[15] Uddin, M. Y. S., Wang, H., Saremi, F., Qi, G.-J., Abdelzaher, T., and Huang, T. (2011). Photonet: A similarity-aware picture delivery service for situation awareness. *Proceedings of the 2011 IEEE 32nd Real-Time Systems Symposium*, pp. 317–326, November 2011.

[16] Uddin, M. Y. S., Amin, M. T. A., Abdelzaher, T., Iyengar, A., and Govindan, R. (2012). "Photonet+: Outlier-resilient coverage maximization in visual sensing applications," *Proceedings of the 11th International Conference on Information Processing in Sensor Networks*, April.

[17] Wang, S., Hu, S., Li, S., Liu, H., Uddin, M., and Abdelzaher, T. (2013). Minerva: Information-centric programming for social sensing. *Proceedings of the 22nd International Conference on Computer Communications and Networks*, pp. 1–9, July.

[18] Dron, W., Leung, A., Uddin, M., Wang, S., Abdelzaher, T., Govindan, R., and Hancock, J. (2013). Information-maximizing caching in ad hoc networks with named data networking. *Proceedings of Network Science Workshop*, pp. 90–93.

[19] Kumar, S., Shu, L., Gil, S., Ahmed, N., Katabi, D., and Rus, D. (2012). Carspeak: A content-centric network for autonomous driving. *Proceedings of ACM SIGCOMM*, pp. 259–270.

[20] Cristescu, R., Beferull-Lozano, B., Vetterli, M., and Wattenhofer, R. (2006). Network correlated data gathering with explicit communication: Np-completeness and algorithms. *IEEE/ACM Transactions on Networking*, pp. 41–54.

[21] Pattem, S., Krishnamachari, B., and Govindan, R. (2008). The impact of spatial correlation on routing with compression in wireless sensor networks. *ACM Transactions on Sensor Networks (TOSN)*, 4(4).

[22] Vuran, M. C., Akan, Zgr B., and Akyildiz, I. F. (2004). Spatio-temporal correlation: theory and applications for wireless sensor networks. *Comput. Network. J.* 245–259, January.

[23] Gupta, H., Navda, V., Das, S., and Chowdhary, V. (2008). Efficient gathering of correlated data in sensor networks. *ACM Transactions on Sensor Networks (TOSN)*, 4(4), January.

[24] Liu, C., Wu, K., and Pei, J. (2007). An energy-efficient data collection framework for wireless sensor networks by exploiting spatiotemporal correlation. *IEEE Transaction on Parallel Distributed System*, pp. 1011–1023.

20

Improvement in Load Balancing Decision for Massively Multiplayer Online Game (MMOG) Servers Using Markov Chains

Aamir Saeed, Rasmus Løvenstein Olsen and Jens Myrup Pedersen

Wireless Communication Networks, Department of Electronic Systems, Aalborg University, Denmark

Keywords: MarKov Chain, load balancing, Hot-spot problem, prediction.

20.1 Introduction

Massively multiplayer online game (MMOG) is a multiplayer online gaming platform that hosts participation of hundreds of thousands of players simultaneously. It provides a persistent game world by maintaining a continuous experience without halting and resetting. The game world has playing characters and non-playing characters (NPC), i.e., enemies, and support characters, which interacts throughout the game play. The movements of game players have a lasting effect, which cannot be reset.

A player connected to the MMOG game server is represented by an avatar in the game world, and the moves carried by remote players are propagated to neighboring avatar. The propagation of player movement is communicated based on area of interest (AOI). Each avatar that has an overlapping AOI with other avatar communicates its status with avatars of overlapping AOI. The MMOG server updates gaming events for playing characters (players) and non-playing characters (NPC) to provide a persistent view of the gaming world to all the players at instant time. Due to physical restriction such as CPU speed, bandwidth requirements, and availability of free memory, a single server cannot host hundreds of thousands of players. Single server experiences an increased message drop rate as analyzed in the study conducted in [1]. To improve the performance of the gaming server, additional servers are added to balance load and maintain high quality of experience (QoE) [2, 3] of game play. Distributed and scalable approaches are employed to divide the virtual world into regions, and regions are then allocated to servers. The servers hosting neighboring regions form a cluster. Load balancing algorithms are adopted to balance load within a cluster.

20.1.1 Hotspot Problem in MMOG

Flocking of players to one area results in a hotspot problem, making it harder for hosting server to manage all the players in that region due to limitation in terms of bandwidth, computation, and free memory. The flocking of players to a hotspot is triggered due to either invitation of players to each other or improving scores to improve profiles. In addition to this, players often move between regions, overburdening server to handle the handover between regions. The server requires excessive computation and communication to handle the hotspot problem and handover problem. The movement of players between regions varies the servers load hosting the region. For example, a player hosted at region R_i moves to another region R_j, affecting the workload of both servers S_i and S_j, respectively. The handover of players between the servers induces additional traffic management and computations, causing lower QoE for players. Load balancing algorithms are employed to balance load between servers in order to tackle the issue of hotspot problem. The approaches available in the literature are both proactive [5] and reactive [6].

20.1.2 Load Balancing Approaches

A number of load balancing approaches exist to equally balance loads between servers, depending on the network resources utilization model, i.e., peer-to-peer model, centralized model, and hybrid model. Locality aware dynamic partitioning [4] uses a decentralized algorithm that employs heuristic-based approach to shed load from an overloaded server. The algorithm maintains a structured approach of status updates by sending periodic updates to communicate its load status and data (consistency maintenance) as shown in Figure 20.1. Each neighboring host adjacent partitions and keeps record of load of its neighbors, and a record of neighbor's neighbors are currently lightly loaded. Under normal circumstances, an overloaded server sheds its load to neighbors in its locally maintained lightly loaded server. In situations when lightly loaded neighbors are not available, loads are shed into neighbor's neighbor (server). Shedding loads into neighbor's neighbors causes locality disruption problem, a condition in which servers host regions that do not belong to the

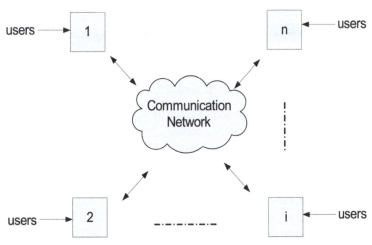

Figure 20.1 A distributed system model.

same locality. In such situation, locality aware partitioning algorithm solves locality disruption problem by employing interneighbor partition aggregation (to avoid excessive computation and communication, interneighbor partition aggregation is carried out during normal load conditions).

Dynamic load balancing architecture discussed in [5] presents a distributed architecture to manage MMOG servers. Normally MMOG has three types of workload, namely process load at a server, communication load due to interaction between entities of multiple servers, and management of load. A number of optimization algorithms for different workload optimization are applied to balance loads between servers. A peer-to-peer-based load balancing model is a novel paradigm of distributed architecture discussed in [6] to divide the region into microcells (a small region in virtual space). All the microcells in a region that are either overloaded with players or have low movement of players are identified and marked. Load balancing is then applied to balance load within a region. Alternate to microcell approach is the KD-tree-based load balancing algorithm that partitions virtual environment, based on the distribution of avatars in the virtual environment [7]. KD-tree-based load balancing algorithm changes the limits of each region by modifying the bounded regions using KD trees. The KD tree data structure allows partitioning at the granular level, resulting in a load assignment closer to the ideal situation.

An important perspective in investigating load balancing approaches for MMOG is to analyze the impact of outdated information that may cause unwanted migration triggers or to-and-fro motion of players within a server due to frequent migration. The Algorithms discussed above does not look into this perspective. In this work, we aimed to minimize the impact of oudated information for load balancing algorithms in distributed systems.

20.1.3 Sharing of Outdated Information

The status information exchanged between servers has an impact on the outcome of the load balancing decisions. Delay in load status information exchange due to network losses, communication delays and information stalling at the remote servers, results in incorrect load balancing decisions. Outdated information often triggers unnecessary load balancing decisions that eventually affect the user experience. Access mechanism for information sharing and delay has an impact on the quality of information being shared. A detailed analysis of how different access schemes and delay impacts the quality of information is analyzed using mismatch probability. The study is supported by analytical and simulation studies for different use cases in [8, 9]. In this chapter, prediction algorithm is presented to minimize the impact of outdated information, by minimizing the number of migration triggers.

20.1.4 Load Balancing Decision Affects Player Response Time

The user satisfaction level of the game play due to load balancing is an indicator to analyze the impact of load balancing decision. Response time of a game player is generally defined as the time a user initiates an activity such as movement of mouse, keystroke, or touch screen, until the system presents the result. Lengthy response time causes lower satisfaction and a high ratio of user stopping using the service. The response time is an important parameter that needs to be considered when developing load balancing algorithms for remote access applications. Remote access application response time is affected by communication delay in the network, application load, and processing rate of applications and devices. Generally, load balancing decisions of MMOG force players to migrate from one server to another server, in order to improve the system performance. The response time during the player migration process is worse, and frequent migration of players further affects the players' response time during the game play.

The remaining part of the chapter is comprised of three sections: Section 20.2 proposes a method to overcome the impact of outdated information (discussed in Section 20.1) for load balancing algorithms; Section 20.3 presents a brief analysis of the impact on the individual user performance; and finally Section 20.4 briefly concludes the chapter.

20.2 Minimizing the Impact of Outdated Information

Ensuring information correctness and minimizing the impact of outdated information improve the decision process. Status information exchanged between game servers is often delayed, either due to network losses, communication delay, stalling of information at remote servers. Figure 20.2 presents a scenario in which the overloaded and underloaded servers communicate load status information. The *"states"* of the servers are exchanged between servers using *"periodic updates."* The circle represents the periodic event, when load status information is provided to the overloaded server. The t_p represents the communication delay between underloaded server and overloaded server. The t_m represents the time spent in migrating the players to an underloaded server from an overloaded server. Communication delay and migration delay may cause a mismatch between shared state and actual state of a server. The impact of outdated information has to be minimized in order to avoid false migration triggers.

20.2.1 Prediction Algorithm

Predicting the load of the servers, based on its shared status and estimated traffic of game play for a particular region, is one of the possibilities to minimize the impact of outdated information. The estimated traffic of the game play for a particular region can be modeled as a birth–death process [10].

The birth–death process is a special case of a continuous time Markov process, where the states $\{1, 2, 3, \ldots, n\}$ represent a current size of population (players), and the transitions are limited to birth and death. The process moves

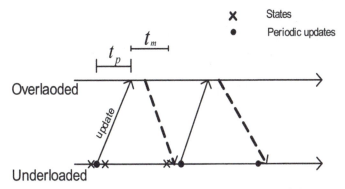

Figure 20.2   Interaction between underloaded and overloaded servers.

from state i to state $i + 1$ when a birth occurs and moves from state i to state $i - 1$ when a death occurs.

The arrival and departure of players at a server are modeled as Markov chain. Transition probabilities are computed using prediction at t_i in future Δt. At time t_i, the future state of a server at t_f (where $t_f = t_i + \Delta t$) is estimated using the probability distribution.

The probability distribution π_0 of server has the i number of players at time t_i, and $\pi_{\Delta t}$ is the probability distribution at Δt. A generator matrix Q is generated as shown in Figure 20.3. Each entry in Q matrix is q_{ij} that represents the different state of the server. The matrix Q has the following properties.

- The row sums in Q are zero.
- The diagonal entries are non-positive, i.e., for all $i, j \in S$,

$$0 \geq q_{ii} = -\sum_{i \neq j} q_{ij}.$$

The probability distribution $\pi_{\Delta t}$ is computed for a generator matrix Q using matrix exponential at Δt when the server is at state i [11].

$$\pi_{\Delta t} = \pi_0 e^{Q\Delta t}. \tag{20.1}$$

The entries in Q matrix represent the arrival and departure rates of the players at the server. The main aim to use the Markov-based prediction is to utilize the current arrival and departure rates as input to Q along with current state of the server to estimate future state.

20.2.2 Accuracy of Arrival ($\hat{\lambda}$) and Departure Rates ($\hat{\mu}$) Estimates

In the literature, numerous approaches are available to analyze the performance of MMOG. One of the anaysis in literature measure network pattern for an MMO game called Lineage II in [12]. It analyzed the game session using network packet sizes, RTTs, session times, and intersession arrival times. Similarly, Ye et al. [13] proposed performance models for MMORPG

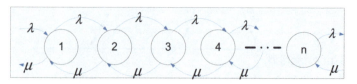

Figure 20.3   Markov chain for clients on a server.

servers and networks based on concurrent player population. Chen et al. in [14] classified packet interarrival times, packet load distribution, and bandwidth utilization of a popular MMORPG game server. In [15], the total population over time was observed using probing-based measurements to try to observe and infer properties of the system. The probes were used to measure populations over time, the duration of each session, the zones each player visited, and the time spent in each zone. These are valued estimates to be used for any server session's arrival and departure rates.

Cramer-Rao bound (CRB) [16] is an effective approach in estimation theory and statistic to express a lower bound on the variance of estimators of a parameter. The inverse of lower bound is called the Fisher information $I_F(h)$. Let the parameters λ and μ and their estimates $\hat{\lambda}$ and $\hat{\mu}$ be represented as h and $\hat{h}$.

Let $y = [y(0) \ y(1) \ y(2) \ldots y(n-1)]$ be a vector of measurements represented by the probability density function (PDF) $f_y(y; h)$, and the measurement y is a function of an unknown parameter h. Let $\hat{h}$ be an unbiased estimator of h and only a function of y. So, if the regularity condition holds, then

$$E[(\hat{h} - h)^2] \geq \frac{1}{I_F(h)},$$

where $I_F(h)$ is the Fisher information

$$I_F(h) = \left[\left(\frac{\partial \ln f_y(y)}{\partial h} \right)^2 \right].$$

The estimator, which achieves the lowest possible mean square error among all the unbiased estimators, is therefore the minimum variance unbiased estimator (MVU). The Fisher method achieves maximum efficiency with the Cramer-Rao lower bound as the sample approaches to infinity.

20.2.3 Use Case Scenarios

The impact of sharing current states of a server and predicted state of a server is analyzed for a load balancing algorithm. A simple scenario of two systems is shown in Figure 20.2, comprising of overloaded and underloaded servers. The information that is exchanged between both the systems is susceptible to be outdated due to communication delay and server processing time (t_p). A problem associated with outdated information is false migration triggers.

The objective of this work was to minimize the impact of outdated information in order to avoid false migration. To achieve this objective, a

Markov-based prediction approach is used that uses current state of a server and estimated arrival and departure rates of players at a server. Therefore, based on these rates, future states are predicted in an estimated time window Δt, i.e., approximately equivalent to the processing time of information at a server and communication delay.

 False migration triggers cause player disruption that directly impacts the user experience of the game play. The load balancing algorithms are designed to minimize user game play disruption. In this work, a load balancing algorithm is designed that uses Markov-based prediction algorithm output as an input for load balancing metric. Markov-based prediction algorithm is compared with normal approach where actual state of server is exchange. The load balancing algorithm for both approaches is analyzed using number of migration triggers.

20.2.4 Results

Unnecessary migration triggers, due to outdated information, affect the QoS of some players. By predicting future states based on the estimated traffic, the number of migration triggers was minimized as shown in Figure 20.4.

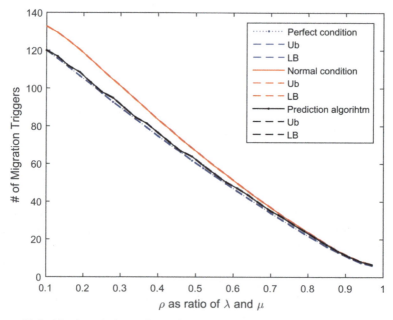

Figure 20.4 Number of triggers for perfect condition, normal and prediction approaches.

The figure shows three different conditions: (1) *Perfect condition:* When the information is provided instantly (means no delay), the time to enable migration takes no time (instantaneous). (2) *Normal condition:* The information shared between neighboring servers may be outdated, and the migration process takes time. (3) *Prediction-based approach:* The future states of the server are predicted based on the estimated value of players at a server for certain duration of time. The prediction-based approach has lesser number of migrations and is much nearer to the perfect condition.

In cases when $\mu = \lambda$ with $\frac{1}{\mu+\lambda} = 1$, it is concluded that the prediction approach gives a similar outcome as normal approach. The prediction approach guarantees better performance for most of the λ and μ values. In order to investigate the reason for equal number of triggers when μ and λ are same, the convergence speed of Markov chain was analyzed as shown in Figure 20.5. This figure shows that when μ and λ are approximately same (the case when $\frac{1}{\mu+\lambda} = 1$), the Markov chain converges slowly to steady state. For any decision process, it is important to know whether it hits the steady state of the system or not within the timeframe (Δt). If this is not the case, then it is more likely that decision is not correct to predict future state and reduces the efficiency of the proposed methodology. Therefore, by looking into the convergence time of the models, we can conclude that similar performance for both approaches (normal and prediction

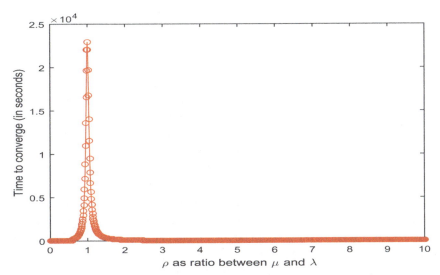

Figure 20.5 Convergence speed of Markov chains for average stay $\frac{1}{\lambda+\mu} = 1$.

approaches) is due to slow convergence of the Markov chain, when μ and λ are equal.

The convergence speed of Markov chain was computed by decomposing generator matrix Q into diagonal matrices A and D.

$$Q = ADA^{-1} \tag{20.2}$$

A detailed description is available in [17] to compute the convergence speed (t_{covg}) of the system using Equation (20.3).

$$t_{covg} = -\frac{\ln(\varepsilon)}{\lambda_2} => t_{covg} = 3 * \ln(10)/\lambda_2 \tag{20.3}$$

Figure 20.5 shows the convergence speed ρ (ratio of $\mu : \lambda$). As discussed earlier, the convergence of the Markov chain to steady state is required to have stable decision. Due to longer time to converge to steady state for ρ equal to 1, the prediction algorithm and normal algorithm do not improve.

20.3 MMOG Server Load Migration Affects User Experience

User experience of game plays is affected by the system performance enhancement algorithms as discussed in the previous section. It is important to understand the impact of load migration decisions on an individual user of MMOG: In what conditions a player benefits and in what conditions a player performance is affected by system performance enhancement algorithms?

User satisfaction of the games depends on various factors such as the response time of game, the expertise level to play the game, and external factors. In [18], QoS parameters were listed for multimedia applications, and in [19], the author summarized the maximum allowable response time a user can tolerate for different MMOG games, i.e., beyond which the user is distracted from game experience. Premature departure of game players is also an indication of user unsatisfactory experience of game play. It refers to the player departure from the game within 10 minutes of experience as discussed in [20]. The premature departure and normal departure are differentiated by the time a user stays in the game. The premature departure time was improved by enhancing processing priorities and buffer management.

In the next section, the impact of migration trigger on a user response time is explained.

20.3.1 System Model and Impact of Migration Decision on a User Response Time

A general distributed system model is considered as shown in Figure 20.1. The system consists of n servers connected by a communication network. We denote N as a set of nodes. Users arrive or send a job at a node i ($i \in N$) according to a Poisson process with parameter λ_i. The job or the user is either served locally or transferred to another node j ($j \in N$). The departure rate of the user is also represented as a Poisson process with parameter μ_i.

Each server i communicates its load with the neighboring server, using periodic updates. The players presence at the server at suitable time intervals was modeled with a stationary birth–death process of Markov chain in our previous work [21]. At the time when a server is overloaded, it adopts load exchange mechanisms to offload extra load.

20.3.1.1 Impact of Migration on User Response Time

A migration decision triggers the transfer of players from the current server to a neighboring server. During this transition process, the response time of a player is affected. Figure 20.6 shows the impact of migration decision on a user response time, due to the migration process. Frequent migrations of a player between servers may be beneficial for the overall system, but it can be an unsatisfactory experience for the players who experience frequent migrations.

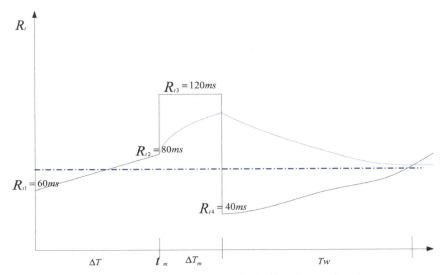

Figure 20.6   Response time for a single client due to migration.

The user response time due to migration is represented in Equation (20.4).

$$R(t) = \frac{1}{\Delta T + \Delta T_m + Tw} \left(\int_{t_m - \Delta T}^{t_m} R(t)dt \right.$$
$$\left. + \int_{t_m}^{t_m + \Delta T_m} R(t)dt + \int_{t_m + \Delta T_m}^{t_m + \Delta T_m + Tw} R(t)dt \right) \tag{20.4}$$

where ΔT represents the time a player spent at a previous server, ΔT_m represents the time a player spent in the migration process, and T_w represents the time a player spent on the new server.

In the next section, we analyze the impact of the load balancing decision on response time of a single user under different server loads. The aim was to analyze when the load balancing decision improves and when it degrades the user response time.

20.3.1.2 Simulation and Results

In this section, a simulation study was conducted with the setting specified in Table 20.1. A single-player response time due to load transfer is to be analyzed in different server traffic conditions. The load transfer decision is defined as a migration triggers when a server has more connected players than a threshold. The player is migrated based on the migrated trigger decision, and during the migration process, it has a lengthy response time. A player that migrates between servers due to a load transfer decision has either an improved or degraded response time.

Servers exchange load information between each other through periodic updates and can be outdated. For example, a server notifies neighboring servers about its load, and based on this information, the neighboring servers infer that a neighbor server has the capacity to provide a better service to the user and a migration is triggered. After having migrated to a new device, the user now experiences a worse response time than the previous device and soon will be in the probable list of being selected for migration. Due to the mismatch of information, the user QoE is affected.

Table 20.1 Simulation parameters

Parameters	Values
Migration trigger threshold	Players ≥ 128
Server utilization	$M/M/1$ Queue with average wait time 1,2,3 seconds
Delay	100 msec
Periodic update interval	300 msec
Simulation time	5,000 seconds

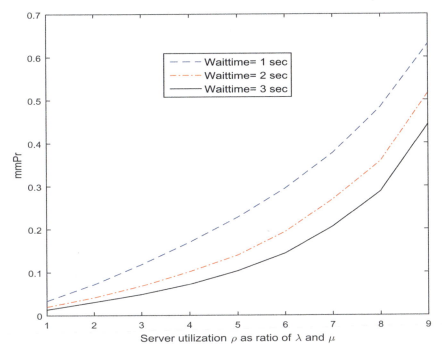

Figure 20.7 Server utilization and mismatch probability.

In Figure 20.7, we show mismatch probability of information between server actual state and received state, for periodic update strategy for different server traffic conditions. Different wait times for the $M/M/1$ queue were introduced to show how fast the server states are changing. For the case when the wait time is 3 sec, the states of the server change slowly.

The player is migrated between servers due to the load balance decision on the servers. The number of migration triggers a player experiences and its respective response time (Equation (20.4)) are shown in Figures 20.8 and 20.9, respectively.

Figure 20.8 shows the number of migration triggers for different rates of servers, and it is shown that as the server utilization is increased, the number of migration triggers is correspondingly higher. For the wait time of 1 sec, the number of triggers is the highest, which is because the server states change very fast, as compared to the wait time of 2 and 3 sec.

Figure 20.9 shows that for a certain increase in migration triggers, the response time improves (lower better), but again the response time starts to degrade. The reason for this phenomenon of degradation is that load balancing

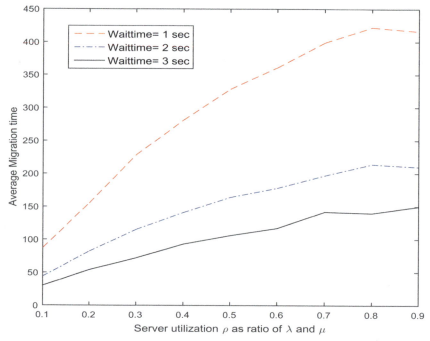

Figure 20.8   Server utilization and migration triggers.

does more harm for these cases. The player spends most of his time in moving between servers, resulting in a high response time. The load balance algorithms need to be designed in such a way that the players' response time is being respected. The user-centric approach of load balancing algorithm improves the user existing load balancing algorithms.

The two approaches discussed in this chapter, namely the (1) prediction approach and (2) user-centric load balancing approach, are two different approaches to improve the overall system performance. The first one uses the estimation technique and prediction algorithm, to minimize the number of migration triggers. The second approach shows how a single-user response time is affected by migration triggers. The purpose of this analysis is to show the importance of user-centric approaches for load balancing decision. A detailed analysis and design of user-centric algorithm is employed, using sigmoid function and probability mass function that updates user selection probability for migration, based on the user arrival at a specific server for being serviced.

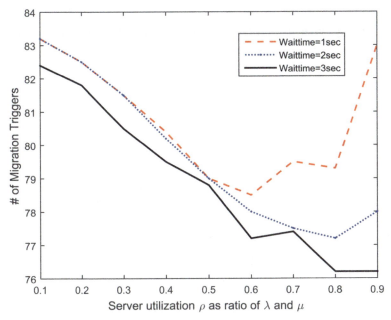

Figure 20.9 Server utilization and single-user average response time.

20.4 Conclusion

Massively multiplayer online game (MMOG) servers facilitate hundreds of thousands of concurrent users. Players flocking to a hotspot in the game world may cause the server to be overloaded. A Markov chain-based prediction approach was proposed to balance load between two servers, with an objective to minimize the impact of the outdated information and false migration triggers. Although the prediction algorithm increased the computational load on server, the benefit it provides is worth to have this cost. The algorithm was analyzed with normal load condition and perfect conditions. The proposed algorithm estimated the future state of a server based on the current state and estimated arrival (μ) and departure (λ) rates of players at a server. The prediction approach showed better performance than the normal approach and was mostly near-to-perfect condition. Furthermore, how the load balancing decision impacts the response time of a single user was analyzed. The analysis showed that load balance transfer decision's impact on a single-user response time depends on the rates (μ and λ) at which players traffic at a server at any instance of time. Load balancing transfer decision improves user response time of a player, when the server has a lower traffic of incoming and outgoing clients. On the other hand, the user response time of a player degrades due to load transfer decision due to frequent migration triggers, when the server has a higher traffic of players. In future, we aim to construct an algorithm that respects a player's arrival time at a server, in order to be selected for next migration transfer.

References

[1] Lu, F. (2006). Load balancing for massively multiplayer online games. In *Proceedings of 5th ACM SIGCOMM NetGames*.

[2] Jain, R. (2004). Quality of experience. In *MultiMedia, IEEE* 11.1, pp. 96–95. ISSN: 1070-986X. DOI: 10.1109/MMUL.2004.1261114.

[3] Kuipers, F., Kooij, R., De Vleeschauwer, D., and Brunnström, K. (2010). Techniques for measuring quality of experience. In *Proceedings of the 8th International Conference on Wired/Wireless Internet Communications.* WWIC'10. Lule&#229. Sweden: Springer-Verlag, pp. 216–227. ISBN: 3-642-13314-2, 978-3-642-13314-5. DOI: 10.1007/978-3-642-13315-2_18.

[4] Chen, J., Wu, B., Delap, M., Knutsson, B., Lu, H., and Amza, C. (2005). Locality aware dynamic load management for massively multiplayer games. In *Proceedings of the Tenth ACM SIGPLAN Symposium on Principles and Practice of Parallel Programming.* PPoPP '05. Chicago, IL, USA: ACM, 2005, pp. 289–300. ISBN: 1-59593-080-9. DOI: 10.1145/1065944.1065982.

[5] Lim, J. Y., Kim, J. R., and Shim, K. H. (2006). A dynamic load balancing model for networked virtual environment systems using an efficient boundary partition management. In *Advanced Communication Technology, 2006. ICACT 2006. The 8th International Conference.* Vol. 1. 2006, pp. 730. DOI: 10.1109/ICACT.2006.206068.

[6] Ahmed, D. T., and Shirmohammadi. S. (2008). A microcell oriented load balancing model for collaborative virtual environments. In *Virtual Environments, Human-Computer Interfaces and Measurement Systems, 2008. VECIMS 2008. IEEE Conference on.* 2008, pp. 86–91. DOI: 10.1109/VECIMS.2008.4592758.

[7] Bezerra, C. E. B., Comba, J. L. D., and Geyer. C. F. R. (2012). Adaptive Load-balancing for MMOG Servers Using KD-trees. In *Comput. Entertain.* 10.3 (Dec. 2012), 5:1–5:16. ISSN: 1544-3574. DOI: 10.1145/2381876.2381881.

[8] Hansen, M. B., Olsen, R. L., and Schwefel, H.-P. (2006). *Probabilistic models for access strategies to dynamic information elements.* Department of Mathematical Sciences, Aalborg University.

[9] Olsen, R. L. Schwefel, H. P. and Hansen, M. B. *Quantitative analysis of access strategies to remote information in network services.*

[10] Taylor, H. M., and Karlin, S. (1998). *An Introduction to Stochastic Modeling.* Academic Press, ISBN: 9780126848878.

[11] Bolch, G., Greiner, S., de Meer, H., and Trivedi, K. S. (1998). *Queueing Networks and Markov chains: Modeling and performance evaluation with computer science applications.* New York, NY, USA: Wiley-Interscience, 1998. ISBN: 0-471-19366-6.

[12] Kim, J., Choi, J., Chang, D., Kwon, T., Choi, Y., and Yuk, E. (2005). Traffic Characteristics of a Massively Multi-player Online Role Playing Game. In *Proceedings of 4th ACM SIG-COMM Workshop on Network and System Support for Games.* NetGames'05. Hawthorne, NY: ACM, 2005, pp. 1–8. ISBN: 1-59593-156-2. DOI: 10.1145/1103599.1103619.

[13] Ye, M., and Cheng, L. (2006). System-performance modeling for massively multi-player online role-playing games. In *IBM Syst. J.* 45.1 pp. 45–48. ISSN: 0018-8670. DOI: 10.1147/sj.451.0045.

[14] Chen, K.-T., Huang, P., and Lei, C.-L. (2006). Game traffic analysis: An {MMORPG} perspective. In *Computer Networks* 50.16, pp. 3002–3023. ISSN: 1389-1286. DOI: http://dx.doi.org/10.1016/j.comnet.2005.11.005.

[15] Pittman, D., and GauthierDickey, C. (2010). Advances in multimedia modeling: 16th international multimedia modeling conference, MMM 2010, Chongqing, China, January 6–8, 2010. Proceedings. In ed. by Susanne Boll, Qi Tian, Lei Zhang, Zili Zhang, and Yi-Ping Phoebe Chen. Berlin, Heidelberg: Springer Berlin Heidelberg, 2010. Chap. Characterizing Virtual Populations in Massively Multiplayer Online Role-Playing Games, pp. 87–97. ISBN: 978-3-642-11301-7. DOI: 10.1007/978-3-642-11301-7_12.

[16] "Estimation Theory". English. In *Identification of Physical Systems.* Ed. by Rajamani Doraiswami, Chris Diduch, and Maryhelen Stevenson. 2014, pp. 117–165. ISBN: 978-1-119-99012-3. DOI: 10.1002/9781118536483.ch3.

[17] Saeed, A., Olsen, R. L., and Pedersen. J. M. (2015). Optimizing the loads of multi-player online game servers using markov chains. In *2015 24th International Conference on Computer Communication and Networks (IC-CCN).* pp. 1–5. DOI: 10.1109/ICCCN.2015.7288445.

[18] Mathy, L., Edwards, C., and Hutchison, D. (1999). Principles of QoS in group communications. English. In *Telecommunication Systems* 11.1-2, pp. 59–84. ISSN: 1018-4864. DOI: 10.1023/A:1019132914996.

[19] Henderson, T., and Bhatti, S. (2003) Networked games: A qos-sensitive application for QoS-insensitive users? In *Proceedings of the ACM SIG-COMM Workshop on Revisiting IP QoS: What Have We Learned, Why Do We Care?* RIPQoS'03. Karlsruhe, Germany: ACM, 2003, pp. 141–147. ISBN: 1-58113-748-6. DOI: 10.1145/944592.944601.

[20] Chen, K.-T., Huang, P., and Lei, C.-L. (2009). Effect of network quality on player departure behavior in online games. In *Parallel and Distributed Systems, IEEE Transactions on* 20.5, pp. 593–606. ISSN: 1045-9219. DOI: 10.1109/TPDS.2008.148.

[21] Olsen, R. L., Saeed, A., and Pedersen, J. M. *Modeling Load Prediction and User Experience to balance Workload for MMOGs.*

Index

603

About the Editors

Dr. Kewei Sha is an Associate Director of Cyber Security Institute at University of Houston, Clear Lake (UHCL). He is also the Assistant Professor of Computer Science at UHCL. Before he moved to UHCL, he was the Department Chair and Associate Professor in the Department of Software Engineering at Oklahoma City University. He received Ph.D. in Computer Science from Wayne State University in 2008. His research interests include Internet of Things, Cyber-Physical Systems, Mobile Computing, Data Analytics, and Network Security and Privacy. Dr. Sha has served as the secretary of Technical Committee on the Internet of the IEEE Computer Society (IEEE-CS TCI). He also served as Editor or Guest Editor of many international journals, and served as Chairs of many conferences and workshops, including Technical Program Chair of ICCCN 2015 and MedSPT 2015. His research has been supported by NSF, CNSF, Oklahoma City University and UHCL.

Prof. Aaron Striegel is an Associate Professor and serves as Associate Chair in the Department of Computer Science & Engineering at the University of Notre Dame. He received his Ph.D. in December 2002 in Computer Engineering at Iowa State University under the direction of Dr. G. Manimaran. Prof. Striegel's research interests focus on instrumenting the wireless networked ecosystem to gain insight with respect to user behavior and global network performance. Further research interests of Prof. Striegel include computer security and the adaptation of low-cost gaming peripherals for rehabilitation. Prof. Striegel has received several best paper awards including USENIX LISA, IEEE Healthcom, and HotPlanet. Prof. Striegel has received various research and equipment funding from NSF, DARPA, Sprint, Intel, Google, and Alcatel-Lucent. He has also been the recipient of a NSF CAREER award in 2004 and has been a recent participant in NAE symposia on Engineering Education and the Informed Brain in the Digital World.

Prof. Min Song served as Program Director with the NSF from 2010 to 2014 and is currently the Founding Director of Institute of Computing and Cyber-systems, Dave House Professor and Department Chair of Computer Science, and Professor of Electrical and Computing Engineering at Michigan Tech.

Through his outstanding contributions in promoting NSF's international leadership, Min received the prestigious NSF Director's award in 2012. Min's research interests include design, analysis, and evaluation of wireless communication networks, cognitive radio networks, network security, cyber physical systems, and mobile computing. His research has been supported by NSF, DOE, NASA, and private Foundations. Min's professional career comprises 26 years in industry, academia, and government, and has held various leadership positions and gained substantial experience in performing a wide range of duties and responsibilities. Min launched and served as Editor-in-Chief of two international journals. He also served as Editor or Guest Editor of 14 international journals, and served as Chairs of many conferences, including General Chair of IEEE INFOCOM 2016 and Technical Program Vice-Chair of IEEE GLOBECOM 2015. Min is currently serving as the IEEE Communications Society Director of Conference Operations. Min was the recipient of NSF CAREER award in 2007.